Dr. J. M. Cherrett.
University College
Department of A
Bangor, Ca
AF745352

C.U.P.
NET IN U.K.
£19·50

ORDER NUMBER

PO NYAATP08

BOOK NUMBER

9194657

BAY NUMBER

2078-0

AUTHOR	Brian, MV (ed)
TITLE	Production Ecology of Ants and Termites (International Biological Programme Synthesis)

DESCRIPTION

Volume 13. This is an ex-library book and may have the usual library/used-book markings inside.This book has hardback covers. Clean from markings. In fair condition, suitable as a study copy. No dust jacket.

Production ecology of ants and termites

THE INTERNATIONAL BIOLOGICAL PROGRAMME

The International Biological Programme was established by the International Council of Scientific Unions in 1964 as a counterpart of the International Geophysical Year (IGY). The subject of the IBP was defined as 'The Biological Basis of Productivity and Human Welfare', and the reason for its establishment was recognition that the rapidly increasing human population called for a better understanding of the environment as a basis for the rational management of natural resources. This could be achieved only on the basis of scientific knowledge, which in many fields of biology and in many parts of the world was felt to be inadequate. At the same time it was recognised that human activities were creating rapid and comprehensive changes in the environment. Thus, in terms of human welfare, the reason for the IBP lay in its promotion of basic knowledge relevant to the needs of man.

The IBP provided the first occasion on which biologists throughout the world were challenged to work together for a common cause. It involved an integrated and concerted examination of a wide range of problems. The Programme was coordinated through a series of seven sections representing the major subject areas of research. Four of these sections were concerned with the study of biological productivity on land, in freshwater, and in the seas, together with the processes of photosynthesis and nitrogen-fixation. Three sections were concerned with adaptability of human populations, conservation of ecosystems and the use of biological resources.

After a decade of work, the Programme terminated in June 1974 and this series of volumes brings together, in the form of syntheses, the results of national and international activities.

INTERNATIONAL BIOLOGICAL PROGRAMME 13

Production ecology of ants and termites

EDITED BY

M. V. Brian

Institute of Terrestrial Ecology, Furzebrook, Dorset

CAMBRIDGE UNIVERSITY PRESS

CAMBRIDGE

LONDON · NEW YORK · MELBOURNE

U.W.B LIBRARY
P.C.B LLYFRGELL

Published by the Syndics of the Cambridge University Press
The Pitt Building, Trumpington Street, Cambridge CB2 1RP
Bentley House, 200 Euston Road, London NW1 2DB
32 East 57th Street, New York, NY 10022, USA
296 Beaconsfield Parade, Middle Park, Melbourne 3206, Australia

© Cambridge University Press 1978

First published 1978

Printed in Great Britain at the
University Press, Cambridge

Library of Congress Cataloguing in Publication Data

Main entry under title:

Production ecology of ants and termites.

(International Biological Programme; 13)

Bibliography: p.

1. Ants. 2. Termites. 3. Insects–Ecology.
I. Brian, Michael Vaughan. II. Series.
QL568.F7P7 595.7'96 76-54061
ISBN 0 521 21519 6

UWB Deiniol Library
QL568.F7 P7
30110005983995

Contents

List of contributors		*page* xiii
Preface		xv
	J. B. Cragg	
1	Introduction *M. V. Brian*	1
2	Empirical data and demographic parameters *C. Baroni-Urbani, G. Josens & G. J. Peakin*	5
3	Production by ants and termites *M. G. Nielsen & G. Josens*	45
4	Food and feeding habits of termites *T. G. Wood*	55
5	Food and feeding habits of ants *D. J. Stradling*	81
	Foraging by seed-harvesting ants *W. G. Whitford*	107
6	Respiration and energy flow *G. J. Peakin & G. Josens*	111
7	Nutrient dynamics of termites *J. P. La Fage & W. L. Nutting*	165
8	Nutrient dynamics of ants *A. Abbott*	233
9	The role of termites in ecosystems *T. G. Wood & W. A. Sands*	245
10	The role of ants in ecosystems *J. Pętal*	293
	Comparison of various biomes *B. Pisarski*	326
Appendixes		
	1 Tables 1, 2 and 3 of ant data by C. Baroni-Urbani and B. Pisarski	333

2 List of IBP projects known to include social insects (ex IBP News 13) plus other information 341

Bibliography 343

Index 401

Table des matières

Liste des collaborateurs — *page* xiii
Préface — xv
J. B. Cragg

1 Introduction — 1
M. V. Brian

2 Données empiriques et paramètres démographiques — 5
C. Baroni-Urbani, G. Josens & G. J. Peakin

3 Production chez les Fourmis et les Termites — 45
M. G. Nielsen & G. Josens

4 Aliments et habitudes alimentaires des Termites — 55
T. G. Wood

5 Aliments et habitudes alimentaires des Fourmis — 81
D. J. Stradling
Le fourragement chez les Fourmis moissonneuses — 107
W. G. Whitford

6 Respiration et flux d'énergie — 111
G. J. Peakin & G. Josens

7 Dynamique nutritionnelle des Termites — 165
J. P. La Fage & W. L. Nutting

8 Dynamique nutritionnelle des Fourmis — 233
A. Abbott

9 Le rôle des Termites dans les écosystèmes — 245
T. G. Wood & W. A. Sands

10 Le rôle des Fourmis dans les écosystèmes — 293
J. Pętal
Comparaison des divers biomes — 326
B. Pisarski

Appendices
1 Tables 1, 2, et 3 de données sur les Fourmis par C. Baroni-Urbani et B. Pisarski — 333

2 Liste des projets PBI connus pour concerner des Insects sociaux (d'après IBP News 13), et information additionnelles 341

Bibliographie 343

Indice 401

Содержание

Список авторов *страница* xiii

Предисловие xv
J. B. Cragg

1 Введение 1
M. V. Brian

2 Эмпирические данные и демографические параметры 5
C. Baroni-Urbani, G. Josens и *G. J. Peakin*

3 Продукция муравьев и термитов 45
M. G. Nielson и *G. Josens*

4 Пища и особенности питания термитов 55
T. G. Wood

5 Пища и особенности питания муравьев 81
D. J. Stradling
Сбор зерен муравьями 107
W. G. Whitford

6 Дыхание и поток энергии 111
G. J. Peakin и *G. Josens*

7 Динамика пищевых веществ термитов 165
J. P. La Fage и *W. L. Nutting*

8 Динамика пищевых веществ муравьев 233
A. Abbott

9 Роль термитов в экосистемах 245
T. G. Wood и *W. A. Sands*

10 Роль муравьев в экосистемах 293
J. Pętal
Сравнение различных биомов 326
B. Pisarski

Приложение 1 Таблицы 1, 2 и 3 данных по муравьям
C. Baroni-Urbani и *B. Pisarski* 333

2 Список тем МБП включающих общественных насекомых (из IBP News 13) и другие сведения 341

Литература 343

Указатель 401

Sumario

Relación de colaboradores *página* xiii
Prólogo xv
J. B. Cragg

1 Introducción 1
M. V. Brian

2 Datos empíricos y parámetros demográficos 5
C. Baroni-Urbani, G. Josens & G. J. Peakin

3 Producción en hormigas y termes 45
M. G. Nielsen & G. Josens

4 Alimento y hábitos alimentarios en los termes 55
I. G. Wood

5 Alimento y hábitos alimentarios en las hormigas 81
D. J. Stradling
Recolección en las hormigas almacenadoras de granos 107
W. G. Whitford

6 Respiración y flujo energético 111
G. J. Peakin & G. Josens

7 Dinámica de los nutrientes en termes 165
J. P. La Fage & W. L. Nutting

8 Dinámica de los nutrientes en hormigas 233
A. Abbott

9 Papel de los termes en los ecosistemas 245
T. G. Wood & W. A. Sands

10 Papel de las hormigas en los ecosistemas 293
J. Pętal
Comparación de distintos biomas 326
B. Pisarski

Apéndice
1 Tablas 1, 2 y 3 de datos sobre las hormigas, por C. Baroni-Urbani and B. Pisarski 333

2 Lista de proyectos del IBP, que incluyen trabajos sobre insectos sociales (tomada de IBP News 13), más otras informaciones 341

Bibliografía 343

Indice 401

List of contributors

A. Abbott, Institute of Terrestrial Ecology, Furzebrook, Dorset, England
C. Baroni-Urbani, Museum d'Histoire Naturelle, Bâle, Suisse
M. V. Brian, Institute of Terrestrial Ecology, Furzebrook, Dorset, England
J. B. Cragg, University of Calgary, Alberta, Canada
G. Josens, Université Libre de Bruxelles, Bruxelles, Belgique
J. P. La Fage, University of Arizona, Tucson, Arizona, USA
M. G. Nielsen, University of Aarhus, Aarhus, Denmark
W. L. Nutting, University of Arizona, Tucson, Arizona, USA
G. J. Peakin, University of London, London, England
J. Pętal, Institute of Ecology, Dziekanów Leśny near Warsaw, Poland
B. Pisarski, Institute of Zoology, Warsaw, Poland
W. A. Sands, Centre for Overseas Pest Research, London, England
D. J. Stradling, University of the West Indies, St Augustine, Trinidad
W. G. Whitford, New Mexico State University, Las Cruces, New Mexico, USA
T. G. Wood, Centre for Overseas Pest Research, London, England and Agricultural Research Station, Mokwa, Nigeria

Preface

At meetings in Poland in 1966 under the chairmanship of Dr François Bourlière, at that time Convener of the Section, the programme of the IBP section concerned with the Productivity of Terrestrial Communities was agreed upon. The social insects were selected for special study because of their major activities as consumers in many ecosystems.

The Working Group in charge of the Theme 'Social Insects', hoped to organise a programme which would include most groups of social insects. However, it soon became clear that the task was beyond the limited resources that could be made available to the Group. The Sectional Committee's original description of the scope of the Theme 'Social Insects' had stated that 'in semi-arid savannah, steppe, prairie and heath habitats (particularly those in the tropics) both ants and termites abound and are unquestionably of great importance in ecosystem functioning'. It was therefore agreed by the Working Group that attention should be concentrated on ants and termites.

Many biologists must regret that the Theme's scope had to be limited. As events have shown, the Working Group made a wise decision. National Committees gave the bulk of their funding to major biome studies and some even ignored the need for intensive studies on special groups of organisms no matter what their functional importance in particular ecosystems. The Theme 'Social Insects' survived, whereas two others concerned with consumer groups – those on invertebrates and large herbivores – disappeared as discrete entities. These other organisms have not been wholly neglected in IBP as will be evident when the Biome synthesis volumes are published.

The task facing the Working Group, even when its activities were limited to ants and termites, was quite formidable. It was Aristotle who said 'Of all insects, one may also say of all living creatures, the most industrious are the ant, the bee, the hornet, the wasp, and in point of fact all creatures akin to these . . .' (*Historia Animalium* Translated by D'Arcy W. Thompson). In addition to their industry, Aristotle could justifiably have commented on their complexity and numerical abundance. For those readers of this volume who are unaware of the tremendous success of social insects, it is enough to quote the first few lines from the chapter on social insects in E. O. Wilson's *Sociobiology* (1975). 'There are more species of ants in a square kilometre of Brazilian forest than all the species of primates in the world, more workers in a single colony of driver ants than all the lions and elephants in Africa'. Figures for the number of species of ants and termites are equally impressive and of the thousands that have been given a binomial label, fewer than 20, perhaps even 10, are known in great enough detail to measure their role in ecosystem functioning.

In spite of the social mode of life, quantitative investigations on social insects are not easy and this volume rightly emphasises the importance of developing reliable techniques. Variations in the composition of individual 'classes' in colonies, the longevity of some individuals and of colonies, are among the problems which have been faced with considerable success by contributors to this volume. As for the measurement of metabolic activities, while some species are ideal subjects for laboratory study, it is no simple matter to extrapolate laboratory findings to field conditions where direct measurements are rendered almost impossible by the presence of other organisms and, in particular, large amounts of symbiotic fungi.

This synthesis of studies on ants and termites is indeed timely. IBP was created to study biological productivity and human welfare. The social insects are among the most successful of organisms when it comes to utilising the resources of their environment. In terms of metabolic activity, many areas of Africa, for example, can be said to belong to the termites. There, and in many other parts of the world, their only serious competitor is *Homo sapiens*. Can *Homo sapiens* outwit them and construct an integrated set of ecosystems in which natural and man-made sub-units can function in some degree of balance? This book does not answer that question but it provides many useful starting points for further study.

Homo sapiens is a very brash newcomer to community living. Ants and termites had solved many of the problems which now face modern man millions of years before even his primate ancestors came into existence. I have no wish to suggest that we turn to the ant for advice on all matters dealing with the division of labour, population control, utilisation of basic resources and conservation of essential food sources. However, there are many parallels which are worthy of examination. One of the features which has made large colonies of social insects possible has been a switch in feeding habits from an animal protein diet to one of plant material. A transformation which Western man is somewhat reluctant to make, as witness the amount of primary production utilised to feed domestic cattle.

Like man, social insects have utilised emigration as a method of dealing with over-population, but they have also shown that it is necessary to establish population limitation systems as well. Whilst the solutions adopted by social insects are not available to man as part of his genetic makeup, the fact remains that man will have to evolve a system of living which encourages population control in some systematised way. In the immediate future, his first task is to develop ways of working with social insects in certain ecosystems to take advantage of their remarkable success as ecosystem 'managers'.

J. B. CRAGG
Killam Memorial Professor
Faculty of Environmental Design,
University of Calgary, Calgary, Canada

1. Introduction

M. V. BRIAN

This contribution to the IBP synthesis series is exclusively concerned with termites and ants. The study of these groups of highly social insects from the production point of view was greatly stimulated by the IBP. As a result, some crude measurements have been made and more important, a new approach to their population biology initiated and a new set of problems formulated. This is one reason why those who dip into this volume will find a certain roughness: a conglomerate rather than a blend. Another reason is that the writers are intensely occupied with their investigations and could only be pursuaded, with great difficulty, to write anything at all between spells in the field. The compensation of course is that their views are fresh. Lastly, they speak, write and think in different languages, but all have without exception written in English. There has been no translation, and only enough editorial alteration to provide reasonable intelligibility and fluency.

The IBP/Productivity of Terrestrial Communities Working Group on social insects

At a meeting in Warsaw in 1966 Professor F. Bourlière who was at that time convener of the IBP/Terrestrial Production (PT) Section set up a number of Working Groups. Some were designed to give special emphasis to biome types like forest or grassland, others, to give emphasis to groups of organisms with obvious importance in ecosystems, e.g. large herbivores, graminivorous birds and social insects. These are listed as 15 themes in IBP News No. 13, and social insects are Theme 12.

The Working Group first met in the headquarters of UNESCO in Paris 1967 during the symposium on 'Methods of study in soil ecology' organised by Dr J. Phillipson. Two of the Group contributed papers on ants and termites to the symposium; one on social insects as modifiers of soil vegetation in grassland (Pętal, Jakubczyk & Wójcik, 1970), the other on the measurement of population and energy flow in these insects (Brian, 1970). The Group also met in 1968 in Moscow during the International Congress of Entomology.

This was followed by a meeting in Warsaw in 1970 under the auspices of the Polish Academy of Sciences. The meeting was arranged by Dr J. Pętal and about 30 ecologists from 10 countries took part and produced a very stimulating exchange of information and ideas. Visits were arranged by our Polish colleagues to the Institute of Ecology at Dziekanów Leśny near Warsaw and the Institute of Zoology Field Station at Ustrzvki in south Poland. The proceedings of this meeting have been published as a collection of 18 papers (Brian & Pętal, 1972) under the general heading 'Productivity investigations on social insects and their role in the ecosystem'.

It was after this that the Group began to collect material for this synthesis volume. The form was decided in the Mølslaboratory of the Zoological Institute of the University of Aarhus during a weeks' session in the summer of 1972, arranged by Dr M. G. Nielsen. In general it follows the IBP Handbook No. 13 on *Productivity of terrestrial animals. Principles and methods* written by K. Petruswicz & A. Macfadyen (1970); however, since most of the authors are either termite or ant specialists separate chapters on each group have often been unavoidable.

The Group would have liked to include bees and wasps in its orbit but decided, nevertheless, to restrict its interests to termites and ants. Bees, of course, play a very vital and still little appreciated role in natural processes as pollinators, and wasps prey on and consume very many and varied insects. Both make nests that are the foci of activity for many other organisms and both are immensely interesting in their social life and evolution. As they live over most of the world, though they have a lower biomass than ants and termites, their ability to maintain nest temperatures well above ambient, their aerial mobility, and their rich and intricate behaviour give them a subtle and pervasive influence on natural systems. The fact that they are the only social insect with a mutualistic relation to man should also not be forgotten.

Nevertheless, it was inevitable that a subgroup of IBP/PT should first concentrate on termites and ants. Termites are almost wholly decomposers of wood, humus, organic matter and foliage. They both nest and live for the most part in or near the soil, where they often attain high densities and considerable microclimatic regulation. Ants, in contrast, are diverse feeders ranging from predators to herbivores; they inhabit all strata of most habitats including the soil in which they usually but not always nest. Termites live almost exclusively in tropical and sub-tropical zones but these may vary from very wet to very dry. Ants range more widely: from the equator to the arctic circle. Even in temperate regions they may live in any habitat between mire and desert if it is sunny enough. Their ecological radiation is astonishing and probably much greater than that of termites; however, both have converged on fungus cultivation though in different parts of the sub-tropical zone.

Termites and ants started their evolution from different initial states, though both were soil insects. Termites probably began as cockroach-like creatures living clandestinely and eating wood or the fungi that decompose wood. Ants are thought to have evolved from wasp-like ectoparasites of soil insect larvae. In termites both sexes lived together and probably had similar behavioural repertoires. They continued so and evolved similar polymorphism comprising a reproductive and a worker caste with, almost invariably, a defensive caste as well. Ants evolved similar differences, but they were restricted to the female sex. This was probably because long before sociality started, there was a profound difference in the complexity of male and female behaviour: the former merely fed themselves and provided sperm, but the latter built earth cells, sought, identified, paralysed and collected hosts, laid eggs on them and stayed to guard the young larvae. Even though males had probably lost the capacity to evolve socially, there are genetic reasons why in an order with sex determined by haplo-diploidy, male sociality is improbable (West Eberhard, 1975). The bi-sexuality of termite societies has been a source of additional polymorphism; in the highly evolved family Termitidae one sex may give soldiers and workers whilst the other gives only workers. Often too, the sexes differ in size and function.

Termites and ants compared with vertebrates

Ants and termites are both highly socialised. Their groupings are, in extreme cases, very intricately differentiated and organised. This enables them to function as large organisms comparable with the bigger mammals, in their influence in ecosystems. Within their temperature, humidity and gas-regulated nests the enzymic processes of digestion, assimilation, growth and reproduction proceed as powerfully as in a large homeothermic vertebrate. Outside, along permanent trunk routes, often underground for protection from predators and weather, hordes of foragers carry fluids (internally) and solid food items (in their jaws). Added together, this collection of a multiplicity of small objects over a long time rather than a few big ones over a short time amounts to a very substantial intake.

Though both groups of social insects have evolved these mammoth communities, one element of the vertebrate's ability is usually lacking: its mobility. The bigger the insect society the less its mobility as a group and the further the individual foragers must walk. This limits them to places where resources are ample, steady, reliable and accessible, without too much struggle and competition. As a result no doubt the species with the biggest colonies tend to be primary consumers and digest plants either unaided or with the help of micro-organisms such as bacteria, fungi or protozoa in their intestine or in the nest cells. A small part of the vertebrate's mobility has been achieved by army ants (sub-family Dorylinae). In this tropical group the whole colony has nomadic phases when they wander in search of food, and rest in different sheltered places each day, often above the ground. The food collected in these periods is fed into growing larval bodies which subsequently transform into adults during a sedentary phase when feeding is light and local (Schneirla, 1971). Whilst such giants have evolved, the vast majority of ant and termite species has remained outside the mainstream of ecological processes; they still live in comparatively simple societies inhabiting inter-territorial parts of the habitat mosaic and collecting only the crumbs left by the main species.

The cost of evolving these societies falls largely on communication and transport. The benefits are physical strength, environmental control, reproductive and population control, and trophic stability arising from a larger storage capacity and the establishment of mutualistic relationships with other species, social and non-social. Such types become more exclusive and able to isolate themselves from the multiplicity of contacts that is probably normal with other organisms in wild conditions. The reduced loss of matter and energy which results from this must, presumably, more than compensate for the increased energy and time devoted to transport and communication.

I wish to thank Eileen Crutchfield for invaluable secretarial help and Andrew Abbott for substantial assistance in checking and collating references.

2. Empirical data and demographic parameters

C. BARONI-URBANI, G. JOSENS & G. J. PEAKIN

The wide geographic distribution of ants and termites and the variety of the environments with which they are associated makes it impossible to recommend a generally applicable method for assessing their populations. Hitherto, investigations have, for the most part, been concerned with population studies with strictly limited objectives. In reviewing them and assessing their contribution to the estimation of energy flow through ants and termites in ecosystems, it is important to consider what attention has been paid to two important attributes of social life, namely nesting habits and organisation.

Concerning their nesting behaviour, one is faced immediately with the difficulty of defining a nest. The word is essentially one of convenience since nests form a continuous spectrum, extending from a completely diffuse system of galleries in the soil, with the members of the society scattered throughout them, to a highly concentrated, architecturally elaborate structure in which the majority of the colony is housed and from which foragers emerge into the surrounding territory. Furthermore, the nest is generally regarded as a more or less permanent feature of social organisation but, of course, the nomadic doryline and ecitonine ants confound this view, though their temporary bivouacs are clearly equivalent to the nest. Despite these difficulties of definition, it is none the less useful to attempt broadly to classify nesting habits, in order to appreciate the nature of the problems facing production ecologists.

Six categories may be used as a foundation for this classification, gathered into two groups on the basis of the biotope utilised. Three categories comprising the first group are associated with the soil: the insects may nest underground without external indication of their presence; they may have surface holes, for the passage of foragers, manifesting their presence; or, they may nest partly underground and partly above ground, forming epigeic mounds containing the majority of the society. The second group includes species that nest in the aerial parts of plants and again there are three categories: they may do so without external sign of their activities or with external evidence of their presence or, most obviously, the nest may be externally attached to the plant. It cannot be emphasised too strongly that the nesting habits of many species do not conform exclusively to one category or another, but may change during the development of the society or in response to the particular characteristics of the environment in which they are found.

The population densities (number of individuals per unit area or unit volume or per nest) are the prime data when demographic studies are undertaken. The evaluation of density, however, presents characteristic problems (in statistical sampling and treatment) due to the very highly clustered (social) distribution of individuals. Any way that can 'normalise' the dispersion of the populations or reduce the effects of their aggregation will increase the precision of the results. It is obviously possible, in the case of several social insect species, to characterise clusters as nests containing an entire society, nests containing a calie (that is, a part of a society), or simply a group of chambers. The clusters may then be taken as individuals in density studies, and the number of termites or ants per cluster studied concurrently.

Several papers have recently treated the problem of sampling and density studies and summarised the relevant literature. It is sufficient to quote here the important reviews by Pętal & Pisarski (1966) for ants, Brian (1969) for ants and termites, Lévieux (1969), for ants, and Lee & Wood (1971*b*) and Sands (1972*a*) for termites. From these works and subsequent papers the following methods of evaluating natural populations can be extracted.

Sampling of nests with epigeic structures

Density of the mounds

For mound nests built by the social insects and obviously visible, several different methods have been used to estimate the nest densities. However, it is important to remember that not every nest of the mound-building species is epigeic; the young ones, for example, are generally subterranean. Moreover, mounds and colonies cannot always be equated.

Aerial photographs

Glover, Trump & Wateridge (1964) studied the characteristic patterns of vegetation on the Loita Plains of north-western Kenya. The patterns (6.5–10 m in width, with an average length of 14 m) were based on low termite mounds of an *Odontotermes* sp.; 650 such patterns were counted in one square kilometre.

Bouillon & Kidieri (1964) used aerial photographs to make a census of the large nests of *Macrotermes subhyalinus* (Rambur) in the Ubangi region (northern Zaire); the large mounds appeared conspicuously on 1/25000 scaled photographs, and the authors suggested that densities of *Trinervitermes* and *Cubitermes* could be studied by means of such pictures at a suitable scale.

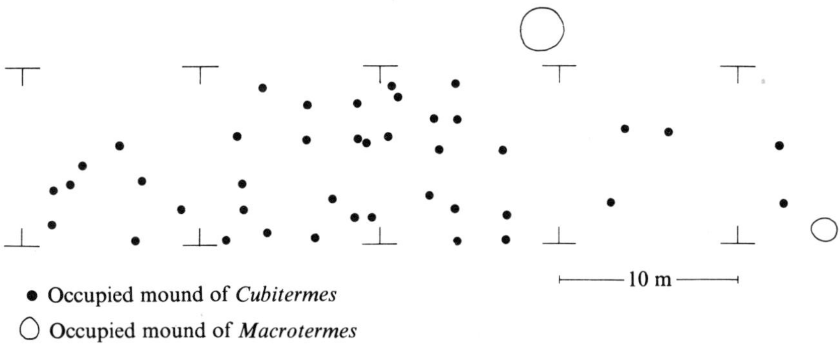

Fig. 2.1. Dispersion of the occupied mounds of *Cubitermes sankurensis* and *Macrotermes bellicosus* (= *Bellicositermes natalensis* before the revision of Ruelle, 1970) on a transect 10 m wide, in a savannah near Kinshasa in Zaire. (After Mathot, 1967.)

Wood & Lee (1971) counted the tall mounds (up to 6 m high) of *Nasutitermes triodiae* (Froggatt) in a tree savannah of northern Australia and studied the dispersion of these nests by means of aerial photographs. Lee & Wood (1971*b*) emphasised the fact that aerial photographs can only be used in savannah countries with rather obvious nests. They are most useful where the density of the mounds is low, and the season and time of day are chosen carefully to maximise contrast with surroundings, but a ground survey is always desirable for completing the study.

Fisser (1970) compared surveys of harvester ant mounds (though he does not mention the species), made from aerial photographs with those from direct counts. The photographs gave a mound density of 6.92 and 7.12 per hectare in 1954 and 1966, respectively. A direct count in 1966 found 7.51 mounds per hectare, though this apparently higher mean density was not significantly different from that established from the aerial photographs. Fisser mentioned a number of sources of error in determining the density from aerial photographs, such as distinguishing between ant mounds and other denuded areas and, of course, the fact that young nests may be overlooked. He emphasised the importance of experience and familiarity with the nesting habits in the correct interpretation of the photographs.

Line transects

Mathot (1967) has observed very large variations in nest densities of *Cubitermes sankurensis* Wasmann: in seemingly homogeneous savannah biotopes, species achieved very high densities (up to 4000 nests per hectare) in certain areas, yet might be absent in adjacent areas. He studied a belt 10 m wide and 210 m in length – in fact a linear succession of 100 m² quadrats – and counted, in the 21 successive plots: 0, 0, 0, 0, 0, 0, 0, 0, 0, 0, 11, 10, 10, 8, 15, 13, 3, 2, 0, 0 and 0 living nests of *C.*

Table 2.1. *Some examples of densities of termite mounds, and the sampling area used in each case*[a]

Species	Density: number of *occupied* mounds per hectare	Size of quadrats used in study	Author
Cubitermes exiguus	Up to 652	100 m^2	Bouillon & Mathot, 1964
Cubitermes sankurensis	Up to 550	100 m^2	Bouillon & Mathot, 1964
Trinervitermes geminatus	Up to 505	405 m^2	Sands, 1965*a*
Amitermes laurensis	217	2800 m^2	Wood & Lee, 1971
Trinervitermes trinervoides	82	8550 m^2[b]	Coaton, 1948*a*
Trinervitermes geminatus	Up to 58	2500 m^2	Josens, 1971*c*
Amitermes laurensis	21	1.71 hectares	Wood & Lee, 1971
Macrotermes subhyalinus (*Bellicositermes bellicosus rex*)	2.3–2.9	244 hectares by aerial photograph (scale 1:25000)	Bouillon & Kidieri, 1964

[a] For more numerous and detailed examples concerning the terminates, refer to Lee & Wood (1971*b*): Tables 5, 8, 9, 10 and 12.

[b] 8550 m^2 approximates the 'Morgen' surface unit used in South Africa.

sankurensis, the last five quadrats contained (or were surrounded by) nests of *Macrotermes* (Fig. 2.1).

Josens (1972), taking a census of the mounds of *Trinervitermes* spp. at the IBP Lamto project (Ivory Coast), used a reel of disposable thread attached to a trip counter which indicates the distance covered. The nests were counted in a belt 6 m wide, and while this method furnishes little information on the dispersion of the mounds, it can provide data on the variations of the specific composition of communities when several biotopes are traversed.

The quadrat method

The quadrat method, used by many research workers, provides the most detailed information on density as well as on dispersion. It is, however, rather time consuming, as the positions of the quadrats have to be randomised and they may be difficult to lay out in tree savannahs or woodlands. Moreover, small nests may escape counting if the vegetation is high. Josens (1971*b*) used this method only after bushfires, when the mounds were readily visible. The size of the quadrats is determined by the density of the nests: as small as possible but large enough to contain (if possible) some tens of nests. Table 2.1 summarises the sizes of quadrats used by various authors in different conditions of mound densities and Table 2.2 compares the densities of mound-building termites in three different

Table 2.2. *Density of termite mounds (only occupied ones) per hectare in West African savannahs: a comparison between three countries*

	Northern Nigeria (Sands, 1965*a*)			Middle Ivory Coast (Josens, 1972)			Southern Ivory Coast (Bodot, 1966, 1967*a*)		
	A	B	C	D	E	F	G	H	I
Amitermes evuncifer	—	—	—	6	1	3.2	51.4	4.3	6.4
Cubitermes severus	—	—	—	—	—	—	45.4	—	—
Macrotermes bellicosus	14.8	—	—	< 1	< 1	< 1	1.85	—	0.6
Odontotermes sudanensis	5.0	—	—	—	—	—	—	—	—
Trinervitermes geminatus	59.2	235	501	11	52	40.5	—	—	—
Trinervitermes togoensis	2.5	27.2	9.9	30	10	14.7	—	—	—
Trinervitermes trinervius	17.3	4.9	12.3	2	—	0.7	9.05	3.0	3.6
Trinervitermes occidentalis	4.9	4.9	—	—	1	0.1	38.3	6.3	8.8
Trinervitermes oeconomus	24.7	7.4	7.4	2	9	2.6	—	—	—
Total	128.4	279.4	530.6	51	73	61.8	146.0	13.6	19.4

Biotopes: A = northern Guinean woodland-savannah with *Isoberlinia*, undisturbed.
B = the same, man-modified but not cultivated.
C = the same, cultivated a year before.
D = open *Loudetia*-savannah with *Borassus aethiopum*.
E = woodland *Hyparrhenia*-savannah with *B. aethiopum*.
F = mean Lamto-savannah with *B. aethiopum*.
G = open *Loudetia*-savannah.
H = open *Anadelphia*-savannah.
I = open *Schizachyrium*-savannah on laterite.

savannahs of West Africa. From these data, it appears that large variations may occur in nest densities, even in adjacent biotopes.

Dispersion of the mounds

The analysis of the spatial dispersion of the mounds is a desirable complement to each density study. The first and easiest expression of dispersion is supplied by mapping out the sites of the mounds and the natural peculiarities of the biotopes on a convenient scale. Mathot (1967) observed the absence of living mounds of *Cubitermes sankurensis* close to those of *Macrotermes* (Fig. 2.1). Josens (1971*b*) noted, on a hydromorphic soil, the concentration of *Trinervitermes* and *Amitèrmes* mounds on small smooth eminences (probably very old and collapsed nests of *Macrotermes*): these microbiotopes avoid the drastic hydromorphic conditions during the wet seasons (Fig. 2.2).

In more homogeneous conditions, mathematical methods become more useful. The quadrats may be divided, if great and populous enough, into smaller plots (some tens at least); after counting the mounds in such plots,

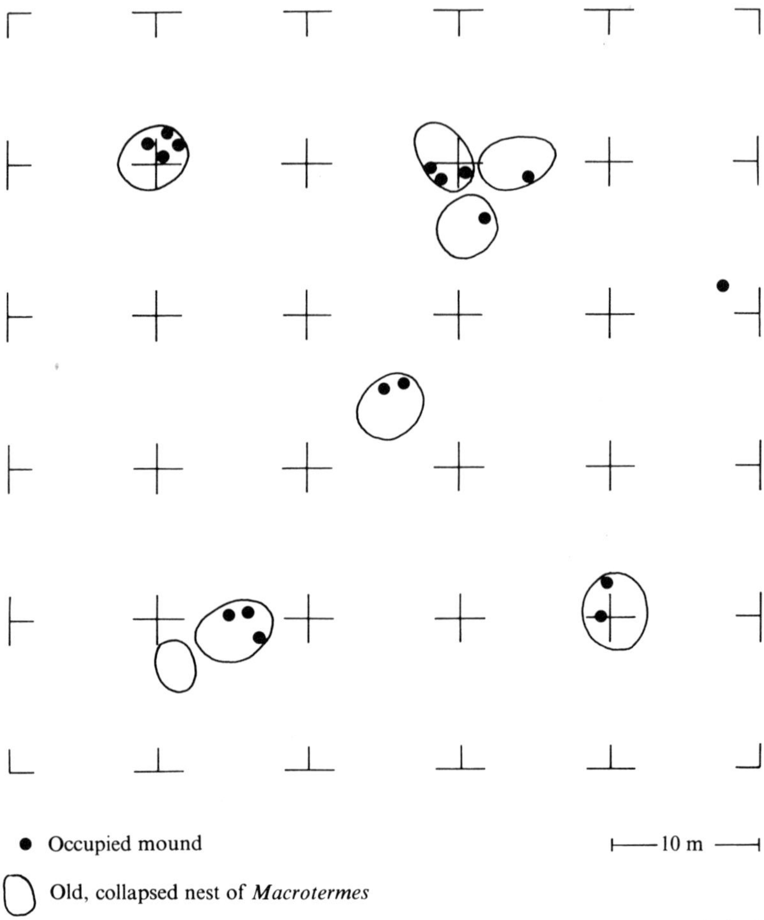

Fig. 2.2. Dispersion of mounds of *Trinervitermes trinervius*, *T. togoensis* and *Amitermes evuncifer* on a hydromorphic soil, at the IBP Lamto project, in the Ivory Coast. (After Josens, 1971*b*).

one may try to fit the observed frequency distribution to a theoretical distribution. One may also make use of the nearest-neighbour technique (Skellam, 1952; Clark & Evans, 1954), as Brian (1956*b*) did for ants in west Scotland. Sands (1965*a*), using this method, detected an aggregative tendency in the dispersion of mounds of *Trinervitermes geminatus* Wasmann, which normally builds polycalic nests. Wood & Lee (1971) discussed spatial distribution of termite mounds in various biotopes of Australia and found out that the nests were generally overdispersed, revealing competitive relations for food and/or space. The nearest-neighbour technique is, however, unable to distinguish different kinds of contagious distributions. Hébrant (1970*a*) mapped out 611 nests in an area of one hectare. To analyse the dispersion, he first used several variants of the nearest-neighbour technique. Second, he divided the area into plots; this gave more accurate results than the first, but he pointed out the importance of the size of the plots. By choosing plots of different sizes (25, 100 and 200 m^2) he noticed in each case a low contagious tendency.

Populations inhabiting the mounds

Estimate by taking a portion of the nest

When colonies are very populous, to count all individuals is virtually impossible. In this case, a measure of the total volume of the colony may be obtained, and the insects counted in small volume samples (Rettenmeyer, 1963). Of course, a dormant period must be selected for excavation, but this needs care as the brood composition of the nest varies. In South Africa, Hartwig (1956) used a soil auger of the pipe type capable of holding a sample of soil 2.54 cm in diameter and 61 cm deep. Though he did not estimate the total population by this technique, he found that the count of samples taken at sunrise or later in the morning differed very little from counts of samples taken at noon or during the afternoon. Sands (1965*b*), in a study of *Trinervitermes geminatus* in northern Nigeria, modified this method by mounting a core sampler of known volume on a club, which could be swung at the termite mound and so take a sample sufficiently rapidly to prevent significant migration of the termites from the sample point. The termites were extracted from the core by flotation or hand sorting and their density estimated. To determine the total population Sands related the estimated density in the sample to the supposed volume of the nest. This was taken arbitrarily as a sphere with a diameter equal to the mean diameter of the base of the nest; in an earlier study Sands (1961*b*) had defined the mean diameter of the base of the nest. He says 'this treatment...gave greater accuracy for comparative purposes than the use of the ordinary geometric mean diameter. How widely this is true must remain dubious. The main problem in using this method to estimate populations is the diurnal variation of the samples'. Operating at 04.00 hours, at 12.00 hours and 16.00 hours Sands (1965*b*) observed that 'distinct movements of considerable parts of the population occurred within the mound during the day and night'. Bouillon & Lekie (1964) have reached the same conclusions after their researchers on populations of *Cubitermes sankurensis*.

Estimate by sampling whole nests

Several authors excavated whole mounds to estimate the populations; the insects were extracted by hand or by flotation, and a proportion counted.

The most important precaution in using this method is to avoid sampling when a significant part of the society is out of its nest!

Greaves (1967) applied dry ice onto the trunks of trees containing nests of *Coptotermes acinaciformis* (Froggatt) and *Coptotermes frenchi* (Hill) so as to concentrate the termites into the nursery before removing the nest. Bouillon & Mathot (1964) sifted the soil around the nests of *Cubitermes exiguus* Mathot, both to collect termites outside the nest and to isolate those in the nest, before excavating the nest itself. Josens (1972) trenched around the mounds of *Trinervitermes geminatus*, to stop termites escaping by the subterranean galleries.

A few authors have tried to calculate correlations between the size of the nest and the number of inhabitants. Bouillon & Lekie (1964) obtained a very weak correlation between the weight of the nests of *Cubitermes sankurensis* and their populations (Spearman's coefficient = 0.580, significant at the 95% confidence level) with 12 nests sampled each time at the same hour. In the case of *Nasutitermes costalis* (Holmgren) and *Trinervitermes geminatus*, Wiegert & Coleman (1970) and Josens (1972) found rather good correlations between the surface of the outer wall of the nest and the number of termites, suggesting a constant adjustment of the size of the nest to their respiratory function, the exchange of oxygen and carbon dioxide through the outer wall.

Sampling of nests with obvious entrances

Density and dispersion of the nests

When each nest of a subterranean species has one obvious major entrance, the nest may be regarded as epigeic, as far as the sampling techniques are concerned.

The quadrat method, described above, is the most useful one in estimating their densities, and both the nearest-neighbour technique and the plot division method can be used in studying their dispersion. However, when subterranean nests have several foraging holes, one should first study their structure and dimensions; otherwise unwarranted suppositions may be made. For example, Bitancourt (1941) assumed that the number of nest entrances is a linear function of the number of workers in *Atta sexdens* (L.). But it must be remembered that the number of foraging holes can vary widely throughout the year. Coaton (1948*b*) described a nest of *Hodotermes mossambicus* (Hagen), a grass-cutting termite, which had 193 foraging holes in May and more than 552 in September. During July, he counted about 15000 foraging holes per hectare in an area densely populated by the harvester termite – 'the whole veld appeared to be covered with large, bare patches separated from one another by narrow strips and patches of semi-denuded grass'. If 550 holes per nest is assumed as an average for July, then the subterranean harvester termite would occur at a density of 27 nests per hectare in the area he described. Nel (1968*b*) studied the density and spatial dispersion of foraging holes and soil dumps made by *H. mossambicus*. Twenty-nine plots of 108.6 m^2 (1.83 m wide), not randomly dispersed, were examined weekly. Large monthly variations occurred, more than 830 soil dumps per hectare were erected during April, against only 16 during September, while the minimum number of foraging holes was open from November to January and the maximum during July. The negative binomial distribution could generally be fitted to the distribution of both foraging holes and soil dumps. Extending this work, Nel (1968*a*) studied the aggressive behaviour of workers collected from one hole and mixed (in a Petri dish or on the ground) with workers from another hole. Making the following assumptions: (i) workers that attack each other come from different societies and (ii) a group of foraging holes with compatible workers are situated on at least part of the foraging territory of one society, he mapped out the territories covered by 'such' holes. From this study in an area of about 0.3 hectare he estimated the nest density at about 45 per hectare.

Sampling the nests

A large number of ants (and a few termites) have nests with obvious entrances so that when a portion of a nest has been removed, the insects tend to escape, sometimes very quickly. Thus, for accurate counting, the whole nest population must be killed with carbon disulphide (Yung, 1900) or by carbon tetrachloride, ethyl acetate, alcohol or any other suitable poison. This method seems to be much the best for common species with small colony populations, so that with a limited sampling effort one can obtain a reasonably exact figure for the mean colony composition. In numerically more complex situations the direct count cannot be recommended at all, both on account of the amount of time involved and the low accuracy of the results.

If whole societies can be collected, the population may be estimated by counting a portion and extrapolating. This method, in ants, was first applied by Schneirla (1957*a*, 1958) who used it to evaluate the brood content of some New World army ants. Nearly complete broods from colonies of different *Neivamyrmex* spp. were collected and a random half of the brood, determined by weight, was carefully counted; an estimate of the total number of individuals of a pre-imaginal stage was obtained by doubling the previously obtained figure.

The same procedure can obviously be used to evaluate the total colony population with more or less large samples according to the degree of accuracy required. This has been done by Rettenmeyer (1963), also on neotropical army ants. Samples of 200–300 ml of ants, etherised or killed by alcohol and tamped repeatedly, were carefully counted. From a measure of the whole colony volume the total population is easy to compute. Though this method is easy to apply to driver ants which do not elaborate true nests, particular care should be taken to sample during the statary phase and in a quiescent period. Accurate volume estimates require the collection of the whole nest, but for a rough evaluation, an estimate of the colony volume in the field has often been made (Rettenmeyer, 1963). The most complete analysis of an ant population by this method has been carried out by Martin *et al.* (1967). A nest of *Atta colombica tonsipes* Santschi was dug up and the total worker population weighed. The weight of workers collected was 6.35 kg and this was estimated to represent between 65 and 85% of the total colony population, which should then have had a total weight of 7.48 to 9.78 kg. A random sample of workers had individual weights varying between 4 and 8 mg, so the total weight represented between one and two and a half million individuals.

Different methods of extrapolation of the total colony population from limited sampling have been proposed. They have been used mostly for very populous species, where direct counting and collecting of all individuals is practically impossible, and this is the only reason for their sporadic use, though the low degree of accuracy is recognised. The first attempt of this type was probably that of Vosseler (1905). In order to evaluate the colony population of an *Anomma* species, this author counted the number of ants leaving the nest per second, and, by multiplying this figure by the total number of seconds in which the migration occurred, he obtained the total number of ants. A more sophisticated method is suggested, for the same genus, by Raignier & van Boven (1955). They observed that the average speed of a migrating colony was 169 m per hour and the average duration of a migration as 56.3 hours, giving a total length of the migrating colony of 9515 m. The average width of a migrating column was found to be 2.98 cm and its average density 7.71 ants/cm^2, so the total population of an average *Anomma wilverthi* colony would be $951\,500 \times 2.98 \times 7.71 = 21\,861\,474$. Rettenmeyer (1963) also used a similar method to study the *Eciton* colony population. He recorded the time necessary for 50 (or 25) ants to pass a given point in a migrating column and, from the total migration period, the population was estimated. Clearly, there are a number of sources of error in such an estimate: fluctuations in the speed of movement of the ants, not all the ants move in the same direction and no account is taken of the ants foraging or those who have already passed the census point. In spite of these drawbacks Rettenmeyer (1963) was of the opinion that this method gave more accurate results than the volumetric technique already described.

Direct sampling from the soil

Civil engineering works

Hartwig (1966) used the sewerage and stormwater excavations made by the municipality of Bloemfontein (Orange Free State) for observing the nests of *Odontotermes latericius* (Haviland). He obtained an estimate of their density by digging at random a trench 0.61 m wide and 15.87 km long in an area of 282 hectares. Knowing the diameter of the nests – from three to six times larger than the width of the trench – the actual nest density was calculated by substituting in the formula:

$$d = (N \times W_t / W_n \times 10\,000)/S_t \qquad (2.1)$$

where d is the nest density per hectare, N is the number of nests encountered, W_t is the width of the trench, W_n is the mean diameter of the nests and S_t is the area of the trench in a square metre. If $W_n < W_t$, W_t/W_n was omitted. Hartwig assumed a random distribution of the nests of *O. latericius* and applied the Poisson distribution for estimating the probability of meeting 0, 1, 2, . . . nests per surface unit.

Bouillon, Lekie & Mathot (1962) also took advantage of human activities (borings, contourings, drainage, foundations, road and railway works) for estimating nest densities of *Apicotermes* spp. not far from Kinshasa (Zaire). They analysed the dispersion of the nests of *Apicotermes gurgulifex* (dead and alive together) and found that a Pascal distribution fitted their data reasonably well, as the nests seemed to be dispersed contagiously without being polycalic. *A. gurgulifex* was found in savannah (of a savannah–forest mosaic). In a total area of 2467 m^2, between 24 and 100 occupied nests per hectare were recorded in different sites in the Amba Mountains. *Apicotermes desneuxi* was found in the same location, but only in forest patches, with a density of 238 occupied nests per hectare (the area examined being 674 m^2).

The quadrat method: pits

Empirical use of the quadrat method

In ants, this method has been widely used by different authors and with different purposes but up to now only Pętal (1972) has been concerned with a measurement of productivity. The use of direct sampling, of course, poses many problems, according to the degree of accuracy required. In production studies where the unit is represented by the individual or classes of individuals matched for sex, size, or age, great accuracy is required.

Oinonen (1956) showed that a sample containing all seven ant species

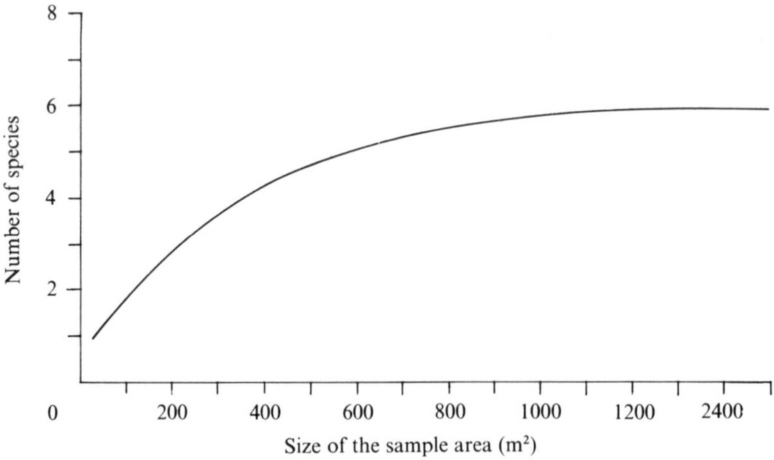

Fig. 2.3. Number of ant species in S. Finland according to the sample size (redrawn after Oinonen, 1956).

present in a Finnish rocky environment should be at least 1200 m^2 (Fig. 2.3); major differences have been found in samples ranging between 25 and 300 m^2.

The best quadrat size depends on the purpose of the study. For this reason, Baroni-Urbani (1968), in order to show spatial competition between ant species in some coastal ecosystems in central Italy, used samples containing no more than 75% of the species present. This condition, in an area containing five to twelve species, was met with quadrats of 14 m^2. Subsequently, Baroni-Urbani (1969), dealing in the high altitude Appennine grasslands with an impoverished ant fauna, both in terms of population and number of species, decided that a sample containing 50% of the total number of species (six) was appropriate and this was accomplished by using 11 m^2 quadrats. Baroni-Urbani & Kannowski (1974) found a 100 m^2 quadrat very useful to study both population density and interspecific competition of the ant *Solenopsis invicta* Buren in Louisiana.

Lévieux (1969) investigating ants in an Ivory Coast savannah (at the IBP Lamto project) used the same graphical method as Oinonen (1956). He found that by plotting the number of 16 m^2 quadrats against the number of species of ants located in the quadrats, a smooth curve was produced (Fig. 2.4), from which it could be deduced that 25 quadrats were adequate for locating 80% of the ant species. Josens (1971*b*), employing the same technique, showed that essentially all of the common subterranean termite species in the same location could be found in the same number of quadrats (Fig. 2.4). Josens (unpublished data) plotted in the same way the results obtained with quadrats of 16 and 1 m^2 and found that a total area

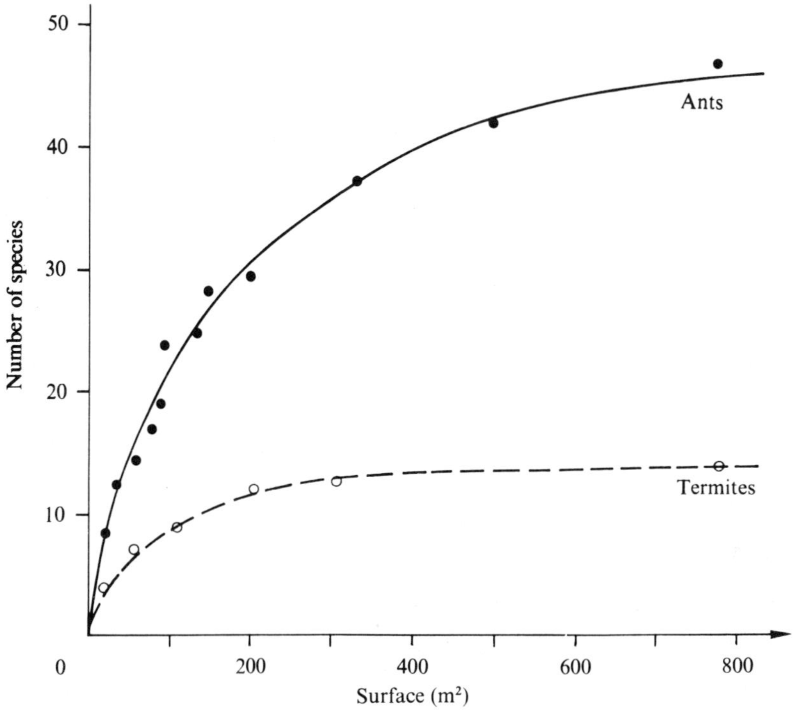

Fig. 2.4. Number of ant and termite species found in terms of the prospected surface of savannah, at the IBP Lamto project in the Ivory Coast. (After Lévieux, 1969 for the ants, and Josens, 1971*b* for the termites.)

of 800 m^2 sampled with large quadrats was required to locate the 14 commonest termites, as against 60 m^2 with the smaller quadrats (Fig. 2.5). However, the large pits were dug out quickly, and the termites were seen only if abundant, while in the small pits, the soil was crumbled by hand and the termites were recorded with more reliability. The costs (in man-hours) of sampling were about the same in the two cases. Given the particular nature of the savannah studied by Lévieux and Josens (frequent and shallow layers of stones), a depth of 25 cm (implying a volume of 4 m^3 of soil per 16 m^2 plot) seems to be sufficient, but in different conditions and with different species (e.g. *Messor*, *Cataglyphis*, etc.) a deeper sample would be necessary.

Pits have been used in some interesting studies on termite densities in Africa. The sizes and depths of the pits were defined by local circumstances and by the actual aims of the work. Kemp (1955) dug out trenches of 5 m length, 1 m wide and 1 m deep 'in sites free from obstacles such as trees and rocks, selected at random'. The method was used essentially

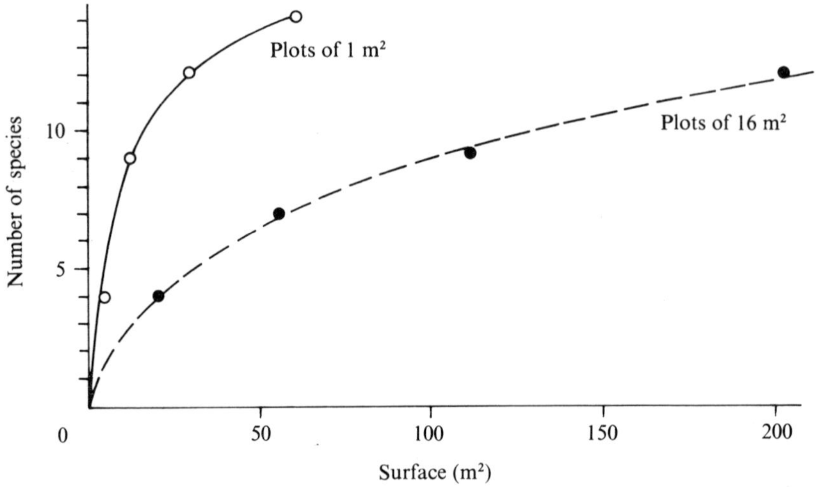

Fig. 2.5. Number of termite species found by means of two different quadrat techniques, in the same savannah, at the IBP Lamto project, in the Ivory Coast (Josens, unpublished data).

in order to compare the relative abundance of termites in the major biotopes of the north-eastern parts of Tanzania. The number of species met with varied from zero in cedar forests (five trenches) up to 12 in a woodland (10 trenches) and 13 in a thicket vegetation zone (10 trenches). Sands (1965*a*) used pits of 1.67 m^2 and 0.91 m deep for counting the termite 'occurrences' in the form of fungus combs of the hypogeic Macrotermitinae. A depth of 91 cm was considered sufficient owing to ironstone concretions frequently met (and the rarity of termites) at that depth. Ten quadrats of 405 m^2 were examined in each area studied and five pits were dug out in each quadrat. The positions of the quadrats and the pits had been selected by means of tables of random numbers. The results of both Sands and Kemp are reported in great detail in Lee & Wood (1971*b*).

Bigger (1966) studied the nest system of four subterranean Macrotermitinae and their fungus comb density on a plot of 111.5 m^2 (1.83 m deep). He worked in the south-eastern part of Tanzania where he encountered deep soils and considered that the termites could have constructed some combs even deeper than 1.83 m. Nevertheless, most combs were found in the superficial layers, about 75% of them occurring between the surface and a depth of 61 cm (Table 2.3). Bigger also dug out two trenches, 1.22 m wide, 0.91 m deep, and 68.28 m in length, cutting across several agricultural plots, to investigate the effects on termite nest density of several insecticides.

Table. 2.3. *Densities of occupied chambers of four subterranean Macrotermitinae in a cultured woodland (maize, soya, sorghum, etc.) in south-eastern Tanzania; after Bigger, 1966*

Species	Number of occupied cells per hectare, extrapolated from a plot of 111.5 m²
Microtermes albopartitus	38 000
Microtermes redemanius	14 900
Ancistrotermes latinotus	4 840
Allodotermes tenax	4 390
Total	62 130

At the IBP Lamto project (Ivory Coast), 79 quadrats of 16 and 25 m^2 and 25 cm deep, were excavated to study the fungus comb density and nest structure of four subterranean Macrotermitinae (Josens, 1972, in preparation). This study was preceded by some preliminary digs to greater depths in pits of 1 to 9 m^2, proceeding 10 cm at a time until two consecutive layers were without any fungus combs. It was found in quadrats totalling 373 m^2 that termites seldom nested deeper than 1 m, owing to a layer of stones limiting penetration of the insects. Moreover, seasonal variations were observed in the vertical dispersion of the combs. (Fig. 2.6.)

As the final part of this study, Josens estimated the densities of termites themselves with a third kind of pit, 1 m^2 by 0.60 m deep. A trench was dug out as quickly as possible around the pit, 60 cm deep, in order to limit the escape of insects; the sandy soil was then crumbled by hand and the termites were collected. In this way, 170 such pits were excavated, their positions in 0.25 hectare quadrats having been determined by means of tables of random numbers. A correlation was calculated for each species between the number of fungus combs and the number of termites found in the pits. The pits also proved very convenient for estimating populations of those species which form diffuse nests, such as soldierless and some humivorous species. Forest galleries and five types of savannah biotope, ranging from open savannah to woodland, constitute the ecosystem of Lamto: vegetation No. 8 on the map of Africa (Keay & Aubreville, 1959). All the savannah biotopes were studied and showed very important differences in their termite populations. Some results given in Table 2.4 are data corresponding with an average Lamto-savannah, where account is taken of the relative proportion of each type of savannah and the density of the termites in each area.

This method of estimating ant and termite populations provides reasonably accurate results, but, although in cold and temperate climates the sample size and number is certainly reduced, it seems to consume too

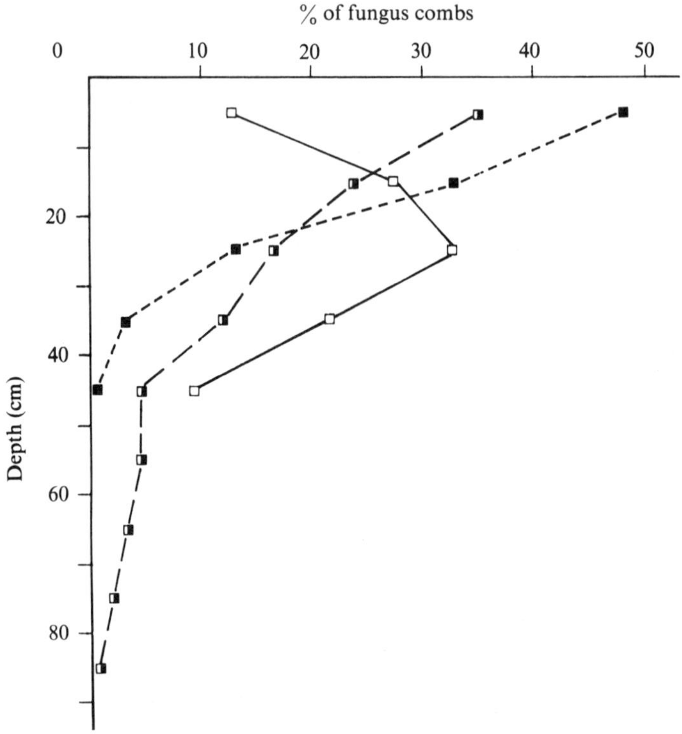

Fig. 2.6. Vertical dispersion of the fungus combs of *Ancistrotermes cavithorax* at three different periods of the year, at the IBP Lamto project, in the Ivory Coast. (After Josens, 1972.) ■—■, wet season; ◧—◧, beginning of the dry season; □—□, dry season.

much time and money to be applied to every situation. Moreover, in restricted and non-uniform habitats, it would imply an unacceptable destruction of a large part of the biome.

The first attempt to produce a quantitative picture of the ant population in a particular environment was that of Nefedov (1930), who studied the forest steppe of Troitsk in Russia. This author dug out 45 sample areas each of 100 m^2 in order to show the difference in the ant population between chernozem, alkaline and saline soils. The depth of the sample is not mentioned, although it would have provided useful information in such a situation.

In order to study the fluctuations of the ant population, Pickles (1940) dug out repeatedly an 880 m^2 area in Yorkshire during five years. The density he obtained (17 ants per m^2) is remarkably low for the first year and decreased during the subsequent years, probably because of the disturbance due to sampling.

Quispel (1941) scraped the humus layer off the soil with a hoe to study

Table 2.4. *Densities of the most common subterranean termite species in a 'mean Lamto-savannah' (Ivory Coast); after Josens, 1972*

	Densities per hectare		
Species	Nests	Fungus combs	Thousands of individuals
Basidentitermes potens	?	—	1390[a]
Adaiphrotermes sp.	?	—	1000[a]
Other humivorous species	?	—	380[a]
Microcerotermes parvulus	75	—	245
Pseudacanthotermes militaris	35	8000	320
Odontotermes sp.	17	3300	850
Ancistrotermes cavithorax	146	21500	2130
Microtermes toumodiensis	44	24400	1170

Total: about 7.5 million subterranean termites per hectare of 'mean Lamto-savannah'.
[a] Those numbers assume that there are less than five million humivorous termites per hectare in the open savannah biotope; however, new studies (e.g. D. Corveaule, unpublished data) show the humivorous termites to be more abundant (about 20 million individuals per hectare) in that biotope. The densities of the humivorous species given here, including *Procubitermes sjöestedti*, might therefore be increased substantially to about 5.8 million per hectare instead of 2.77 million individuals in the 'mean Lamto-savannah'.

the ant population in the national park *De Hoge Veluwe* in Holland. He studied a total of 54 samples of 50–100 m² each, finding a range of ant density between zero and 0.44 nests per m² according to the sample, although none of the three major environmental types considered (sandy soil, oak stands and pine plantations) is regularly destitute of ants. Westhoff & Westhoff-de Joncheere (1942) determined the minimum area necessary to have a representative sample of the ant population in a Dutch forest. They studied the *mierencombinatie* in different quadrats of 10, 25, 50 and 100 m² and found that the 50 m² quadrat was representative, while that of 100 m² offered very slight advantages. Nevertheless, the authors employed the 100 m² unit when taking 126 samples in 22 forest types.

Headley (1943, 1952) preferred to dig small areas out completely in an Ohio wood; perhaps because of this much more intensive collecting he obtained one of the highest figures of ant density, 11.1 ant colonies per m² in a total sample area of 27 m². However, one should also remember the well-known fact that nearctic ant fauna is much richer than the palaearctic both in the number of species and of individuals at the same latitude.

Headley excavated, 1 m² at a time, a total area of 27 m². In this respect, his work differs from that of Talbot (1953, 1954, 1957*b*, 1965) who studied larger areas by examining regularly distributed samples of 1 m^2 each. According to Talbot (1953), in such an area it may be hard to detect all the nests present, but the nests of at least a good proportion of the species were located by distributing cake crumbs and following the foragers returning to the nest with a particle in their mandibles. Later Talbot (1957*b*) examined smaller plots of a square yard (0.836 m^2) in preferentially distributed sites (i.e. easier to work in) within a large homogeneous area. These units were dug out to a depth of 15–43 cm, according to the season, in order to allow for the species going deeper in the soil in dry, summer conditions. In a subsequent paper (Talbot, 1965) the same size quadrat was used but it was regularly distributed within the study area.

The greatest sampling effort in absolute terms seems to be represented by the work of Oinonen (1956) already mentioned who, in south Finland, studied a total of 153 quadrats varying in size between 25 and 5600 m^2. Each sample was systematically searched for ants with a mattock. These samples included eight different environmental types and the ant demographic data were positively correlated with the growth of the forests.

Brian (1956*b*), instead of fixing a standard area unit, counted the total number of nests in whole glades of a pine plantation in Scotland. The glade area was measured by considering the vertical projection of the canopy. A somewhat different method has been used in a later paper (Brian, Hibble & Stradling, 1965) where small homogeneous areas of different size were used for the study of some selected ant species populations. Only one of the species proved to be regularly distributed within all these areas (*Tetramorium caespitum* L.) and its nest density was determined by accurate counting of 24 circular samples of 4 m radius. An evaluation of the average nest population by the mark–recapture method (see later) enabled some conclusions on the density of individuals per unit area to be made. The nest distribution of *T. caespitum* was found to approach that of the Poisson distribution.

Odum & Pontin (1961) evaluated the number of ants per mound in *Lasius flavus* (F.) by the mark–recapture method and the total number of mounds in an area of 277 m^2. By combining these figures the worker density was found to be 485 per m^2, corresponding to a biomass of 0.6 g per m^2. Golley & Gentry (1964) based their work on the bioenergetics of *Pogonomyrmex badius* Latreille on similar but unavoidable extrapolations. The number of ants per colony was determined as the average result of repeated excavations in different colonies and was found to be relatively constant at between 4000 and 6000. The number of ants hills was counted over an area of several hectares and the total number of active hills multiplied by an average population of 5000 was taken as the actual number of ants in the study area.

Francoeur (1965, 1966*a*, *b*) combined some of the sampling techniques already described for Headley and Talbot. Twenty-five 1 m^2 samples were taken to include all the micro-environmental types and the ants searched for on and below the ground surface to a depth of 20 cm. However, the assumption that all the micro-environmental types must be sampled within a fixed number of random samples, implies voluntary disregard of the principle of random selection. This must distort the proportions of species represented in the result, as well as invalidating any subsequent statistical analysis.

Another remarkable sampling effort has been produced by Lévieux (1966, 1967*a*) in the Ivory Coast. Three pedologically homogeneous biotopes were studied by cutting the grass and digging in each of 16 quadrats of 16 m^2 to a depth of 30 cm. These results can be regarded as very accurate. They reveal considerable homogeneity, the ant density varying only between 0.55 and 0.69 nests per m^2 on iron sand and black soils, respectively.

Hocking (1970) attempted to determine the density of three arboricolous ant species in East Africa. This author found that the three commonest species (*Crematogaster* (*Acrocoelia*) *mimosae* Santschi, *C.* (*Acrocoelia*) *nigriceps* Emery and *Tetraponera penzigi* (Mayr)) associated with *Acacia drepanolobium* (Harm sex) Sjöstedt have an average density of 11.7×10^6* individuals of all stages (eggs excluded) per acre (0.4 hectare). However, in spite of much field work, this figure inevitably involves many assumptions and approximations concerning the number of ants per stipular swelling, the size of stipular swelling and the number and distribution of trees.

Mathematical estimates of the desirable number of quadrats

The two main problems involved in evaluating the demographic parameters of social insects, namely the aggregation of individuals within colonies and the simultaneous presence of individuals of very different productive potential, physiology and biomass, seem to be best solved if attacked separately. This has been done recently by Pętal (1972) who evaluated the composition of ants in nests at different periods by excavation of one to two nests each week. From these data she extrapolated to the total number of ants in the biotope by sampling the nests only. The determination of the number of samples needed to represent the area studied has been done by a graphical method similar to that of Lévieux (1969) and also by a statistical method derived from the Student's *t* distribution (see Healy, 1962).

With both these methods Pętal determined the most convenient number of samples for the area studied as 12 to 15 of 12.5 m^2 each. The difference in sampling effort compared with that of Lévieux is striking, although Pętal studied the population of one species only.

It must be emphasised that all these methods based on Student's *t*

* Given as 10^{-6} by Hocking (1970); an error surely.

distribution are applicable only if the number of colonies per sample has a distribution approaching the normal. The only case in which a normal distribution has been demonstrated in ants up to now is that of the Appennine grassland communities studied by Baroni-Urbani (1969), while other students (see, for example, Yasuno, 1963; Brian, 1965) describe situations recalling the negative binomial distribution. Nevertheless, such extrapolation from small, natural samples still seems to be the most promising way for the future.

Soil cores

In most quantitative studies, numerous small plots provide better results than a few large ones of the same total area. However, in the case of subterranean ants and termites, the smallness of the pits seems to be limited by the necessity of preventing the insects from escaping. Very small pits can be simulated by means of soil cores, though the borer has to be driven into the soil very quickly and, as a result, the samples will generally be rather shallow. Brian *et al.* (1965) estimated *Lasius alienus* (Foerst.) in this way.

In fact, few quantitative studies have been done specifically for social insects, but some authors, as part of a more general survey of the soil fauna, have mentioned any ants and termites they have found. A few examples of this type of research deserve mention.

Berwick (1934) (quoted by Strickland, 1944) did not mention any termites but cited very high figures for subterranean Hymenoptera, and estimated 42 and 62 million individuals per hectare in cacao and sugar cane plantations in Trinidad, derived from 60 samples per biotope. Williams (1941) extracted the soil microfauna of a Panama natural rainforest by means of a Berlese apparatus. He took 23 samples of litter and calculated a density of 130000 termites and 16.6 million ants per hectare. Strickland (1944, 1945) studied several biotopes in Trinidad and found very different densities of termites. There were none in his samples from the Tortuga Estate but an estimated population of 90 million per hectare in the Tupuma Reserve. The aggregation of termites and the limited sampling were undoubtedly responsible for such extreme results. Ten samples were taken in the Tupuma Reserve and the fauna was extracted by flotation. Five cores with an area of 66 cm^2 taken to a depth of 22.9 cm gave an estimated population of 1.82 million per hectare, while five more of the same area but taken to only 7.6 cm gave an estimated population of 180 million per hectare.

Salt (1952) also used soil cores, 81 cm^2 in area, froze the samples and later washed and extracted the animals by flotation. Again, the author, working in various biotopes in Uganda and Tanzania, was making a general survey of soil fauna but mentions that of the 23 cores taken, termites were found in 10, with a maximum of 38 animals in one sample and that ants were found in 18 samples, with a maximum of 389 individuals in one sample.

Each of the results referred to above was obtained by means of soil cores totalling less than 0.5 m^2 in a given biotope, which is clearly too little. The method could, however, be applied successfully in areas with dense populations if the number of samples is calculated after an adequate transformation of the data in order to normalise the distribution.

During their exploration of the Garamba reserve (north-eastern Zaire), De Saeger and his colleagues (quoted by Harris, 1963) collected various nest samples and took 432 core samples of 15 cm in depth and 75 cm^2 in area (thus totalling 3.24 m^2) from eight characteristic biotopes, with the aim of producing a comprehensive inventory of the fauna of the reserve. The animals were extracted by means of a Berlese funnel. Harris (1963) analysed the termite material and identified 34 species, 16 of them being present in the soil cores, totalling 5976 individuals. If one calculates the mean density from these data (which was not done by Harris) a figure of more than 18 million individuals per hectare is arrived at. However, eight biotopes were covered by the sampling, from a grassy savannah on a sandy soil, in which no termite was found in more than 60 cores, to a sloping wooded savannah, where termites were present in each of the 10 groups of six cores taken (see Table 13 in Lee & Wood, 1971*b*). Since the relative surface of each biotope was not taken into account and since the samples were not collected in a random pattern (the aim of the study was to construct an inventory of animal species), the mean of 18 million termites in the top 15 cm of the soil must be accepted with reservations.

Nests in wood without external signs of habitation

Several wood-boring termites can be detected by their superficially constructed galleries. However, some species, especially among the lower termites – 'dry-wood termites' and 'damp-wood termites' – confine their activities to the interior of the wood. Some of these species constitute pests in so far as they attack cultivated trees, or timber in houses, and considerable efforts have been devoted to detecting them without cutting down the trees.

In his bibliography of termites, Snyder (1956, 1961, 1968) reviews the detection methods. Microphones and audio-amplifying apparatus have been used for detecting sounds produced by termites, and X-rays have been used to search for signs of damage within the timber, but these methods do not always give satisfactory results. If damage is detected it is possible to recognise the activities of dry-wood or damp-wood termites by their characteristic faecal pellets.

Very large societies of termites, which have invaded big trees, may be located by temperature measurements. Greaves (1964) studied temperature in living trees infested by colonies of *Coptotermes acinaciformis* and *C. frenchi* and he noted temperatures higher by 13–20 °C in infested trees, adjacent to the brood chambers of the nest, as compared with uninfested trees. If small societies have invaded wood, then there seems to be no alternative to random sampling and laboriously sawing and splitting the wood. Even then it is usually only the foraging galleries of larger societies which are met with. Nutting (1970) used this method and obtained much information on both composition and size of societies of wood-boring termites from the Sonoran desert (Arizona and Mexico). He concluded 'that mature drywood termite colonies seldom contain more than a few thousand individuals'. He suggested that the proportion of soldiers was influenced by the size of the society (both by its population size and by the number of openings to be guarded) and that the production of alates was increased dramatically under conditions of water stress.

The mark–recapture method

The general method of marking, reviewed by Southwood (1966) can in most cases be used for ants and termites, though if external marks are used, they must be able to withstand grooming. This has been found to apply, for example, to external radioactive markers applied by dipping the insects into solutions. Contamination is reduced by drying them before replacing them in the field (Nielsen, 1972*a*, *b*). Even internal radioactive markers taken with the food are transmitted by food-exchange from one individual to another, both within and outside the colony (Chauvin, Courtois & Lecomte, 1961). The most commonly used isotope is ^{32}P (Kannowski, 1959; Odum & Pontin, 1961; Brian *et al.*, 1965; Stradling, 1970; Nielsen, 1972*a*, *b*). Nielsen, (1972*a*, *b*) took advantage of the short half-life of ^{24}Na to estimate worker turnover. Other kinds of marks successfully used in ants include very thin copper belts around the petiole (Dobrzański, 1966) and removal of an epinotal spine (Brian, 1951, 1972). In spite of the considerable effort necessary in marking a great number of ants, the last method seems to be the most useful, being free from secondary effects on behaviour and impossible to destroy or transmit. Of course, the method is inapplicable to the numerous species without epinotal spines. Sanders & Baldwin (1969) combined physical marking with radioactivity by gluing small pieces of iridium (Ir) wire (12.7 μm diameter, 130 μm in length), irradiated to enhance the proportion of ^{192}Ir, to the alitrunk of the ants. While this method was not used in population estimates, there seems no reason why it should not be, though it is obviously only feasible with ants of considerable size.

After marking a sample, the ants are released into the field, given time to mix with the rest and then recaptured. The total population (*P*) is calculated according to Petersen (1889) (quoted by Le Cren, 1965) as follows:

$$P = (a \times n)/r \tag{2.2}$$

where *a* is the total number of marked individuals, *r* the number of marked individuals recaptured, and *n* the total number recaptured. This method, originally developed for fish, was first applied to ants by Chew (1959). Bailey (1951), Leslie (1952) and others demonstrated the advantage of calculating the total population by the formula

$$P = a(n+1)/(r+1) \tag{2.3}$$

This allows the calculation of variance and other measures of dispersion.

Notwithstanding its general popularity, its use in ant population studies has been severely criticised by Ayre (1962), Golley & Gentry (1964), Pętal & Pisarki (1966), Lévieux (1969, 1972*c*) and Erickson (1972). It has also been used by many authors (Odum & Pontin, 1961; Brian *et al.*, 1965; Dlussky, 1965, Brian, Elmes & Kelly, 1967; Stradling, 1970; Brian 1972; Nielsen, 1972*a*, *b*) mostly because it still seems to represent the only practical approach to several situations. Of the numerous criticisms of the applicability of the Petersen Index in ant population studies, we shall deal especially with that of Ayre (1962) which is by far the most complete and which is based on exhaustive experimental data. According to this author, there are numerous sources of error depending on the interval of time between marking and recapture, and the variation of foraging activity in time. The main source of error is represented by the fact that in the species studied (*Formica fusca* L., *Formica exsectoides* Forel and *Camponotus herculeanus* (L.)), the foragers outside the nest never exceed 20% of the total colony population. This proportion is still greater than that ascertained for *Pogonomyrmex badius* and *Camponotus nylanderi*, where never more than 10% of the total colony population has been observed to forage outside the nest at a given moment and the daily average number of foragers is less than 1% of the total colony population (Golley & Gentry, 1964; Baroni-Urbani, 1965). The last author, in another paper, gives comparable figures for *Lasius alienus* and *Tetramorium caespitum*, where the maximum number of foragers per unit time is 8% and 7%, respectively (Baroni-Urbani, 1969). Evidently, by this method and in normal conditions, we can obtain an estimate only of the foraging workers and not of the whole colony population. This statement is based on the assumption (never strictly demonstrated) that the foragers, because of polyethism, are a fixed, small minority within the colony population. However, Brian (1969) maintains that in sampling an ant population it is essential to take individuals from the nest and make suitable preparations to this end. This

Table 2.5. *Foraging in* Camponotus aethiops: *the relationship between number of foragers and colony size in laboratory-bred colonies*

Sample no.	Number of larvae	Number of pupae	Biomass larvae and pupae (mg)	Number of workers	Biomass of workers (mg)	Total biomass	Number of foragers
1	1	0	9	23	240	249	23
2	50	0	120	12	187	307	9
3	17	4	204	23	453	657	22
4	5	5	293	19	400	693	17
5	123	1	626	22	290	916	18
6	0	20	733	12	273	968	12
7	0	0	0	92	1134	1134	71
8	150	7	789	40	410	1199	15
9	70	7	528	57	1027	1555	53
10	118	138	3307	107	1723	5130	77

method avoids the error due to marking an ethologically separate subset of the colony; it is only essential that the samples be unbiased. The idea is evidently supported by the paper of Stradling (1970) who obtained good results with the mark–recapture method using an adequate sampling of the nest. It is of interest to report in detail here the main results obtained by Stradling in this paper which should be useful for those who use the mark–recapture method. Stradling used ^{32}P as a marker, but found a high degree of contamination between marked and non-marked workers after dipping the first in the radioactive solution, even if they were subsequently dried on filter paper. Internal marking by feeding the ants with a solution containing 25% honey in water and 1 mCi of ^{32}P as orthophosphate in 5 ml of solution was then tried. However, because of the liquid food transmission between individuals, non-marked workers showed a degree of contamination even greater than those present in the dipping experiment. Contamination became negligible if the marked ants were starved for a while before their re-introduction into the colony; this allows for full assimilation of the radio-isotope. The required starvation period needs to be determined in the laboratory because it apparently differs from species to species. The longest period observed was in *Lasius niger* (L.) (three days). Even so, the radioactivity was detectable for at least nine days after the cessation of feeding, allowing an adequate time for release and capture.

More recently, Erickson (1972) found again the inadequacy of the mark–recapture techniques when proper safeguards are ignored in evaluating a population of *Pogonomyrmex californicus* (Buckley). He compared the results obtained by this method with direct counts of the total

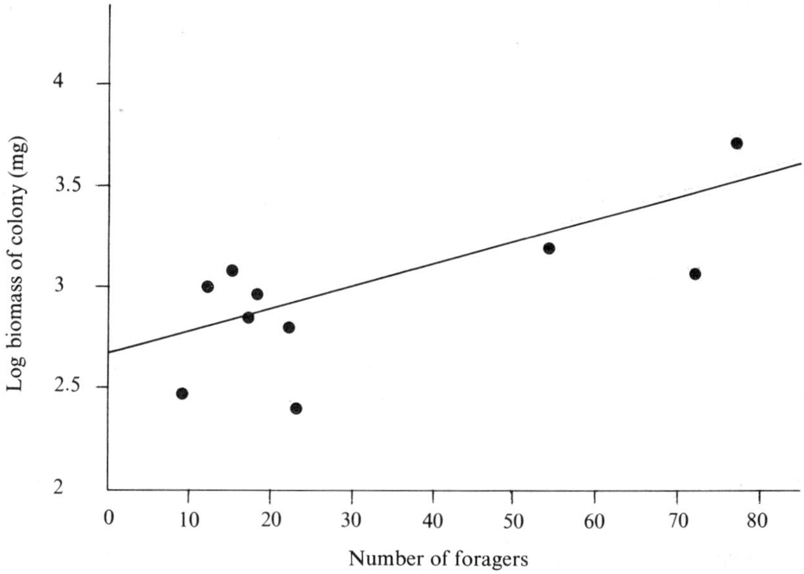

Fig. 2.7. Relationship between colony biomass and the number of foragers in *Camponotus aethiops*.

population after excavation in the field, itself a very unreliable method. Moreover, he used the dipping technique which, as he himself noted, would reduce the estimated number of ants because of secondary tagging. Furthermore, only ants collected on the ground outside the nest were marked, with the result that the estimated population represented about 10% of the true ant population.

To test the reliability of the hypothesis that foragers constitute a well-defined numerical minority in the nest, a simple series of laboratory observations have been made. Ten colonies of *Camponotus aethiops* with known composition were bred in artificial nests with defined foraging territories. Observations on foraging workers were made on each colony at intervals of four hours, day and night. Every forager was marked and the observations were continued until only marked workers in the foraging territory were encountered on at least five consecutive occasions. The total observation period was never longer than eight days. The results are given in Table 2.5. One can easily see how, although the number of foraging workers observed at any moment remains within the prescribed limits, the total number of foragers varies between 40 and 100% of the total colony population. In two small samples (colonies 1 and 6), the whole worker population participated in foraging activity. These data support the hypothesis that all workers, including nurses, may come out to a bait placed close to the nest. In this case a population estimate made by the mark–

recapture technique in the proper conditions could give a reasonable result. Moreover, a statistically significant correlation ($r = 0.721$, $p < 0.05$) exists between the logarithm of the total colony biomass and the number of foragers (Fig. 2.7).

Another example where the mark–recapture method has provided useful results involves the ant population of a Danish grassland (Nielsen, 1974*b*). This author was not interested in determining the colony population, but only the total density per unit area of two species (*Lasius niger* (L.) and *Tetramorium caespitum*). The normal mark–recapture method was used on samples of the ants found under specially distributed cement plates. The proportion of ants under these plates is much higher in comparison with the foragers observed on the ground, but the reliability of the results obtained by the mark–recapture method with such samples was tested by counting the real number of ants per unit volume of soil in the same grassland and the results are quite good. However, the use of the cement plates tended to overestimate the number of ants present in the deeper strata of the soil, so a correcting factor to compensate for this source of error, based on field experience, was calculated. The good results obtained in this study by mark–recapture and by this particular method of sampling, seem to depend upon several environmental and behavioural characteristics of this particular ecosystem. Nielsen himself considers his results a special case; but then all sampling occasions are, the appropriate method requires careful selection.

Determination of biomass and energy content

We have seen, in considering the methods that have been employed in determining population densities for ants and termites that if the implications of social organisation are ignored, serious errors may arise and the value of the results for production studies be materially limited. Transformation of population data into biomass and subsequently into energy content is similarly fraught with pitfalls for the unwary. The importance of sampling an accurate cross section of the population, to include all castes in their correct proportions has been emphasised. (See also: annual variations in composition). The polymorphism characteristic of ants and termites has an obvious bearing on biomass but, in addition, recent studies have shown that considerable variation in the energy content of individuals relates not only to their biomass but also to their body composition.

Table 2.6. *Fresh weights (mg per individuals) of termites in some* Cubitermes *spp. When a percentage is in parentheses, it represents the proportion of that caste in a population without eggs and without nymphs or sexuals*

Caste	*Cubitermes fungifaber*[a]	*Cubitermes sankurensis*[b]	*Cubitermes exiguus*[c]	*Cubitermes exiguus*[d]
Queen	—	50	40	—
King	—	8.5	5.0	—
Egg	—	0.06	0.04	—
Larva I }		0.26	0.38	—
	10.0	(15.0%) }		
	(42.5%)		(36.2%)	
Larva II }		1.25 }	0.94	—
		(10.8%) }		
Worker }		4.85 }	3.87	—
	12.4	(71.5%) }		
	(56%)		(62.3%)	
White worker }		1.68 }	1.25	—
		(1.6%) }		
Soldier	17.45	5.18	3.89	—
	(1.5%)	(0.7%)	(1.4%)	
White soldier	—	3.69	2.90	—
		(0.3%)	(0.2%)	
Nymph I	—	1.25	0.94	0.5
Nymph II	—	1.22	0.91	0.8
Nymph III	—	2.53	1.82	1.8
Nymph IV	—	7.48	5.47	3.7
Nymph V	—	9.66	8.12	7.1
Alate	—	8.94	5.43	6.4

[a] (Maldague, 1967).
[b] (Hébrant, 1970*a*); percentages calculated from Hébrant's Table 20 (12 societies).
[c] (Hébrant, 1970*a*); percentages calculated from Hébrant's Table 18 (18 societies after flight).
[d] (Bouillon, 1974); weights estimated from Bouillon's Fig. 4.

Biomass

Though biomass is readily determined from the samples taken during population estimates, the literature contains relatively few figures for ants and termites and these generally express only the order of magnitude of the biomass rather than details of its partition amongst the population components.

Thus Lee & Wood (1971*b*) suggest 3 g per m^2 for *Nasutitermes exitiosus* (Hill) in dry, sclerophyll forest of south Australia, where the population density was 600 per m^2. The same authors (Lee & Wood, 1971*b*) combine Sand's (1965*a*) and Bodot's (1967*a*) data to give values of 5–50 g per m^2 for populations of 1000–10000 per m^2 of termites in northern Nigeria and

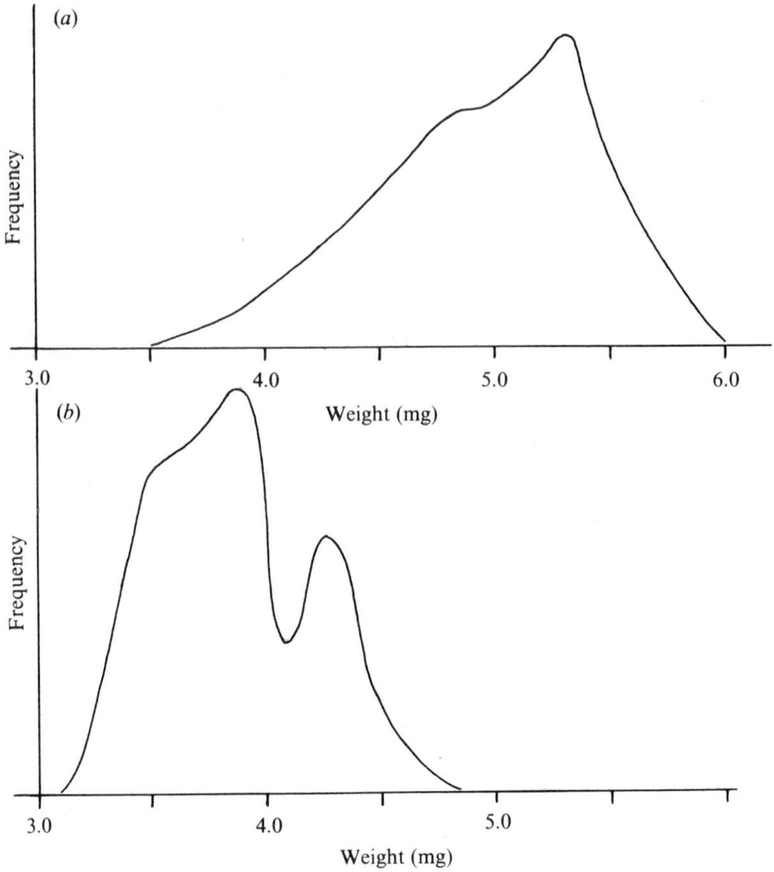

Fig. 2.8. Distribution of individual fresh weights of (*monomorphic*) workers in two *Cubitermes* species. Smoothed curves after Hébrant's data (1970*a*). (*a*) *C. sankurensis*, after 236 weighings of 10 individuals each; (*b*) *C. exiguus*, after 582 weighings of 100 individuals each.

the Ivory Coast. The highest figure should perhaps be treated with some reservation since, for example, Bodot found high densities in only one area out of the three. More detailed figures in which attention is paid to the proportional representation of castes and developmental stages, have been provided by Maldague (1967) for *Cubitermes fungifaber* (Sjöstedt), by Hébrant (1970*a*) for two species of *Cubitermes*, and Josens (1972) for various species of Macrotermitinae, *Microcerotermes parvulus*, *Trinervitermes geminatus* and a number of humivorous species.

Thus Maldague (1967) arrived at a value based on the estimated density of termites (in a *Brachystegia* forest in Zaire) of 8.4×10^6 individuals per hectare with a mean fresh weight of 11.45 mg per individual (96 kg per hectare). In fact, the complete estimate was based on the population (40% of the total) inhabiting the mounds of *Cubitermes fungifaber*. Hébrant

Table 2.7. *Caste proportions (%) and fresh weights (mg per individual) in the societies of the common termites at the IBP Lamto project (Ivory Coast). After Josens (1972)*

The percentages have been calculated without the eggs or the nymphs and alates having been taken into account. The biomasses are in parentheses.

Species	Larvae	Minor workers	Major workers	Minor soldiers	Major soldiers
Humivorous					
Basidentitermes potens	8 (0.47)	80 (3.01)		12 (1.81)	
Adaiphrotermes sp.	16 (0.29)	84 (1.73)		0	
Aderitotermes sp.	23	77 (3.37)		0	
Astratotermes sp.	10	90 (11.7)		0	
Wood-eating					
Microcerotermes parvulus	24.5 (0.34)	74 (0.82)		1.5 (0.89)	
Fungus-growing					
Ancistrotermes cavithorax	46.9 (0.55)	18.0 (0.83)	26.6 (2.10)	6.5 (0.85)	2.0 (1.92)
Microtermes toumodiensis	19.5	30.2	40.9	9.4	—
Pseudacanthotermes militaris	34.5 (0.90)	32.0 (1.25)	23.5 (4.64)	5.5 (1.84)	4.5 (12.5)
Grass-cutting					
Trinervitermes geminatus	7 (0.99)	0.5	74.5[a]	12 (1.40)	7 (3.66)

[a] The (major) workers of the *Trinervitermes* spp. moult several times and increase their weight with their age (see Table 2.9).

(1970*a*) studied *C. sankurensis* and *C. exiguus* and he observed a range of biomasses (fresh weights) from 2.48 to 83.35 g per society of *C. sankurensis* (12 societies sampled completely) and from 3.82 to 53.95 g per society for *C. exiguus* (38 societies sampled). This last species, with 510 living nests per hectare represented a living biomass of 11.5 kg per hectare. Furthermore, it is clear from Hébrant's data that besides the considerable variation between societies in their total biomass there are also differences in the proportions of castes, which should not be ignored (Table 2.6). Moreover, Hébrant provided much information on the distribution of weights of several castes in the two species studied; for example, from Fig. 2.8 it appears that large variations occur in the weight of the monomorphic workers of *Cubitermes*. He showed that these variations were influenced both by the society and by the physiological state of the workers.

Josens (1972) gives details of the proportions of castes and develop-

Table 2.8. *The biomass of workers and sexuals in* Tetramorium caespitum (*after Brian, Elmes & Kelly, 1967*)

Number of colony	Territory (m^2)	Population of workers (±S.E.)	Weight of workers (g)	Population of males	Weight of males (g)	Population of queens	Weight of queens (g)	Total weight of colony (g)	Biomass (g per m^2)	Worker density (per m^2)
					1963					
1	62	14470±678	8.23	485	2.88	336	3.46	14.57	0.235	233
2	2	2603±132	0.90	0	0	0	0	0.90	0.410	1183
3	59	26284±2279	15.38	208	1.31	28	0.27	16.96	0.287	445
4	54	15905±1163	9.37	241	2.29	76	0.88	12.54	0.232	295
5	36	25268±1668	11.95	164	1.08	388	4.04	17.07	0.474	702
6	12	6735±505	3.23	0	0	0	0	3.23	0.269	561
7	43	10087±749	6.41	49	0.51	19	0.25	7.17	0.167	235
8	34	9952±552	5.63	289	2.54	321	3.42	11.59	0.341	293
9	63	11629±876	6.43	145	0.96	256	2.53	9.92	0.157	185
10	35	29571±2444	12.01	0	0	0	0	12.01	0.343	845
11	45	13763±684	6.55	320	2.41	24	0.27	9.23	0.205	306
12	30	9393±913	5.37	228	1.69	276	3.09	10.15	0.338	313
13	68	16989±1285	12.06	364	2.14	120	1.16	15.36	0.226	249
14	29	22467±1259	11.53	38	0.24	272	2.41	14.18	0.489	775
15	50	8563±439	4.10	259	2.47	437	5.04	11.61	0.232	171
16	56	12407±836	9.02	194	1.22	113	1.05	11.29	0.202	222
17	21	8619±1583	3.44	0	0	0	0	3.44	0.164	410
18	72	15561±1000	11.50	117	0.82	27	0.26	12.58	0.175	216
19	57	10128±610	6.01	1467	10.56	778	8.77	25.34	0.445	178
20	75	11666±1062	9.29	304	2.06	90	0.95	12.30	0.164	156
21	23	16576±805	10.69	321	2.41	248	2.58	27.64	0.709	918
22	16	19227±1455	11.96							
23	25	5718±715	2.75	39	0.32	74	0.87	3.94	0.158	229
Mean	42	14068±1247	7.99	277	1.64	169	1.79	11.42	0.272	440

1964

1	81	7421±956	5.27	641	3.89	283	2.90	12.06	0.149	92
2	4	1395±166	0.56	0	0	0	0	0.56	0.140	349
3	49	5081±340	3.35	444	2.65	105	1.07	7.07	0.144	104
4	81	7195±355	4.32	124	0.81	142	1.35	6.48	0.080	89
5	32	12156±1129	6.88	322	2.05	246	2.71	11.64	0.364	380
6	17	3636±272	1.95	104	0.56	53	0.51	3.02	0.178	214
7	31	3209±209	2.19	55	0.33	2	0.01	2.53	0.082	104
8	34	12873±1192	8.12	294	1.96	290	2.76	12.84	0.378	379
9	67	30943±2954	19.80	247	1.21	94	0.84	21.85	0.326	462
10	33	6943±376	3.28	167	0.86	41	0.41	4.55	0.138	210
11	36	4625±421	2.70	0	0	0	0	2.70	0.075	128
12	46	6322±654	3.81	30	0.20	23	0.22	4.23	0.092	137
13	60	7021±463	5.27	315	2.00	127	1.24	8.51	0.142	117
14	9	7446±422	4.25	84	0.40	86	0.86	5.51	0.612	827
15	33	12676±1539	8.92	259	1.74	265	2.71	13.37	0.405	384
16	30	5439±407	4.12	216	1.22	91	0.77	6.11	0.204	181
17	19	8626±696	4.68	0	0	0	0	4.68	0.246	454
18	69	5948±800	4.87	211	1.22	117	1.15	7.24	0.105	86
19	57	2731±258	1.66	452	3.08	181	1.60	6.34	0.111	47
20	71	9738±1072	7.69	630	4.51	287	2.89	15.09	0.213	137
21	36	5559±395	3.81	281	1.74	222	2.37	7.92	0.220	154
22	8	5607±392	3.85	174	1.10	41	0.41	5.36	0.670	700
24	23	12102±1090	6.73	29	0.15	151	1.36	8.24	0.358	526
25	43	3426±431	2.16	82	0.52	262	2.50	5.18	0.120	79
26	43	8929±667	5.70	172	1.02	275	2.63	9.35	0.217	207
Mean	40	7881±1150	5.04	213	1.32	135	1.35	7.71	0.193	261

mental stages and the situation he found in some species is summarised in Table 2.7.

Essentially the same conditions obtain in the ants, so that while many estimations of populations and population densities have been reported (see Appendix 1: Tables 1 and 2) relatively few take account of the biomass and its partition between the castes and the brood. One investigation is that of *Tetramorium caespitum* populations in a southern English heath by Brian *et al.* (1967) in the years 1963 and 1964. Their results are summarised in Table 2.8. The striking inter-colonial variation in total biomass and the partition between castes that was seen in termites is again very evident. For example, not all colonies produced sexuals, yet in one colony in 1963, 76% of the biomass was due to them. Similarly, the worker biomass (this species is essentially monomorphic in the worker caste) varied from 1.66 g to 19.80 g fresh weight,* with a mean of 5.04 g fresh weight in the second year (1964) in which measurements were made. Not surprisingly, biomass per unit area also varied over an order of magnitude, though it should be noted that it did not correlate with the absolute values. Furthermore, variations from one year to the next were well marked, thus in 1963 the mean biomass of the colonies (workers and reproductive castes) was 11.42 g or 0.272 g per m^2, while in 1964 it was 7.71 and 0.193 g per m^2. Censuses of these colonies continued to be taken until 1968 and the data obtained have recently been analysed (Brian & Elmes, 1974). They showed that underlying the great variation described above there was during this period an increase in the foraging area, in worker size and, most interesting from our present point of view, an increase in biomass, in which the sexuals played an important role.

Pętal (1967), investigating energy flow through a population of *Myrmica* spp. in a *Stellario-Deschampsietum* in Poland, estimated the biomass in 1964 and 1965 at 0.21 g per m^2 dry weight and 0.14 g per m^2, respectively. In the most abundant species, *Myrmica laevinodis* Nyl., with a density of 0.073 colonies per m^2, she shows that in 1964 the mean dry weight of the brood (larvae) in each colony increased from 8.5 mg at the start of the growing season to 1933 mg at the end, of which 234 mg comprised the reproductive castes. In the 1965 season the increase was from 44.6 mg to 1262 mg, with 352 mg due to sexuals. More recently, Nielsen (1972*a*, *b*) has presented data on worker production of *Lasius alienus* in a Danish heath. He examined in considerable detail the population of a colony whose nest covered an area of 3.9 m^2, and showed that the worker biomass fluctuated between 1.75 g and 3.24 g dry weight during the summer. These examples relate only to temperate ants. As far as tropical ants are concerned there is a dearth of information on biomass, which is perhaps a commentary on the difficulties associated with their study.

* (Ignoring the obviously immature colony.)

Martin *et al.* (1967) collected 6.35 kg of ants from an *Atta colombica tonsipes* nest and they suggest that this weight represented 65–85% of the total colony, since many of the smaller workers were undoubtedly lost. Lockwood (1973) has determined the density of *A. colombica* and *Atta cephalotes* (L.) in different kinds of tropical forest in Costa Rica and found a maximum density of 2.5 nests per hectare, so that by combining the data from these two sources, it can be deduced that the biomass of *Atta* spp. may be of the order of 1.59 g per m^2 fresh weight.

Energy content

Turning now to the question of energy content, it will have been noted that the authors cited above expressed their data in terms of either fresh weight or dry weight. Clearly, the transformation of biomass into energy content will have to take account of the very different calorific values of the individuals in these different conditions. Furthermore, even if the data on biomass were standardised by expressing them as dry weight values, allowance has to be made for the variations in body composition which has an important bearing on the calorific value of the individuals. Josens (1972) measured the calorific content of the castes and developmental stages of a number of termite species and clearly demonstrated its variability. Table 2.9 illustrates the state of affairs in two species which may be taken as representative of the range which exists in the termites. It will be seen from the table that young larvae of *Trinervitermes geminatus* have the lowest calorific value per g fresh weight, while the alates have the highest, and that a factor of almost six separates them. Clearly, a great deal of the difference is due to the fact that the larvae contain 88% water while the alates are only 50% water. A much reduced but still important variability remains even if the biomass is expressed in terms of dry weight. Thus the dry matter of the alates is about 1.5 times as energy rich as that of the workers, presumably because of variation in fat content, for the most part, since Wiegert & Coleman (1970) have shown that the calorific value of termites is correlated highly with fat content.

Another important factor, particularly in the humivorous species, is the proportion of inorganic material in the digestive system of the termite. Thus the workers of *Basidentitermes potens* have an apparently low calorific value for the dry matter, but because 71% of this is due to the soil in the intestine, the calorific value of the dry tissues is about the same as that of the workers of *Trinervitermes geminatus*. There is no doubt that the greatest importance attaches to a detailed determination of the calorific values of each class of individuals in the colony. The same is true for ants, though the importance of the ash content is not as great as in the humivorous termites. For example, Peakin (1972), investigating growth of

Table 2.9. *Comparative weights and energy content of the castes of* Trinervitermes geminatus *and* Basidentitermes potens (*after Josens, 1972*)

Species and caste	Fresh weight mg per individual	Dry weight,[a] mg per individual	Water content % of fresh weight	Energy content J per individual	Energy content J per dry weight (with ashes)	Energy content J per dry weight, (ashfree)	Ashes % of dry weight
Trinervitermes geminatus							
Eggs	0.067	0.016	76	0.374	23390	24650	5.1
Larvae (1+2)	0.99	0.11	88	2.405	21890	23020	4.9
Workers 1	3.97	0.72	82	14.40	20000	22390	10.5
White workers[b] 1–2	3.95	0.58	85.5	12.97	22350	23270	4.0
Workers 2	5.18	1.03	80	20.42	19840	22100	10.2
White workers[b] 2–3	6.14	0.89	85.5	19.92	21260	22350	4.7
Workers 3–7	7.23	1.48	80	28.52	19270	20760	7.2
Major soldiers	3.66	1.00	73	25.03	25030	25950	3.6
Minor soldiers	1.40	0.28	80	5.775	20650	22680	8.9
Young nymphs[c]	3.30	0.49	85	13.02[e]	26600[e]	27710[e]	4.0
Old nymphs[d]	31.5	13.3	58	353.6[e]			
Alates	28.7	14.4	50	409.3	28420[f]	29170[f]	2.3
Basidentitermes potens							
Eggs	?	?	?	?	24570	26370	6.9
Larvae	0.47	0.07	85	1.340	19110	21470	11.0
Workers	3.01	1.17	61	6.570	5610	19170	71[g]
Soldiers	1.81	0.56	69	13.98	24940	25950	4.5
Old nymphs[d]	3.41	1.26	63	33.94	26950	27750	3.0

[a] Drying was done at 70–75 °C.
[b] Close to their moulting, the workers become whitish and, clear out their gut.
[c] Nymphal instar with first visible wing rudiments.
[d] Last nymphal instar with long wing rudiments and pigmented eyes.
[e] Large variability in each instar.
[f] Comparable values were obtained by Wiegert & Coleman (1970): 28980 J per g (ashfree) for alates of *Nasutitermes costalis*; and even higher values were observed by Josens (1972): 30930 J per g (with ashes) and more than 32400 J per g (ashfree) for alates of *Ancistrotermes cavithorax* and *Microtermes toumodiensis*.
[g] This rate of ashes may exceed 80% in certain other humivorous species.

the reproductive castes in *Tetramorium caespitum*, has shown that there are substantial changes in the water content of individuals during development (Table 2.10).

Overwintering, third-instar larvae hydrate at the start of the growing season and then accumulate dry matter in rather greater proportions as development proceeds. The most spectacular changes are, however, reserved for the postmetamorphic growth period. They are particularly well marked in the queens as they prepare for the rigours of claustral colony foundation by increasing their stores of fat. The effect on the calorific

Table 2.10. *Variations in body composition and energy content, during the course of development, in sexuals of* Tetramorium caespitum (*after Peakin, 1972*)

Developmental stage	Mean fresh weight (μg)	Mean dry weight (μg)	Water content (%)	Calorific value[a] (kJ per g) fresh weight	Calorific value[a] (kJ per g) dry weight
Overwintering larvae[b]	568	205	64	9.43	26.12
Small, sexual larvae[c]	3153	622	80	4.78	24.22
Medium, sexual larvae[c]	9328	1923	79	5.19	25.17
Full-grown sexual larvae[c]	16183	4295	73	6.66	25.09
Prepupae[c]	12983	3447	73	6.60	24.87
Young, male pupae	11530	2934	75	6.32	24.83
Old, male pupae	9806	2235	77	5.24	22.97
Young, male adults	6770	2094	69	6.72	21.74
Old, male adults	5050	1515	70	6.19	20.63
Young, queen pupae	13067	3672	72	6.97	24.82
Old, queen pupae	11769	3199	73	6.52	23.97
Young, adult queens	9799	3130	68	7.81	24.46
Old, adult queens	10793	6023	44	16.38	29.35

[a] Calculated from the body composition, on the assumption that 1 g of fat is equivalent to 39.33 kJ and the remaining dry material has a calorific value of 18.83 kJ per g.
[b] May include some worker potential larvae, which are indistinguishable from sexual potential larvae at this stage.
[c] Male and queen potential taken together.

values when expressed both in terms of fresh weight and dry weight is evident. It is interesting to note the contrast in energy content between queens and males in this species as the time of the nuptial flight approaches. It is, of course, only the queens that are concerned with colony foundation, unlike the termites in which both sexes are involved and where both males and queens are energy rich (Table 2.9). Peakin (1972) has also shown that there are significant variations in the calorific content of the workers of *Tetramorium caespitum*, apparently related to the growth of the sexual brood. Similar fluctuations have been noted in *Formica* spp. Kirchner (1964), investigating the annual cycle in the body composition of workers of *Formica polyctena* Foerst., *Formica rufa* L. and *Formica nigricans* Em., noted that fat accumulated strikingly in the first week in the life of newly emerged workers of all the species to reach a maximum in preparation for overwintering. In the spring there was a sharp fall coinciding with brood growth. While Kirchner (1964) was not primarily concerned with energy flow and calorific content of the workers it will be clear from Wiegert & Coleman (1970) that the variations in fat content of

Formica workers will have a bearing on their calorific content. Similar changes in worker-body composition, and by implication, in calorific value, to those in *Formica* are known to occur in *Lasius flavus* (Peakin, unpublished). Incidently, this species shows essentially the same growth pattern for the sexual brood and the same postmetamorphic distinctions between males and queens as have been described for *Tetramorium caespitum* (Peakin, unpublished).

Population dynamics

Longevity of societies

The longevity of termite colonies is often considerable in stable ecosystems and some remarkable, though admittedly isolated, examples have been cited. For example, Grassé (1949) refers to a mound of *Macrotermes* sp. said to be 80 years old and the 60-year-old colony of *Nasutitermes triodiae* described by Hill (1921) is an often-quoted instance of colony longevity. Similarly, in ants the longest surviving colony that has been recorded is one of the *Formica rufa* group which was known to have existed for 56 years (Forel, 1928). Colonies of *Messor capitatus* Latr. and *Liometopum microcephalum* (Panzer) have been observed for 20 and 18 years, respectively (Baroni-Urbani, unpublished). In all probability these are exceptional cases and, generally, it would appear that colonies survive for shorter periods. Thus, Noirot (1970) estimates that the mounds of *Cubitermes fungifaber* are occupied for no more than five years, and Ernst (personal communication) concluded from his observations in Tanzania that *Cubitermes* nests survive 20–25 years. Furthermore, termite colonies seem to be quite heavily preyed upon, often by ants, to judge from the number of 'abandoned' mounds that have been recorded by several investigators. Thus, Bodot (1966) found that 85 out of 131 mounds of *Cubitermes severus* in the *Loudetia* – savannah of the southern Ivory Coast were abandoned, and no less than 66 out of 68 mounds of *M. subhyalinus*. Similarly, 177 out of 486 *Tumulitermes hastilis* (Froggatt) mounds were abandoned in the savannah woodland of Northern Territory, Australia (Lee & Wood, 1971*b*). Josens (unpublished) found that of 90 mounds of *Trinervitermes geminatus*, observed throughout a year, six were killed by doryline ants and four were abandoned for unknown reasons. It should be noted that since each colony produces between three and six mounds, the longevity of the colony will generally be longer than that of the individual mounds.

While colony longevity in ants will also be determined to a degree by predatory pressure, the situation is generally rather more complex than in termites. This is because of the habit of queen supersedure, which in theory should result in the immortality of the colony. In the diffuse, highly

Table 2.11. *Composition of the societies of* Cubitermes severus *in terms of their age (after Bodot, 1969)*

	Immature society	Mature society	Senile society
Number of individuals	< 10000	10000–40000	> 40000
Number of examined societies	25	45	12
Mean population	5600	30000	50000
Percentage of workers	47	63	85
Percentage of larvae	52	36	14
Percentage of soldiers	0.85	0.75	1

polygynic forms one probably finds the closest approximation to immortality and even in the modest colonies of *Myrmica* spp. survival for many years is probable. Thus Brian (1972) has presented continuous records for colonies of *Myrmica sabuleti* Meinert, which were in general still growing after six years and in which, at least in some of the colonies, queen supersedure was demonstrated. Even in monogynic colonies, without queen replacement, survival for up to 20 years is a possibility since there are several instances of queens living in the laboratory for up to this time. For example, Donisthorpe (1936) kept a *Stenamma westwoodii* queen for 17–18 years. More recently, Kutter & Stumper (1969) have recorded survival for 14 years in queens of *Formica rufibarbis*, 18 years in *Lasius flavus*, 20 years in *Formica sanguinea* and 28 years 8 months in a single queen of *Lasius niger*.

King and queen replacement is well known in termite societies too. Some of the most reliable data are given in Bouillon's recent work (1974): 11% of the sexuals found in 110 nests of *Cubitermes exiguus* were replacement ones.

Variations in the composition of the societies

Apart from the spectacular flight of alates, very little is known about fluctuations in the composition of termite societies. A percentage of soldiers has been established in many species of termites (Bouillon, 1970). The percentages of soldiers, workers, larvae (and 'nymphs' or larvae of alates), however, vary with the seasons. The work of Sands (1965*b*) is now well known, having been quoted by Bouillon (1970) and Lee & Wood (1971*b*), and relates to the fluctuations in population density in mounds of *Trinervitermes geminatus*.

Much information is available in Bodot's work (1966, 1967*b*, 1969) also: she observed a rise in the proportion of workers from about 45% in

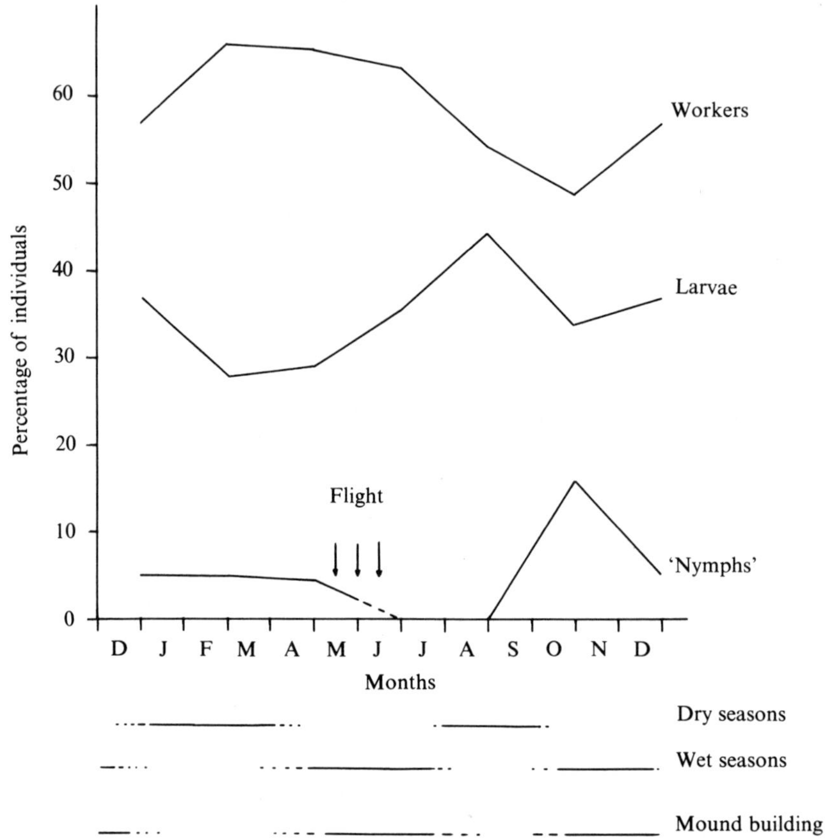

Fig. 2.9. Annual fluctuations of the percentages of workers, larvae and nymphs in societies of *Cubitermes severus*; the soldier proportion always stays beneath or near 1%. Smoothed, simplified curves after P. Bodot (1966, 1969).

immature societies of *Cubitermes severus* (societies with less than 10000 individuals and which do not produce alates) to 85% in senile societies (more than 40000 individuals, Table 2.11). Furthermore, she shows annual variations in the caste proportions correlated with the rain cycle and the production of alates (Fig. 2.9).

Features which all ants and termites seem to have in common are the following:

(*a*) Newly founded colonies, after the initial, workerless stage, pass through a phase in which only sterile castes are produced. In termite societies, soldiers appear with the first workers; in ant societies, while workers may be present after a few weeks or months, one or more years are usually necessary for the production of 'soldiers'; in both cases, sexuals are produced much later.

(*b*) The first workers, which develop when the food input into the colony is lacking or very limited, are much reduced in size and weight in comparison with the species average.

(*c*) The production of sexuals in most of the species studied is limited to a short and constant period of the year, but there are species that seem to be able to produce sexuals through the whole year (Kannowski, 1963; Markin *et al.*, 1971).

(*d*) Mature termite societies produce sexuals of both sexes together, but in ants both males and females may be produced or one sex may be virtually absent. This phenomenon does not seem to be strictly species specific (Marcus, 1949; Kusnezov, 1949) and may perhaps reflect the nutrient status of the colony at the time of sexual production (Brian, 1965; Brian & Elmes, 1974).

The worker/larva ratio could be an intrinsic species character; if sufficient experimental evidence were obtained for this hypothesis a very useful simplification in production studies might result. In fact, Brian (1950, 1953*b*) has shown that such a constant ratio between worker and larval weight exists in one species of *Myrmica*. Also, different studies have shown that the growth rate both in ant and termite colonies is well approximated by the logistic curve (Kalshoven, 1930; Bodenheimer, 1937; Bitancourt, 1941; Brian, 1965). From all these data, Brian (1965) has been able to construct a mathematical model with valuable predictive properties. The potential power of such an instrument in productivity studies should be self-evident. Elmes (1973) has developed a queen/worker system on similar lines.

Longevity of individuals

Little evidence is available concerning intra-colonial mortality in the population. In termites, Buchli (1958) found that workers of *Reticulitermes lucifugus* rarely lived more than one and a half years, though he suggests that in large, more homogeneous colonies survival may last five years. Pseudergates of *Calotermes flavicollis* (Fabr.) survived for nearly two years (Grassé, 1949) and workers and soldiers of *Coptotermes lacteus* (Froggatt), *Coptotermes acinaciformis* (Froggatt) and *Nasutitermes exitiosus* for about the same time (Gay, Greaves, Holdaway & Wetherly, 1955). Williams (1959*b*) calculated that the workers of *Cubitermes ugandensis* Fuller survived for between 196 and 339 days. All these estimates have been derived from laboratory studies, and under similar conditions Brian (1951) has shown that the workers of *Myrmica rubra* (L.) live for a maximum of two years. The difficulties of determining longevity in the field need no emphasis, though some progress has been made in the last few years with regard to free ant colonies. For example, Brian (1972) found that only 20% of the workers of *M. sabuleti* survived for up to a year and less than 1% for two years. *M. rubra* survival was lower, only 6% living for a year or longer, while none survived for 2 years. Nielsen (1972*b*) found that in *Lasius alienus* 34000 workers produced during the summer were able to sustain a colony population of about 10000, which implies survival of less than a year for a large proportion of the workers.

3. Production by ants and termites

M. G. NIELSEN & G. JOSENS

In the IBP terminology (Petrusewicz, 1967; Petrusewicz & Macfadyen, 1970) production is defined as: the net balance of food transferred to the tissue of an organism or a population for a defined time period, and described by

$$P = A - R = C - (R + FU) \tag{3.1}$$

Where P is production, A is assimilation, R is respiration, C is consumption and FU are the rejecta (faeces and urine). In other words the production is the amount of biomass which is elaborated by an organism or a population in a specified time period, including elimination (removal), and excluding weight losses of members of the population:

$$P = \Delta B + E \tag{3.2}$$

where ΔB is change in biomass, E is elimination of biomass, e.g. by predation, death, exuviae, emigration and organic deposit (mammals milk, elaborated saliva of termites, etc.).

Production can also be expressed as

$$P = P_r + P_g \tag{3.3}$$

where P_r is the production due to new born organisms and P_g is the production of body growth. The ecological parameters (e.g. P, A, R, E, B) are normally expressed in Joules per unit time and unit area or volume. The symbol 'Λ_x' will be used for the energy content of x.

Another expression of formula 3.3 is based on age specific longevity and may be used in studies on social insects:

$$P = Tn_e\Lambda_e + T\sum_{i=1}^{k}[(\Lambda_{i+1} - \Lambda_i)N_i/t_i] + E \tag{3.4}$$

where n_e is the number of eggs laid daily, Λ_e is the average energy value of an egg, T is time measured in days, N_i is the number of individuals of stage i, t_i is the average duration of stage i in days, Λ_i is the average energy content of an individual of stage i, and E is elimination of biomass, sometimes important in termites as will be seen further on. This is due to Winberg, Pečen & Šuškina (1965) (quoted by Petrusewicz & Macfadyen, 1970: p. 94). A simplified version of this formula will be proposed and used further on.

Production is one of the parameters which can easily be measured for a single organism, but at the population level it often involves huge problems. The literature about production in ant and termite societies is very sparse; this is no doubt due to the special difficulties of working with social insects, mainly the intricate degree of coordination and huge size of the populations.

Production of eggs

The number of eggs produced per queen varies enormously from species to species and with the age of the queen. The young queen generally lays a few eggs some days after her foundation, and does not start again until just before the development of the first workers. In well-developed societies, the egg production is roughly regular, seasonal or even periodic.

Eciton hamatum Fabricius produces a huge amount of eggs every five weeks (Schneirla, 1956) and *Myrmica ruginodis* Nyl. lays two batches a year (Brian, 1957*b*). The queens of *Amitermes hastatus* (Haviland) produce several hundreds of eggs in a few days and then rest for some days or weeks before they start again; this is only during the summer season (Skaife, 1955). In tropical climates, egg laying is more regular, but seasonal fluctuations are probably frequent, as was shown by variations in the larval rate of *Trinervitermes geminatus* Wasmann (Sands, 1965*a*), and of *Cubitermes severus* Silvestri (Bodot, 1969) populations. Table 3.1 gives some figures of egg production for ants and termites.

The number of eggs produced per queen is dependent on several factors, e.g. number of queens in the nest (Gösswald, 1951*a*, *b*), food restrictions (Brian, 1965, 1957*b*) and temperature (Brian, 1957*b*). The rate of egg production under laboratory conditions might be quite different from the rate in the field, and therefore the values in Table 3.1 must only be taken as an order of magnitude. When egg laying is rather regular, the egg production is observed generally by putting a queen (just caught) in a vial and counting the eggs some minutes or hours later.

In some cases (among tropical species), the daily egg production (P_e) may be approximated from several independent observations. If egg laying is estimated regularly during a period exceeding the hatching time plus the duration of stage 1 (first larval stage), either knowing the average hatching time (t_e) and the number of eggs present in the nests (N_e), or knowing the number of individuals of stage 1 (N_1), and the average duration of stage 1 (t_1), then

$$P_e = N_e/t_e = N_1/t_1 \qquad (3.5)$$

This either assumes no mortality among eggs (and stage 1), which is very unrealistic, or that it is known.

This method provides a useful verification under humid, tropical climates, where fluctuations in populations of ants and termites may be of

Table 3.1. *Production of eggs for some queens of different ant and termite species*

Species	Number of eggs produced per queen	Author
Ants		
Monomorium pharaonis	400–500 per year	Peacock, 1950
Myrmica rubra	400–800 per year	Brian & Hibble, 1964
Formica polyctena	2.1×10^3 per year	Gösswald, 1951*a*, *b*
Formica rufa	6.3×10^4 per year	Gösswald, 1951*a*, *b*
Eciton burchelli	2×10^5 per month	Schneirla, 1971
Dorylus nigricans	$3\text{–}4 \times 10^6$ per month	Raignier & van Boven, 1955
Termites		
Some Calotermitidae	200–300 per year	Grassé, 1949
Cryptotermes havilandi	8 per day	Wilkinson, 1962
Apicotermes gurgulifex	400–500 per day	Bouillon, 1964
Apicotermes desneuxi		
Trinervitermes geminatus	500–1 000 per day	Josens, unpublished
Microcerotermes arboreus	1 690 per day	Emerson, 1938
Nasutitermes surinamensis	3 917 per day	Emerson, 1938
Macrotermes subhyalinus	36 000 per day	Grassé, 1949
Odontotermes obesus	26 208 per day	Arora & Gilotra, 1960
O. obesus	86 400 per day (= one per second)	Roonwal, 1960
Cubitermes severus		
Immature society (< 10 000 individuals)	50–200 per day	Bodot, 1966
Mature society (10 000–40 000 individuals)	up to 600 per day	Bodot, 1966
Senile society (> 40 000 individuals)	80 per day	Bodot, 1966

relatively little magnitude. Josens (unpublished data) observed that laying by captive queens of *Trinervitermes geminatus* gradually decreased and stopped some days later, even when about 2000 workers and soldiers accompanied the queen; a graphical extrapolation suggested a laying of about 30 eggs per hour, and a verification using N_1 and t_1 in formula 3.5 showed that laying had to be *at most* equal to 21 eggs per hour.

Egg production by workers is known in many ant species, but very little has been done to quantify this problem. Brian (1953*a*, 1957*a*, 1969) has studied the egg production by workers of the genus *Myrmica*, and he found that workers, on average, lay 16 eggs per year. Passera (1966, 1969) has studied the production of worker laid eggs in *Plagiolepis pygmaea* Latr. He found that the workers laid two types of eggs, *viz.* small eggs which are used as food for queens and larvae, and large eggs which develop into males.

Brood growth and survival

No satisfactory methods for field conditions have been used to estimate production in the larval stage. The main difficulty is to measure the production of the larvae which die before the imaginal state. In laboratory experiments, mortality and growth rate for larvae can be measured without real difficulty, but there is no evidence that information obtained in the laboratory is applicable to natural conditions.

Several methods can be used to measure production, see for example, IBP Handbook No. 13, Chapter 4 (Petrusewicz & Macfadyen, 1970), but most of the methods assume knowledge of parameters which are impossible to obtain in field experiments.

Pętal (1967, 1972) dug up two or three nests of *Myrmica laevinodis* Nyl. (= *M. rubra* (L.)) each week and measured the weight of 12 size classes of larvae, pupae and young imagoes. This method gives the possibility for estimating the mortality in the larval and pupal stages, and by knowing the average weight of each size class, the biomass of the dead larvae and pupae can be calculated. The production is then found by adding the biomass of the eliminated larvae and pupae to the biomass of the workers produced.

This method assumes an equal distribution of the different size classes in all nests at the same time, and it has the disadvantage that it is destructive and requires a lot of nests. If the nests are widely spaced, the different age classes will generally not be randomly dispersed; therefore the method will be applicable only when the nests are dense. Moreover, daily fluctuations may occur in dispersion of age classes, even in concentrated nests, as was shown by Sands (1965*a*) in the case of *Trinervitermes geminatus*.

Worker production

Most effort has been directed towards measuring the production of new workers by the population. Golley & Gentry (1964) have a minimum estimate of worker production in nests of *Pogonomyrmex badius* Latreille, which they found by excavating two nests in May and three nests in October, and then by taking the difference of the mean populations. This estimate is very rough, primarily because the elimination of workers between May and October is not included and also because of the small number of samples. They have a further estimate which applied the information of Cole (1934) that it takes about 30 days for workers of *Pogonomyrmex occidentalis* (Cresson) to develop. By taking the maximum number of larvae found in a nest and multiplying it by the number of 30-day periods in the growing season of each year the production was estimated as 3200 workers, whereas the first method gave an estimate of 2000 workers produced per year per nest.

The method used by Pętal (1967, 1972) on *Myrmica laevinodis* has the advantage that it could estimate the production of eliminated larvae and pupae, and in addition to this it could distinguish young imagoes from aged ones, so the elimination of workers could be estimated too; the latter ranged between 20–50% of the worker population for the season.

Brian (1972) has studied the worker turnover for *Myrmica sabuleti* Meinert and *Myrmica rubra* (L.). Marking the workers by removal of an epinotal spine and using Petersen's mark–recapture technique to estimate the population density, he found that the replacement rate for workers of *M. sabuleti* and *M. rubra* was 80% and 94% per year, respectively. The ant nests were only investigated once a year, so the workers that had hatched and died between the yearly samples were not included in the production.

By using durable radioisotope marks Nielsen (1972*a*, *b*) has estimated the production of workers in a nest of *Lasius alienus* (Foerst.). The population density was estimated weekly by using Petersen's mark–recapture technique, and the ants were labelled with ^{24}Na (half-life 15 hours, γ-emission). Besides the ^{24}Na-labellings, workers were marked with ^{32}P (half-life 14 days, β-emission) which could be detected more than six weeks afterwards. The ants were dipped for two minutes in a solution containing about 200 μCi per ml. Afterwards the ants were kept for four to six hours on filter paper in a 'drying cage' which was ventilated by an air flow. In this way contamination with isotope from labelled to non-labelled ants was eliminated. The mortality of the ants during marking was normally very low, about 1%. By taking the samples from different places in the nest in a reproducible way the proportion of young and old workers was the same at each sampling.

From the change in the proportion of ^{32}P-labelled ants, together with the population density, the production of workers can be calculated (Nielsen 1972*b*). It is equal to the increase in the population plus the number of animals which leave the population (mainly by dying) during the same period. To cover the whole season three ^{32}P-labellings were carried out, and the total production of workers during the season amounted to 1.89 of the maximum population density in the nest. The method is rather time consuming and involves all the mark–recapture technique problems just discussed, but it is not destructive, so all measurements can be carried out on the same nest year after year.

Table 3.2. *Number of worker stages known for some termite genera of the family Termitidae (after Noirot, 1955)*

Genus	Minor workers[a]	Major workers[a]	Workers[b]
Amitermes	—	—	4
Microcerotermes	3	4	—
Termes	—	—	3
Cubitermes, *Noditermes*, *Orthotermes*, *Pericapritermes*	—	—	1
Acanthotermes, *Ancistrotermes*, *Bellicositermes*, *Microtermes*, *Odontotermes*, *Protermes*, *Pseudacanthotermes*, *Sphaerotermes*	1	1	—
Eutermellus	—	—	1
Leptomyxotermes, *Coarctotermes*	1	2	—
Nasutitermes	2	4	—
Trinervitermes	1	7	—

[a] Minor and major workers develop independently; in some genera the minor workers moult into soldiers.
[b] Undifferentiated workers.

Total brood and worker production

The method of Winberg, Pečen & Šuškina, used by Pętal (1967, 1972), is very precise, but it too is rather time consuming. A simplified version of this method may provide some limited estimates of total production, using the mean egg production and the mean worker mortality in societies of stationary population – neither increasing nor decreasing – provided the egg production is regular or annual mean data are available. If the mean egg production is lower than the mean worker mortality, it seems logical to suppose that an experimental error, or an artefact, has occurred. If the mean egg production and the mean worker mortality are equal or close, mortality occurs almost only in the worker stages, and the total brood and worker production is maximal. Formula 3.4 then becomes:

$$P = n_e \times \Lambda_w \times T + E \qquad (3.6)$$

where Λ_w is the energy value of a worker.

Josens (1973) gave a first rough estimate of production in two termite species *Trinervitermes geminatus* and *Ancistrotermes cavithorax* (Sjö-

Table 3.3. *Production of the larval and the worker stages in societies of* Trinervitermes geminatus *(after Josens, 1972)*

Stage	Fresh weight (mg per individual)	Dry weight (mg per individual)	Energy content (J per individual)	Production (J per individual)
Egg	0.067	0.016	0.372	Larval: 14.04[a] (from egg to the first worker stage)
Larvae (two stages mixed)	0.99	0.11	2.406	
Worker (first stage)	3.97	0.72	14.41[b]	
Worker(secondstage)	5.18	1.03	20.43[b]	Worker: 14.11[c]
Worker (third and following stages)	7.23	1.48	28.52[b]	

[a] Overestimated because the energy content of the first worker stage is a mean datum; it would have been necessary to use the energy content of the freshly moulted workers for calculating the actual larval production.
[b] Gut content is 10 to 20% in these data.
[c] Underestimated, for the same reason given in (*a*), and again, a mean energetic content has been used for workers from the third to the last stage.

stedt). Using the total population number, the percentage of the larvae (field data), and their mean duration (laboratory data), he calculated the mean number of individuals produced daily. Assuming a stationary population and a negligible mortality in young stages, he then obtained the annual production taking into account both the energy values of the workers and the soldiers and their respective proportions in natural populations. In this first rough estimate, a very important difference appeared between the two species considered: the annual rate of production/biomass was about one in *T. geminatus* but about 14 in *A. cavithorax*.

Worker growth

In ant societies, the weight increase of workers during the adult stage is normally very small, and for some species it is negligible (Nielsen 1972*a*). The newly hatched workers can often be distinguished from older workers by their white or pale colour, and therefore a comparison of weights is possible. It might be necessary to compare weights frequently during the season because the mean weight of newly hatched workers often fluctuates. In some species, e.g. *Myrmica*, older workers forage more than younger workers and the crop content can increase their weight substantially; on the other hand, younger workers produce more eggs, the presence of which must augment their weight.

In some termite genera, the workers moult several times after their larval stages: (Table 3.2); sometimes too their production may be more important during their worker life than during their larval stages (Table 3.3).

Elimination of biomass

Apparently nothing is known about the biomass of the exuviae of ants and termites; in termite societies, the exuviae are very difficult to collect for they are eaten by the workers. Another biomass loss is very difficult to estimate but may be very important in some termite societies: namely the elaborated saliva of the workers. In the termite societies, the sexuals, the larvae, the nymphs (and in some genera also the soldiers) are nourished with an elaborated saliva produced by the workers and comparable with the milk fed to new-born mammals and honey bee larvae. In some species, the salivary glands become so developed that they invade the thoracic cavity and even part of the abdominal cavity in the worker body. No work has been done to quantify directly this biomass loss. Nothing is known, in terms of production, about other biomass losses such as the venom of ants or the secretion of nasute termite soldiers.

Production as food for the society

As already mentioned above, production loss, the elaborated saliva of the termite workers, has a feeding function. Moreover, in termite societies, the whole of this production (the defensive secretions and the hard, chitinised structures excepted) is in fact consumed: the dead individuals are eaten by the living ones. In some cases, cannibalism also occurs and both eggs and larvae are regularly devoured. This consumption of dead individuals is consistent with economy in nitrogen, phosphorus and other mineral cycles of the societies; the scarcity of these elements (found in litter, decayed wood) in their food is usually striking. In ants, the brood is regularly eaten in times of starvation and quite a number of species are now known to produce non-viable trophic eggs.

Production of sexuals

The production of sexuals in ant nests is correlated with a number of factors, e.g. the number and size of workers, the age of the queen, food resources and climatic factors. A more detailed discussion of this problem is given by Gösswald & Bier (1957) and Brian (1965).

The production of sexuals has been studied in several species. Most investigations have been carried out by estimating the number of sexuals which leave the nest. Gösswald (1957*a*, *b*) has studied the production of sexuals in *Formica rufa* L. nests by building a tent over the nest so that the sexuals were trapped when they took off for the mating flight. The same method has been used for collecting sexuals of *Lasius alienus* and *Lasius flavus* (F.) (Nielsen, unpublished).

Brian *et al.* (1967) collected sexuals of *Tetramorium caespitum* Latr. under slates which the ants had adopted for nesting. By sampling frequently, until no more sexuals appeared under the slates, most of the sexuals were collected. For this species the production of sexuals ranged from 43 to 53% of the average worker biomass. The same method has been used by Pontin (1961, 1963), who studied the queen production by *Lasius flavus* and *Lasius niger* (L.) and correlated the production with the territory of the nest. He calculated that the production of queens in *L. flavus* was 15 per m^2 where the worker population was 1130 per m^2. The number of winged sexuals in *Myrmica sabuleti* nests has been estimated by Brian (1972) using mark–recapture technique. This method can only be used on species where all sexuals emerge before the first sexuals take off for the mating flight.

Using his core sampling method Sands (1965*a*) found 5% of nymphs (potentially alates) in the nests of *Trinervitermes geminatus*; this observation was performed in northern Nigeria. By entire mound samplings, Josens (1972) observed a production of only 1% of nymphs and alates in the nests of the same species, in the Ivory Coast, at the border of the semi-deciduous tropical forest. These two observations are probably not contradictory for production must vary with the distance from the optimal geographical and climatic area. In the Ivory Coast, at the border of the forest, *T. geminatus* is at the border of its geographical distribution, whereas the optimal condition for its existence probably occur in northern Nigeria.

Swarming as an energy loss

Whilst the production of neutral castes is consumed by the termite societies after the death of the individuals, the swarming of alates represents an important energy loss. Whereas the societies of *T. geminatus* produce only 1% of alates, these sexuals contain more than 15% of the energy of the society (Josens, 1972). This would exceed 50% of the energy of the societies where the production of alates reaches 5% of the populations, as in northern Nigeria. Such an important percentage (more than 50% of the energy) was also observed in the Ivory Coast, as in the alate production in societies of *Ancistrotermes cavithorax* (Josens, 1972).

The energy loss from ant nests through alate swarming exceeds the same values in termites. Brian *et al.* (1967) found that for *Tetramorium caespitum* the biomass of the sexuals ranged from 43 to 53% of the worker biomass, which represents a loss of energy of about 50%. More recently the production of sexuals by this species has been shown to have a bienniel periodicity (Brian & Elmes, 1974).

4. Food and feeding habits of termites

T. G. WOOD

Food supplies for the termite colony are collected by the workers who feed themselves and the dependent castes: larvae, nymphs, soldiers and primary and supplementary reproductives. This chapter is concerned with the qualitative and quantitative aspects of food and foraging and with digestive processes in so far as they determine utilisation of food. The effects of food collection, storage, consumption, digestion and defaecation on the cycling and spatial distribution of organic matter and nutrients in the ecosystem is discussed in Chapter 7.

Food and feeding

Range of foods and feeding habits

The range of food materials eaten by termites has been reviewed by Adamson (1943), Noirot & Noirot-Timothée (1969) and by Lee & Wood (1971*b*). Their basic food is plant material: living, recently dead, dead but in various stages of decomposition and soil rich in organic matter (so-called humus). Only under unusual circumstances are other substances, such as leather or plastics, attacked. The trophic levels of termites are those of primary consumers (i.e. herbivores) and decomposers. Although they are cannibalistic within the colony, as far as is known termites are not predators although they occasionally attack vertebrate corpses, consuming skins, feathers and dried-out tissues. Several categories of food can be recognised and the classification indicated below, has been selected to emphasise trophic level.

Living vegetation

(i) Aerial (i.e. tree trunks, branches) and subterranean (i.e. thick roots) woody tissues.

(ii) Aerial (i.e. grasses, herbs) and subterranean (i.e. slender roots) non-woody tissues.

Fresh, dead vegetation

(i) Standing, dead trees, fallen trees and branches, thick roots.

(ii) Standing, dead grasses and herbs, slender roots.

(iii) Plant litter lying on the soil surface, consisting of dead leaves and small twigs of trees and shrubs, grasses, herbs, seeds, etc.

Decomposing vegetation

(i) Rotting wood either standing, fallen or subterranean.
(ii) Rotting plant litter on the soil surface.
(iii) Dung.

Humus

Soil organic matter consisting of intimately mixed organic and mineral particles: see Lee & Wood, (1971*b*).

Fungi

(i) Fungi cultivated in special regions (fungus gardens) of the nest: exclusive to and universal within the Macrotermitinae.
(ii) Fungi ingested as a significant component of the diet (e.g. in decomposing vegetation).

Special or incidental foods

(i) Other termites of the same colony (cannibalism and oophagy).
(ii) Fungi, algae, lichens, etc.
(iii) Organic-rich portions of nests of termites.
(iv) Skins, etc. of vertebrate corpses.

These categories are not mutually exclusive. Some species are almost polyphagous herbivores, such as the Australian *Mastotermes darwiniensis* Froggatt (Ratcliffe, Gay & Greaves, 1952), the north African and Arabian desert-inhabiting *Psammotermes hybostoma* Desneux (Harris, 1970). Some of the Macrotermitinae consume, in addition to a wide variety of living plants (wood and non-woody), many forms of dead plant material but they do not seem to attack decomposing vegetation; this may be related to their habit of cultivating specific fungi on faecal material. Wood-feeding is regarded as a primitive habit (Grassé & Noirot, 1959*b*). However, although the majority of the lower termites (i.e. families other than Termitidae) is wood-feeders, the wood-feeding habit is retained by many of the higher termites (Termitidae): at least 33 of the 53 species of Australian Termitidae* considered by Lee & Wood (1971*b*: Table 1) fed on wood. The majority of wood-feeding termites feeds on the dead material, relatively few feeding on living wood. Grass, herbs, dung and plant litter are resources for a wide range of termites which either forage in the open or under the protection of runways or sheets of soil constructed over the food. Some of these species tend to be polyphagous (e.g. Macro-

* The sub-family classification of the Termitidae adopted here is that of Sands (1972*a*). The four sub-families recognised are Termitinae, Apicotermitinae, Nasutitermitinae and Macrotermitinae. The name 'Amitermitinae' has become a junior synonym of the Termitinae.

termitinae, such as *Macrotermes subhyalinus* (Rambur)), while others are specialised grass-feeders (e.g. Hodotermitinae, some Termitinae and Nasutitermitinae). Humus-feeding is a specialised habit and can be regarded as an extension of the habit of feeding on decomposing vegetation, although it is sometimes difficult to distinguish the incidental ingestion of soil particles along with decomposing plant material from the more specialised habit of ingestion of mixed mineral and organic matter (i.e. humus). Humus-feeding is found only among the Termitidae and is associated with morphological specialisation of the mouthparts of the workers, which has occurred convergently in the Termitinae, Apicotermitinae and Nasutitermitinae. The significance of fungi in the diet of termites varies, from an apparently symbiotic relationship in Macrotermitinae, to incidental and insignificant relationships in other species, but they are an important and often essential component for many species feeding on decomposing vegetation. Extreme specialisation in food requirements is rare, but does exist, as in the lichen-feeding *Hospitalitermes* in Indonesia (Kalshoven, 1958) and in the Australian genera *Ahamitermes* and *Incolitermes* which are obligate inquilines in the nests of *Coptotermes* (Gay & Calaby, 1970) where they feed exclusively on the carton of the nest material (i.e. structures made from faeces) of their hosts. Cannibalism is thought to be partly a means of sanitation and partly a means of conserving and recycling nitrogen. Recent work (Benemann, 1973; Breznak, Brill, Mertins & Coppel, 1973) indicates that some termites can fix nitrogen in low amounts (up to 566 μg nitrogen per g fresh weight of termites).

Observation of feeding in the field

Some termites, particularly primitive wood-feeding species, construct their entire nest system within their food so that determination of feeding preferences in the field depends on locating colonies and tracing galleries. This may prove difficult in the case of species with small colonies within the upper branches of tall trees: Nutting (1965) described a colony of *Zootermopsis laticeps* (Banks) at a height of 11 to 13 m. Other endophagous nest systems, such as those of *Porotermes adamsoni* (Froggatt) in rotting logs (Ratcliffe *et al.*, 1952), may be more accessible but pose similar problems in that there is often no external evidence of the presence of the termites, and sampling is, of necessity, destructive and unrewarding. Similar problems are posed by those subterranean species that forage entirely within the soil. Many of these species are humus-feeders while others may feed on roots and it has not yet been possible to make direct observations of their feeding behaviour. Some indication may be given by extracting termites from hand-dug pits or from soil cores and noting their range of vertical distribution within soil, as done by Williams

(1966) with *Cubitermes*, or their location in relation to pedological features, such as the structure, state of decomposition and distribution of organic matter. Again, sampling is destructive and often unrewarding and other methods (see below) may be more satisfactory. However, the food and foraging habits of many termites can be observed in the field by direct observation of foraging parties, observation of removal of marked food, examination of food stored in nests and galleries and tracing of foraging galleries to the source of food.

Direct observations of foraging parties can be readily made on species which forage in the open without the protection of covered runways. These species usually forage during the night, early morning and evening, or occasionally during the day if the sun is obscured by heavy cloud. Foraging in the open has been adopted by many different groups of termites: among the Hodotermitinae are *Hodotermes* and *Microhodotermes* in Africa (Coaton, 1958) and *Anacanthotermes* in Asia (Harris, 1970); among the Rhinotermitinae is *Schedorhinotermes derosus* (Hill) in Australia (Watson, 1969); among the Termitinae are various *Drepanotermes* in Australia (Gay & Calaby, 1970; Watson, personal communication); among the Nasutitermitinae are *Nasutitermes* in Australia (Gay & Calaby, 1970), *Trinervitermes* and *Fulleritermes* in Africa (Bouillon, 1970), *Hospitalitermes* and *Lacessititermes* in Indo-Malaya (Roonwal, 1970), *Syntermes* in South America (Emerson, 1945) and *Tenuirostritermes* in North America (Nutting, 1970). It should be possible to quantify food collection and foraging of these surface-active species as had been done many times with ants (e.g. Holt, 1955; Ayre, 1958; de Bruyn & Kruk-de Bruin, 1972; Nielsen, 1972*c*) by estimating duration of foraging, intensity of foraging (i.e. traffic per unit time) together with estimates of food collected and returned to the nest. Some species (e.g. *Hodotermes*, *Microhodotermes*, *Trinervitermes*, *Drepanotermes*) climb tall grasses, often to a height of 2 m (Sands, 1961*b*), cut down pieces of grass which may or may not be cut into smaller pieces which can be carried back to the nest; other species appear to collect only plant litter lying on the ground. Also, some observations (Kalshoven, 1958; Sands, 1961*b*) show that workers returning to the nest with fragments of grass or other plant material have full crops, indicating consumption of food while foraging. The only quantitative studies of this nature on termites are for *Hodotermes mossambicus* (Hagen) (Nel & Hewitt, 1969, Nel, 1970) which is one of the principal grass-harvesters in southern Africa and for *Trinervitermes geminatus* Wasmann (Ohiagu & Wood, 1976) in Nigeria. However, for a complete assessment of the impact of such species on the vegetation, in addition to measurements of termite populations and the duration and intensity of foraging, the following measurements (made at suitable intervals of time) are required:

(i) standing vegetation (biomass and production);

(ii) litter (biomass and production);
(iii) vegetation cut down by termies;
(iv) vegetation cut down but not collected;
(v) vegetation collected and transported to the nest;
(vi) ingestion of food by workers whilst foraging.

Similar observations, though more time consuming and requiring the development of methods for assessing duration and intensity of foraging, could be made on species feeding on living or dead trees and concentrated sources of food such as fallen trees, branches, dung, etc. on the soil surface. Studies of foraging activity are more readily made on species which construct covered runways above ground level than on those which have an almost completely subterranean system of foraging galleries. Using a line-transect sampling method, Lepage (1972) made quantitative estimates of food consumption in the field of *Macrotermes subhyalinus* which is often noted for the abundance of its covered runways. He measured the length occupied by runways along 5 m line transects, measured the weight of herbaceous vegetation and litter on ground covered with runways and on ground where there was no foraging, and thereby calculated the amount of vegetation removed by the termites. Food selection in species with largely subterranean galleries is not readily studied by tracing the galleries. Often the only evidence of feeding by this group is holes in the soil surface leading to the subterranean galleries, or frequently, short, covered runways or sheets of soil originating from a foraging hole. This group includes the majority of subterranean Macrotermitinae, which in the tropics, except for South America and Australia where they do not occur, is the most important group of termites causing damage to timber, forestry trees and agricultural crops. There has therefore, been a considerable amount of field work devoted to assessing their food preferences.

The most widely used method of studying the feeding habits of subterranean termites, particularly the polyphagous and wood-feeding species, is by the use of baits presented in varying situations – on the soil surface or completely or partially buried. The baits are usually laid out close together in a systematic layout (the so-called graveyard tests) and baiting materials vary, depending on the aims of the experiments, but usually consist of various timbers, timber treated with preservatives or insecticides, plastics or other materials subject to attack by termites (e.g. Coaton, 1947; Kemp, 1955; Butterworth & MacNulty, 1966; Williams, 1973; Gay, Greaves, Holdaway & Wetherley, 1975); toilet rolls were used by Nutting, Haverty & La Fage (1973). A similar technique employing nylon mesh bags for holding leaf litter is widely used in the field of soil ecology to assess the rate of decomposition of leaf litter and the significance of different groups of organisms in decomposition processes (e.g. Heath, Edwards &

Arnold, 1964; Madge, 1969; Wood, 1971, 1974; Anderson, 1973). Although the basic principles of baiting for termites and litter bags for studies of decomposition are similar, in that the aim of both techniques is to assess rate or amount of feeding by different groups of soil organisms, the methods used are somewhat different. The aims of baiting studies have usually been based on the economic necessity of determining the resistance of various timbers to attack by termites. The widely accepted procedure is to pre-bait for approximately three months with timber susceptible to termite attack and then, once the termite population has built up (whether it is by increase in abundance of species normally abundant on the site or by attraction of termites that would not normally be abundant is not clear), to lay out the timbers to be tested. Thus the termite population is presented with an unusual abundance of food which, by the very nature of the experiment, may change the relative abundance of species of termites in the study area. In contrast, litter bag experiments usually present decomposer organisms with their naturally-occurring food, in naturally-occurring situations (i.e. on the soil surface) and in amounts which correspond to the natural accession of leaves, branches, etc. in the particular habitat under investigation. Methods of assay also differ. Feeding by termites is usually scored subjectively (three to five classes of damage are frequently selected) and in some cases attack by fungi is also scored. Williams (1973) adopted the method, which is generally used in litter bag studies, of measuring loss in weight of baits exposed to attack by termites and showed that such subjective assessments of termite feeding are inaccurate. It should be possible to apply some of the more objective methods used in litter bag studies to baiting experiments, in order to assess, qualitatively and quantitatively, the food and foraging habits of many groups of termites.

Experiments along these lines are being carried out on species feeding on woody litter, as part of the Centre for Overseas Pest Research/Ahmadu Bello University (COPR/ABU) Termite Research Project in Nigeria. Measurements are made of the natural rate of wood decomposition with termites present and the rate in the absence of termites. Wood baits are used to measure the relative consumption by different species of termites and, assuming that these species have the same proportional consumption of baits as natural, woody litter, it is possible to calculate their natural rates of consumption.

Three recent modifications of the baiting method extend the possibilities of studying food selection and food consumption. The method developed by Josens (1971*a*) to study food consumption by subterranean Macrotermitinae suffers from some of the disadvantages of baiting noted above, but offers the possibility of estimating turnover of food within the colony. Macrotermitinae use their excreta to construct fungus combs which support the growth of specific fungi. The old parts of the comb, which have been subjected to decomposition by the fungus are consumed by the termites while fresh excreta is continuously added to the comb. Josens (1971*a*) provided sawdust baits impregnated with soot and measured the length of time it took the newly deposited, soot-stained faeces, to be completely replaced in the fungus combs. Consumption was calculated from total weight of combs and their rate of turnover. Haverty & Nutting (1975) estimated consumption by the deadwood-eating *Heterotermes aureus* (Snyder) in the Sonoran desert by a simulation method. They used data on temperature-dependent functions of foraging intensity (number of termites at toilet roll baits) (Haverty, La Fage & Nutting, 1974), daily turnover of foragers, laboratory-determined values of wood consumption and relative rates of consumption of different species of wood. These calculations took account of seasonal variation in foraging but incorporated two untested assumptions. Firstly, that intensity of foraging at palatable baits was similar to that at their natural food; experience of baiting procedures indicates that this would overestimate foraging intensity. Secondly, that there was continuous movement to and from the baits; as foraging termites spend a considerable amount of time in mutual grooming and other activities, such as construction of galleries and carrying soil particles, this assumption also overestimates consumption. Williams (1973) placed wooden baits in a recess cut in the soil surface, covered the bait with a glass plate so that the plate was just below the level of the soil surface and covered the glass with a hollow brick to provide shade and insulation. Foraging could then be observed without disturbing the bait, by lifting the brick. There is a possibility that this method could be adapted to study foraging and feeding habits of completely subterranean termites such as the humus-feeders.

Observation of feeding in the laboratory

The methods of rearing or maintaining termites in the laboratory are outside the scope of this book and have been adequately summarised by Becker (1969). They vary from maintaining a small number of individuals in petri dishes for a short period of time to maintaining entire breeding colonies. Most of the problems encountered in maintaining laboratory colonies are related to the control of temperature, moisture supply and humidity at optimum levels and in preventing death of the termites through pathogenic organisms or deterioration in quality of the food.

Most of the qualitative and quantitative laboratory studies of feeding have been conducted with non-breeding colonies and principally, because of the relative ease of handling their food and also because of their economic importance, with wood-feeding species (e.g. work by Kofoid

& Bowe, 1934; Gay *et al.*, 1955; Seifert & Becker, 1965; McMahan, 1966*a*). These and other experiments with wood-feeding species have shown that various parameters of the relationship between the termites and their food (amount of wood eaten, percentage survival, gain or loss in weight) vary according to the species of wood offered, as well as with environmental conditions. However, few of these experiments, although providing information on consumption of food, were designed to determine food preferences. Of those that were, only that of Haverty & Nutting (1974, 1975) integrate laboratory and field determinations of foraging intensity on different foods. Determining preferences of many wood-feeding species is also complicated by the fact that fungal attack, often by specific species, may determine whether or not a particular type of wood is suitable and in fact may be an important factor in determing survival and growth of the termites (Becker & Kerner-Gang, 1964; also reviewed by Sands, 1969). In contrast with the field testing (graveyard) methods, laboratory evaluation of consumption of or attack on wood by termites has invariably been done gravimetrically (e.g. Gay *et al.*, 1955; Becker, 1965*b*; Seifert & Becker, 1965). In laboratory feeding trials it is possible to exclude all other organisms except micro-organisms and to know, with some degree of accuracy, how many termites (usually determined gravimetrically) of a particular species have been feeding on the food offered.

Non-breeding colonies have been used to study some grass-harvesting species of *Trinervitermes* (Sands, 1961*b*), using moistened nest material as a matrix. Foraging commenced within 24 hours and direct observations could be made on feeding in relation to various species of grasses placed in the foraging chamber. Nel (1968*a*) adopted Dropkin's (1946) method of refrigeration, to allow newly paired alates to be mixed with workers, soldiers and juveniles collected from a field colony to form a breeding colony in the laboratory. This technique, which could have more general application, was subsequently used by Nel, Hewitt & Joubert (1970) to study food selection in the grass-harvesting *Hodotermes mossambicus*. Many grass-harvesting species build small mounds and in Nigeria, Ohiagu (personal communication) has removed whole mounds of *Trinervitermes geminatus* from the field for observation of feeding habits in the laboratory.

There have been few attempts to study feeding habits of Macrotermitinae or of litter-feeding and humus-feeding termites in the laboratory. Methods, such as those developed by Light & Weesner (1947) for various scavenging termites, and Ausat *et al.* (1962) for the macrotermitine *Odontotermes obesus* (Rambur), using artificial, prepared diets of agar and sawdust, are not suitable for experiments on food selection. Humus-feeding *Cubitermes* spp. have been studied in the laboratory by Alibert (1963) and

Williams (1959*a*, *b*) but these studies did not include food selection and foraging which in humus-feeders has yet to be studied in the laboratory. It is likely that laboratory studies of litter-feeding and humus-feeding termites can be more readily undertaken by using whole colonies (e.g. of mound-building species) removed from the field as done by Skaife (1955) with *Amitermes hastatus* (Haviland) which feeds on decomposing plant litter.

Observation of food stored in nests

The best-known instances of food storing are by grass-harvesting termites which often pack the galleries of their mounds (e.g. *Trinervitermes* in Africa and India, certain *Drepanotermes*, *Nasutitermes* and *Tumulitermes* in Australia and *Cornitermes* in South America) or subterranean nests (e.g. *Syntermes* in South America and *Drepanotermes* in Australia) with fragments of stored grass. Often the grass is stored in fragments up to 10 mm or more in length which can be measured (Sands, 1961*b*; Wood & Lee, 1971) although identification of the species of grass is difficult, particularly in species such as *Tumulitermes tumuli* (Froggatt) in Australia (Lee & Wood, 1971*b*) which store small pellets of very finely comminuted plant debris. Certain fungus-growing termites (Macrotermitinae) store comminuted food in special chambers. These accumulations (*boules de sciure* or *amas de sciure*) have been described by Grassé (1944, 1945) in *Macrotermes*, *Acanthotermes* and *Pseudacanthotermes* in Africa and by Kalshoven (1956) in *Macrotermes* in Indonesia. These food stores appear to be soaked with salivary secretions by the termites which consume the food once it has dried. The fungus combs of Macrotermitinae are a special case of food storage but as the material used in their construction is finely comminuted and partially degraded the components are not readily identified (see *Examination of faeces* below).

Examination of gut contents

Examination of gut contents is a widely used method for determining the feeding habits of many groups of soil animals such as Acari (Schuster, 1956), Collembola (Anderson & Healey, 1972), Diptera (Healey & Russell-Smith, 1971) and others, but has not been applied to termites since Adamson's (1943) study of the gut contents of various termites from Trinidad. Her observations indicated that plant remains were largely unidentifiable due to their advanced stage of comminution and degradation. Observations in a weakly ferrallitic, sandy clay near Mokwa (Wood, unpublished) indicate that *Adaiphrotermes* sp. is found predominantly in the topsoil and its alimentary system is invariably full of dark, organic-rich soil which appears to contain a greater proportion of organic matter than the topsoil in which the species is found. In contrast, the gut of *Anenteotermes* sp. is invariably packed with red soil containing little humified organic matter which, microscopically, is similar to the subsoil in which the species is predominantly found. Microscopic examination of the gut contents of humus-feeding termites, supported by suitable chemical and physical analyses of soil in the gut and soil from the termites' habitat, should enable some contribution to be made to our scanty knowledge of this group of termites. Such observations and analyses have yet to be made.

Examination of faeces

Examination of faecal pellets is merely an extension of analysis of gut contents but as faecal pellets can be counted, weighed and analysed without damaging individual animals, feeding can be quantified and estimates made of assimilation. These techniques have been used in both field and laboratory for many groups of animals and are too well known to be repeated here. Among termites, faecal production has been measured only for wood-feeding species, largely for lower termites such as *Calotermes* (Seifert, 1962) and *Zootermopsis* (Hungate, 1943) but also for higher termites such as *Nasutitermes exitiosus* (Hill) (Lee & Wood, 1971*a*). Measurement of faecal production in termites is made difficult by the fact that the majority of species does not produce discrete faecal pellets but produce semi-liquid faeces which are used to construct certain portions of the nest system, or to line the walls of galleries, or are fed to dependent castes or re-ingested by the termites themselves. Microscopic examination of faeces can be achieved by preparing thin sections of nests of galleries embedded in polymerising resins (Stoops, 1964; Lee & Wood, 1971*b*; Sleeman & Brewer, 1972). In general, the food is so finely comminuted and degraded to an advanced stage of decomposition by the termites' digestive processes that it is impossible to identify the origin (e.g. wood or grass) of the faeces which are almost amorphous in appearance. The pelleted faeces of certain primitive termites (e.g. *Mastotermes* and *Porotermes*, Lee & Wood, 1971*b*) and the fungus combs of certain Macrotermitinae (e.g. *Microtermes*, Wood, unpublished) contain fragments of plants with cellular structure still visible.

Examination of morphology of mouthparts and alimentary system

In many groups of animals, specialisation in feeding habits is accompanied by corresponding specialisation of mouthparts. It has already been noted that extreme specialisation of feeding habits is rare in termites but one might expect to see morphological differences in mouthparts and alimentary systems between, for example, wood-feeders, grass-feeders and humus-feeders. As far as mouthparts are concerned the best example amongst termites are the humus-feeders which can be characterised (Sands, 1965*c*; Deligne, 1966) by the loss of transverse grinding ridges on the molar plate of the worker's mandibles (particularly the right mandible) and their transformation to crushing cusps by the development of rounded flanges on both sides. This adaptation occurs convergently in the subfamilies Termitinae, Apicotermitinae and Nasutitermitinae. In grass-eating termites such as *Hodotermes* (Hodotermitinae), *Trinervitermes* (Nasutitermitinae) and certain *Amitermes* (Termitinae), workers have mandibles with dissimilar dentition, as do the three genera of Termopsinae (*Archotermopsis*, *Hodotermopsis* and *Zootermopsis*) which feed on rotting wood (Ahmad, 1950).

Noirot & Noirot-Timothée (1969) have discussed the anatomy and functions of the alimentary system of worker termites in relation to feeding habits. In general, lower termites, which are principally wood-feeders, have a short, relatively simple alimentary canal as do the Macrotermitinae (Termitidae). Among other Termitidae the recognised variations in gut anatomy are only poorly correlated with feeding habits; for example, although humus-feeders have a well-developed hindgut their members exhibit both of the two major types of alimentary system recognised by Noirot & Noirot-Timothée (1969) in the Termitinae. Thus, in the present state of knowledge the morphological features of mandibles and alimentary systems of termite workers provide some indication of feeding habits, but precise deduction cannot yet be based solely on anatomical features.

Food Selection

Mechanisms of food selection

The physiological and behavioural basis of food selection in termites is poorly understood (McMahan, 1966*a*; Abushama, 1967) and although the mechanisms of food selection are largely beyond the scope of this book it is apparent that the olfactory sense operates only within a few centimetres of the food. Searching for food is therefore, both extensive and intensive – a feature of termite populations which is at once obvious to those studying termites in the field.

Selection of certain species of plant

Most of the available information relates to wood-feeding termites as, for economic reasons, there have been many studies of the susceptibility and resistance of various timbers to attack by termites. Occasionally, resistance of timber to attack by termites is due to specific chemicals in the wood which repel certain termites. For example, in Australia, *Pinus* is attacked by a wide range of termites but not by *Nasutitermes exitiosus* (Ratcliffe *et al.*, 1952) and resistance to this species appears to be due to the fact that essential oils in the timber contain α- and β-pinenes which are a dominant constituent of the termites' alarm pheromone (Moore, 1965). Resistance is commonly more general as in the case of *Eucalyptus marginata* Donn ex. Sm. which is resistant to most Australian termites except *Nasutitermes exitiosus*; in contrast, hoop pine (*Araucaria cunninghamii* Ait. ex D. Don) is readily attacked by *Coptotermes* but is resistant to *N. exitiosus* (Ratcliffe *et al.*, 1952). Resistance is often due to a complex of chemical substances, as in *Eucaluptus microcorys* F. Muell. (Rudman & Gay, 1967), and in various coniferous woods showing resistance to *Coptotermes formosansus* Shiraki (Saeki, Sumimoto & Kondo, 1971). Many of the indications of feeding preferences noted in the field are likely to be due to the presence or absence of certain chemicals in the wood but few of these observations have been followed by biochemical investigations. Even the preference for (or avoidance of) certain tissues within the same piece of timber may be due to differences in the distribution of certain chemicals, as was shown for inner and outer heartwood of *Eucalyptus marginata* (Rudman & Gay, 1967). Two other factors appear to be of some significance in determining susceptibility, one being hardness of the wood and the other being the previous experience of the termites themselves. Soft, large-celled, fast-growing tissues are preferred by *Reticulitermes* to denser, small-celled, slow-growing tissues (Snyder, 1948), and *Coptotermes* in Africa attacks only the spring wood of rubber trees (Grassé, 1936). However, relative hardness does not appear to affect all wood-eating termites, as *Cryptotermes* attacks both spring and autumn wood equally. McMahan's (1966*a*) laboratory studies with *Cryptotermes brevis* (Walker) showed that conditioning of the termites had a significant effect on food selection – the termites preferred the wood (poplar) on which they had been previously fed to another (maple) to which they were unaccustomed, even though maple was preferred to poplar when the termites had not been conditioned to either. There is no information as to whether or not conditioning is of any significance in the field.

Information on the feeding preferences of grass-eating termites is limited to a very few species. Measurement of the relative abundance of grasses and shrubs and of the incidence of attack on these by *Hodotermes mossambicus* in South Africa enabled Nel & Hewitt (1969) to rank 16 species of plants in order of preference as food for the termites. These field observations were followed by laboratory studies (Hewitt & Nel, 1969*a*; Nel *et al.*, 1970) and the combined field and laboratory work showed that *H. mossambicus* preferred certain species of grasses and preferred grasses to shrubs, and that selection was influenced by thickness and width of leaves and by their physiological state – dry leaves were preferred to green leaves. Chemical factors were not analysed except for demonstrating the presence of substances in *Chrysocoma tenuifolia* (Berg) (Compositae) which were toxic to the termites. Sands' (1961*b*) laboratory experiments with *Trinervitermes geminatus* showed that many different species of grasses were accepted although smaller, fine-leaved species were preferred to larger, coarse-leaved species and that both *T. geminatus* and *T. trinervius* (Rambur) showed little inclination to eat *Aristida kerstingii* Pilger or *Cymbopogon citratus* (Hochst.), the latter (known as lemon grass) being reputed to be repellent to insects. It is worth noting that the species of grass most readily eaten by *Hodotermes and Trinervitermes* were not the most abundant or frequently-occurring grasses in the particular habitats where these termites were studied. There is no information on the food preferences of the many grass-eating termites which are partially (e.g. certain *Drepanotermes* in Australia) or largely (e.g. *Macrotermes subhyalinus* in Africa) litter-feeders, consuming, in addition to dead grass lying on the soil surface, fragments of leaves, seeds and other plant debris. However, it may be expected that differences between species of plants will be of less significance in determining food selection by litter-feeders than herbivores and even less by termites feeding on decomposing vegetation.

Food selection is, as in all animals, effected by availability of the various sources of food so that in the absence of the preferred food alternative sources may be exploited. This was demonstrated by Lepage's (1974*a*) observations on *Trinervitermes trinervius* in Senegal. This species is normally a grass-harvester but during 1972–73 when, due to drought, little grass was available it fed on deadwood.

Selection of stage of decomposition

The stage of decomposition is an important factor in determing food selection by decomposer invertebrates and although its significance to termites has often been acknowledged there is little precise information available and published work largely refers to wood-eating termites and their relationships with fungi. This subject was reviewed by Sands (1969). It appears that rotting wood sustains better growth and longevity of certain termites, such as *Zootermopsis angusticolis* (Hagen) (Hendee, 1934, 1935). Certain fungi produce substances which positively attract some termites but repel others, while some fungi are toxic. Various beneficial effects have been attributed to the presence of wood-rotting fungi, such as provision of nitrogen-rich tissues, vitamins or growth factors and breakdown of repellent or toxic chemicals, but these effects have rarely been elucidated. However, Williams (1965) demonstrated that resins and turpentine in the heartwood of *Pinus caribaea* Morelet were both repellent and toxic to *Coptotermes niger* Snyder but that when these substances were broken down by a brown-rot fungus *Lentinus pallidus* Bark. and Curt. the rotting heartwood was readily attacked.

The symbiotic relationship between the Macrotermitinae and fungi belonging to the genus *Termitomyces* is a special case of dependence on decomposition of plant material. Josens (1971*a*) estimated that the time between deposition of faeces on the fungus comb and their consumption by termites was approximately five to six weeks in *Odontotermes pauperans* (Silvestri) and two months in *Ancistrotermes cavithorax* (Sjöstedt) and *Microtermes toumodiensis* Grassé but it is not certain whether or not this time interval is related to degree of decomposition by the *Termitomyces* or to the nutritional requirements of the colony.

The effects of microbial decomposition are likely to be of some significance in determining food preferences of termites feeding on decomposing vegetation and also for humus-feeding termites but there is virtually no information available to enable this topic to be discussed satisfactorily. Skaife (1955) showed that *Amitermes hastatus* had a distinct preference for partially decomposed stems of Restionaceae, the preferred state being soft stems still retaining their original form. Leaf litter and humus were rejected but moist, rotten wood and dung were accepted if their preferred food was not available. Obviously, the stage of decomposition of their food is an important factor in determining food selection. Many termites eat dung (Ferrar & Watson, 1970) but the available records do not enable any conclusions to be made concerning preferences for dung in various stages of decomposition. Food selection in humus-feeders has not been studied. Adamson (1943) examined the gut contents of several humus-feeders in Trinidad and concluded that soil appeared to be swallowed without selection of organic particles. However, this seems unlikely and one would expect that, like other humus-feeding soil animals (such as certain earthworms), there would be selection for organic-rich components and that stage of decomposition would be an important factor influencing selection.

Selection of size of food source

The influence of the dimensions of particular sources of food on food selection has received little attention, but for certain species observations indicate that a minimum size is required to initiate foraging. For example, in dry sclerophyll forest in south Australia, *Nasutitermes exitiosus* is the dominant consumer of dead *Eucalyptus* wood in the form of tree stumps, logs and large branches but it does not attack small *Eucalyptus* twigs measuring only several millimetres in diameter – this source of food is exploited by *Heterotermes ferox* (Froggatt) and *Microcerotermes* spp. (Wood & Lee, unpublished). Williams (1973) found that feeding by *Pseudacanthotermes militaris* was more intense on large wood baits than on small ones. On the other hand, large food sources are unlikely to be attacked by termites that specialise in feeding on plant litter, twigs, seeds, etc. lying on the soil surface, as many of these species either forage in the open or under specially constructed sheetings or covered runways.

Size of food source, therefore, should be considered when planning field experiments to study food selection by baiting, nylon bag or similar techniques. For instance, Madge (1969) used leaf discs 2.5 cm in diameter enclosed in nylon mesh bags to study decomposition in forest and savannah in Nigeria. He found no evidence of termites feeding on the leaves, whereas previous work by Hopkins (1966) on the same sites showed that the rate of disappearance of pieces of wood laid on the soil surface was largely dependent on the activities of termites.

Foraging

Extent of foraging

All termites construct a system of galleries or runways, either covered or uncovered, along which they travel in search of requisites such as food, moisture and soil particles. Some termites, notably the wood-feeding Calotermitidae, Termopsinae, Stolotermitinae and Porotermitinae, excavate their nest and galleries solely within their food supply. Also in this category are *Ahamitermes* and *Incolitermes* (Termitinae) which are inquilines in the nests of *Coptotermes* in Australia (Gay & Calaby, 1970), and possibly some humus-feeding termites also fall into this category. These termites are characterised by having relatively small colonies (less than 10000 and often only 2000–3000 – see Lee & Wood, 1971*b*; Table 17) and a compact nest and gallery system – galleries of *Zootermopsis laticeps* excavated by Nutting (1965) extended for approximately 2 m within the branch where the nest was located.

Most termites construct special foraging galleries or runways leading from the central nest system to sources of food and often these are very extensive. Greaves (1962) excavated the gallery systems of two colonies of *Coptotermes acinaciformis* (Froggatt) in southern Australia. The central nests were located within the trunks of living trees and one colony had six main galleries leaving the tree below ground level, the other colony had 10 main subterranean galleries. The galleries branched and extended to and entered logs lying on the ground, and also living trees. Galleries from the first colony were traced to nine trees (two species) while galleries from the second colony were traced to 15 trees (five species). In both colonies the gallery system covered an area of approximately 0.16 hectare and for both colonies the maximum length of any particular gallery was approximately 48 m, although because the galleries did not travel in straight lines the distance from the central nest was only 30 m. Similar studies on *Nasutitermes exitiosus* and *Coptotermes lacteus* (Froggatt), both mound-building, wood-eating species constructing subterranean galleries, in south-eastern Australia (Ratcliffe & Greaves, 1940) showed that there were between 9 and 36 main galleries leaving each mound and that their maximum extent (in a straight line) was approximately 30 m.

Many grass-harvesting and litter-feeding species construct subterranean galleries which terminate in foraging holes from which the termites emerge to forage in the open. In certain *Drepanotermes* spp. in Australia, subterranean galleries extend up to 20–25 m from the mound or subterranean nest and the termites forage for up to 0.5 m from each foraging hole (J. A. L. Watson, personal communication). *Nasutitermes triodiae* (Froggatt) constructs subterranean galleries for distances up to 30 m from its mound before emerging to feed on grass (Gay & Calaby, 1970). At Mokwa, Nigeria, I have observed *Trinervitermes geminatus* foraging in the open for up to 2 m from their foraging holes, the latter which may be 3–4 m from the nearest mound – the maximum foraging distance from the breeding colony has not yet been ascertained as this species is polycalic (Sands, 1961*a*), a single colony occupying several interconnected mounds some of which are used solely for storing grass. Sands (1961*b*) observed foraging parties of *T. geminatus* spreading out over areas up to 3 m^2 in area and climbing tall grasses up to heights of 1 m, whereas the sympatric species, *Trinervitermes oeconomus* (Tragardh), mainly foraged near its mounds, often emerging directly from the mound. Mounds of various species of *Macrotermes* in Africa have been estimated at densities of 2–4 per hectare (see Lee & Wood, 1971*b*: Table 5) and thus the average area occupied by a colony would be 0.25–0.5 hectare and galleries would extend 30–45 m from the central part of the nest system. In one species (*Macrotermes subhyalinus*) studied at Mokwa, Nigeria (Wood & Ohiagu, 1976), a colony appeared to forage over an area of 0.31 hectare. This species forages on the surface (consuming plant litter of all types) for short distances from its foraging holes and consequently, in areas of the territory where foraging is active there are many foraging holes (up to 30 per m^2 but more often 10–15 per m^2). Other species, such as *Trinervitermes geminatus* and *Hodotermes mossambicus* which forage on the surface for some distance from their foraging holes, have fewer foraging holes – Nel (1968*b*) estimated 0.08–0.56 per m^2 for *H. mossambicus* and Ohiagu & Wood (1976) estimated 0.015–0.133 per m^2 for *T. geminatus*.

Certain arboreal-nesting species, particularly those foraging in the open, travel great distances from their nests but rarely follow the shortest distance from the nest to source of food. Emerson (1938) described how a covered runway built by *Nasutitermes guayanus* (Holmgren) led from a nest, which was 9.1 m up a palm tree, down some vine stems to the ground and to a small, dead tree; in all a distance of approximately 49 m although the dead tree was only 10.7 m in a straight line from the palm tree. Kalshoven (1958) has observed the open-foraging columns of *Hospitalitermes* at distances exceeding 60 m from the nest.

Dynamics of foraging

Diurnal periodicity

Many observers have noted the occurrence of diurnal patterns of foraging in termites and, not surprisingly, most observations relate to species foraging in the open. Bouillon's (1970) generalisation that foraging tends to occur at times when environmental conditions in the diurnal cycle depart the least from conditions in the nest requires more data than are at present available for it to be generally acceptable and there are, as Bouillon admits, some notable exceptions. Foraging in the open, either on the soil surface or on living plants, appears to be largely nocturnal although it may occur during daylight hours if the sky is cloudy; observations on *Hospitalitermes* sp. (Kalshoven, 1958), *Trinervitermes* spp. (Sands, 1961*b*), (Bodot, 1967*b*), *Tenuirostritermes tenuirostris* (Desneux) (Nutting, 1970) and *Drepanotermes* spp. (Gay & Calaby, 1970; J. A. L. Watson, personal communication) confirm this generalisation. Observations at Mokwa, Nigeria (Ohiagu & Wood, 1976) on *Trinervitermes geminatus* show that between December and June, foraging parties emerge twice during a 24-hour period, from 06.00–09.30 hours in the morning and from 16.30–18.30 hours in the early evening i.e. foraging occurs largely during daylight, and the last half-hour of the morning period and first half-hour of the evening period it is carried out in bright sunlight. Low temperatures early in the morning (e.g. during the harmattan) restrict foraging to the evening period. In contrast, during the same period, *Macrotermes subhyalinus* never started foraging before 21.00 hours. The most detailed studies of foraging have been made on *Hodotermes mossambicus* which differs from the foraging crepuscular or noctural species, in habitually foraging during the day (Coaton, 1958; Nel, 1968*b*). It forages mainly in winter and emerges from its foraging holes in the morning or early afternoon as air temperatures increase. Nel's (1968*b*) observations showed that activity commenced between 08.45 and 15.30 hours and ceased between 10.30 and 16.30 hours although on any particular day foraging from different foraging holes was not synchronised, for example, on 1 March activity started between 09.30 and 15.30 and ceased between 10.30 and 15.30 hours. Foraging usually started at temperatures of 22–27 °C but had been observed to start at temperatures of 12 °C. Foraging holes were in use for an average of 1.9 hours per day, with a minimum of 7.5 hours and at any one time an average of 9–14 workers were out of the foraging hole and each worker spent 1.5–3.2 min out of the hole.

Very few detailed observations have been made on diurnal foraging patterns in species with more cryptic habits. Observations in southern temperate regions of Australia on the mound-building, wood-eating *Nasutitermes exitiosus* (Holdaway & Gay, 1948) showed that temperatures in the central nursery chamber were always higher than soil or air temperatures and showed less variation so that the greatest difference between mound and air or soil surface temperatures occurred early in the morning. At this time of day the temperature difference may be as great as 16 °C (Lee & Wood, unpublished) and correlated with this is the greatest concentration of workers in the mound: as air temperatures increase during the day workers leave the mound via subterranean galleries to feed in logs, tree stumps and similar large pieces of wood. In contrast, in the tropics, nocturnal temperatures are relatively high and even cryptic species appear to forage during the night. Observations by Bouillon & Lekie (1964) and Mathot (1964) on the humivorous, mound-building *Cubitermes sankurensis* Wasmann showed that the number of termites in the nest was greatest between 13.00 and 15.00 and least between 23.00 and 02.00 hours; the degree of repletion of the workers gizzards was greatest during the night and least during the day.

There is little information on the speed of and energy cost of foraging (see Chapter 6) but species with long galleries (up to 48 m) must spend a considerable amount of energy searching for, obtaining and transporting food. Skaife (1955) observed that workers of *Amitermes hastatus* took two hours to travel 23 m and back (i.e. 46 m) to their nests at temperatures of 25 °C – a speed of 38.3 cm per min. Much faster rates were noted for the surface-active *Hodotermes mossambicus* (Nel, 1968*b*) which moved at 396 cm per min at 35 °C and at almost twice this rate at 38 °C.

Seasonal periodicity

Seasonal variations in foraging activity have been studied in some detail for certain species and, together with observations scattered throughout the literature, enable certain generalisations to be made. Bouillon (1970) suggested that the seasonal foraging cycle would be more evident in areas where the climate is most variable (e.g. in temperate as opposed to tropical climates and tropical savannah as opposed to tropical rainforest) and where the food supply shows most seasonal variation (e.g. in grasslands as opposed to forests). In some areas of temperate, southern Australia, temperatures during the day are often low (approaching 0 °C) and the foraging of *Nasutitermes exitiosus* which has subterranean galleries is curtailed. In South Africa, *Hodotermes mossambicus* forages primarily in winter, not during the nights which are too cold, but during the day on the surface and often in bright sunlight (Nel, 1968*b*); in regions where the soil surface becomes too hot in summer it may forage during early morning or late evening. In this species both daily and seasonal foraging patterns appear to be determined largely by temperature. In contrast, the daily foraging pattern of *Trinervitermes geminatus* appears to be largely determined by temperature, but the seasonal pattern (Sands, 1961*b*) is determined partly by availability of food and partly by climate. Foraging (in northern Nigeria) commences in September when there is abundant grass of high nutritive content and is curtailed by the onset of the rainy season in May. Stored grass enables this species to survive throughout the wet season with little or no foraging activity whereas nearby in the Ivory Coast *Trinervitermes trinervius* (Bodot, 1967*b*), which in that area does not store grass in its mounds, forages all the year round.

N. M. Collins (personal communication) observed seasonal variation in foraging among several species feeding on woody litter in savannah woodland in Nigeria. *Microcerotermes* and *Odontotermes* foraged mainly in the dry season, *Microtermes* and *Ancistrotermes* mainly in the wet season, while *Macrotermes bellicocus* (Smeathman) and *M. subhyalinus* foraged almost all the year with least activity at the end of the dry season. In the drier, Sahel savannah, Lepage (1974*a*) observed seasonal variation in foraging of *M. subhyalinus* and also differences in mean annual activity which were directly related to rainfall.

Some species may be adapted to foraging over a wide range of temperatures. Nutting *et al.* (1973) observed *Heterotermes aureus* and *Gnathamitermes perplexus* (Banks) foraging in the Sonoran desert of Arizona at temperatures ranging from 7–49 °C; an obvious adaptation to the desert environment which enabled foraging to be carried out all the year round.

Quantitative aspects of feeding

Amount of food eaten or collected

There is very little information on the amount of food eaten by termites. Laboratory measurements of food intake, particularly in the case of termites which are often maintained as non-breeding colonies lacking primary reproductives, are always open to the uncertainty of relating laboratory conditions to those prevailing in the field. On the other hand, measurement of food intake in the field presents many difficulties and with termites there is the added difficulty of estimating abundance of the termites themselves (Lee & Wood, 1971*b*: Chapter 4 and Appendix).

Some laboratory measurements of consumption rates are shown in Table 4.1. The wide range for wood-feeders (2.0–90.8 mg per g per day) reflects differences between species of termite, differences between palatability of wood species and probably differences in culture methods. All these determinations were made on non-breeding groups of individuals, whereas the single value for a grass-eating termite (*Hodotermes*) was made on breeding colonies. The majority applies to lower termites (non-Termitidae) which with few exceptions (e.g. *Mastotermes* in northern Australia) have little impact in tropical ecosystems, although some may be important in arid and semi-arid regions (e.g. *Heterotermes aureus* in the Sonoran desert and *Hodotermes mossambicus* in South Africa).

Measurement of consumption rates in the field is very difficult. Estimates of natural rates of consumption (Table 4.2) have in some cases relied on extrapolation from laboratory data while others have relied on direct measurement in the field.

Lee & Wood's (1971*a*) data for *Nasutitermes exitiosus* in dry sclerophyll *Eucalyptus* forest in Australia was based on laboratory-determined consumption rates combined with estimates of population density and biomass; no allowance was made for seasonal variation in foraging. Haverty & Nutting (1974) estimated consumption by *Heterotermes aureus* in the Sonoran desert by using laboratory-determined consumption rates combined with field observations on foraging intensity at baits; they allowed for seasonal variation in foraging but assumed that foraging intensity at baits was similar to that at natural foods and that all termites observed in the baits were consuming food and not occupied in other activities. Hébrant's (1970*b*) figures for consumption by the soil-feeding *Cubitermes exiguus* Mathot were based on measurements of metabolic rate, calculations of the quantity of cellulose and hemicelluloses required to maintain this metabolic rate, assumed assimilation efficiency of 15% and certain assumptions on the quantity of these substances in soil.

Direct estimates of consumption have relied on either measurement of removal of naturally-available food or removal of food in the form of baits. The two grass-harvesting species *Hodotermes mossambicus* and *Trinervitermes geminatus* forage on the surface and carry grass back to their nests. Observations on *H. mossambicus* were made at intervals during a full year and measurement of mean weight of food collected, numbers of pieces of grass per foraging hole and the daily and seasonal duration of foraging resulted in food intake being estimated at 1.5 g per m^2 per year. This rate of feeding gave a good correlation with an estimated damage (percentage removal of canopy) of 1% of the grass harvest of 1664 kg per hectare in 1967. However, the much greater incidence of damage in 1966 (33% of the grass harvest of 1107 kg per hectare, Nel & Hewitt, 1969) indicated that food intake may reach 36 g per m^2 per year. Abundance

Table 4.1. *Food consumption by termites under laboratory conditions* (*mg food* (*dry weight*) *per g termites* (*live weight*) *per day*)

Species	Food	Termite live weight (mg)	Consumption (mg/g/day)	Temperature (°C)	Author
Mastotermitidae					
Mastotermes darwiniensis	Wood (*Eucalyptus*)	41.7	11.5	26	Gay *et al.*, 1955
Calotermitidae					
Cryptotermes cavifrons	Wood (*Pinus*, *Tilia*)	4.1	16.0	25	Hrdý & Zelaný, 1967
Cryptotermes dudleyi	Wood (mean of many species)	5.5	28.6	26	Becker, 1967
Calotermes flavicollis	Wood (mean of many species)	12.9	26.8	26	Becker, 1967
C. flavicollis	Wood (*Pinus*, *Fagus*)	7.0	23.5	26	Seifert, 1962
Neotermes castaneus	Wood (*Pinus*, *Tilia*)	17.5	31.0	25	Hrdý & Zelaný, 1967
Neotermes jouleti	Wood (*Pinus*, *Tilia*)	11.4	22.0	25	Hrdý & Zelaný, 1967
Paraneotermes simplicicornis	Wood (mean of three species)	7.4	45.7	26	Haverty & Nutting, 1974
Hodotermitidae					
Zootermopsis angusticollis	Wood (mean of many species)	70.7	10.1	26	Becker, 1967
Hodotermes mossambicus	Grass (*Themeda*)	7.5	49.0	30	Nel *et al.*, 1970
Rhinotermitidae					
Coptotermes acinaciformis	Wood (*Eucalyptus*)	3.6	17.8	26	Gay *et al.*, 1955
Coptotermes amanii	Wood (mean of many species)	2.8	26.8	26	Becker, 1967
Coptotermes lacteus	Wood (*Eucalyptus*)	3.7	12.2	26	Gay *et al.*, 1955
Heterotermes aureus	Wood (mean of four species)	1.9	19.0	26	Haverty & Nutting, 1974
Heterotermes indicola	Wood (range of many species)	1.9	13.0–90.8	26	Becker, 1962
Reticulitermes lucifugus	Wood (mean of many species)	2.8	19.3	26	Becker, 1967
Prorhinotermes simplex	Wood (*Pinus*, *Tilia*)	3.3	8.0	25	Hrdý & Zelaný, 1967
Termitidae					
Nasutitermes costalis	Wood (*Pinus*, *Tilia*)	2.1	2.0	25	Hrdý & Zelaný, 1967
Nasutitermes exitiosus	Wood (*Eucalyptus*)	4.8	12.2	26	Gay *et al.*, 1955
N. exitiosus	Wood (*Eucalyptus*)	5.0	10.6	26	Lee & Wood, 1971*a*
Nasutitermes ripperti (Rambur)	Wood (*Pinus*, *Tilia*)	5.4	9.0	25	Hrdý & Zelaný, 1967

The data from the papers by Becker (1967) and Hrdý & Zelaný (1967) were kindly supplied by Dr M. I. Haverty.

Table 4.2. *Food consumption by termites in the field*

Species	Natural food	Estimated consumption (g per m² per year)	Termite biomass (g per m²)	Daily consumption (mg per g per termite)	Author
Hodotermitidae					
Hodotermes mossambicus	Grass	1.5	—	—	Nel, 1970
Rhinotermitidae					
Heterotermes aureus	Wood	7.9	—	—	Haverty & Nutting, 1975
Psammotermes hybostoma	Grass/wood	0.7	0.14	13.7	Lepage, 1974*b*
Termitidae					
Macrotermitinae (four species)	Wood/grass	130.0	0.63	565.0	Josens, 1972
Ancistrotermes cavithorax	Wood/grass	16.1	1.59	27.7	Collins & Wood, unpublished
Macrotermes bellicosus	Wood/leaves	24.1	0.47	140.5	Collins & Wood, unpublished
Macrotermes subhyalinus	Grass	7.2	0.60	32.9	Lepage, 1974*b*
Microtermes spp.	Wood	30.9	1.02	84.7	Collins & Wood, unpublished
Odontotermes spp.	Wood/grass	12.4	0.78	43.6	Collins & Wood, unpublished
Odontotermes smeathmani	Grass/wood	2.1	0.12	48.0	Lepage, 1974*b*
Pseudacanthotermes militaris	Wood	540–1508	—	—	Williams, 1973
Nasutitermes exitiosus	Wood	11.6	3.00	10.6	Lee & Wood, 1971*a*
Trinervitermes geminatus	Grass	8.1	0.80	27.7	Ohiagu & Wood, unpublished
Cubitermes exiguus	Soil	11300.0	1.20	4110.0	Hébrant, 1970*a*

of termites in the field was not estimated so that no comparison can be made between food intake in the field and in the laboratory (see Table 4.1). The observations on *T. geminatus* were made in the latter half of the dry season. Daily rates of grass-harvesting by a termite population of approximately 200 per m^2 (1.0 g per m^2) varied from 11 mg per m^2 to 83 mg per m^2 with a mean of 42 mg per m^2. As there is distinct seasonal variation in foraging intensity by *T. geminatus*, annual harvesting rates cannot be calculated from this figure but it was suggested (Wood & Ohiagu, 1976) that it would not be more than 15 g per m^2 per year. These figures did not include data on grass consumed whilst foraging.

The measurements on *Ancistrotermes*, *Macrotermes*, *Microtermes* and *Odontotermes* indicate that Macrotermitinae have higher weight-specific consumption rates than other species, undoubtedly due to the fact that their consumption is required to maintain their symbiotic fungi in addition to themselves. The figure of 84.7 mg per g per day for *Microtermes* is probably a reasonable estimate as these species feed almost exclusively on woody litter of the type with which they were baited. However, as *Ancistrotermes* and *Odontotermes* also consume grass litter and bark of trees and *Macrotermes bellicosus* also consumes leaf litter and bark of trees the figures for these species are underestimates. Likewise, the figures for *M. subhyalinus* and *Odontotermes smeathmani*, which are based solely on removal of surface litter, are likely to be under-estimated.

Estimates of consumption by *Pseudacanthotermes militaris* (Hagen) and Macrotermitinae (four species) were based on baiting with palatable food over a short period (two to three months) after first clearing the study area of alternative food. The disadvantages of these methods has already been discussed. Not surprisingly, these studies gave high figures for consumption (540–1508 g per m^2 per year for *P. militaris*) and weight-specific consumption (594 mg per g per day for Macrotermitinae). Although indicating potentially high, short-term rates of consumption, in natural situations, limiting factors, such as climatic restrictions on foraging and the fact that natural sources of food are less palatable and less concentrated (necessitating a longer time spent in searching for food) than baits, mean that these rates are never achieved.

Utilisation of food

Food storage

It has already been noted that many termites, particularly grass- and debris-feeding termites, store food in their nests. There are no published estimates of the quantity of food stored but unpublished observations by C. E. Ohiagu on *Trinervitermes geminatus* and N. M. Collins on *Macrotermes bellicosus* at Mokwa, Nigeria, give some indication of the quantities involved. *T. geminatus* stores grass in its mounds but no more than 60–70% of the mounds contain stored grass. The quantities varied from a few fragments amounting to 2–3 g to maxima of 320 g per mound with a seasonal variation of 15–143 g per mound (0.3–2.9 g per m^2). *Macrotermes bellicosus* stores finely macerated wood and other food materials in loose aggregations above the fungus combs. Examination of 11 nests showed that the weight of food store varied from 0.068 to 4.226 kg. It is not known if stored food is used entirely by the termites or whether other organisms, particularly micro-organisms, are able to make use of this local concentration of energy and nutrients. Stored grass may have some function in insulating the nest (Coaton, 1958; Lee & Wood, 1971*b*) and its turnover rate may be regulated in relation to its usefulness for this purpose as well as for providing a reservoir of food for the colony.

Digestion and assimilation of food

Digestive processes of termites and their ecological significance were discussed in some detail by Lee & Wood (1971*b*). Most of their energy is derived from the breakdown of polysaccharides (particularly cellulose and hemicullulోses) and there is evidence that some species can digest lignin. Details of digestive processes are beyond the scope of this review and interested readers are referred to the review by Lee & Wood cited above. The intense degradation of plant materials in the termite gut appears to be accomplished by the activities of symbiotic micro-organisms, largely flagellate protozoa in the lower termites and bacteria in the Termitidae (higher termites). Many details of these processes have not been worked out but the net effect, the extremely efficient digestion of structural polysaccharides, sets termites apart as a particularly significant group of soil animals.

The majority of quantitative studies of digestion has been carried out with lower termites that feed on fresh deadwood (Becker & Seifert, 1962; Seifert, 1962; Seifert & Becker, 1965; Lee & Wood, 1971*a*), there is one study on a grass-feeding species (*Hodotermes mossambicus*) and none at all on detritivorous, humus-feeding or fungus-growing termites. Assimilation efficiency and digestion of components of plant tissue by various wood-feeding and one grass-feeding termite is shown in Table 4.3. Assimilation is high (range 54–93%) compared with other invertebrates feeding on similar substrates (e.g. Diplopoda, 30–40% (Striganova, 1971); Isopoda, 33% (White, 1968); Orthoptera, 27% (Smalley, 1960); Acari-Oribatei, 14% (Berthet, 1964)). In the case of the wood-feeding species coprophagy could contribute to some of the very high values for assimilation efficiency to a greater extent than might be expected under natural conditions. Without comparable data on several grass-feeding, litter-feeding, humus-feeding and fungus-growing termites it is impossible to generalise on the

Table 4.3. *Assimilation efficiency and digestion of food by laboratory colonies of termites*

Species	Food	Assimilation efficiency (% dry weight)	Digestion (% of original dry weight degraded)			Author
			Cellulose	Hemicelluloses	Lignin	
Calotermes flavicollis	*Fagus* wood	59	90	70	3	Seifert, 1962
C. flavicollis	*Pinus* wood	59	95	65	4	Seifert, 1962
C. flavicollis	Several woods	54–64	74–91	—	2–26	Seifert & Becker, 1965
Heterotermes indicola	Several woods	62–69	78–89	—	14–40	Seifert & Becker, 1965
Reticulitermes lucifugus	Several woods	86–93	96–99	—	70–83	Seifert & Becker, 1965
Nasutitermes ephratae	Several woods	75–85	91–97	—	42–52	Seifert & Becker, 1965
Nasutitermes exitiosus	*Eucalyptus* wood	54	—	—	—	Lee & Wood, 1971*a*
Hodotermes mossambicus	Red grass (*Themeda*)	61	—	87	0.3	Nel *et al.*, 1970

fate of ingested food. There are indications from the examination of faeces that the fungus-growing Macrotermitinae do not digest their food as efficiently as other termites; published work on lignin/cellulose ratios of fungus combs is not very useful as no account has been taken of the comb age (Sands, 1969; Lee & Wood, 1971*b*). However, it is likely that the fungus comb can be considered as an extension of the termites digestive system, but in the absence of adequate analytical data these ideas cannot be substantiated.

Production and use made of faeces

Because of their high assimilation efficiency termites appear to excrete not more than 40–50% and possibly as little as 30% of their ingested food. The properties of this faecal material have been discussed by Lee & Wood (1971*a*, *b*) and certainly for those species for which information is available the organic fraction appears to have the following chemical properties: low nitrogen (less than 1%), high carbon/nitrogen ratio (20–25 in grass-feeding species, more than 50 and approaching 100 in wood-feeding species), high lignin content (40–65%) and low carbohydrate content (8–29%); the lignin/cellulose ratio approaches 5:1. In one species, *Nasutitermes exitiosus* (Lee & Wood, 1968, 1971*b*) the faecal material contains significant quantities of humic acids and is toxic to micro-organisms.

All termites make use of their faeces for building purposes – lining their galleries and walls of nest chambers or for constructing special portions of their nest system. The ecological implications of this behaviour are discussed later on (see also Lee & Wood, 1971*b*); here, it is sufficient to note that unlike other soil invertebrates their faecal material is not immediately available for use by other soil organisms. This behaviour appears to have been extended by the Macrotermitinae (fungus-growers) to the exclusive or near-exclusive use of their faeces for building fungus combs. The controversy as to whether fungus combs are constructed from macerated plant material or from faecal material has been reviewed by Sands (1969) and this author's conclusion that the combs are made from faeces is supported by Josens (1971*a*) studies and by direct observation of incipient colonies of *Microtermes* by R. A. Johnson at Mokwa, Nigeria. Further studies are required to confirm whether or not this behaviour applies to all Macrotermitinae. There is little doubt, however, that the cultivation of fungi on the fungus combs, followed by re-ingestion of old portions of the fungus comb containing material degraded by fungi leads to complete or near-complete utilisation of food by the termite–fungus system. In view of the efficiency of this system it is not surprising that many Macrotermitinae consume a wide range of plant materials. The dynamics of this fascinating system have yet to be studied in detail.

5. Food and feeding habits of ants

D. J. STRADLING

Ant food and feeding: objectives and principles

The main purpose of feeding studies in ecology is to estimate the quantitative feeding characteristics of whole populations. More meaning is given to estimates of food consumption if they can be related to habitat area, particularly where ants maintain feeding territories (Elton, 1932; Brian *et al.*, 1965; Brian *et al.*, 1967).

Measurements of food consumption by social insect colonies can lead to an independent check on estimates of gross production where net production (P) and respiration (R) have been measured as $A = P + R$ where A = assimilation. However, it is also necessary to measure the rejecta in order to measure assimilation as $A = C - (F + U)$ where C = consumption, F = faeces and U = urine, and in the case of ant colonies consumption must be carefully defined. Quantitative measures of food consumption, ingestion or assimilation of whole ant colonies enables the ratios of these with production to be obtained, making possible the calculation of ecological and production efficiencies which are important indices of the effect of the population on subsequent food chains in the community. This information is thus valuable in assessing the relationships of ant populations to other trophic levels. Efficiencies of consumption and assimilation by field populations are determined by the nature and selection of food both of which may vary temporally and spatially. The effects of food sharing by social insect groups are also important and will be dealt with later.

The single colony which because of its integrity is the most conveniently and frequently studied unit shows certain trophic analogies to a single organism. The worker population which is concerned with food collection, selection and processing is primarily a consumer of energy producing carbohydrates, although it may in some cases also be involved in the assimilation of protein and lipid for storage (Dlussky & Kupianskaya, 1972; Peakin, 1972). The larvae which represent the active stage of tissue production are the chief consumers of protein. Sexual production demands high protein and lipid intake in the larval stage and also after metamorphosis in the case of gynes (Peakin, 1972). Food intake is therefore related to brood development and reproduction, being often markedly seasonal for both extrinsic and intrinsic reasons. Schneirla (1957*a*) has shown that the nomadic phase in *Eciton* is one of intensive food collection and also of maximum larval growth.

Consumption of food: qualitative aspects of food selection

Diets

Highly selective consumers tend to achieve a high assimilation/consumption ratio because their food contains more nourishment and because they are better adapted to digest it (Petrusewicz & Macfadyen, 1970). Ants show extremes of dietary specialisation and generalisation but are selective of foods with high protein and carbohydrate contents even when the diet is catholic. Thus, predation is selective for high protein content, honeydew and plant exudate collection for carbohydrates and seed collection, while not as remunerative as predation, is selective for the most nutritious of plant tissues. The ability of the attines inhabiting tropical rainforest to use a fungus to convert material rich in cellulose from a wide range of plant species into a single, highly nutritious food source is of obvious ecological advantage in such floristically complex habitats. Selection continues after food collection, invertebrate cuticles and parts of seeds being rejected before ingestion. Most ant species are omnivores which combine predation, scavenging and the collection of plant materials. It is therefore not possible to allocate most species to either primary or secondary consumer trophic levels. It is, however, possible to recognise items of primary and secondary production in the diet.

Primary consumption

The fungus-cultivating attini are unique among the ants in that some feed probably exclusively on their symbiotic fungi, a relationship parallelled in the Macrotermitinae. In most of the attine genera the fungus is cultivated on a substrate of vegetal origin. Such substrates fall into five main categories; insect faeces, insect cuticle, rotting wood, plant debris and freshly cut plant fragments. Martin & Martin (1971), Weber (1972*a*) and Martin *et al.* (1973) have reported that the primitive genera *Cyphomyrmex*, *Mycocepurus* and *Myrmicocrypta* incorporate some insect carcasses in the fungal substrate. However, the bulk of the substrate is vegetal in origin. All the insect faecal material recorded as used in fungal substrate comes from phytophagous species and is thus principally vegetal material. The higher attines collect fresh plant material and thus exert the same ecological effect as primary consumers.

In most attine genera the fungus culture takes the form of a dense mycelium ramifying over the surface of the substrate which has been carried to the nest, malaxated, and built into a loose cellular structure by the ants. The fungal hyphae produce swellings termed gongylidia, either in the centre or terminally and these are used as food by the ants. The gongylidia frequently occur in clusters called staphylae. Interspecific

Table 5.1. *Nutrients derived from the fungus cultured by* Atta colombica tonsipes. (*After Martin, Carman & MacConnell, 1969*)

Nutrient	mg per g dry fungus
Trehalose	160
Glucose	47
Mannitol	51
Arabinitol	14
Protein-bound amino acid	130
Free amino acid	47
Lipid (major sterol, ergosterol)	2
Total weight identified	451

variation in both the structure of the fungal culture and the form of gongylidia has been described by Weber (1972*a*). In the primitive *Cyphomyrmex rimosus trinitatis* the fungus takes the form of a cheese-like bromatium of yeast-like cells which were found by Weber to form hyphae in culture. The sporophores of ant-fungi are of extremely rare occurrence rendering identification difficult. Möeller (1893) described the production of sporophores from the fungal culture of *Acromyrmex* (*Atta*) *discigera* and named the fungus *Rhozites* (= *Pholiota*) *gongylophora*, a basidiomycete. Weber (1957*a*) reported the production of sporophores by cultures of the fungus obtained from colonies of *Cyphomyrmex costatus*. This fungus was initially named *Lepiota* n. sp. but Heim (1957) considered that it was the same species as that described by Möeller (1893) and proposed the name *Leucoprinus gongylophorus* (Möeller). Weber (1966*a*) reported the production of a sporophore of matching description from cultures of the fungus of *Myrmicocrypta buenzlii*. The implication seems to be that the same fungus is cultivated by a wide range of attines. This is supported by the experiments of Weber (1972*a*) in which a wide acceptance of a range of free-living *Lepiota* spp. by a number of attine species was found. That many attine species will accept the fungus cultured by others as food also suggests a close similarity.

The nutrient value of the fungus of *Atta colombica tonsipes* Santschi has been investigated by Martin, Carman & MacConnell (1969). Fungal hyphae were removed from the ants and cultured on Sabouraud's dextrose medium. Samples of the fungal mycelium were subsequently separated from the substrate, freeze dried and analysed. The results of these analyses are summarised in Table 5.1.

More than 50% of the fungus dry weight was found to be soluble nutrient. Approximately 27% was carbohydrate, 13% protein-bound amino acid, 4.7% free amino acid, 0.2% lipid. A total of 19 amino acids were identified. Among the carbohydrates no polysaccharides were found and the distribution of sugars was found to vary with the age of the culture and the carbon source of the substrate. With the exception of the low lipid content the fungus appears to provide a rich and complete diet. Although analyses were of necessity carried out on artificially cultured material it is probably reasonable to assume that the nutrient content did not depart greatly from that of the fungus cultivated by the ants themselves. At the present time this is by far the most detailed analysis of an ant diet. Many authors have considered that the fungus is the sole source of food for the attines but it now seems probable that the workers obtain water and some nutrients directly from plant sap during the collection and malaxation of substrate.

The use of symbiotic micro-organisms by herbivores to assist in the digestion of cellulose is widespread and it is therefore not unreasonable to consider the attines as primary consumers. No other ant species has been conclusively shown to be totally herbivorous although plant materials predominate in the diets of some grain-harvesters.

Granivorous ants of the genera *Messor*, *Veromessor* and *Pogonomyrmex* make extensive use of plant seeds in their diet. The seed-harvesting habit seems to be most highly developed in desert habitats although it is practised to a lesser degree by related and unrelated genera in more hospitable temperate environments. Whole *Calluna* seeds are reported to constitute a large proportion of the larval food of *Tetramorium caespitum* L. by Brian *et al.* (1967). The seeds of certain plant species are usually favoured and preferences have been reported by Tevis (1958), Brian *et al.* (1965) and Clark & Comanor (1973). Selection continues even after harvesting in that only parts of some seeds are consumed. *T. caespitum* collects graminaceous seeds but eats only the embryos, *Lasius niger* (L.) and *Lasius alienus* (Foerst.) collect the seeds of *Ulex minor* Roth. but eat only the caruncles (Brian *et al.*, 1965). Bequaert (1912) recorded *Anomma* chewing the oily pericarp of palm nuts.

Other ants whose diet is composed largely of primary production are found in the genera *Azteca* and *Pseudomyrma*. These form special associations with certain neotropical trees and obtain an unknown proportion of their diet from 'Mullerian' and 'Beltian' food bodies produced by the plant. Rickson (1971) has found that the 'Mullerian' bodies of *Cecropia* contain glycogen having the same composition as animal glycogen. An extensive account of these ant–plant associations has been given by Wheeler (1942) and the nature of the association has been discussed more recently by Janzen (1973*b*).

Plant exudates and sap are consumed extensively by the majority of ant species. These are obtained from both floral and extrafloral nectaries, glands and wounds. Soft plant tissues are sometimes chewed to extract the sap. These materials provide a ready source of sugars, water and probably some amino acids and proteins.

The most widely exploited source of plant sap is undoubtedly that provided by the honeydew excreted by plant-feeding Homoptera, principally Aphidae and Coccidae. The Homoptera probe the phloem of the plant and thus tap a rich supply of plant nutrients. Excess plant sap passes through the alimentary canal of the insect and is voided from the anus in the form of droplets. In the absence of ants these are flicked away and fall either to the ground or onto lower leaves. Wheeler (1910) records that some *Leptothorax* spp. will lick the fallen honeydew from leaves. More usually, however, the ants solicit the honeydew directly from the Homoptera.

Mutualism between ants and Homoptera has been discussed by Nixon (1951) and comprehensively reviewed by Way (1963). Honeydew has been shown to contain high concentrations of a wide range of sugars (20%, to quote Zoebelein, 1956*a*). These may differ in composition from those present in plant sap, having been partly converted by the homopterans to melezitose which may be harmful when consumed by other insects (Zoebelein, 1956*a*, *b*). The extensive consumption of honeydew by ants, however, indicates that harmful effects must be negligible. Many free amino acids, amides, minerals and B vitamins have also been found in honeydew. Thus 22 free amino acids comprising 13.2% of the dry weight, and including all those known to be needed by animals were identified in the honeydew of one aphid species by Maltais & Auclair (1952). Strong (1965) records the presence of small concentrations of lipids and sterols. Analyses of honeydew have been extensively reviewed by Auclair (1963). The implications are that honeydew may be a complete food for ants but it has not been shown that any species of ant relies on it as a sole source of food. Collection of this liquid food is associated with elaborate modifications of the proventriculus to facilitate the filling of the crop without midgut absorption or evacuation (Eisner & Brown, 1958). In this way the capacity for transport, trophallaxis and storage is enhanced. The number of ant, aphid and coccid species indulging in these associations is large and although some ants show preference for certain species of Homoptera many will tend a wide range, for example *Lasius niger* and *Lasius flavus* (F.) (Pontin, 1963). Conversely, some aphid species such as *Forda formicaria* v. Heyd., are found only in association with ants (Sudd, 1967) while others are scarce in the absence of their attendant ants (Way, 1963).

Secondary consumption

Most ant species are predaceous to some degree. Purely carnivorous species are found among the Ponerinae and Dorylinae and represent the primitive ancestral forms. Animal flesh is a source of high quality protein and is thus a desirable food although predation often involves the collection of substantial quantities of undigestible cuticle.

Selection of prey may be due to specialisation, as in *Myopias* which feed largely on Diplopoda and *Leptogenys* which are isopod feeders (Wilson, 1959). Many ponerines effect raids on termitaria and probably constitute an important link in the food web of tropical ecosystems. This merits more investigation. The doryline genus *Aenictus* and the ponerine Cerapachyini are specialised ant-feeders (Wilson, 1958*c*; Sudd, 1960). *Lasius* spp. prey on other ants a great deal and even on their own species (Brian *et al.*, 1965). Selection may be less specific and due to the temporal and spatial availability of prey as well as the ant's ability to capture it. Defensive mechanisms protect many insects from ant predation, such as the hirsute nature of some lepidopterous larvae, the heavy armour of Coleoptera and the repugnatorial substances produced by Coccinelidae and some Hemiptera. Investigations of predation in *Formica rufa* L. have been made by Wellenstein (1954). Many suitable prey animals are ignored due to their immobility and this is particularly true of egg and pupal stages. Whereas inertness tends to prevent predation the movements of weakened and dying animals attract the attention of foragers. Aphids are prized by ants as sources of honeydew but Pontin (1958) has shown that they are also taken as a source of solid protein.

A great many records of the prey of ants exist and these indicate that most insect orders are attacked and that Arachnida, Myriapoda, Crustacea and Oligochaeta are regularly taken. Both vertebrate and invertebrate carrion are also utilised. Among the attines the primitive genera mentioned in the previous section make use of some insect material as substrate for their fungal culture. Martin *et al.* (1973) have shown that these genera, as well as some species of *Sericomyrmex* and *Atta* produce faecal material containing a chitinase which it is suggested is involved in the breakdown of insect cuticle when this is included in the fungal substrate. The primitive attines therefore appear to use some secondary production as a source of nutrients for their fungus.

Composition

The composition of the diet of an ant colony may be a reflection of its own specialisation or the relative availability of different food materials in its environment.

Although the attines grow their fungal food within the nest, the collection of substrate is ecologically analogous to the collection of food by other ants and the range of materials utilised is appropriately considered here. The five principal categories of substrate material have already been listed. The more primitive attines, such as *Cyphomyrmex rimosus trinitatis*, use the frass of caterpillars as the principal substrate material. *C. costatus* and *Myrmicocrypta ednaella* combine insect frass with plant debris. *Apterostigma* spp. usually nest in or beneath rotted wood which is combined with the frass of wood-boring beetles to form the fungal substrate. *Trachymyrmex* spp. combine the use of plant debris and insect faeces with freshly cut plant material and are considered by Weber (1958) to be transitional to the higher attine genera, *Sericomyrmex*, *Acromyrmex* and *Atta*. The latter are almost completely dependant on fresh vegetation for substrate. The type of material used varies between species according to location and availability. Several species of *Atta* which inhabit the savannah regions of South America have adapted to the exploitation of Graminae. The most notable of these are, *Atta capiguara* Gonçalves (Amante, 1967*b*), *Atta vollenweideri* For. (Bucher & Zuccardi, 1967) and *A. laevigata* (Jonkman, 1971). *Acromyrmex landolti* and several members of the sub-genus *Mollerius* have also specialised in this habit (Labrador, Martinez & Mora, 1972).

Werner (1973) has studied the selection of substrate by *Acromyrmex versicolor* in the southern USA and has presented data showing the relative proportions of different materials foraged during nine consecutive months. For the months of July and August the Graminae accounted for just over 50% of foraging records. In September 35% of records were grasshopper faeces and in November 60% of the material foraged were leaves of the leguminous tree *Prosopis juliflora*. It is evident from these data that *A. versicolor* can adapt to the collection of different substrates when available. The attines which inhabit neotropical rainforest are surrounded with a flora of high diversity and are well known to use a very wide range of plant species for substrate. In Guyana, Cherrett (1968) recorded 72 plant species from an area of 1194 m^2 surrounding an *Atta cephalotes* (L.) colony. During a 10-week period of observation 36 of these species were exploited for substrate. This is a wide range of food species for a phytophagous insect but on closer examination Cherrett's lists suggest that the ants are being selective and are concerned to avoid those species which produce repellent resins and latex which impedes cutting. Plants with hirsute leaves and stems are not favoured as substrate probably because the hairs impede the cutting process. Many neotropical plants bear accessory glands at the bases of leaves and flowers and these produce secretions which attract ponerine and formicine sentinels which repel attine foragers. Arboreal-nesting ants such as *Azteca* also appear to repel attines. Much attention has been focussed on the apparent preference of attines for crop plants. Many of these plants, e.g. *Citrus*, have been introduced to the neotropics and lack protective devices.

The higher attines show some selection for particular plant parts, flowers of *Theobroma*, *Hibiscus* and *Ixora* being notably favoured by *Acromyrmex octospinosus* and *Atta cephalotes*. Fallen fruits of *Psidium guajava* are particularly attractive to *Acromyrmex octospinosus* and the white albedo of citrus fruit is attractive to both *Atta* and *Acromyrmex*. Cherrett

& Seaforth (1970) have shown that fruit rind, old and young leaves of *Citrus* and the flowers of *Hibiscus* contain both attractive and repellent substances to attines. These no doubt play an important role in substrate selection. Cherrett (1972) studied the selection of substrate by a colony of *Atta cephalotes* in Guyana. Of 22 101 workers returning to the nest 42.05% were carrying burdens in their mandibles. This material comprised 49.1% newly cut leaves, 44.83% flowers, 3.56% dead leaves, 0.34% twigs, 0.34% other ants and 2.82% were unidentifiable. Of the flower and leaf material 59% by fresh weight were leaves. Differences were also detected between the leaf material carried by the ants and samples taken at random from the habitat. Ants tended to cut leaf material which was thinner, less tough, of higher moisture content and of lower weight per unit area. Apart from the first, all these characters were even more accentuated in ant-cut flower material. These results suggest that there are physical features which make young leaves more attractive to cut than older ones.

The diet of *Formica rufa* was recorded by Wellenstein (1952) as being composed of the following: honeydew (62%), insect prey (33%), sap and resin (4.5%), fungi and carrion (0.3%) and seeds (0.2%). Sap was collected most early in the year when insect prey was least available, and Ayre (1959) states that honeydew is collected more in late summer than in spring. Honeydew undoubtedly ranks high in the diet of *F. rufa* but has a large water component. Insect prey constitutes about one-third of the diet but it is not all digestible protein. The composition of the insect prey is variable; Chauvin (1966) found that 50% of the prey of *Formica polyctena* Foerst. were Diptera, whereas Gösswald (1958) reported an immense intake of lepidopterous larvae by *F. rufa* . By contrast, items collected by *Veromessor pergandei* (Mayr) were shown by Tevis (1958) to comprise seeds (68.2%), insect prey (10.3%) and miscellaneous (21.5%) during the period from July to March, and seeds (91.9%), insect prey (0.9%), flowers (6.7%) and miscellaneous (0.5%) for the period April to June. Clark & Comanor (1973) investigated the food intake of this species and showed that 34% of the total biomass collected was composed of fruit and seeds and 66% was stems and leaves. In terms of numbers of items, fruit and seeds comprised 88% and stems and leaves 12%. The stems and leaves are presumably taken for their sap and water content. Miscellaneous non-food items were also carried to the nest, including small gravel. The effect of an unbalanced diet has been investigated by Brian (1956*a*) who found that when fed a sugar-rich, protein-short diet, cultures of *Myrmica rubra* (L.) survived but reared workers rather than sexuals. The development of small larvae ceased, although they remained healthy until more protein was provided. Vanderplank (1960) found that protein-starved colonies of *Oecophylla longinoda* (Latr.) produced more small, timid workers compared with large aggressive workers produced on a protein-rich diet.

Food collection and storage

Foraging

Most ants do not possess very well-developed eyes and recognition of food is probably for the most part chemosensory. Barrer & Cherrett (1972) have shown that in *Atta cephalotes*, foragers are more attracted to cut or damaged leaves than to intact ones, more chemical information being available where cells are ruptured. This is probably also the reason for the attractiveness of injured and dying prey to predaceous species.

Apart from the honeydew excreted by root-feeding aphids and coccids within the nest, all items of food must be collected at some distance and carried home. Transport of food is achieved in three ways, according to its physical state and size. Liquid foods are imbibed and transported in the crop. Small prey may be consumed at the site of capture and the nutrients carried in the crop, or carried whole in the mandibles and maxillae. Larger items of food are either carried by groups of workers or dismembered at the site of capture and the pieces carried by individuals. In the higher attines the larger workers use their mandibles to cut pieces from living plant tissues. These are then grasped in the mandibles and carried hoisted over the head so as not to impede locomotion.

Much work has been published on the foraging patterns and behaviour of ants (Sudd, 1967). Foraging may be limited to a small proportion of the total worker population, as in *Formica* (Ayre, 1962) and *Lasius niger*, or it may involve most of the population, as in *Myrmica* (Stradling, 1968) and probably most ponerines. Small proportions of foragers are particularly associated with liquid feeding and high trophallaxis rates. It is generally accepted that foragers in such species are older individuals (Sudd, 1967). In *Formica fusca* L. the readiness to forage is correlated with hunger (Wallis, 1962) as well as the state of the brood and the temperature. Hunger increases ant activity and returning foragers boost the activity of the colony, stimulating more foraging. This must lead to increased exploitation of available food supplies as it brings about the recruitment of more foragers. Sudd (1960) has demonstrated the effectiveness of recruitment in *Monomorium pharaonis* (L.).

Simultaneous mass recruitment with a 'guide ant' has been observed in the ponerine, *Leptogenys kitteli* (Mayr) by Baroni-Urbani (1973). The Dorylinae have very large colonies which have no fixed nest site. They practise well-defined systems of mass group raiding which in *Eciton* vary according to statory and nomadic phases (Schneirla, 1957*b*).

Where sources of food are large and recruitment occurs there is a tendency for clearly defined trails to develop. These are marked by scents laid down by the workers and the nature of the trail substance in *Atta texana* has been investigated by Moser & Blum (1963) and Moser (1967).

In the genera *Atta* and *Acromyrmex*, trail utilisation to areas of substrate gathering may persist for long periods of time and as debris is cleared the trails become physically discernible (Weber, 1972*a*). Removal of debris clearly aids by speeding up the traffic of foragers and these genera frequently take obstacle-free routes along fallen logs, branches and even concrete where available. The speed of attine foragers has been investigated by Weber (1972*a*) and Pollard (1973).

Physical environmental factors such as temperature and humidity may impose rhythms of foraging activity on ant colonies. In temperate species low night temperatures may restrict nocturnal activity and in some desert habitats diurnal activity may cease during the heat of midday. Werner (1973) found that in the southern USA the foraging rhythm of *Acromyrmex versicolor* responded to changing environmental factors during the course of the year. The general pattern of foraging was nocturnal in summer and diurnal in winter. More difficult to explain is the well-marked diel foraging rhythm of the tropical attines. Whilst there was general agreement on the existence of a rhythm, reports on the timing were very conflicting until Pollard (1973), by long-term observations, showed that *Atta cephalotes*, whilst predominantly nocturnal, undergoes periodic phases of diurnal activity. Nocturnal and diurnal phases may persist for several weeks and would not be recognised by the short-term observer. Trails from the same nest site were sometimes found to indicate asynchronous patterns of activity which hence could not be correlated with physical environmental factors (Lewis, Pollard & Dibley, 1974). An intrinsic control, perhaps of a similar nature to that found in *Eciton*, seems to be indicated although the possibility of interaction with other ant colonies must not be overlooked. The role of hunger as a stimulus to forage in *Formica fusca* (Wallis, 1962) cannot explain the attine rhythm as clearly, as the material foraged is for the most part used as substrate for the fungus culture and not eaten directly. However, the workers do show interest in the plant sap exuding from cut surfaces and some of this is probably consumed.

Territory

In most ants the nest site remains in the same place except in the event of massive disturbance. Foragers then operate within a certain surrounding area, the extent of which will depend on a number of factors, such as the size of the population, size of workers, abundance of food, and the nature of the terrain. Brian & Brian (1951) suggested that insolation and temperature also affected the area foraged. The foraging area – trophoporic field of Holt (1955) – may or may not be defended against other ants (Brian *et al.*, 1965). The food territory of *Formica rufa* was investigated by Elton (1932). Defence of food territories is usually associated with the collection of large food items and honeydew, which cannot be transported by a single forager. In particular, the attendance of colonies of Homoptera was noted. The monopolisation of such food sources by different ant species has been discussed by Brian (1955). Wellenstein (1952) has shown that *F. rufa* will defend aphid colonies at a distance of 100 m from the nest.

Good communication, rapid recruitment and aggressiveness are necessary for the effective defence of food territories. Many ant species utilise permanent tracks through their territory which result in a more efficient retrieval of food from the area (Holt, 1955). In this way the area is divided into sectors radiating from the nest and these are searched by scouts which leave the permanent tracks throughout their length. The discovery of prey and recruitment of foragers by *F. polyctena* have been investigated by de Bruyn & Mabelis (1972). Pickles (1935) expressed the extent of the food territory as a circle with the nest at the centre and whose radius was the greatest distance at which foragers were observed. More recent studies indicate that food territories are usually irregular in shape and rarely approach circular form. The extent of these territories may be investigated in more detail by the use of sugar baits (Brian *et al.*, 1965) and by the use of radioactive tracers (Mortreuil & Brader, 1962), and maps prepared. This procedure enables what Elton (1932) termed the 'economic density' of the population to be calculated. This quantity has more ecological significance as it is more closely related to potential food supply.

Storage

Differences in the time of availability and the colony demand for food and water have lead to the evolution of storage behaviour by ants. The collection of seeds by granivorous species in desert habitats is usually followed by storage. Stores of seeds must be prevented from germinating and must be protected against attack by micro-organisms. Schildknecht & Koob (1971) found that the metathoracic glands of myrmicine ants, including the seed-storing species *Messor barbarus* L., produce a secretion containing the substance D-β-hydroxydecanoic acid (myrmicacin). This substance was shown to have a germination-inhibiting effect with both seeds and fungal spores. *Tetramorium caespitum* stores seeds of *Calluna* in late summer and these are used for larval feeding in the following spring when food supplies are less abundant (Brian *et al.*, 1967). Liquid food is stored by *Myrmecocystus* in the crops of repletes. These ants are found in arid habitats and probably use this method to store water as well as nutrients. The liquid food stored in the crops of *Prenolepis imparis* Say was found by Talbot (1943) to be kept during the winter and fed to the brood early in the following summer when foraging ceased.

Dead insects may be stored by some ants. Bohart & Knowlton (1953) found dead bees stored in the nests of *Pogonomyrmex occidentalis* (Cresson) but gave no indication of the durability of the store. Generally, animal protein does not keep well and Dlussky & Kupianskaya (1972) considered that the variability of daily intake by *Myrmica rubra* when supplied with excess protein was accounted for by this. Protein can, however, be stored after assimilation in the form of ant tissue by workers and queens and made available at a later stage by the laying and eating of eggs. This will be dealt with in Chapter 6.

Direct observations

The removal of burdens from incoming foragers and their subsequent identification is the most direct method of determining the quality of the diet and most authors have adopted this approach. Shortcomings are due to observer bias towards the larger, more conspicuous items and failure to consider crop contents. In the higher attines where burdens consist of fragments, identification may be difficult. However, it is usually possible to follow the foraging trail to the site of cutting and identify the source of the material directly. Where liquid foods are taken, observations are limited to the site of consumption. This is practicable with plant secretions and honeydew where the site is repeatedly visited by the ants and is epigaeal. Hypogaeal ants are not so easy to observe and their attended root-feeding Homoptera may represent a major food source. Pontin (1963) lists 16 species of underground Homoptera being tended by *Lasius niger* and *L. flavus* in an area of less than 300 m^2.

Solid food may be broken up to a greater or lesser degree within the nest by workers and fed to the larvae. Examination of clumps of larvae inside the nest frequently reveals fragments of solid food which can be collected and identified. This approach has been adopted by Pontin (1961). In temperate regions, nest inspection is facilitated when nests are located under flat stones; slates and glass can be utilised effectively for this purpose (Stradling, 1970).

Identification of non-assimilated remains

The procedure of identifying types of food consumed by faecal analysis which is so useful in vertebrate studies is not practicable with adult ants which, being liquid feeders, do not produce faeces with identifiable contents. Furthermore, social behaviour renders the collection of worker faecal samples extremely difficult. Many species, however, dispose of the non-ingestible cuticular parts of their prey and the husks of seeds on distinctive middens outside the nest or in refuse chambers within the nest. Examinations of ant refuse have been made by Went *et al.* (1972) to determine the seed types collected by *Veromessor pergandei*. Chauvin

(1967) used this technique to determine the types of prey eaten by *Formica rufa* and went on to devise apparatus for collecting the refuse. In the field the ants' rejecta are frequently rapidly dispersed making collection difficult.

Gut content analyses

These have been used widely in the study of vertebrate diet, where large identifiable fragments are obtainable. The principles are the same as those for faecal analysis but have the advantage that food may be identified more easily when digestion has not proceeded so far. Worker ants may be persuaded to regurgitate the liquid contents of their crops in response to mechanical pressure and sometimes to carbon dioxide anaesthesia. Ant larvae defaecate only just prior to metamorphosis and faecal material accumulates and is stored in peritrophic membranes within the midgut. The faeces are expelled within the cocoons of species which spin them and are more easy to collect. Dissection of mature larvae to obtain the gut contents is not impracticable. This material represents the residues of the entire larval life and although workers may remove urine from the larvae, the analysis of faecal pellets may yield much valuable information. Modern micro-analytical techniques have been used by a number of authors to determine the composition of aphid honeydew, (see Auclair, 1963) and there is no reason why worker crop contents and larval faeces should not be treated in a similar manner. Identification of carbohydrates and amino acids can be carried out using paper chromatography. The most satisfactory method of determining amino acid composition is ion exchange chromatography. Most laboratories conducting amino acid analysis use modern automatic amino acid analysers employing the ion exchange principle originally developed by Moore & Stein (1951).

The use of immunological techniques for identifying the source of proteins in the gut has been advocated by Pickavance (1970). These micro-techniques are quite suitable for the investigation of small samples from ants. The analysis of fats is more difficult although they can be separated by solution in organic solvents.

Some indication of the diets of worker ants has been obtained by gut enzyme analysis. Ayre (1967) carried out an analysis for digestive enzymes on the workers of five ant species and found that differences in the activity of the enzymes in different species tended to reflect their feeding habits.

The quantitative measurement of consumption

Laboratory methods

Surprisingly, few studies have been made on the quantitative aspects of food consumption in the laboratory although as long as the ants are easily cultured no great difficulties are presented. Laboratory studies are particularly useful in elucidating the effect of different components of the diet on productivity. Intake, assimilation, rejecta and production can be measured precisely in relation to larval stage and differing queen/worker/larval ratios. Altering the dosage of different dietary and nutritional constituents whilst providing an excess or deficit of others can reveal the requirements for production. Data from field measurements can be utilised in the design of such experiments to obtain a closer approximation to field conditions.

Ayre (1960) investigated the effect of varying the protein intake of *Formica polyctena* cultures on larval, worker and gyne production. Sugar and water were provided in excess and protein in the form of dipteran larvae was supplied in a series of daily rations by weight. This form of protein supply was chosen because of the thin cuticle and hence relatively low proportion of non-digestible material provided. Dlussky & Kupianskaya (1972) adopted a similar procedure for investigating protein consumption in three species of *Myrmica*. Numbers of queens and workers, the protein supply and the duration of trial were all varied and effects were measured in terms of changes in egg, larval, pupal and adult numbers and biomass. Protein was supplied in the form of *Lasius* and *Formica* larvae which, like dipteran larvae, have thin cuticles. The rate at which food was taken from an unlimited supply was very variable and thought to be due to the ants' attempts to store perishable material. Under natural conditions more heavily chitinised prey are transported to the nest where the non-digestible parts are rejected. Thus care must be taken in the collection and interpretation of field data.

Laboratory studies on the feeding and growth of *Myrmica rubra* were carried out by Brian (1973*a*). Food intake by ant cultures was measured in terms of the weights of *Drosophila* adults eaten, sugar solution drunk and cuticular remains rejected. The proportions of queens, spring larvae, sugar and water supply were varied and productivity was assessed in terms of numbers of eggs laid, larval weight gains and the weights of queens and workers. The presence of both queens and actively growing larvae increased the consumption of flies, sugar and water. Except in the absence of larvae or queens, provision of sugar reduced the number of flies taken. Concentrated sugar solutions (40%) were avoided but more dilute solutions (10 and 20%) were attractive. With flies available in excess their efficiency of utilisation was decreased by the absence of sugar and the consumption of water was reduced. Gösswald & Kloft (1963) have used radioactive tracers to study trophallaxis in laboratory cultures. This will be considered elsewhere.

The attines are readily cultured in the laboratory but because of the intimate relationship between the ants and their fungus culture and the physical structure of the latter, quantitative measurements of food intake are rendered extremely difficult and so far have not been attempted. The feeding of isolated groups of ants and larvae with artificially cultured fungal material might be contemplated although the attines do not thrive when separated from their own culture. The fungus forms an intermediate stage in the transfer of nutrients from foraged substrate to ant tissue, and the input of substrate to laboratory cultures is easily monitored. Weber (1972*b*) measured the total fresh weight of substrate per month taken in by laboratory cultures of three attine species over a period of two years. Monthly measurements of the volume of fungus culture were also made. Although in some instances there is an indication that increased intake of substrate is coupled with increasing volume of fungus culture this is not always so and some knowledge of the ant population and the rates of nutrient turnover and refuse ejection is necessary to clarify this relationship. There is scope for much more work in this field.

Field-based estimations

Food collected outside the nest by foragers is carried either in the mandibles and maxillae or in the crop, and hence different approaches are needed for quantitative assessments. Laden foragers may be sampled at the nest entrance although it must be remembered that some ants use underground tunnels for foraging and tend hypogaeal Homoptera.

Most quantitative data recorded have been obtained by direct observation and the collection of samples. Holt (1955), by marking individual *Formica rufa* foragers, carried out a census at a number of points on ant trails. He was able to determine the harmonic mean duration of all-round foraging trips and the number of trips made per unit time. This work is a valuable contribution to the quantitative study of food collection.

Attempts to overcome the limitations of human observation have been made by the use of photo-electric devices and electronic recorders to monitor foraging activity. Total numbers of ingoing and outgoing foragers can be obtained in this way although the precision dwindles at higher ant densities. The separation of inward and outward streams of foragers in laboratory cultures was attempted by Siddorn (1962). Systems of glass tubes in which the flow of traffic was facilitated in one direction at two points served to channel the two streams of foragers in front of two photosensitive devices. This system operates adequately under controlled laboratory conditions where ants are provided with small pieces of solid food. Attempts to use this technique in the field by Stradling (1968) were less successful due to obstructions caused by large pieces of prey in the confines of the narrow tubes. Equipment based on this principle has been adapted by Dibley & Lewis (1972) and used with some success for counting the leaf fragments carried by attines.

Solid food

The rate of solid food intake may be readily assessed by taking samples of laden foragers, removing their burdens and replacing them. Variation in foraging rates and rhythms must be taken into account. This approach is good for measuring seed intake and has been used by Tevis (1958) and Clark & Comanor (1973) to estimate seed intake by *Veromessor pergandei* colonies. Clark & Comanor recorded an average of 0.05 g of seeds per five-min interval during the morning foraging peak and 0.16 g during the afternoon period.

Wellenstein (1952), Holt (1955), Gösswald (1958) and Pętal (1972) have used this method to assess solid prey intake by species of the genera *Myrmica* and *Formica*. Holt (1955) found that 19% of returning *F. rufa* foragers were carrying solid food. The mean weight of a burden was found to be approximately 0.008 g and hence the daily intake of approximately 57000 items of solid food weighed 456 g. Gösswald (1958) calculated that a large colony of *F. rufa* had a daily intake of between 65000 and 100000 caterpillars. Pętal (1972) does not quote separate data for estimates of solid food intake. The problem of observer error has already been mentioned in the section *Direct Observations*. Holt (1955) reduced this error by using an electronic device operated by push buttons. A portable tape recorder could be used for this purpose. The frequency and duration of this kind of sampling will be ultimately determined by observer stamina. Chauvin (1966) devised an automatic method for retrieving the prey of *F. polyctena*. A barrier was erected around the nest and the returning foragers were forced to pass through a box impregnated with diesel oil. The fumes of the oil caused an alarm response and solid burdens were dropped.

Care must be taken in evaluating data obtained by these methods as the ants may react to this robbery by increasing or decreasing their foraging efforts.

Estimates of substrate intake by attine colonies have been attempted by a number of authors and these have been summarised by Lugo *et al.* (1973) and are presented in Table 5.2. Most of these data have been obtained either by periodically making direct counts of the number of ants carrying burdens past a point on a foraging trail during a timed interval, or making periodic counts of laden ants on a measured length of trail.

Table 5.2. *Rates of leaf fragment input into various leaf-cutter ant nests as reported in the literature. After Lugo* et al. (*1973*)

Species	Location	Date	Nest size	Leaf input g per hour[a]	Time period	Author
Atta sp.	Costa Rica	July–August 1969	Medium–Large	80.0	0700–2200 hours	Harris, L. D., 1969
Atta sexdens	Brazil	—	—	47.7	77 months	Autuori, 1940[b]
Atta cephalotes	Costa Rica	February–November 1966	37.8 m^2	64.0	24 hours	Markham, 1966
A. cephalotes	Costa Rica	December1969–March 1970	4 trails	290.0	—	Gara, 1970
A. cephalotes	Guyana	10 weeks 1963	56 m^2	13.1	24 hours	Cherrett, 1972
Atta colombica	Costa Rica	February–March 1967	3 trails	41.8	0830–1230 hours	Emmel, 1967
A. cephalotes	Panama	July–August 1955	2 trails	85.2	0700–1800 hours	Hodgson, 1955
Atta sp.	Costa Rica	June–July, 1967	—	118.0	0700–1500 hours	Lloyd, 1967[c]
Atta sp.	Costa Rica	—	—	42.0	0230–2400 hours	Ferrand, 1966[c]
Atta sp.	Costa Rica	—	—	16.5	0730–2230 hours	Wood, 1966[c]
A. cephalotes	Costa Rica	July–August 1966	Various nests	17.2	0530–1300 hours	Parsons, 1968
A. cephalotes	Guyana	10 weeks 1963	56 m^2	10.9	0100–2400 hours	Cherrett, 1968
A. colombica	Costa Rica	August 1971	44.8 m^2	47.5	50 hours	Lugo *et al.*, 1973

[a] Assumed 45% dry weight and/or 8.0 mg dry weight per leaf fragment of 1 cm^2.
[b] As cited by Weber (1966*a*, *b*).
[c] As cited by Harris, L. D. (1969).

Samples of burdens removed from foragers are used for mean fresh and dry weight determinations. The information must subsequently be coupled with that on the diel foraging rhythm before extrapolation to an estimate of daily intake. The inherent weakness in most of these estimates due to the short duration of observations has been pointed out by Pollard (1973), who made continuous records over a period of several months using the apparatus described by Dibley & Lewis (1972). The intake of substrate was found to vary considerably in time and between different trails serving the same colony. Cherrett (1968) also reported a wide variation in the utilisation of six foraging trails by an *Atta cephalotes* colony which was under observation for 58 days. Numbers of laden ants recorded for the same periods in each day ranged from 2 to 359. An inspection of Table 5.2 suggests that besides the variation in estimates produced by extrapolation, there is a general lack of standardisation in other parameters. A good example of this is the measurement of nest size which different authors have recorded in terms of unit area, numbers of trails, subjectively or not at all. Even where comparable units have been used wide differences in estimates occur. Amante (1967*b*) estimated that the grass-cutting *Atta capiguara* cuts 52.5 kg per hectare per day and Cherrett, Pollard & Turner (1974) estimated that another grass-cutting species *Acromyrmex landolti* cut 566 g per hectare per day. Undoubtedly more standardised and comprehensive data on attine substrate intake are required.

Liquid food

The rate of liquid food intake by a colony of *Formica rufa* was investigated by Holt (1955). To assess the weight of honeydew carried by foragers, samples of ants arriving and departing from a pine tree with tended aphids were collected and weighed. The weight difference of these samples gave the mean weight of one load of honeydew as 0.0019 g. As the census indicated that 19% of returning foragers were carrying solid food it was assumed that the balance of 81% were carrying honeydew. That foragers carrying solid burdens were also carrying honeydew was not considered. Using these figures the daily intake of food by this colony was approximately 918 g of which half was honeydew. Herzig (1938) estimated that a colony of *Lasius fuliginosus* Latr. could collect as much as 6 kg of honeydew in 100 days and Zoebelein (1956*b*) found that a *Formica rufa* colony could collect as much as 500 kg of honeydew in a year.

F. rufa normally carries prey whole in the mandibles so that it is fair to assume that the bulk of the crop contents will be sap or honeydew. Other species of ants consume more prey at the site of capture and therefore crop contents are likely to contain substantially lower percentages of sap and honeydew. Pętal (1967, 1972) estimated the intake of liquid food by

Myrmica laevinodis Nyl. colonies using weight differences between samples of ingoing and outgoing foragers. Live and dry weights of samples were measured and an attempt was made to allow for errors due to differences in worker size, but the equation quoted is not satisfactory.

Indirect methods

Measurements of the prey available to ants and its rate of disappearance in the field can give an index of the food intake by ant colonies. This approach has been used by Wellenstein (1957) and Kajak, Breymeyer, Pętal & Olechowicz (1972). Problems arise because the ants are rarely the sole predators and the losses from the prey population due to other predators and parasites must be taken into account. Elton (1932) realised that it was desirable to relate the population of an ant colony to the area foraged and expressed this as an 'economic density'. The production of harvestable seeds by plants within the foraging territory of *Tetramorium caespitum* was measured by Brian *et al.* (1967), and it was found that they provided an annual potential food source with a mean of 214 g per territory. The quantity of seeds consumed is unknown. Clark & Comanor (1973) measured territory size of the granivorous *Veromessor pergandei* and found it to be highly variable.

Where food supply is restricted seasonally and stores are made it might be possible to estimate intake by measuring the amount of food stored. Golley & Gentry (1964) excavated four hills of *Pogonomyrmex badius* Latreille and found stores of seeds weighing 16, 73, 130 and 303 g.

The difficulties of obtaining comprehensive, detailed, quantitative data of food intake by omnivorous ant colonies and relating this to population and productivity have been stressed by Baroni-Urbani (1972) who has suggested that some of these difficulties may be overcome by the use of computer simulation. The model presented takes into account seven different food sources whose availability changes temporally, the relative preference for these of four ant species and the populations of these ants. This sort of simulation is useful in that the results of experiments with small samples in the laboratory, and intermittent sampling of field ant populations and food supplies can be related to the complex system which actually exists. The programme can of course be revised to include additional results as they become available.

Interactions between foodstocks and the organisms feeding on them

It has long been realised that ants are an important factor in the flow of energy in many ecosystems. Ant feeding may lead to either increase or decrease in different food supplies and affect ecological stability. Long-term selective feeding by certain species of ants can lead to a change in the balance of competition between remaining species. Holt (1955) showed that with *Formica rufa* there was an intensification of foraging in regions of high prey density. Undoubtedly, communication and recruitment make for a more effective exploitation of food supplies, particularly where prey is aggregated. Ant predation can certainly have dramatic effects on the populations of some prey species. Kajak *et al.* (1972) found that 32% of newly emerged Diptera, 43% of leaf hoppers, 49% of Lycosid spiders were removed by *Myrmica* predation in a meadow ecosystem. Some species of *Formica* have long been known to reduce the populations of forest pests (Otto, 1958*c*; Pavan, 1959*a*). Defoliating larvae of certain Lepidoptera are particularly susceptible to attack and Gösswald (1958) estimated that large colonies of *F. rufa* could collect from 65000 to 100000 caterpillars per day. Wellenstein (1954) has shown that population densities of defoliating Lepidoptera are significantly lower inside *Formica* territories than outside and has investigated the relative rates of disappearance of caterpillars when they were artificially introduced to foraged and ant-free areas. This heavy predation of pest insects by wood ants has led to their cultivation for biological control purposes.

The ability of most ants to exploit a wide range of different food supplies undoubtedly helps them to maintain an adequate continuous food intake, and as predation on aggregations of prey is preferred through recruitment, declining populations will receive less attention as they become uneconomic to collect.

The depletion of prey populations by ants is in marked contrast with their effect on tended Homoptera. Herzig (1938) suggested that the effect of ant attendance was to stimulate increased feeding and excretion rates by Homoptera. This has been confirmed for *Aphis fabae* Scop. attended by *Lasius niger* by Banks & Nixon (1958). Homoptera which are kept underground on roots in the nest, or within ant-built shelters must receive some measure of protection. Nixon (1951), whilst considering that the protection conferred by ant attendance was much exaggerated, admitted that some advantages were gained in the form of reduced attack by natural enemies, and improved hygiene in the removal of surplus honeydew. Way (1963) has reviewed this subject extensively.

The effects of collection on seed supply have been little studied. Golley & Gentry (1964) have calculated that of the seeds produced in an old-field system, between 5 and 10% were stored by *Pogonomyrmex badius* and just over 50% were taken by mice and birds. Tevis (1958) showed that the collection of desert ephemeral seeds by *Veromessor pergandei* constituted a very small proportion of the seed production of the plants. The rate of seed intake by ant colonies did not noticeably diminish during the long period between flowerings, suggesting that the supply of seeds is not significantly reduced by ants. High densities of seedlings after rain support this conclusion. It is suggested, however, that as seed collection is selective, there might be an overall long-term change in the balance of the flora.

Perhaps the closest interaction between ant populations and their food supply is shown by the attines. This occurs at two levels, the fungus culture and the source of substrate. The ecological advantage of tapping the cellulose from a wide range of plant species as a source of carbohydrate has already been stressed. Few higher organisms can produce the necessary cellulase for this process and most herbivores resort to the use of endosymbiotic micro-organisms. The fungus cultivated by *Atta colombica tonsipes* has been shown by Martin & Weber (1969) to provide the necessary enzymes to bring about the degradation of this material, thus providing an enriched supply of assimilable carbohydrate through the fungal tissue (Martin, Carman & MacConnell, 1969). Curiously, the fungus is deficient in proteolytic enzymes but this is overcome by pre-treatment of the substrate by the ants and the liberal application of rectal secretions which have been shown by Martin, J. S. & Martin, M. M. (1970) and Martin & Martin (1971) to have significant protease activity. Furthermore, nitrogenous excretory products such as allantoin are abundant in the rectal fluid and serve to disperse the substrate protein and render it more susceptible to hydrolytic breakdown (Martin, M. M. & Martin, J. S., 1970).

In the presence of the ants their own fungus species is maintained as a monoculture, a practice as beset with difficulties for ants as for man. The culture is continuously vulnerable to invasion by competing alien fungi and bacteria which rapidly take over on the removal of the ants. Analyses of the metathoracic gland secretions of *Atta sexdens* have revealed the method by which the ants control their culture. Maschwitz, Koob & Schildknecht (1970) demonstrated the presence of the antibiotic phenylacetic acid in these secretions. The ability to suppress bacterial growth implicates the use of this substance in protecting the fungus culture from bacterial invasion. Schildknecht & Koob (1970) found that these secretions also contained the heteroauxin 3-indolyl acetic acid which has growth promoting effects on fungal mycelia. Schildknecht & Koob (1971) reported the discovery of a third component D-β-hydroxydecanoic acid (myrmicacin) in the metathoracic gland secretion. This substance was demonstrated to be a strong inhibitor of germination in fungal spores and no doubt serves to keep the ant's culture free of alien fungi. Martin *et al.* (1973), who demonstrated the presence of a chitinase in the faecal material of five attine genera, suggested that this enyme may be involved in the suppression of competitors by selective attack on chitinous fungi. The interactions between attines and their symbiotic fungus, as at present understood, are summarised in Fig. 5.1.

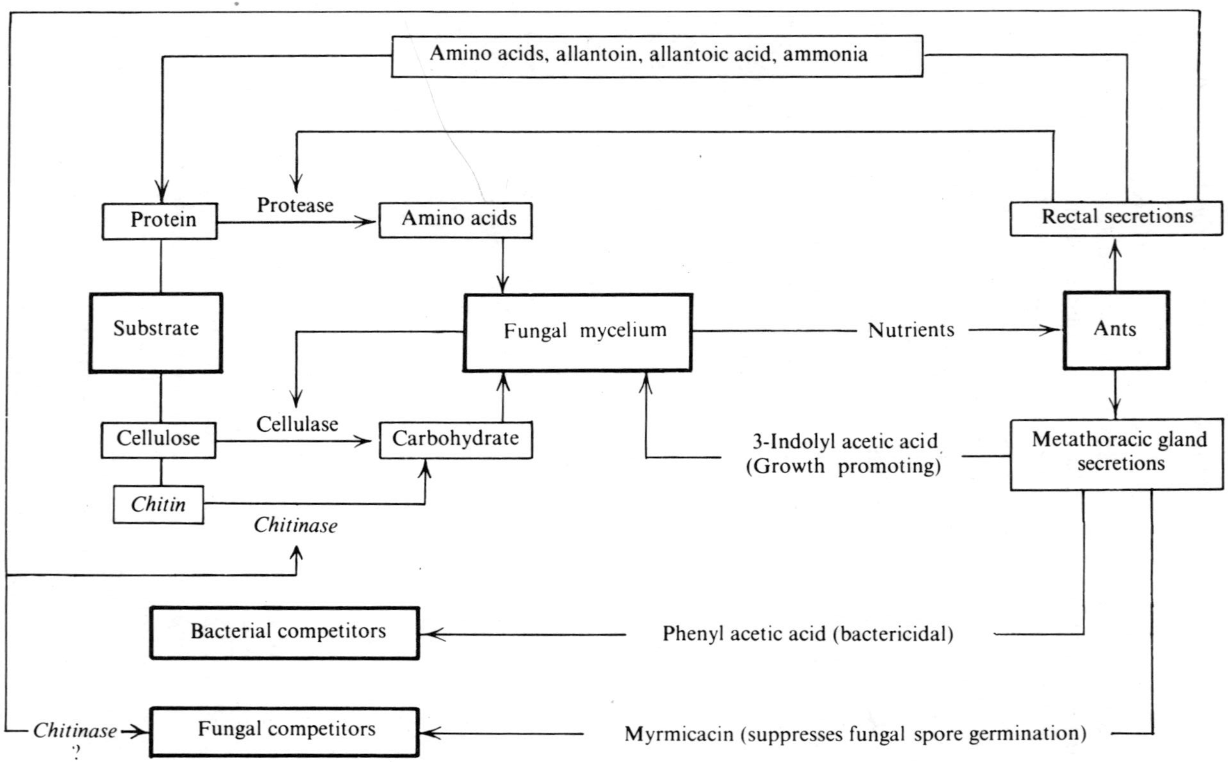

Fig. 5.1. Chemical interaction between an attine ant population and its symbiotic fungus culture.

The relationship between attines and the plants serving as substrate shows ecological interaction at another level. The ability of large *Atta* colonies in tropical rainforest to defoliate whole trees is well known, and it might be expected that repeated defoliation in the immediate vicinity of the nest site would result in the destruction of the forest habitat. Although there is a tendency for the nest site to be free from undergrowth the mature trees do not appear to be affected. Cherrett (1968) found that a colony of *A. cephalotes* in Guyana exploited the forest trees more intensively in a zone between 31 and 47 metres from the nest site, and suggested that the foraging pattern acted as a conservational grazing system which prevented over exploitation of the habitat. Weber (1972*a*) reports that trail formation in attines starts shortly after establishment of the colony and that the trails increase in length as the colony grows and ages. The clearing and maintenance of these trails is a continuous energy-consuming process (Lugo *et al.*, 1973) and as might be expected they are not transient structures but may persist for many months and even years. As the sites of leaf cutting, and probably also scouting, are at the ends of these trails there will be a natural tendency for the exploitation of vegetation to gradually become more distant from the nest site. The use of fallen trees and branches as trail routes has already been mentioned and in forest species the trunks and branches of growing trees will serve as easy routes to large supplies of foliage. It is evidently more rewarding for the forest-dwelling attines to exploit this supply of material as the undergrowth is much less used. Weber (1972*a*) noted the contrast between the forest attines pattern of relatively few broad trails and the many narrow ones used by grassland species. Cherrett (1968) has pointed out that the quantity of available foliage in plantation crops such as citrus and cocoa is very much less than in rainforest and therefore the grazing by attines is more likely to be detrimental in cultivated land. This effect is no doubt accentuated as the crop does not consist of a mixture of palatable and unpalatable plants. Lugo *et al.* (1973) have studied the energetics of *Atta colombica* in Costa Rica and have discussed the impact of its grazing on the forest ecosystem. Although the removal of photosynthetic material can reduce the potential productivity of the forest by 5.92 kJ per m^2 per day it is suggested that this is compensated for by the ant's effect of accelerating the flow of energy and cycling of nutrients, particularly phosphorus.

U.W.B LIBRARY P.C.B LLYFRGELL

Digestive efficiency and the relation of assimilation to consumption and production

Principles and examples

The distinction between 'digestive efficiency' and 'assimilation efficiency' has been made by Petrusewicz & Macfadyen (1970). The second ratio, which is the more desirable one to measure from an ecological point of view, is not easily measured for ants. The principal difficulty arises in the interpretation of consumption. Substantial quantities of material collected and carried to the nest are rejected before ingestion, and may or may not be regarded as consumption. If consumption is to be limited to the food which is ingested by workers and larvae the rejecta must be carefully retrieved in order to assess how much has been ingested. Faecal material may well be deposited with non-ingested rejecta making the separation of these quantities impossible. Whilst the collection of faecal residue may be feasible from full term, final-instar larvae, the quality and quantity of food ingested, including that supplied directly from worker crops, is difficult to assess. The problem of measuring rejecta of ant cultures quantitatively has resulted in most authors avoiding the issue by providing food with a high ingestibility potential (ant and fly larvae with thin cuticles) and equating ingestion with assimilation. A high assimilation efficiency is assumed, faecal material is ignored and ingestion is related to productivity.

Brian (1973*a*) supplied *Myrmica rubra* cultures with whole *Drosophila* adults and took the difference between the dry weight of flies taken and residues left as a measure of food ingested. Faeces were not regarded as a major component since food is filtered of insoluble particles before ingestion. Ayre (1960) found that protein intake was directly related to worker production in *Formica polyctena* cultures although gyne production was less at higher levels of protein supply. Similarly, Dlussky & Kupianskaya (1972) found that in three species of *Myrmica*, production (P) was related to protein consumption (C_{pr}) in a curvilinear manner. Relative productivity was calculated from $P/C_{pr}\times 100$, and distinguished from general productivity $P/C\times 100$, where the whole consumption of food is taken into account. There is no doubt that relative productivity is one of the most important indices for the assessment of the role of ants in ecosystems. Values for *Myrmica* spp. were found to range from 10–40%. Using Ayre's (1960) data in conjunction with those of Dlussky & Kupianskaya (1972) *Formica rufa* is found to have a relative productivity of 32% and a general productivity of 4–10%, which falls within the range for *Myrmica*. General productivity calculated from the whole consumption of food of course gives a much lower index, as insects collected under natural conditions contain larger proportions of non-digestible cuticle. Thus the general productivity of field populations of *M. laevinodis* (= *M. rubra* L.) was calculated by Pętal (1967) to be 2.6% and that of *Pogonomyrmex badius* was calculated by Golley & Gentry (1964) to be about 2%. With an increase in daily protein consumption, Dlussky & Kupianskaya (1972) found that relative productivity values of *Myrmica* decreased and then stabilised. Evidently the stabilised level of productivity is optimal for the species and is thought to occur under natural conditions. The highest correlation between protein intake and productivity occurs at the time of maximum larval growth.

Production/ingestion (*P*/*I*) ratios were calculated by Brian (1973*a*) for cultures of *M. rubra*. The production was taken as the increase in larval weight and eggs layed. The ratio varied little for large or small, spring larvae, irrespective of the presence of a queen and was found to have a mean value of 0.45±0.01, somewhat higher than the value of Dlussky & Kupianskaya (1972). With summer larvae the ratio was lower and more variable due to the tendency for larvae to diapause. Higher levels of protein supply had a tendency to reduce the ratio. In the absence of both queens and sugar, the highest ratios were attained at 0.56 and the effect of high sugar concentration in the absence of queens reduced the ratio to 0.04. The effect of increasing the supply of lower concentrations was also to depress the *P*/*I* ratio.

All of these results are expressed in terms of weight ratios which though a reasonable approximation should more correctly be based on energy measurements.

The effect of environmental factors on consumption, assimilation and efficiency

The metabolic rates of poikilothermic animals are directly dependent on the ambient temperature, above and below levels of which no feeding occurs. Ants, however, have the power to choose nest sites with favourable temperature regimes and to move the brood to selected parts of the nest in response to changes in temperature, thus keeping the brood in the optimum conditions available. Many soil-nesting ants in temperate regions make use of the thermal properties of surface stones in nest construction and build in situations where insolation is high (Brian & Brian, 1951; Steiner, 1929). In tropical regions high temperatures are avoided by deeper penetration of the soil or by building in the shade. Brian (1956*b*) has shown how the closure of a tree canopy with its accompanying reduction of ground insolation caused the regression of *M. rubra* colonies by increasing the distance between the ants and their attended Homoptera and reducing the temperature, thus rendering the capture of prey more difficult. Holt (1955) investigated the effect of temperature on the walking speed of *Formica rufa* foragers and came to the conclusion that the rate of food collection was dependent on ambient temperature. Hodgson (1955) observed that rainfall interrupted the collection of substrate by attines.

Cultivation of Homoptera within the nest, the construction of shelters and hypogaeal habits undoubtedly assist in providing a more constant food supply. Interruptions to foraging, such as heavy rain and extremes of temperature, are thus considerably reduced. Where seasonal availability of food or rigours of the physical environment reduce the time available for food collection, ants adapt by resorting to food storage.

Foraging by seed-harvesting ants

W. G. WHITFORD

Because of the high colony densities and habit of denuding areas around nests, the activities of seed-harvesting ants in the semi-arid regions of the United States have received considerable attention. Studies of seed-harvesting ants in other arid regions have been limited and few data are available on which to base generalizations concerning foraging activities.

Most studies on foraging ecology in seed-harvesters deal with the larger-sized species, i.e. *Pogonomyrmex*, *Veromessor* and*Novomessor* with only scattered references to the smaller, numerous *Pheidole* spp. Hence, this discussion will be mainly confined to the large North American forms.

Several studies have addressed the question: do harvester ants remove sufficient quantities of seeds to effect the seed reserves and hence viability of a particular plant community? (Tevis, 1958; Rogers, 1974; Whitford & Ettershank (1974). Tevis (1958), concluded that *Veromessor pergandei* harvested less than 8% of the seeds produced in a sandy Sonoran Desert area in California. Rogers (1974) estimated that in short-grass prairie in northeastern Colorado, *Pogonomyrmex occidentalis* remove approximately 2% of the available seed supply. Whitford (1975) has estimated that three species of *Pogonomyrmex* in the Chihuahuan desert in southern New Mexico removed less than 10% of the total seed production in one year and less than 3% the following year. It was concluded that variation in the impact of harvester ants on the total seed reserves was a function of the immediate past climatic history of the area. Foraging activity was intense and numbers of active colonies were large in a year of high productivity which had been preceded by an 18-month drought and low seed production. The following year, foraging activity was greatly curtailed despite high seed production by preferred forage species. This decrease was attributed to a reduction in foraging drive due to the granaries being nearly full from the previous year's activity.

Based on studies of foraging in harvester ants over a four-year span, it was concluded that both colony satiation, i.e. degree to which granaries were filled, and climate (soil surface temperature, saturation deficit and light intensity) regulated the foraging activity in harvester ants. In an area supporting three sympatric *Pogonomyrmex* spp. and *Novomessor cockerelli* (André), two species were exclusively diurnal, *Pogonomyrmex californicus* (Buckley) and *Pogonomyrmex desertorum* Wheeler and two both diurnal and nocturnal, *Pogonomyrmex rugosus* Emery and *N. cockerelli.* Within these constraints, foraging was regulated by saturation deficit

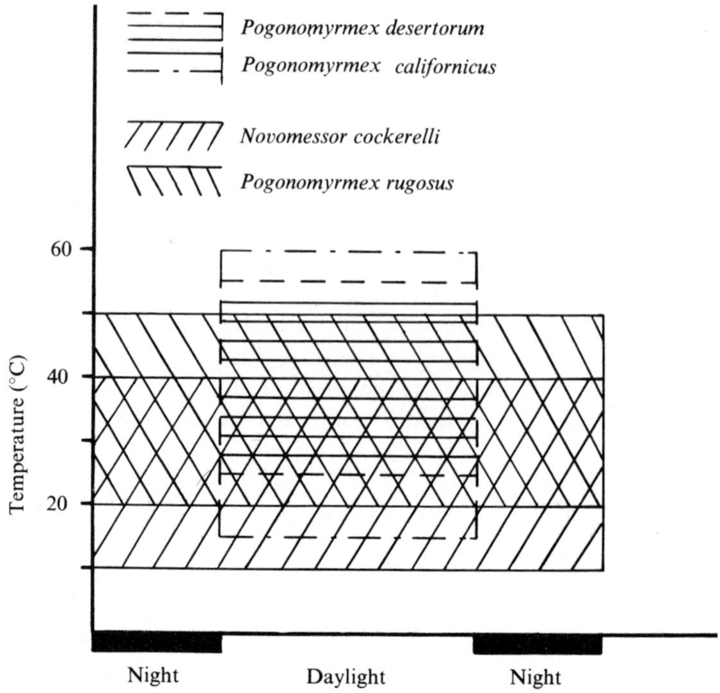

Fig. 5.2. The effect of soil surface temperature and time of the day on the distribution of foraging activity in four species of seed-harvesting ants from the Chihuahuan Desert in southern New Mexico.

and soil surface temperature with higher saturation deficits and soil surface temperatures resulting in cessation of foraging in *N. cockerelli*, *P. rugosus*, *P. desertorum* and *P. californicus* in a successive series (Fig. 5.2). Seasonal differences in peak foraging were recorded for the two diurnal *Pogonomyrmex* spp. with *P. desertorum* exhibiting peak activity in summer and *P. californicus* in spring and autumn (Schumacher & Whitford, 1974). These studies suggest that seasonal constraints and forage availability determine initiation of foraging activity in harvester ants and that saturation deficits, soil surface temperature and light act as regulators of daily foraging patterns.

Several investigators have reported the influence of light and surface temperature on daily foraging in harvester ants. Box (1960) reported that *Pogonomyrmex barbatus* foraged at night, the only period when ambient temperatures fell below 35 °C during his studies in south Texas. *P. barbatus* in New Mexico and Arizona are strictly diurnal exhibiting an early morning and late afternoon peak. It was concluded that light was less important than temperature in regulating activity in *P. barbatus*. Rogers

(1974) showed that the peak activity of *P. occidentalis* occurred at soil surface temperatures between 40–45 °C which are identical to the soil surface temperatures at which peak activity occurred in the four *Pogonomyrmex* spp. studied (Whitford, 1975). Clark & Comanor (1973) showed that *Veromessor pergandei* ceased foraging at a soil surface temperature of approximately 39 °C which is similar to the maximum foraging temperature in *N. cockerelli*. Although these studies indicate the importance of temperature and possibly light, they do not consider saturation deficits which are an important factor regulating foraging activity in ants in arid and semi-arid regions.

Although harvester ants remove only a small percentage of the total seed production of an area, they do not select from the seed pool at random. Forage selection by harvester ants and the factors affecting forage selection have been studied by several investigators (Tevis, 1958; Box, 1960; Willard & Crowell, 1965; Eddy, 1970; Went *et al.*, 1972; Clark & Comanor, 1973; Rogers, 1974; Whitford, 1975). These studies suggest that harvester ants select seeds which are (1) abundant and (2) available in high densities over much of the foraging season. *Pogonomyrmex* and *Pheidole* seem to prefer the seeds of annual grasses and forbs (Willard & Crowell, 1965; Clark & Comanor, 1973; Rogers, 1974; Whitford, 1975). Costello (1944) found that colony density of *Pogonomyrmex occidentalis* was closely correlated with the successional stage of the vegetation with high densities in the annual grass-forb stages in abandoned field succession. Our studies have shown that three sympatric species of *Pogonomyrmex* exhibit high preference for one or two species of forbs. When spring annuals were available these ants preferentially collected seeds of only one of the five most common annuals.

In summer, they selected seed of three annuals that exhibit continuous production of seed throughout the season. However, in late summer when seeds of an annual grass *Bouteloua barbata* appeared, their foraging preference switched to that grass. Regardless of season, *Pogonomyrmex* spp. removed more than 50% of the total seed production of the preferred forage species. Heavy impact on the seeds of preferred forage species was also reported by Tevis (1958), Willard & Crowell (1965), Eddy (1970) and Rogers (1974). Although harvester ants do not appear to have an effect on total seed reserves, their activities could greatly affect the success and distribution of species which are preferred forage.

The generalization that harvester ants selectively forage on fruits or seeds of the most numerous plants and/or those species which produce fruits over extended periods does not account for the failure of these species to harvest seeds of some equally abundant or productive plants. Since all seeds or fruits are not foraged in proportion to their relative abundance, forage preference must be related to some characteristic of the seed species. We have examined size and spatial distribution as factors affecting forage selection. *Pogonomyrmex* spp. selected seeds from a wide spectrum of sizes rejecting others of a similar size. Seeds distributed in piles around plants or dispersed were equally selected for or not taken. It was concluded that seed selection was based on some chemical properties of the seeds. However, not all foragers within a colony forage on prefered species since a sample of foraged material contains some seeds of most of the plants in the area.

Although seed-harvesting ants are dependent on seeds as stored food, whenever subterranean termites are available at the surface, all harvester ants prey heavily on them. In the Chihuahuan desert, species of *Amitermes* and *Gnathamitermes* forage on dead annuals when saturation deficits are close to zero and soil surface temperatures are low. Under these conditions *Pogonomyrmex*, *Novomessor* and *Pheidole* prey heavily on termites which may account for nearly 50% of the forage brought to the colony. Termites are the only animals which appear to be preyed upon by *Pogonomyrmex*. Other arthropods brought to the colony appear to have been collected as carrion. Termites transported to nests are often alive and struggling, but other arthropods are immobile. When termites are not abundant, carrion makes up less than 2% of the forage.

Plant seeds are less important as forage for *Novomessor* and *Veromessor* (Tevis, 1958; Went *et al.*, 1972; Whitford, 1975). These species transport a variety of materials such as leaves, stems, floral parts, fecal material, parts of exoskeletons to the nest. We found that over half of the forage of *Novomessor* was miscellaneous plant and animal parts. Nocturnal-foraging *Novomessor* spp. prey on a variety of arthropods in addition to termites which are a favourite prey item. When the large seeds of mesquite *Prosopis glandulosa* were available (average weight 0.04 g), *Novomessor* collected large quantities of these seeds. Foraging activity in *Novomessor* is more dependent on climate and less on seed availability than *Pogonomyrmex*, suggesting that this species is a facultative seed-harvester in contrast with the *Pogonomyrmex* which are obligate seed-harvesters.

6. Respiration and energy flow

G. J. PEAKIN & G. JOSENS

The difficulties of determing the energy requirements of maintenance are well known to production ecologists. They have been discussed in general terms by Macfadyen (1967), Petrusewicz & Macfadyen (1970) and Grodzinski, Kłekowski & Duncan (1975). It is clear that the problems these authors draw our attention to certainly apply to ants and termites, as Brian (1971) has observed.

Ideally, measurements of the costs of maintenance should be carried out in the field. Attempts to do this have been made for soil animals in general and Macfadyen (1971) describes some of the methods that have been employed and draws attention to their limitations. The principle of these is, of course, that they apply to the whole soil ecosystem and it is generally impossible to partition the energy flow manifested as respiratory activity amongst the individual species within the ecosystem. Most ants and termites are associated with the soil so that such difficulties certainly apply, especially with those species nesting diffusely. However, the tendency to aggregate into elaborated nests holds out the possibility, as Brian (1971) has suggested, of measurement of colony respiration in the field, with at least some of these sources of error removed. It is clear that as the degree of isolation inherent in nest building increases, so greater confidence can be placed in the assumption that the level of respiratory activity determined is that of the ants or termites, rather than extraneous organisms. It should not be forgotten that in many cases metabolic activity other than that due to the ants or termites is associated with nests. A conspicuous example is furnished by those species which harbour symbiotic fungi, where a large proportion of 'colony respiration' must be due to the fungi, which, while constituting part of the total energy flow due to the social insects, will confuse comparison of laboratory studies of isolated insects with field determinations. According to Ruelle (1962), 1 l of fungus garden from a *Macrotermes bellicosus* (Smeathman) nest produces about 80 ml of carbon dioxide per hour; that is to say that about 20 l of carbon dioxide evolve daily from the fungus gardens of a medium-sized nest. Where isolation is practically complete, as in arboreal- and aerial-nesting species, the difficulties of measuring metabolic activity in the field may be fewer and in this connection the continuous flow, gas exchange method described by Woodwell & Botkin (1970) for measuring respiratory activity in trees may lend itself to adaptation to similar measurements in these species.

However, again, it should not be forgotten that termites deposit their faeces upon the walls of their galleries and nest chambers, accumulating substantial organic material as food for bacteria and fungi; the walls themselves may be made of carton. During recent experiments, Josens (unpublished data) measured the carbon dioxide evolution from a whole nest of *Trinervitermes geminatus* Wasmann (isolated from the soil); subsequently the termites were hand-sorted and therefore the nest broken into small pieces. All the nest material was then put back into the respirometer and the new carbon dioxide evolution was higher than before! This artefact was attributed to microbial and/or fungal respiration stimulated by the crumbling of the nest. But at least it shows how substantial the respiration of the nest itself can be.

Hébrant (1970*a*) observed a circadian rhythm in the respiration of whole nests of *Cubitermes exiguus* Mathot; thus he tried to establish correlations between his different measurements of oxygen consumption and the population biomass. He obtained the best correlation using the maximum respiration and the worst one with the minimum respiration. In this case, it may well be that a substantial proportion of oxygen was consumed by the micro-organisms of the nest.

Since measurements in the field must always be imprecise, useful, indeed, necessary confirmation of the values determined may be obtained from measurements in the laboratory and the most widely practiced procedure is, in fact, to determine metabolic rates in controlled laboratory conditions and then extrapolate them to the conditions which are assumed to obtain in the field. Measurements in the laboratory certainly result in obvious improvements in precision and sensitivity because of the great variety of methods that have been developed in recent years. However, whereas investigators in the field, while having legitimate doubts about the precision of their measurements, may be reasonably sure that the organism being investigated is behaving normally in the physiological sense, those in the laboratory have justified confidence in the accuracy of their readings but are acutely aware of the artificiality of the environment in which the experimental animals are confined (Macfadyen, 1967; Petrusewicz & Macfadyen, 1970). The effects on metabolic activity are often unknown and where they are, these doubts are often confirmed. Nevertheless, the combination of field and laboratory estimates provide mutual support in a preliminary estimate, though perhaps only to within an order of magnitude of this component of energy flow.

However, it is all too common for preliminary data of this kind to become incorporated into the conventional wisdom unless caution is continually exercised. It would seem prudent therefore, at the start of this review of the present state of knowledge of respiratory activity in ants and termites to consider in some detail the nature of the limitations of the methods available, so that existing information may be properly assessed, the degree to which it may be extrapolated to field conditions determined and, perhaps most important, subsequent work may be fruitfully directed to elucidate this most intractable problem.

Determination of metabolic costs in the laboratory

Two approaches may be adopted to measure the costs of maintenance in the laboratory, the direct and the indirect. Direct determination is effected by calorimetry, since all metabolic activity ultimately expresses itself in the generation of heat energy. Indirect determination (indirect calorimetry), since the units employed are the same as in the direct method, utilises respirometry and an oxycalorific equivalent, which equates the respiratory quotient (RQ) and the quantity of oxygen consumed (or carbon dioxide evolved) with the energy released in metabolism.

Indirect determination

Respirometry has, hitherto, been overwhelmingly the preferred method employed by production ecologists and this is, of course, reflected in the accepted terminology and symbolism for this component of the energy budget (Petrusewicz, 1967; Petrusewicz & Macfadyen, 1970). Of the three synonyms – cost of maintenance, metabolism and respiration – it is the last which is most commonly used, symbolised by R.

There is no doubt that the principle reason for this choice is the great variety of adaptable respirometers available to research workers (Petrusewicz & Macfadyen, 1970), with which respiratory gas exchange can be measured with great ease and precision. For example, when oxygen consumption is measured there is a choice between physical, analytical and manometric methods. Physical methods include the use of oxygen electrodes and the utilisation of oxygen's paramagnetic properties, while the best-known analytical methods, which are perhaps in general the least precise of the three, include Winkler's method and the use of pyrogallol. The greatest variety and the most familiar techniques are to be found in the manometric field. The Warburg and Gilson respirometers depend on the constant volume principle, as do the more recently developed electrolytic types (Macfadyen, 1961; Phillipson, 1962) and the ultra-respirometric variants developed by Gregg (1947, 1950, 1966) and Gregg & Lints (1967). The ultimate in sensitivity is achieved with the Cartesian diver technique in which there have been a number of important developments in recent years (Løvlie & Zeuthen, 1962; Zeuthen, 1964; Løvtrup & Larsson, 1965; Lints, Lints & Zeuthen, 1967).

Where carbon dioxide evolution is measured, then again physical (infra-red analysers, gas chromatographs), analytical (absorption in alkali, etc.)

and manometric methods are available. In the latter instance, where carbon dioxide evolution is measured contemporaneously with oxygen uptake, then a value for RQ may be calculated and this in turn is of value in determining the conversion factor (the oxycalorific equivalent) to energy units. Thus with an appropriate choice of apparatus it is possible to measure, with precision, oxygen consumption ranging from nanolitres per hour to kilolitres per hour, a range of 12 orders of magnitude (Kleiber, 1961).

The very sophistication of respirometric techniques has tended to blind investigators to their inherent shortcomings in relation to studies of energy flow in ecosystems. The artificiality of the environment for many animals is fully recognised. For example, the lack of carbon dioxide and a less than saturated atmosphere are two factors which may affect the respiration of the experimental animals and steps have been taken in the development of electrolytic and continuous flow respirometers to reduce their effect. And, of course, respirometers can only measure aerobic activity and cannot take into account the temporary or permanent occurrence of anaerobic activity in animals.

Further, the fact that it is an indirect method implies the making of assumptions when converting respirometric data to energy units whose margin of error can only be guessed at, unless a great deal is known about the physiology of the organisms. The quantity of energy released as a result of the consumption of unit volume of oxygen depends on the substrate oxidised. If glucose is utilised then each cubic centimetre of oxygen at standard temperature and pressure (STP) will release 21.02 J; if the respiratory substrate is palmitic acid a cubic centimetre of oxygen consumed will release 19.54 J; or if alanine, then 19.06 J. In fact, thermal equivalents ranging from 17.56 kJ per g for carbohydrates to 39.73 kJ per g for fat have formed the basis of tables for converting oxygen consumption to units of energy (Brody, 1945), so it follows that the minimum volume of oxygen required to produce a joule is about 90% of the maximum volume. For carbon dioxide evolution the proportion is 78% (Southwood, 1966). The difficulty of determining the respiratory substrate is alleviated somewhat by a knowledge of the RQ since it is possible to estimate the proportion of fat and carbohydrate being oxidised and so apply an appropriate thermal (oxycalorific) equivalent derived from the tables mentioned above. However, even this has its drawbacks since other metabolic processes may disproportionately influence the RQ. For example, the synthesis of fats from carbohydrates has an RQ of 8.0 (Kleiber, 1961). It is clear that the indirect method of estimation is not without error, and in precise studies of energy flow in populations of animals the effects of other metabolic processes may be large enough to invalidate the results.

Direct estimation

It is surprising that direct methods for determining the cost of maintenance, obviating the difficulties mentioned above, have not been employed. Yet calorimetry has been almost totally neglected by production ecologists, despite its long and distinguished history as a physical method applied to varied biological problems, where much useful physiological information may be derived from the measurement of the direct manifestation of biological activity. It is all the more curious that production ecologists, who are primarily concerned with energy flow, should not make use of the phenomenon, especially as modern micro-calorimeters rival many respirometers in their sensitivity. It would seem that ecologists have been unduly deterred by the apparent difficulties of correcting the heat output of small poikilotherms for such things as heat of vaporisation of water (Southwood, 1966; Petrusewicz & Macfadyen, 1970). Yet it is a simple matter to avoid evaporation, especially as many animals of ecological interest naturally avoid the physiological stress inherent in such loss, by arranging for measurements to be carried out in an hermetically sealed vessel. Even where evaporation can be tolerated and does take place it can be measured and corrected for (Peakin, 1973).

A number of reviews and accounts of the general application of calorimetry to biological problems have been published in the last decade or two and testify to the general interest aroused by the technique (Świętosłaski, 1946; Calvet & Prat, 1956, 1963; Kleiber, 1961; Skinner, 1962; Brown, 1969; Pullar, 1969; Wadsö, 1974) and it is perhaps not out of place to consider briefly the history of the method, particularly with reference to smaller poikilotherms, in an effort to redress the current neglect by ecologists of what could be, by virtue of the simplicity of the principles involved, a powerful addition to the techniques used in the study of energy flow through organisms and ecosystems.

Pullar (1969) has reviewed briefly developments in calorimetry up to the end of the nineteenth century. During the course of these experiments Lavoisier coined his well-known aphorism *La respiration est donc une combustion* and the applicability of the Law of Constant Heat Sums to living animals was established. The result was the development of increasingly sophisticated macro-calorimeters for use in human physiology and agricultural zoology.

Meanwhile, progress in the opposite direction towards the development of micro-calorimeters was stimulated by Hill (1911) whose early, adiabatic calorimeter, in which, amongst other things, he determined the heat output of a number of poikilotherms (frogs, earthworms) led to the astonishingly sensitive calorimeters with which he carried out his classical experiments on muscle and nerve physiology. Bialaszewicz (1933) used a variant of

Hill's calorimeter to determine the heat output of larvae and pupae of *Lymantria dispar* L. The animals were enclosed individually in an hermetically sealed vessel, together with moistened filter paper, thus obviating the problem of heat loss due to evaporation. Heat production at 25 °C ranged from 0.596 J per hour for a 36-mg male larva to 24.15 J per hour for a 1625-mg female larva. Expressed as heat output per unit weight per hour, it was found that the larvae at all stages and in both sexes produced about the same energy, namely 15.75–16.80 J per g per hour. Balzam (1933), in the second part of this investigation, compared direct and indirect methods of calorimetry. Larvae and pupae of *L. dispar* and *Bombyx mori* L. of between 230 and 734 mg formed the experimental material. He demonstrated that the mean oxycalorific equivalent was slightly higher in the larval stages and much lower in the pupal stage than is accepted today (Petrusewicz & Macfadyen, 1970), when it is assumed that the major part of the heat evolved is due to the oxidation of fat and carbohydrate in the proportions reflected in the RQ. In other words, he demonstrated that a discrepancy existed between the two methods of assessing energy flow, for which he was unable to offer an explanation. Taylor & Crescitelli (1937), in determining the heat evolved by pupae of *Galleria mellonella* (L.), in a differential calorimeter similar to Hill's (1911) one, also found discrepancies between direct and indirect estimates. The indirect measurement was calculated in the orthodox way by assuming that all respiratory activity was due to oxidation of fat (the RQ had been shown in previous work to be 0.69 throughout metamorphosis (Taylor & Steinbach, 1931; Crescitelli, 1935)) and they recognised that the discrepancy could arise as a result of any or all of the following factors: (i) The role of protein metabolism in heat production; (ii) syntheses occurring during histogenesis and (iii) evaporation of water from the pupa. This last factor undoubtedly played a part, since the Dewar vessel was only plugged with wool, but the other two should not be discounted entirely since they are metabolic activities associated particularly with pupation and which are not necessarily reflected in respiratory activity.

Extension of these earlier studies on heat production by small poikilotherms awaited the further development of calorimeters. A considerable advance was made with the appearance of the Calvet micro-calorimeter, whose rapidity of response and great sensitivity, allied with continuous recording techniques, allowed detailed study of the qualitative nature of thermogenesis. Prat (1954, 1956), as part of more general studies of these phenomena, investigated heat production in a number of insects (e.g. *Melanoplus differentialis* Thomas, *Galleria melonella*, *Drosophila melanogaster* Meigen, *Periplaneta americana* (L.)) and revealed the complexity of this manifestation of physiological activity. Roth (1964), also using a Calvet micro-calorimeter, was primarily concerned with comparative heat production in worker honey bees (*Apis mellifica* L.) in isolation or in groups of up to 50, rather than comparing direct and indirect methods. He showed that heat production by each bee was consistently lower in groups at temperatures below 35 °C, and at 10 °C individual heat production was about halved when in groups of 25 compared with isolated workers. Peakin (1973) found discrepancies between direct and indirect measures of energy dissipation in the pupae of *Tenebrio molitor* L. Using Ivlev's (1934) oxycalorific equivalent he showed that the indirect measure underestimated, though not significantly, the energetics of the first five days of pupation at 25 °C. On the sixth and seventh days there was almost complete agreement between the two methods but on the final day, indirect measurement apparently overestimated energy flow. Of course, Ivlev's oxycalorific equivalent assumes oxidation of a mixture of fat and carbohydrate. That *T. molitor* pupae do oxidise a mixture of these compounds during metamorphosis is supported by Russell's (1955) and Moran's (1959) findings. However, Krogh (1914) found that the RQ of pupae was 0.7, suggesting that fat is the only respiratory substrate and if the oxycalorific equivalent is calculated on this basis then the discrepancy between direct and indirect estimation becomes greater. More recently, using hermetically sealed vessels to obviate the need for correction for evaporative losses, larger discrepancies have been found with workers of *Myrmica sabuleti* Meinert (G. J. Peakin, unpublished).

In conclusion, there is evidence that if experimental difficulties are overcome, and there seems to be no reason for assuming that they are insuperable, then the direct method of measuring energy dissipation due to metabolic activity gives a better estimate than the indirect, which despite the sophistication of the methods developed contains inherent sources of error. As we shall see, the characteristics of ant and termite social life pose particular difficulties for respiratory investigations; these emphasise their inherent disadvantages and argue for further consideration of calorimetry as the preferred technique.

The characteristics of social life which bear on the study of metabolic costs

The foregoing remarks have sought to emphasise the general limitations, advantages and disadvantages of the methods for determining metabolic costs. It is now necessary to turn to a consideration of those features of social organisation in ants and termites which are likely to affect their respiratory physiology.

The partition of metabolic costs amongst the individuals of the society

Consideration of the way in which metabolic costs are partitioned in the society is primarily determined by the fundamental differences in the phylogeny of ants and termites. Ants are, of course, holometabolous, with immobile developmental stages which play no active role in social organisation, while termites are hemimetabolous, with active immature stages which, at least after the earliest period of development, play an increasingly important role as they differentiate into each caste. In both groups the temporal pattern of energy flow in the society will be substantially affected by the cyclical nature of development determined by such things as environmental conditions; for example, seasonal temperature fluctuations in temperate climates and the alternation of wet and dry seasons in the tropics. In other cases endogenous rhythms are responsible, as is so well known in the army ant *Eciton burchelli* Westwood. As a consequence certain members of the society are more or less transitory, but can, nonetheless, contribute significantly to metabolic costs where they constitute a substantial proportion of the biomass.

Eggs in both groups, whether layed in batches or continuously, will make the smallest contribution to metabolic costs.

Larvae, in ants, will clearly begin as a minor component of the society, but with growth their metabolic costs will rise until they form a substantial and more or less persistent part of the total for the colony. At this time, caste is differentiated so that in mature societies production of sexuals as well as replacement workers may result in larval biomass and, by implication, metabolic costs exceeding that of the workers. For example, Brian & Elmes (1974) showed that some *Tetramorium caespitum* L. societies produced sexuals exceeding the weight of the workers by two or three times. The so-called larvae of termites, since they constitute the initial stages of development and caste determination, will be quite numerous but with relatively low metabolic costs.

The pupal stages of ants find no parallel in termites. The preceding remarks on the metabolic costs of larvae in mature societies apply equally to the pupae, except that it is a relatively short-lived stage. As a resting stage, their only contribution to energy flow through the society is by their respiration.

The nature of caste determination in termites is both more complex and more variable than is the case in ants. It is, therefore, not so easy to compartmentalise the costs of maintenance of each stage. Beyond the 'larval' stage the castes become more distinct and in a sense the majority remains immature as pseudergates or workers and soldiers, though they are functionally equivalent to the sterile adult castes of ants. These clearly constitute the major proportion of the society and consequently account for the bulk of the energy cost. The individuals proceeding toward sexual maturity pass through a nymphal stage which forms a distinct component of the society population. As with the sexual larvae and pupae of the ants, the nymphs may form a considerable (though relatively transitory) proportion of the total society so that their metabolic costs cannot be discounted.

The adult castes in mature societies of ants comprise workers, which may be either mono- or polymorphic, soldiers and the sexuals, including the reigning queen or queens. The workers and, to a smaller extent, the soldiers are, of course, continually present in the society and as a general rule must be the most important component of its metabolic costs. Their contribution will vary with environmental conditions and with the metabolically costly activities of foraging and nursing during the peaks of brood development. The sexuals represent the ultimate expression of the brood development already alluded to and will continue to dissipate substantial quantities of energy until they emerge from the nest during the nuptial flight.

While the respiration of the reigning queen in monogynous colonies will be a negligible proportion of the total, the same is not necessarily true for polygynous species. For example, Elmes (1973) found a worker/queen ratio of about 76:1 in an extensive survey of *Myrmica rubra* (L.) colonies, with some in which the ratio fell to a much lower value. Similar ratios have been reported in *Pseudomyrmex venefica* Wheeler (Janzen, (1973*b*), in *Formica yessensis* Forel (Ito, 1973) and *Formica polyctena* Foerst. (Gösswald, 1951*a*), and, no doubt, exist in other species. Brian (1973*a*) has shown that queens of *M. rubra* require about twice the energy of workers to maintain themselves so that they would contribute about 2%, on average, to the colony metabolic costs. With smaller queen/worker ratios, their contribution may rise to 10–12.5% of the total.

As in ants, the culmination of brood development is the metamorphosis of the nymphs into primary reproductives. They, while maturing before the nuptial flight, may continue to contribute not inconsiderably to colony metabolic costs.

Group effects and respiration

While many of the behavioural attributes of *l'effet de groupe* have been described, little is known of the physiological effects on the individual. Hangartner (1969) found that elevated concentrations of carbon dioxide induced digging behaviour in *Solenopsis geminata* (Fabricius) and, by implication, raised the metabolic rate. Chen (1937*a*, *b*) demonstrated that workers which dug most actively and acted as 'leaders' in groups were more susceptible to starvation, drying and asphixiation by chloroform and ether, again suggesting a higher metabolic rate. These effects also apply to the general activities of ants and termites. For example, Grassé & Chauvin (1944) found that ant and termite workers survived longer in groups than singly, even when supplied with abundant food. Maximum longevity was obtained with groups of 10 or greater. *Neivamyrmex carolinensis* (Emery) workers also survive longer in the presence of the queen (Watkins & Rettenmeyer, 1967). These effects also apply to the reproductive castes. Founding queens of *Lasius flavus* (F.) survive better and use a smaller proportion of the reserves when cooperating in groups (Waloff, 1957; Peakin, 1965) and the same is true of *Solenopsis saevissima* (F. Smith) (Wilson, 1966).

Thus there is some circumstantial evidence which suggests that social activation will have an effect on the physiology of the individuals which might be reflected in their respiration. It is perhaps surprising, therefore, to find that when the respiration of worker ants has been compared singly and in groups no significant difference was evident. For example, Brian (1973*a*) found in *Myrmica rubra* that the respiratory rate determined with single workers or groups of 10 was essentially identical. However, it would be unwise to dismiss group effects in the absence of more exhaustive studies of the respiratory physiology of interacting ants. It is even not evident that a group effect can be attained with only 10 individuals in all species of social insects. In some termite species, such as *Trinervitermes geminatus*, an abnormally high mortality cannot be avoided when working with groups of less than 200 individuals (except in society foundations of course).

Intrinsic factors affecting respiration

Day (1938) has already reported daily fluctuations of carbon dioxide concentration in nests of *Eutermes exitiosus*. A first minimum, close to 1% carbon dioxide happened at 9.00 hours and the highest peak of about 1.8% was reached at 16.00 hours; it was followed by a slight fall at midnight (about 1.3%) and a rise at 4.00 hours (about 1.4%). Day could not explain these variations: no agreement was found between the concentration of carbon dioxide and the temperature within the nursery. So he concluded 'it is obvious that the factors influencing the variation in carbon dioxide concentration in the mound are more complex than those which operate in the laboratory colonies'.

Ruelle (1964) also detected variations in carbon dioxide concentration within nests of *Macrotermes bellicosus*: the peak occurred between 15.00 and 18.00 hours and was correlated with the temperature of the external wall of the nest. Hébrant (1970*a*) noticed a rhythmicity in carbon dioxide evolution from nests of *Cubitermes exiguus* kept at a constant temperature. This rhythmicity was obtained only when the termites were allowed to stay

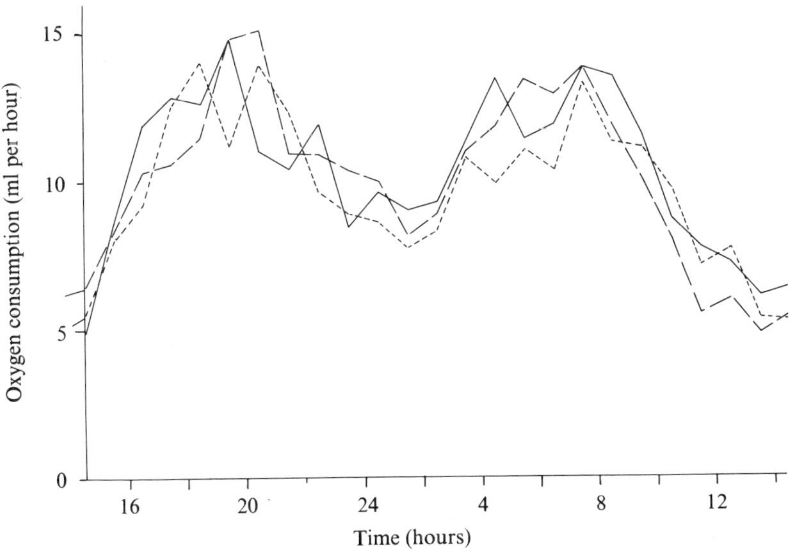

Fig. 6.1. Circadian rhythm of respiration observed in a whole society of *Cubitermes exiguus* within its nest, during a three-day experiment. Redrawn from Hébrant (1970*b*). ——, First day; — — —, second day; - - - -, third day.

within a part of their nest or at least in contact with some material from their nest. Two peaks (8.00 and 20.00 hours) and two falls (2.00 and 14.00 hours) were observed and occurred regularly at the same time day after day (Fig. 6.1).

Fluctuations in carbon dioxide production were also observed in large groups (more than 1000 individuals) of *Trinervitermes geminatus* which were allowed to dig into a sandy soil and were placed in normally fluctuating temperature (Josens, unpublished data). Only one peak of carbon dioxide production was observed daily (between 16.00 and 18.00 hours) but it did not coincide with the daily maximum of temperature (between 22.00 and 02.00 hours).

No similar phenomenon has been described for ants but there seems no reason to suppose that they do not exist since social activities are often marked by cyclical phenomena, as, for example, in foraging behaviour. Saunders (1972) found that peak foraging activity in early June by *Camponotus pennsylvanicus* (De Geer) and *Camponotus noveboracensis* (Fitch) was in the afternoon, when litter temperature was highest, while *Camponotus herculeanus* (L.) peaked in the evening at about 20.00 hours, with a subsidiary peak in the morning, when the temperature of the litter was lower by about 5 °C. In August, *C. pennsylvanicus* foraged most actively during the night with the peak at about midnight. Whitford & Ettershank (personal communication) studied foraging in four sympatric harvester ants, *Pogonomyrmex rugosus* Emery, *Pogonomyrmex desertorum* Wheeler, *Pogonomyrmex californicus* (Buckley) and *Novomessor cockerelli* (André). They found that *N. cockerelli* was primarily nocturnal in midsummer, while the three species of *Pogonomyrmex* were diurnal. It is clear, then, that the relationship between foraging and physical factors is complex and that there is no simple correlation with temperature. It seems not improbable that rhythmic behaviour such as that demonstrated in termites also exists in ants, and certainly the possibility should not be discounted until the matter is researched further.

Influence of abiotic factors on respiration

Temperature

The most important physical factor determining respiratory rate is temperature, as is the case in any poikilotherm. Furthermore, one of the advantages of social life is that it allows a measure of control of micro-climate both by active choice of optimum conditions and by the alteration of the environment implicit in nest building. The importance of this facility is most marked in temperate species where the general environmental temperatures often fall below the optimum level, but is not entirely absent in tropical forms where the principal problem is to avoid excessively high temperatures.

While consideration of temperature conditions is not necessarily the prime determinant of nest architecture it is clear that it often plays an important part. Even where nests are at their simplest, as in a diffuse set of galleries in the soil, the inhabitants gain from the relative stability of the environment and can, of course, still exercise their freedom of choice as to which of the galleries they occupy at any time. As the architecture becomes more elaborate so the effect on the micro-climate becomes greater. Thus those temperate species which show a propensity for nesting under stones are taking advantage of the differential warming of the stone by the sun. Mound-building ants achieve further control of the micro-climate in a number of ways: by, for example, increasing insolation so that the temperature of the mound may be raised several degrees above that of the surrounding soil (Steiner, 1929; Scherba, 1959; Peakin, 1960). Where the mound is constructed from plant debris, then its spongy nature, accentuated by the galleries in it, will improve its thermal properties by virtue of the stagnant air enclosed. The metabolic heat generated by the inhabitants will be retained and have the effect of raising the temperature beyond that achieved purely by insolation. Thus Steiner (1924) in *Formica rufa* L. and Raignier (1948) in *F. polyctena* and Kâto (1939) in *F. yessensis* found that at a depth of 15–30 cm below the summit of the mound the temperature was 23–29 °C day and night. Temperature may also be controlled positively by the opening and closing of apertures which regulate the flow of air through the nest. Metabolic heating is supplemented in the leaf-cutting ants (Attini) by fungus gardens and in *Atta cephalotes* (L.). Stahel & Geijskes (1939) found that the fungus chambers were 15 °C or more above that of the soil. The possibility of the regulation of the temperature by air flow in *Atta sexdens* (L.) has been suggested (Jacoby, 1953, 1955). Metabolic heat is also used by non-nesting species, such as *Eciton hamatum* Fabricius, where the temperature in the centre of a bivouac is not only several degrees above ambient but also more stable (Schneirla, Brown & Brown, 1954).

The restricted distribution of termites to more equable climates has removed the need to acquire heat from the environment but the advantages of stability at an optimum level remain and in a number of species the nest architecture has resulted in increased control of internal conditions, temperature and others besides. Lüscher (1956*b*, 1961*b*) has demonstrated the varying ability of termites, through their nest structure, to regulate temperature. Species which build small, thin-walled nests (for example, *Amitermes evuncifer* Silvestri) possess virtually no thermo-regulating properties, whereas species with large, thicker-walled nests show much improved thermo-regulation (for example, *Cephalotermes rectangularis* (Sjöstedt)). Finally, nest homeostasis reaches its peak in the very large mounds of *Macrotermes natalensis* (Haviland), where not only is temperature controlled within fine limits in the centre of the colony amongst the fungus gardens but its variation elsewhere in the nest also produces the convection currents so important in the air conditioning of the nest.

Whatever the degree of homeostasis achieved through elaboration of the nest, the insects within still possess the ability to choose which parts they inhabit. Holdaway & Gay (1948) reported that the temperature inside mounds of *Nasutitermes exitiosus* (Hill) fluctuated in relation to the surrounding air temperature. That of occupied nests was 8 to 10 °C higher than that of unoccupied nests, and the presence of alates resulted in another increase of 5.5 to 7 °C. When they were faced with weather changes, the 'termites, probably as a result of movement into the mound from galleries away from the mound, were capable of buffering the effects of sudden falls in air temperature'. In *Trinervitermes geminatus* nests, the outer crust reaches temperatures as high as 50–60 °C and decreases to 20 °C (or less) at the end of the night; in the hypogeic part of the mound, however, the temperature is damped between 26 and 28 °C, in the dry season as well as the wet season (Fig. 6.2) (Josens, 1971*a*, *b*). The brood and all the moulting individuals of the society stay in this deep part of the nest (Josens, 1972), whereas active workers, soldiers and some young move up and down daily; these young stages, however, 'are more strongly

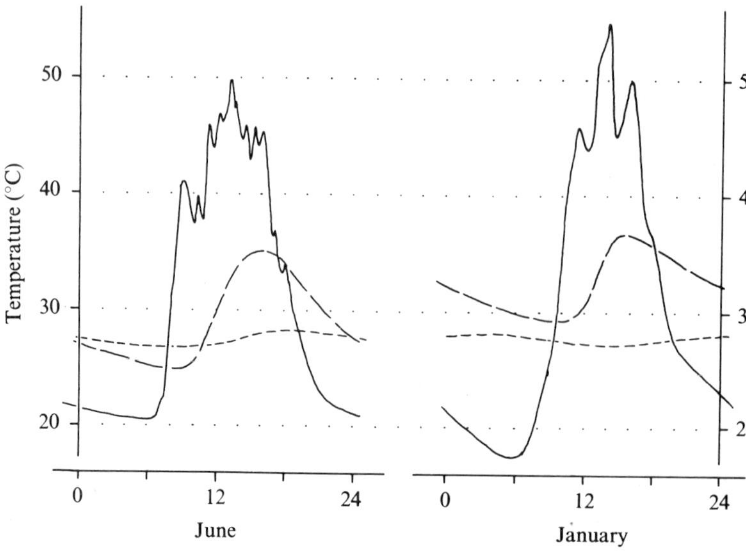

Fig. 6.2. Variations of temperature in a nest of *Trinervitermes geminatus* in June (rainy season) and in January (dry season), at the IPB Lamto Project (Ivory Coast). Recordings made at the top of the epigeic part of the nest, in the centre of the nest (at soil level) and in a subterranean chamber 40 cm deep in the soil. Redrawn from Hébrant (1970*a*). ——, Top of the nest; — — —, centre of the nest; - - - -, subterranean chamber.

affected by the stimuli causing the population movements than the mature workers and soldiers, and either move rapidly or are carried in response to these' (Sands, 1965*a*). In termites, virtually all members of the society (save eggs and newly hatched larvae) possess the mobility necessary to put their choice into effect. In ants, of course, the developmental stages are incapable of independent locomotion and depend on the workers choosing conditions appropriate to their needs. That they do this is generally supposed from the segregation of pupal and the various larval stages into individual chambers in the nest, where, because of the thermal stratification, differing temperature regimes will be experienced. In *Megaponera foetens* societies, living in or at the border of African tropical forests, pupae are brought out of the nest and exposed to the sunshine (Lévieux, 1971*a*). Such segregation presupposes the ability of workers to respond to temperature and exhibit a preference. This has been demonstrated in *Aphaenogaster fulva* Roger, *Camponotus pennsylvanicus* and *Lasius latipes* Walsh (Fielde, 1904), *Formica rufa* (Herter, 1924, 1925; Gösswald & Bier, 1954) and *Myrmica rubra* (Brian, 1973*a*).

Other abiotic factors

There are two other characteristics of the abiotic environment which have an important bearing on the respiratory physiology of ants and termites. The first, the concentration of carbon dioxide in the atmosphere has an almost unknown effect, but, bearing in mind the remarks made earlier concerning the inherent artificiality of conditions in respirometers, deserves examination. The second, the humidity or, perhaps better, the saturation deficit of the atmosphere, is known to affect the respiration rate of ants (Ettershank & Whitford, 1973; Kay & Whitford, 1975) and termites (Hébrant, 1965, 1970*b*).

The concentration of carbon dioxide in the atmosphere of the nests of ants and termites will be determined by the metabolic activity of the colony and the rate at which carbon dioxide can diffuse into the general atmosphere. Probably in most cases where the nest is of moderate size the level, while above that of the air, will not become excessively high (Table 6.1). In the largest nests it might become much more concentrated and it has been shown in termites (Lüscher, 1956*b*, 1961*b*) that the nest architecture has been substantially determined by the need to increase the rate of loss of excess carbon dioxide. Lüscher (1955*b*) measured the oxygen consumption in *Calotermes flavicollis* (Fabr.) and *Zootermopsis nevadensis* (Hagen) and obtained a figure close to 500 μl per mg per hour under 'optimal conditions'. Assuming the same respiration rate for *Macrotermes bellicosus*, a large nest of two million individuals, weighing on an average 10 mg each, must need about 240 l of oxygen – or about 1200 l of air – per day. So he concluded an active mechanism was necessary to provide such an exchange of gas. His measurements of carbon dioxide concentrations (see Table 6.1) and temperatures in *Macrotermes* nests led him to present his well-known hypothesis of air circulation based on convection currents.

This hypothesis has been criticised by Grassé & Noirot (1958*b*) who maintained that the air inside the nest was quiet and that the gas exchanges occurred by diffusion only. Ruelle (1962) collected some data unfavourable to Lüscher's hypothesis and assumptions: first of all, his measurements of oxygen consumption are far lower: 0.289, 0.276 and 0.169 μl per mg per hour in large workers, small and large soldiers. Secondly, his moulding of whole *Macrotermes* nest did not agree very well with Lüscher's diagram: it appeared more complicated, with numerous communications at different levels between the nest and the narrow channels in the external ridges; this seems unfavourable to the idea of ventilation by a thermosiphon. Moreover, the external walls may be warmer than the nest during the afternoons (Ruelle, 1964). However, by means of a sensitive anemometer designed by Loos (1964), it was possible to detect within a nest

Table 6.1. *Carbon dioxide and oxygen concentrations within the nests of some termite species*

Species	Region of nest	Carbon dioxide concentration (%)	Oxygen concentration (%)	Author
Amitermes atlanticus	Epigeic part	4–5; up to 15 before swarming	—	Skaife, 1955
Cubitermes intercalatus	15.5 cm above ground, 15.2 cm down inside base of regenerated mound	0.052	—	Emerson, 1956
	3.5 cm above ground, 15.24 cm down inside base of regenerated mound	0.162	—	
	15 cm above ground, 7.5 cm down in normal nest	0.920	—	
Cubitermes exiguus	10–12 cm deep in unshaded nest	0.6–1.2[a]	—	Hébrant, 1970*a*
	4 cm down in unshaded nest	Less than[a] 0.4	—	
	11 cm deep in shaded nest	0.4–0.8[a]	—	
	4 cm down in shaded nest	About 0.2[a]	—	
Macrotermes bellicosus	Central part with fungus combs and brood	2.6–2.7	18.6–18.2[a]	Lüscher, 1955*b*, 1956*b*, 1961*b*
	'Attic'	2.9	18.7[a]	
	Lower part of ridge channel	0.8	20.6[a]	
	'Cellar'	1.3	19.6[a]	
Macrotermes subhyalinus rex	Central part	Up to 2.8	Down to 18.2	Grassé & Noirot, 1958*b*
Macrotermes bellicosus	Central part	Up to 1.4[b]	—	
Macrotermes bellicosus	Central part Normal nest	Peaks of[c] about 2.0, up to 2.9	—	Ruelle, 1962, 1964
	Central part, nest covered with a plastic sheet	3.5–4.0	—	
Nasutitermes exitiosus	Central part, about 38 cm down in nest	1.00–1.89	19.91–19.10	Day, 1938

[a] Deduced from author's graphs.
[b] 0.6% hydrogen was also detected.
[c] Lower rates in dry season than in wet season (these data were collected during the wet season).

of *Macrotermes bellicosus* irregular air movements which were related to external wind and the porosity of the external walls. When the outside wind was negligible, Loos observed a steady air flow inside the nest: its speed and direction were variably and the 'preliminary results seemed to indicate that the steady flow would be directed upwards above the nest, while in the peripheral galleries leading to the cave a downward flow was found in the morning'. A speed of air flow of 12 cm per min was reported. Nevertheless, 'the results seem to show that the thermosiphon would not work in the afternoon'. Finally this agrees very well with the observations of the carbon dioxide concentration in the nest, which reaches a peak of about 2% each afternoon (Ruelle, 1964).

Little is known about the effects of variable gas concentration on ants and termites. Low tension of oxygen and high carbon dioxide concentration probably cause no respiratory distress, at least in lower termites. Cook (1932) measured the oxygen consumption of *Zootermopsis nevadensis* placed in different concentrations of oxygen: 100, 21, 10, 5, 2 and 0.8%. The respiration in each concentration was compared with, and expressed as, the percentage of respiration in normal air (21% oxygen). It appeared that pure oxygen was slightly inhibitory since the termites consumed 93% of their normal oxygen consumption. Cook suggested that this decrease might have been caused by the inhibition of the termites' gut micro-organisms. In oxygen concentrations lower than normal, the respiration of the termites was depressed but not proportionally: in 5% oxygen, the respiration was still 84.5% of the normal value; in 0.8% oxygen, the termites became motionless and their respiration fell to 28% of normal. Subsequently, Cook tested the ability of *Zootermopsis nevadensis* to tolerate high concentrations of carbon dioxide: concentrations of 5, 10, 20, and 40% carbon dioxide (with 20% oxygen) were used. Up to 20% carbon dioxide, 'the net gas exchange was approximately the same as that of termites in air and the appearance of the termites otherwise was perfectly normal'. But at 40% carbon dioxide 'the termites soon became immobile'.

Williams (1934) also tested the sensitivity of two termite species to carbon dioxide: *Reticulitermes tibialis* was inactivated after 50 to 60 hours in 30% carbon dioxide whereas *Reticulitermes hesperus* Banks was inactivated after the same period in 40% carbon dioxide; the latter lives in wetter soils, with worse gas exchange, and therefore its lower sensitivity to carbon dioxide was expected.

More recently, Lüscher (1955*b*) confirmed that lower termites were relatively independent of abnormal oxygen and carbon dioxide concentrations. Some *Calotermes flavicollis* were placed in two closed vessels; after eight days, the oxygen concentrations were reduced to 2.7 and 6.2% and the carbon dioxide concentrations had risen to 18 and 15.4%. The oxygen consumption under these conditions was, however, lower (0.403 and 0.392 μl per mg per hour at 26 °C) than in air (0.540 μ per mg per hour), 74.6 and 72.6%, respectively, of normal respiration.

Much less experimental data are available for higher termites. Wiegert & Coleman's (1970) observations strongly suggest that the size of the nest of *Nasutitermes costalis* (Holmgren) is related to its ability for gas exchange (and not to the number of termites living in it). A relation was obtained:

$$y = 0.24x^{0.80} \qquad (6.1)$$

where y is the calculated consumption of oxygen (ml per hour) of the whole society and x is a volume index of the nest. This relation shows that the 'oxygen consumption is somewhat more closely related to changes in surface area than to changes in weight'.

In the big nests of *Macrotermes*, a concentration of carbon dioxide higher than 2.9% has never been recorded (see Table 6.1). During his investigations on *Macrotermes bellicosus* Ruelle (1962, 1964) covered a whole nest with a plastic sheet; soon the carbon dioxide concentration within the nest rose from 2.0–2.5 to 3.5–4% and new porous parts were built by the termites, suggesting a high sensitivity to this change in their atmospheric composition.

In nests of *Trinervitermes geminatus*, the number of individuals is best correlated with the area of the outer walls of the nest (Josens, 1972). If such a nest is covered by a plastic jar and shaded in order to avoid the greenhouse effect, the termites will soon build new parts with porous walls and even leave some openings (Josens, unpublished data). Again these observations agree with a relatively high sensitivity of the termites to the composition of their atmosphere. Thus there seems to be a contrast in the tolerance to high carbon dioxide concentrations (or low oxygen concentrations, or both) between the lower and higher termintes. Alternatively, the contrast is between epigeic nest-building termites and others.

In his very detailed work on *Cubitermes exiguus* respiration, Hébrant (1970*a*, *b*) tried to measure the influence of carbon dioxide at concentrations of 0.5, 1, 1.5 and 2%, by means of Pardee's solutions. His measurements suggested a smaller consumption of oxygen when some carbon dioxide was present but he was not very confident of his results because of the possible errors involved in this method.

Although it has been shown in Ettershank & Whitford's (1973) study of desert ants that oxygen consumption does vary with vapour pressure deficit it should be borne in mind that the climatic conditions experienced by these ants are extreme. It is probable that in most cases such conditions are not encountered by foragers, particularly in temperate zone species where soil moisture will keep the atmospheric humidity adjacent to the soil high. Within the nest humidity conditions are more equable and, indeed, have to be so because of the sensitivity of the brood to desiccation. The spiracular anatomy of larval ants precludes control of evaporation, so that water vapour loss in less than saturated air is considerable (Peakin, 1960). The propensity of termites to forage underground or in covered galleries probably also reflects their inability to control evaporative loss.

It is clear from Hébrant's experiments (1965, 1970*a*, *b*) that the respiration of workers of *Cubitermes exiguus* depends slightly on the air humidity. Though a decrease in oxygen consumption was noted only when the relative humidity (RH) was lowered to about 15%, the termites were obviously influenced by the water pressure deficit. The amount of water they lost during the measurements (lasting five hours) varied from 7% at 25 °C and 92% RH to 22% at 35 °C and 15% RH.

Nest architecture again often plays a part in determining humidity levels and the behaviour of the inhabitants also ensures that even in the most extreme conditions the inhabitants are always in contact with virtually saturated air. This has been shown to be true of *Lasius flavus* (Peakin, 1960), and Stahel & Giejskes (1940) found that the RH in the nest of *Atta cephalotes* was always above 90%. This species extends its large and complex nests to a depth of 3 m and may even go down as far as 6 m. The propensity of ants living in dry environments to extend their nest chambers to considerable depth clearly reflects this search for equable humidity conditions. Thus *Pogonomyrmex badius* Latreille may burrow 2 m down (Wray, 1938), *Veromessor pergandei* (Mayr) 3 m (Tevis, 1958) and *Myrmecocystus hortideorum* McCook as much as 5 m (Wheeler, 1910).

Termites, similarly, take steps to ensure that the colony experiences high humidity at all times. Thus *Cubitermes* vacates the above ground part of the nest during the dry season in West Africa (Grassé & Noirot, 1948). *Ancistrotermes cavithorax* (Sjöstedt), which lives in scattered subterranean nests, also tends to desert the upper layers of the soil during the dry season (Josens, 1972). *Macrotermes bellicosus* maintains a high humidity in the central part of the nest by the water content of their fungus gardens or by secreting watery saliva. According to Bodot (1966), these termites fetch water from the water table, the presence of which (in the southern Ivory Coast) is obligatory for the survival of these termites. *Psammotermes* is said to burrow to the water table, 50 m deep, in order to bring up moisture to humidify the nest (Elliot, 1904). Fyfe & Gay (1938) suggest that in *Nasutitermes exitiosus* the structure of the nest and the chemical composition of the different parts of the wall ensure that the humidity remains high in the nursery (96–98%) and that the minimum of water is lost by evaporation.

Modes of expression of respiratory rates

Keister & Buck (1964) discussed the bases of expressing metabolic rate and concluded that as good a way as any is per unit whole body live weight per unit time. For the purposes of studying energy flow through populations this has a number of drawbacks. Energy accumulation or production is, of course, determined by measuring the energy content of the increase in biomass in unit time. As this is done by bomb calorimetry, dry material is used, so it is sensible to express the energy per unit dry weight of tissue. In any case, as water is one of the products of combustion, its existence in fresh material is of no consequence in the determination of the potential energy of the tissues. It would seem that the estimated cost of metabolising the tissue (i.e. the respiratory costs) would also be best expressed per unit dry weight of tissue per unit time. This has the advantage of eliminating the effect of considerable variations in water content observed during the development of both ants and termites (Josens, 1972; Peakin, 1972). Strictly, the dry weight should be exclusive of ash content, but in most cases a simple measure of dry body weight is a more than adequate approximation. However, in the case of humivorous termites the soil in the gut may form a considerable proportion of the fresh weight, and, of course, an even larger proportion of the dry weight and cannot be ignored.

Whatever the form in which respiratory rates are expressed they have to be converted to units of energy and extrapolated to field conditions. Furthermore, it is conventional, where respiratory data are incomplete, to establish some general relationships to aid extrapolation. For example, the RQ is used as a basis for determining the energy equivalent of respiratory rate by making assumptions concerning the respiratory substrate. The limitations of these assumptions have been alluded to above.

The relationship between respiratory rate and body weight is another popular predictive tool. Keister & Buck (1964) have drawn our attention to its limitations, since not only is there variation between species but also variation during the life cycle of insects, as, for example, during metamorphosis, when there may be a ten-fold variation in respiration but, of course, very little change in weight. The usual form of the relationship is given by the equation

$$y = ax^b \tag{6.2}$$

where y = oxygen consumption, x = weight and a and b are constants supposedly characteristic of the species. Schmidt (1968) found that the value of b for the pupae of the three castes of *Formica polyctena* apparently varied both with temperature and the stage of development (Fig. 6.3 and Table 6.2), just as might be expected from the remarks made above. The differences are, however, not significant with respect either to temperature or to stage of metamorphosis and the mean value for b is 1.1. In the adults of the same species Kneitz (1967) found that for the female castes, workers, young and old queens, the exponent had a value of 1.0. The males did not conform to this relationship (Fig. 6.4). Both Schmidt and Kneitz are dealing with a special case, in so far as the polymorphism of the female caste gives the characteristic dispersion of the points plotted, with obvious effects on the determination of the regression line connecting them.

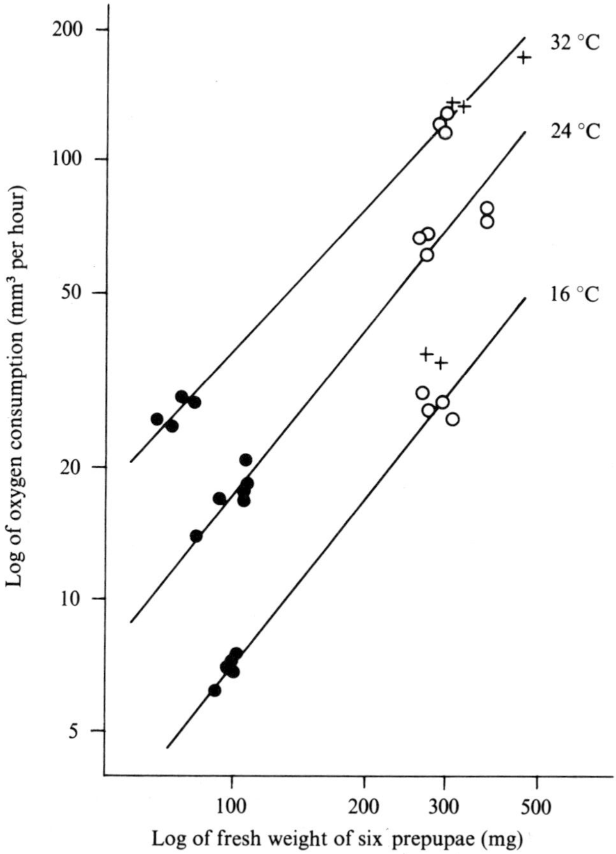

Fig. 6.3. The relation between oxygen consumption and body weight in prepupae, without coccoons, of male, queen and worker *Formica polyctena* plotted using a log scale. (After Schmidt, 1968.) Solid circle, worker prepupae; open circle, queen prepupae; '+', male prepupae.

Table 6.2. *Value of b in* $y = ax^b$, *where* y = *oxygen consumption* (mm^3) *and* x = *fresh weight* (*mg*), *for pupae of* Formica polyctena (*after Schmidt, 1968*)

Developmental stage	Value of b at 16 °C	24 °C	32 °C
Prepupae, without coccoon	1.28	1.28	1.11
Pupation	1.43	1.33	1.15
Beginning of eye pigmentation	1.11	1.11	1.07
Well-pigmented eyes	0.97	1.04	1.00
Beginning of body pigmentation	0.90	1.07	1.11
Deeply pigmented body	1.04	1.11	1.15
Adult emergence	0.97	0.97	1.11

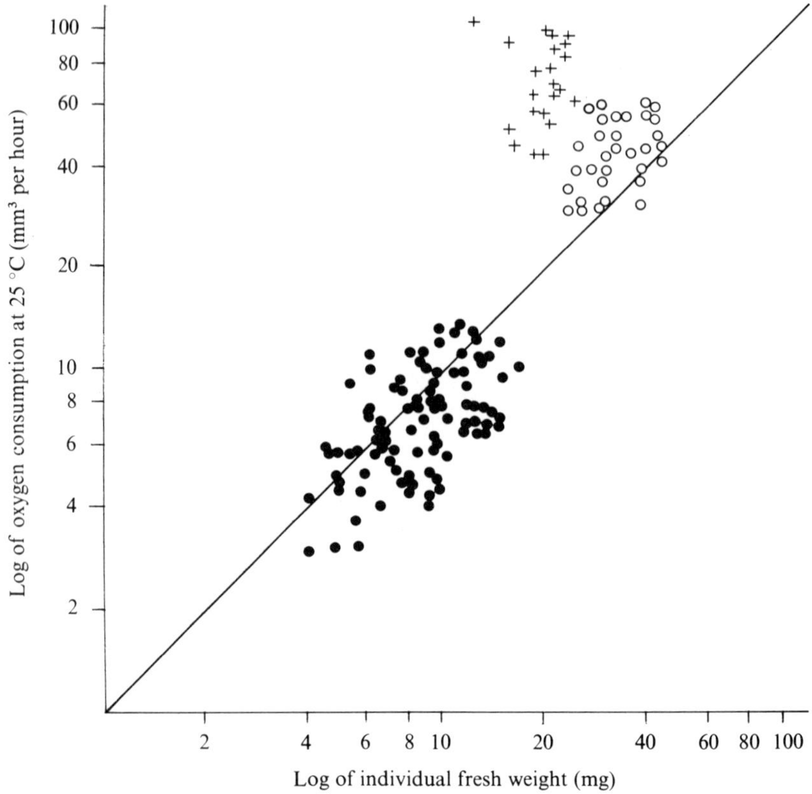

Fig. 6.4. The relation between oxygen consumption and body weight in the adult castes of *Formica polyctena* plotted using a log scale. (After Kneitz, 1967.) Solid circle, workers; open circle, queens; '+', males.

More interestingly, interspecific relationships can be established from two sets of recent data for a number of species of myrmicine and formicine ants. Jensen & Nielsen (1975) determined respiratory rates in nine species of temperate ants, two myrmicine and seven formicine, while Ettershank & Whitford (1973) and Kay & Whitford (1975) have done the same for seven species of desert ant, one formicine and six myrmicine. Their results are drawn together in Table 6.3, and the linear nature of the relationship is clear from Fig. 6.5 where the values have been plotted on a double logarithmic scale. On a weight for weight basis the formicine species respire at a significantly higher rate than the myrmicines. In both groups the relationship between body weight and respiratory rate is significant (for the formicine species, $r = 0.952$, $p < 0.001$ and for the myrmicines $r = 0.901$, $p < 0.01$). The relationship in the formicines is best described by the equation

Table 6.3. *Oxygen consumption and dry body weight at 25 °C in the workers of 16 species of ant from two sub-families*

Species	Code no. (see Fig. 6.5)	Dry weight (mg)	Oxygen consumption (mm^3 oxygen per hour)	Authors
Formicinae				
Formica perpilosa	1	1.01	2.02[a]	Kay & Whitford, 1975
Formica exsecta	2	1.35	3.89	Jensen & Nielsen, 1975
Formica pratensis	3	2.39	5.57	Jensen & Nielsen, 1975
Formica fusca	4	1.47	1.71	Jensen & Nielsen, 1975
Lasius alienus	5	0.46	1.07	Jensen & Nielsen, 1975
Lasius niger	6	0.58	1.08	Jensen & Nielsen, 1975
Lasius flavus	7	0.86	2.04	Jensen & Nielsen, 1975
Camponotus herculeanus	8	6.71	13.56	Jensen & Nielsen, 1975
Myrmicinae				
Myrmica rubra	9	0.92	1.21	Jensen & Nielsen, 1975
Novomessor cockerelli	10	3.91	3.09[a]	Kay & Whitford, 1975
Tetramorium caespitum	11	0.23	0.43	Jensen & Nielsen, 1975
Pogonomyrmex rugosus	12	4.95	3.04[b]	Ettershank & Whitford, 1973
Pogonomyrmex maricopa	13	3.69	4.98[b]	Ettershank & Whitford, 1973
Pogonomyrmex californicus	14	1.39	2.38[a]	Kay & Whitford, 1975
Pogonomyrmex desertorum	15	2.21	1.75[a]	Kay & Whitford, 1975
Trachymyrmex smithi	16	1.07	0.68[a]	Kay & Whitford, 1975

[a] Value for the lowest vapour pressure deficit in Fig. 1, Kay & Whitford (1975).
[b] Value for the lowest vapour pressure deficit in Figs. 1 and 2, Ettershank & Whitford (1973).

$$\text{Log } y = 0.9783 \log W + 0.3162 \tag{6.3}$$

where y = oxygen consumption in mm^3 per hour and W = body dry weight (mg). For the myrmicine species the equation is

$$\text{Log } y = 0.7403 \log W + 0.0729 \tag{6.4}$$

Interesting though these results are, the disparate environment from which the ants have been drawn may be a more important factor than interfamilial differences. Seven of the eight formicine ants were from a temperate climate and six of the eight myrmicine ants from a desert habitat.

Information available on termites is not as abundant. Using respiration data from all castes, Wiegert & Coleman (1970) reported a correlation for *Nasutitermes costalis*:

$$y = 1.07x^{0.73} \tag{6.5}$$

where y is the oxygen consumption in μl per individual per hour and x is the body dry weight (mg).

In his work on *Cubitermes exiguus* and *Cubitermes sankurensis* Wasmann, Hébrant (1970*a*) compared the respiration of groups of workers of

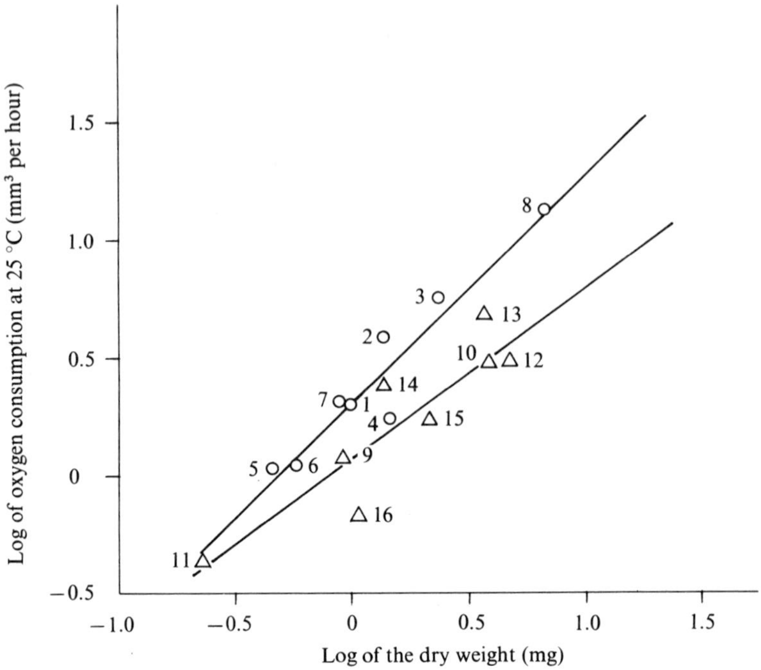

Fig. 6.5. The relation between oxygen consumption and body weight in the workers of two sub-families of ants plotted using a log scale. Open circles, formicine ants; open triangles, myrmicine ants. See Table 6.3 for the code number of each species.

different average weights and obtained contradictory results. In a first series of 59 groups of *C. exiguus* workers, the relation of respiration to body weight fitted the surface law. In a second series of measurements of 36 groups of *C. sankurensis* workers, this relation fitted a linear regression instead of the surface law, and finally, a third series of 88 groups of *C. sankurensis* did not show any significant correlation! In the case of humivores, such contradictions may be expected because of the inert gut content which may fluctuate largely in quantity.

Internal factors, such as the society the workers come from, genetical and historical factors, may also influence the respiration rate; moreover, quite unexpected external factors may be important (see the subsequent discussion about the rate of increase of respiration with temperature).

It has been suggested that the rate of increase of respiration with temperature (Q_{10}) is sufficiently predictable to be used in determining respiratory rates at all temperatures for observations at just a few. While this may be true for simple reactions, it cannot be extended to whole organisms with impunity as Q_{10} tends to decrease with increasing tem-

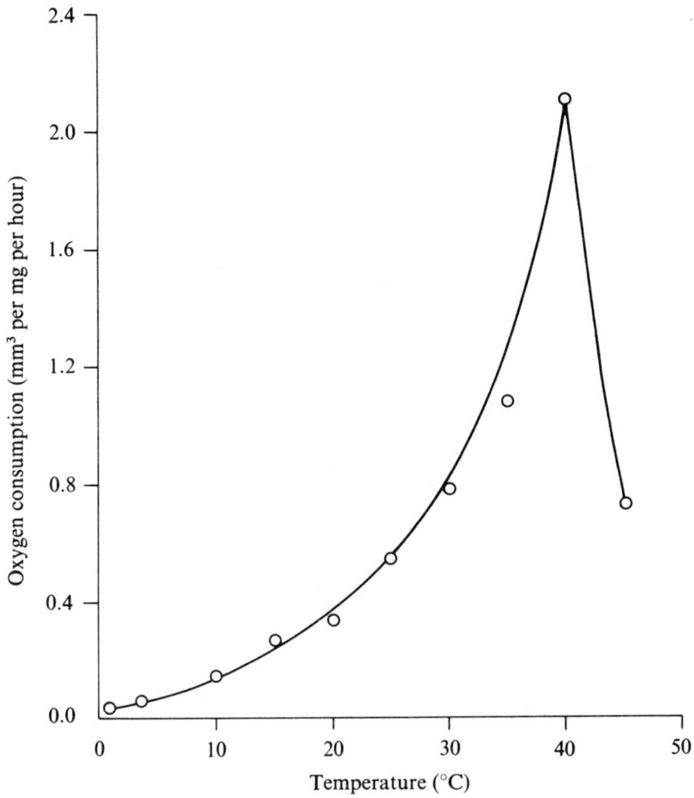

Fig. 6.6. Oxygen consumption in relation to temperature in the overwintering workers of *Formica polyctena*. (After Kneitz, 1967.)

perature and, unlike the value of two for simple reactions, can be as high as 50 in some insects (Keister & Buck, 1964). Where it has been determined in ants, inter- and intraspecific variability has been confirmed. For instance, in *Lasius alienus* (Foerst.) Nielsen (1972*a*) found it to be constant at 1.62 for the temperature interval 5–30 °C but inspection of Kneitz's (1967) curve relating oxygen consumption to temperature (Fig. 6.6) suggests a Q_{10} of 2.50–2.60 for workers of *Formica polyctena* in winter. We are indebted to Schmidt (1968) for his detailed study of the respiratory physiology of the pupae of *F. polyctena*. In it he determined the Q_{10} at six stages of pupation in the three castes at 2 °C intervals from 16–32 °C and observed that the Q_{10} varied from 0.51 to 7.10. In spite of this there is generally a high degree of correlation with temperature. Males and queens at pupation and workers in the last two stages of metamorphosis were the only stages where there was no significant correlation. At all other stages there was a decrease in Q_{10} with increase in temperature, ranging

Table 6.4. *The rate of increase of respiration with temperature (Q_{10}) of five desert ants at various vapour pressure deficits in the temperature range 15–35 °C. (After Kay & Whitford, 1975)*

Temperature interval (°C)	Vapour pressure deficit[b]	Q_{10}				
		Pogonomyrmex californicus	*Pogonomyrmex desertorum*	*Novomessor cockerelli*	*Trachymyrmex smithi*	*Formica perpilosa*
15–25	12.8–23.7	4.2	2.3	1.9	1.4[a]	2.0
	4.5–7.1	5.8	2.1	1.8	1.5	1.7
	2.3–2.4	3.8	2.5	1.9	1.8	2.4
25–35	23.7–42.2	1.7	2.4[a]	3.3	3.7[a]	1.9
	7.1–14.8	1.5	2.3	2.7	2.1	2.1
	2.4–6.3	1.8	2.2	2.1	2.1	1.7

[a] Doubtful values from chambers containing weak ants.
[b] In mm Hg.

from 3.83 in the temperature interval 16–18 °C to 1.45 in the interval 30–32 °C. There were no significant differences amongst the castes.

Very few data are available for termites. The respiration of *Cubitermes exiguus* was tested at 15, 20, 25, 30, and 35 °C by Hébrant (1970*a*); at 40 °C the mortality was too high and the results were discarded. The Q_{10} decreased with temperature from 2.98 to 2.05 and the oxygen uptake (y) (in μl per individual per hour) satisfied the relation

$$\log y = 0.553 + 0.0393T \tag{6.6}$$

with T in °C. In *C. sankurensis*, tested at 20, 25, 30 and 35 °C, the Q_{10} fluctuated between 2.18 and 2.43, and the relation was:

$$\log y = -0.321 + 0.036T \tag{6.7}$$

where the oxygen consumption (y) is expressed in μl per 10 individuals per hour and T in °C.

No other actual determination of Q_{10} has been made for termites. Hébrant (1970*a*) calculated some Q_{10} values for *Zootermopsis angusticollis* (Hagen), which he deduced from Cook & Smith's (1942) graphs. Their measurements had been made at 4, 8, 12, 16, 20, 24, 28 and 32 °C. With termites previously reared at 29 °C, the Q_{10} was 2.52; at 19 °C it was 2.44; at 9 °C, 2.26; and at 4 °C, 2.97. This shows the possible influence of conditions prior to laboratory measurements.

Table 6.4 summarises the findings of Kay & Whitford (1975) and it will be seen that the Q_{10} is sometimes unaffected by temperature, or can rise or can fall in an entirely unpredictable fashion. Table 6.5, based on Jensen & Nielsen's (1975) investigations of respiration in temperate ants presents

Table 6.5. *The rate of increase in respiration with temperature (Q_{10}) of five temperate ants in the temperature range 5–30 °C. (After Jensen & Nielsen, 1975)*

Temperature interval (°C)	Q_{10}				
	Myrmica rubra	*Tetramorium caespitum*	*Lasius niger*	*Formica pratensis*	*Formica fusca*
5–10	—	—	—	2.21	—
10–15	4.45	7.53	4.20	6.88	1.65
15–20	—	—	—	2.56	—
20–25	1.47	1.65	2.20	3.30	3.55
25–30	—	—	—	3.57	—

much the same confused picture. It should be remembered that the respiration rate may vary, following a circadian rhythm, even at constant temperature. As a result the oxygen uptake may be more than twice as high at a given moment than some hours later (or sooner) (Fig. 6.1). Thus Hébrant (1970*b*) estimated the Q_{10} of single, whole nests placed successively at 20, 25, 30 and 35 °C for at least 24 hours. Using *Cubitermes exiguus* he carried out two surveys during the dry season and obtained Q_{20} values of 1.61 and 1.99. Hébrant thought that the first imposed temperature might influence the results at subsequent temperatures and that the climatic conditions at the time the nests were collected might be important too, not to mention the physiological condition of the societies (starvation, water balance, etc.) which must have changed over the four days of measurements. Nevertheless, such experiments should be carried out again with other species, for they will probably provide much more realistic figures than those obtained on isolated individuals in a respirometer, however sophisticated the apparatus.

Besides the possible error introduced in Q_{10} determinations by internal factors, such as circadian rhythms, external factors may play an unexpected part; thus Damaschke & Becker (1964) pointed out the influence of electromagnetic phenomena (see Fig. 6.7).

Thus it may be argued that prediction of respiratory rates from a few measurements is fraught with difficulties and if such predictions are used then they will provide only a rough approximation of the true costs of maintenance.

As to the conditions obtaining in the field, very little is known. In this respect the conditions in the tropics probably make extrapolation from laboratory measurements easier. Seasonal environmental temperature fluctuations are smaller than in temperate regions, so that even if the nest

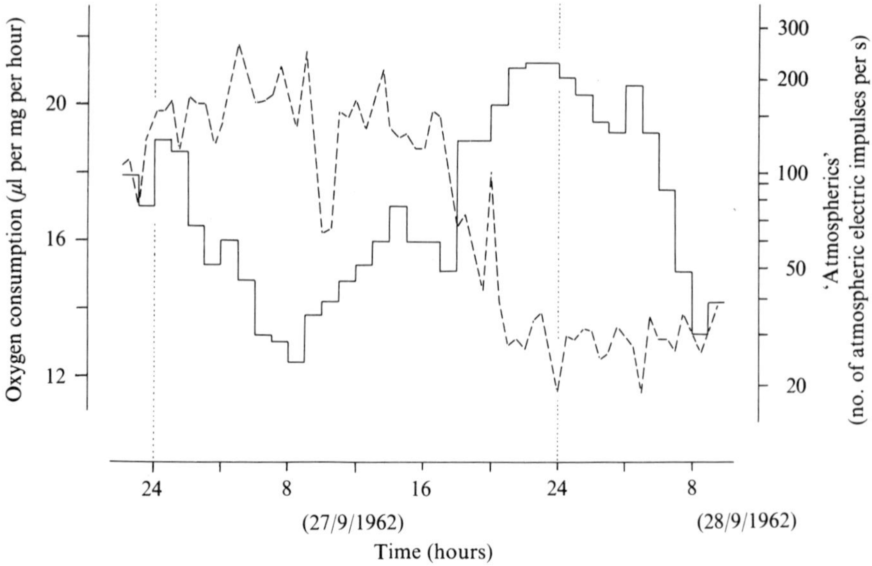

Fig. 6.7. Simultaneous variations of the respiration of *Zootermopsis nevadensis* workers and of the 'atmospherics', or number of atmospheric electric impulses per s. Redrawn from Damaschke & Becker (1964).

architecture is such that it provides little insulation the temperature fluctuations experienced may still only be 4–5 °C, as was shown for termites by Lüscher (1961*b*). In hypogeic species the soil itself will damp the temperature fluctuations as already mentioned. The homeostatic properties of the bivouac in army ants have already been referred to and are sufficiently effective to reduce the fluctuations experienced by the brood to about half those in the general environment (Schneirla *et al.*, 1945).

In temperate climates the problem facing the ants is to maximise the time during which the insects may enjoy optimum conditions. We have already remarked on the ways in which location of the nest and its architecture is designed to achieve this end.

Only two attempts have been made to estimate annual respiratory costs for ants (Golley & Gentry, 1964; Nielsen, 1972*a*). Neither took account of the behavioural adaptations of the ants. Instead, Golley & Gentry (1964) divided the respiratory costs of *Pogonomyrmex badius* into two parts, namely costs above ground for foraging ants when they were active and costs below ground for the majority of ants, i.e. the brood and the foragers, during the time of the day when they were confined to the nest. Temperature above ground was measured in February, May, July and October and it was assumed that these conditions were the same for the

Table 6.6. *Estimate of the monthly cost of maintenance of the workers of a colony of* Lasius alienus. (*After Nielsen, 1972a*)

Month	Average soil temperature (°C)	Population of workers (thousands)	Oxygen consumption (l)	Energy equivalent (kJ)
January	1	11.2	1.75	35.15
February	0	10.8	1.45	29.16
March	1	10.4	1.62	32.64
April	6	10.0	1.94	39.04
May	12	9.6	2.50	50.21
June	18	13.0	4.87	97.74
July	23	14.5	6.34	127.28
August	19	13.5	5.32	106.90
September	16	12.8	3.96	79.58
October	9	12.2	2.90	58.32
November	4	11.8	2.12	42.68
December	4	11.4	2.12	42.59

90-day period about these dates. Below-ground temperature was measured in summer and winter and interpolation between the two figures was used for estimating respiratory rates for the rest of the year. The respiratory rates determined over a range of temperatures in the laboratory were applied to the values found in the field. The wide range in their estimates of costs of maintenance, 59.8 kJ per m^2 to 199.6 kJ per m^2, reflect these approximations.

Nielsen (1972*a*) took as the basis of his estimate of respiratory costs the assumption that the average soil temperatures at a depth of 2.5 cm were those experienced by the ants (*Lasius alienus*). By multiplying oxygen consumption at these temperatures by the numbers of workers present each month throughout the year he arrived at a figure for the annual respiration of the workers (Table 6.6). Using an oxycalorific equivalent of 4.8, as Golley & Gentry (1964) did, the total costs of maintenance for workers were 741.4 kJ. The contribution of the larvae to the total metabolism of the colony was calculated by multiplying the mean larval weight by the number of workers produced in the year and the mean larval developmental period of four weeks at 25 °C. The figure of 152.6 kJ was arrived at in this way.

Table 6.7. *Respiration and equivalent energy expenditure in pupae during metamorphosis in* Formica polyctena. (*After Schmidt, 1968*)

Temperature (°C)	Duration of metamorphosis (hours)	Fresh weight (mg)	Carbon dioxide evolved (mm³)	Oxygen absorbed (mm³)	RQ	Energy expended (J)
			Males			
20.5	934	42.7	5943	7561	0.786	151.33
24.5	554	42.5	5314	6582	0.807	136.78
28.0	396	46.7	5500	6582	0.836	133.38
32.0	360	44.8	6359	7717	0.824	155.13
			Queens			
20.5	934	41.7	5068	6435	0.788	128.84
24.5	554	43.5	4760	6106	0.780	122.10
28.0	396	41.6	4523	5696	0.794	114.19
32.0	360	43.3	5960	7085	0.841	143.95
			Workers			
20.5	934	11.8	1453	1714	0.848	34.84
24.5	554	14.4	1628	1860	0.875	38.08
28.0	396	11.6	1225	1299	0.943	27.03
32.0	360	11.6	1641	1740	0.943	36.20

Laboratory measurements of respiration: ants

Developmental stages

There are apparently no estimates of the respiration of ant eggs.

Nielsen (1972*a*) provides the only determination of larval respiration to date. He found that the worker larvae of *Lasius alienus* consumed 2.8 mm³ O_2 per mg dry weight per hour. For an oxycalorific equivalent of 4.8, this gives 56.23 mJ per mg dry weight per hour.

Schmidt (1966, 1967, 1968), in his exhaustive studies of metamorphosis in *Formica polyctena*, has provided much valuable information on its metabolic costs. He has shown that they are apparently minimal at 28 °C (Table 6.7) but closer examination of his data underline the complexity of metamorphosis and emphasise the difficulties of measuring energy flow by indirect means. It will be noted in Table 6.7 that the RQ, which is used to calculate the energy expenditure on the assumption that it is a measure of the proportion of fat and carbohydrate utilised, varies significantly between the castes and with temperature. Schmidt (1968) has also shown that it varies, also significantly, at each stage of metamorphosis (Table 6.8), which is, perhaps, not surprising bearing in mind the complexity of the metabolic processes at this stage of development. It does, however,

Table 6.8. *The respiratory quotient (RQ) of male, queen and worker pupae of* Formica polyctena *at each stage of metamorphosis and at various temperatures. (After Schmidt, 1968)*

Temperature °C	Prepupae			At pupation			Beginning of eye pigmentation			Well-pigmented eyes		
	Male	Worker	Queen	Male	Worker	Queen	Male	Worker	Queen	Male	Worker	Queen
16	0.79	0.96	0.72	0.78	1.13	0.75	0.83	0.76	0.81	0.75	0.73	0.78
20	0.78	0.80	0.80	0.86	0.92	0.83	0.84	0.85	0.84	0.77	0.87	0.72
24	0.77	0.97	0.78	0.83	0.94	0.85	0.79	0.87	0.79	0.85	0.82	0.77
28	0.87	0.96	0.82	0.79	0.96	0.83	0.81	0.98	0.75	0.95	0.94	0.86
32	0.82	0.97	0.83	0.89	1.06	0.89	0.82	0.99	0.87	—	0.93	0.84
	Beginning of body pigmentation			Well pigmented body			At emergence					
16	0.75	0.76	0.80	0.80	0.83	0.72	0.75	0.91	0.74			
20	0.77	0.82	0.75	0.64	0.82	0.80	0.80	0.87	0.76			
24	0.82	0.84	0.77	0.79	0.86	0.74	0.82	0.87	0.76			
28	0.79	0.90	0.79	0.82	0.97	0.74	0.82	0.89	0.82			
32	0.78	0.89	0.82	0.84	0.92	0.81	0.85	0.90	0.83			

emphasise the importance of not taking an isolated measure of RQ as a basis for the oxycalorific equivalent.

Taking the minimum values of oxygen consumed during metamorphosis it can be shown, as is indicated in Table 6.7, that a male pupa expends 133.4 J, a queen pupa 114.2 J and a worker pupa 27.0 J. A measure of confirmation of these values is provided by Schmidt's (1967) observation that loss of fat and carbohydrate with a potential energy of 124.3 J in males, 120.0 J in queens and 20.4 J in workers takes place during metamorphosis.

Adult castes

Reproductives

Gösswald & Schmidt (1960) and Kneitz (1967) have measured respiration in the reproductive castes of *Formica polyctena* and Brian (1973*a*) has measured that of the queens of *Myrmica rubra*.

Gösswald & Schmidt (1960) found that newly flown queens respired at the rate of 17.07 mm^3 oxygen per hour at 24 °C. The RQ was 0.89 so this respiratory rate is equivalent to 0.35 J per hour. Kneitz (1967), measuring respiration at 25 °C, found that old queens absorbed 0.610 mm^3 oxygen per mg fresh weight per hour, while newly flown queens were more active at 1.063 mm^3 oxygen per mg fresh weight per hour, almost twice the value

Table 6.9. *The respiration and its equivalent energy dissipation of ants at various temperatures. (After Slowtzoff, 1909)*

Temperature (°C)	Carbon dioxide evolved (g per 100 g ants per 24 hours)	Q_{10}	Energy equivalent[a] (mJ per mg per hour)
0	0.106	—	0.47
10	0.240	2.26	1.07
20	2.016	8.40	9.00
24	2.112	1.12	9.43
27	2.112	1.00	9.43
34	2.112	1.00	9.43
44	4.224	2.00	18.85
48	4.707	1.29	21.01
52	0.106	—	0.47

[a] One milligram of carbon dioxide evolved is equivalent to 10.7 J at an RQ of 1.0.

found by Gösswald & Schmidt (1960). It would seem that the respiratory rate of queens at the time of the nuptial flight is subject to much variation, a view supported by Kneitz's (1967) observations, where individual rates of oxygen consumption seem to range from about 10 mm^3 oxygen per hour to 30 mm^3 oxygen per hour. The respiratory rate of the males at 1.236 mm^3 oxygen per mg fresh weight per hour, is also high, but perhaps not surprisingly so, bearing in mind the hyperactivity usually exhibited by males at the time of the nuptial flight.

Brian (1973*a*) found that single queens of *Myrmica rubra* emitted 0.40 ± 0.017 mm^3 carbon dioxide per mg fresh weight per hour, at 19 °C. Since an average queen weighs 4.6 mg fresh weight this amounts to 1.84 mm^3 carbon dioxide per hour for each individual. He argues that this volume of carbon dioxide is equivalent to the breakdown of 1.68 μg of sucrose, which would release 0.0391 J per hour.

Thus the three independent measurements present a picture of some complexity in which comparison is made difficult because of the differing modes of measurement. Even within the same species great variation is evident and even though the respiratory rate of *Myrmica* queens is apparently lower than that of *Formica*, it certainly is within the range found by Kneitz (1967). On the other hand, the differences between myrmicine and formicine workers already noted may also hold for queens. Undoubtedly, queen respiration will vary considerably with physiological condition.

Workers

The bulk of the investigations of respiration in ants have been concerned with workers, but as with the developmental stages and the reproductive castes, few of them have been concerned with the question of energy flow. They do, however, throw valuable light on the problems of estimating metabolic costs in these insects and deserve scrutiny.

Slowtzoff (1909) was apparently the first worker to investigate respiration in ants, though he does not specify which species he was working on – or even whether he used workers. He was primarily concerned with the effect of temperature on respiratory rate. He showed that the RQ of the ants lay between 0.9 and 1.0 and that the Q_{10} was very variable, since respiration was apparently activated between 0 and 20 °C, assumed a constant rate until 34 °C, only to resume its rise up to 48 °C. At 52 °C respiration was severely depressed, at what must have been an essentially lethal temperature. His results are summarised in Table 6.9, together with the energy equivalents calculated from them on the basis of an RQ of 1.0 and the usual assumptions concerning the oxidation of carbohydrate. It will be noted that the heat evolved at 20–24 °C agrees well with the values determined by Gösswald & Schmidt (1960) and Brian (1973*a*) for the queens of *Formica polyctena* and *Myrmica rubra*, respectively.

Holmquist (1928), primarily concerned with the nature of hibernation in *Formica ulkei* Emery, compared the respiration of active ants that had been kept at 20–25 °C with hibernating ants at 3–6 °C. One hundred ants were confined in the respirometer for 24 hours to determine the gas exchange. Six determinations were made with active ants at room temperature (20–25 °C) and he found they absorbed 3.603 cm^3 of oxygen and evolved 3.176 cm^3 of carbon dioxide, giving an RQ of 0.88. Hibernating ants at 3–6 °C used 0.987 cm^3 of oxygen and evolved 1.039 xm^3 of carbon dioxide, indicating an RQ of 1.05. Since he used workers ranging in weight from 7–12 mg these respiratory rates imply a heat output of between 2.57 mJ per mg fresh weight per hour and 4.40 mJ per mg fresh weight per hour at 20–25 °C and 0.73 mJ per mg fresh weight per hour to 1.26 mJ per mg fresh weight per hour at 3–6 °C. It will be noted that the heat output of hibernating ants is in the same range as Slowtzoff's (1909) measurements but that active ants apparently expended only about half the energy of maintenance.

Dreyer (1932) repeated Holmquist's (1928) work with more refined apparatus. Using 25 ants at a time he measured their respiratory rate throughout the year at ambient temperatures and his results are summarised in Table 6.10, together with the usual estimates of heat output and the Q_{10} exhibited at each temperature interval. It is interesting to note the variations of RQ with temperature, a seasonal phenomenon which could

Table 6.10. *Seasonal changes in the respiratory physiology of* Formica ulkei. (*After Dreyer, 1932*)

Ambient temperature (°C)	Respiration of 25 workers (cm^3 per 0.192 g[a] per 105 min)		RQ	Respiratory rate (mm^3 per mg per hour)		Q_{10}	Energy equivalent (mJ per mg per hour)
	Carbon dioxide	Oxygen		Carbon dioxide	Oxygen		
4	52	104	0.50	0.155	0.310	—	5.71
7	62	103	0.60	0.185	0.307	—	5.77
10.5	104	162	0.64	0.310	0.482	4.49	9.19
15.5	135	191	0.71	0.402	0.568	2.36	11.01
18.5	171	217	0.79	0.509	0.646	3.79	12.77
22	247	285	0.87	0.735	0.848	3.75	17.08

[a] Fresh weight of 25 workers.

be induced by subjecting ants to low temperature at any time. Thus, ants kept at 22 °C would behave as hibernating ants if tested at 4 °C and would return to normal after one to two hours acclimation at 22 °C. Dreyer (1932) also found that if ants emerging from hibernation were starved the RQ remained low even at higher temperatures, but rose as soon as the ants had excess of food. The nature of the food also influenced the respiratory rate and the RQ. Workers were starved for two weeks, then fed exclusively on carbohydrate or fat or protein for a further week before their respiration was observed over a total of five weeks. Those on the carbohydrate diet maintained a normal respiratory rate and a high RQ of 0.98, but those on fat and those on protein both exhibited a depressed respiratory rate and a low RQ. Respiration on a diet of fat was only 72.5% of that on a carbohydrate diet, with an RQ of 0.58, while respiration on a protein diet was only 52.3% of that on carbohydrate, with an RQ of 0.60.

Apart from these observations, it will be noted that the heat output of Dreyer's ants was much higher than those of Holmquist and of Slowtzoff, by a factor of five. Dreyer suggests that overcrowding, resulting in formic acid intoxication, and insensitive manometry (the shortcomings of which Holmquist recognised), may account for Holmquist's low estimates.

Kennington (1957) investigated respiration in the major workers of *Camponotus modoc* Wheeler collected at two sites, one at an altitude of 2057 m and the other at 3231 m. Both samples were tested at each altitude at two temperatures, 10 and 25 °C. Curiously, ants from the first site, when tested at the same altitude showed no response to temperature change. Oxygen consumption at 10 °C was 0.58 ± 0.154 mm^3 oxygen per mg fresh weight per hour and at 25 °C only 0.60±0.059 mm^3 oxygen per mg fresh weight per hour. Ants from the second site tested at the same altitude were more responsive, the oxygen consumption rising from 0.61±0.219 mm^3 oxygen per mg fresh weight per hour to 2.79±0.263 mm^3 oxygen per mg fresh weight per hour, a Q_{10} of 3.04. When the site of collection and the site of testing were reversed a more typical response was observed in the 2057 m ants, since their respiratory rate rose from 0.68±0.111 mm^3 oxygen per mg fresh weight per hour at 10 °C to 1.89 mm^3 oxygen per mg fresh weight per hour at 25 °C, a Q_{10} of 1.86. Workers from 3231 m up were less sensitive to temperature change when tested at the lower temperature. Their respiration rose from 0.59±0.101 mm^3 oxygen per mg fresh weight per hour at 10 °C to 1.48±0.208 mm^3 oxygen per mg fresh weight per hour at 25 °C, a Q_{10} of 1.67. It is difficult to compare these results with previous work but it may be noted that the respiratory rate of *C. modoc* is rather higher at 10 °C and, at the highest rates at 25 °C almost three times that of *Formica ulkei* (Dreyer, 1932).

Up to the early 1960s then, the few studies on the respiratory rates of worker ants had been carried out with diverse ends in view and had produced results which are difficult to compare. They were carried out with relatively unsophisticated respirometers and the inescapable conclusion is that the variability of the data for respiration per unit weight in unit time is not entirely due to specific differences.

As has already been noted, Golley & Gentry (1964) were the first workers specifically concerned with the construction of an energy budget for ants. However, their results have been criticised (Engelmann, 1966; McNeill & Lawton, 1970; Petrusewicz & Macfadyen, 1970) since they obtained an extremely high respiratory rate for workers, about 100 times that for the related species examined by Ettershank & Whitford (1973) and Kay & Whitford (1975). If the estimates of Golley & Gentry (1964) of the annual costs of maintenance are reduced by two orders of magnitude, a much better fit to McNeill & Lawton's (1970) curve relating production and respiration is obtained, at least in the case of their lower estimate.

More recently, Kneitz (1967) has investigated seasonal variation in respiration in workers of *Formica polyctena* (Table 6.11). He found that about one-third of the overwintering workers were significantly larger, with substantial reserves of fat. Such workers exhibited a depressed respiratory activity at 25 °C compared with the rest of the population, and an unusually high RQ of 1.2, suggesting, perhaps, fat synthesis. The conspicuously larger workers were associated with the innermost chambers of the nest. Other workers found there were also larger, but with a rather higher oxygen consumption, while the smallest workers, with the highest respiratory rate, were found in the outer parts of the nest. During

Table 6.11. *Seasonal variation in the respiration of workers of* Formica polyctena (*After Kneitz, 1967*)

Season	Worker type	Mean fresh weight (mg)	Oxygen uptake (mm^3 per mg per hour)	Equivalent heat output (mJ per mg per hour)
Winter	With large fat reserves, from the inner part of nest	15.20	0.738	16.34[a]
Winter	From the outer part of nest	10.52	0.901	
Winter	From the inner part of nest, but without large reserves	12.81	0.805	
Summer	From the outer part of nest	7.36	1.121	23.67[b]
Summer	From the inner part of nest	8.60	1.082	22.85[b]

[a] RQ = 1.20, heat output calculated by extrapolation from heat equivalent in the RQ range 1.00–0.70.
[b] RQ = 1.00.

the summer, workers from inner chambers were still rather larger than ants from other parts of the nest, but only about half the weight of the overwintering forms. The respiratory rate per unit weight at 25 °C of all summer workers was higher than overwintering ants and their RQ was at the more normal level of 1.0. Kneitz's (1967) establishment of the existence of seasonal variation in respiration paralleling obvious changes in the physiological condition of the workers is consistent with the observations of the earlier work carried out by Holmquist (1928) and Dreyer (1932). As far as the average level of oxygen consumption in workers of *F. polyctena* is concerned, it will be noted that it is of the same order as that described in *F. ulkei* (Dreyer, 1932).

Stradling (1968), in a study of the energetics of ants in a dune system, found that workers of *Myrmica rubra* respired at 0.10 mm^3 oxygen per mg fresh weight per hour at 10 °C, 0.25 at 15 °C, 0.55 at 20 °C and 0.60 at 25 °C and workers of *Lasius niger* (L.) respired at about the same rate. These values agree well with those of Brian (1973*a*) who found that workers of *M. rubra* evolved 0.50 mm^3 carbon dioxide per mg fresh weight per hour at 19 °C, but are rather lower than those found by Dreyer (1932) in *Formica ulkei.*

Nielsen (1972*a*), when constructing an energy budget for a colony of *Lasius alienus*, found that the oxygen consumption of the workers conformed to the equation

$$\log y = 0.0209T - 0.5685 \qquad (6.8)$$

where y = oxygen consumption in mm^3 per mg fresh weight per hour and T = temperature in °C. This also gives a rate which is in fair agreement with that found for *F. ulkei*.

The most recent observations on respiration in worker ants (Ettershank & Whitford, 1973; Jensen & Nielsen, 1975; Kay & Whitford, 1975) have added substantially to our previous knowledge and some remarks have already been made on the light they throw on relative respiratory rates in different sub-families of ants. Comparisons with previous work are not easy since the rates are expressed in terms of dry weight and it is unfortunate that the RQ was not determined, to allow calculation of the energy dissipated during respiration.

Ettershank & Whitford (1973) and Kay & Whitford (1975) have been concerned, in the species of desert ants they have investigated, not only with the influence of temperature on respiration but also with the drying effect of air, a factor of particular importance in the environment in which these ants live. In all except one of the seven species they have studied, oxygen consumption increases exponentially with temperature. The exception is *Pogonomyrmex californicus* (Buckley), where oxygen consumption (y) is best described by the equation

$$y = 0.1849T - 2.1587 \qquad (6.9)$$

where T = temperature in °C.

They have found that the value of the exponent in each of the six remaining species varies and that in four, *Pogonomyrmex maricopa* Wheeler, *P. desertorum, Trachymyrmex smithi neomexicanus* Cole and *Formica perpilosa* Wheeler, respiratory rate is significantly effected by the vapour pressure deficit of the air in which they were confined, such that oxygen consumption increased, again exponentially, with increasing dryness. In all cases, they have derived equations which describe well the relationship of oxygen consumption to those factors which affect it significantly.

Jensen & Nielsen (1975) have examined respiration in nine species of temperate ants at 25 °C and, again, as rates are expressed in terms of dry weight, comparison with earlier work is not generally possible. However, the relation between respiratory rate and body weight referred to earlier is of interest in that it links their observations with those of Ettershank & Whitford (1973) and Kay & Whitford (1975) in a consistent fashion. The values they determined have already been listed in Table 6.3.

Concluding remarks on ants

There has been recently an encouraging increase in the number of investigations of respiration in ants, particularly with reference to its ecological significance. Thus in the last two years reports have been published (Brian, 1973*a*; Ettershank & Whitford, 1973; Jensen & Nielsen, 1975; Kay & Whitford, 1975) on 17 species of ants, compared with the eight or nine species investigated in the previous 64 years. Table 6.12 summarises the

Table 6.12. *Respiration in ants*

Species	Fresh weight (mg)	Dry weight (mg)	Temperature at measurement °C	Oxygen consumption (mm³ per mg per hour)	RQ	Energy equivalent[a] (mJ per mg per hour)	Author
Formicinae							
Formicini							
Formica polyctena							
Worker pupae	11.6	—	28[b]	0.28[c]	0.94	5.89	Schmidt, 1968
Queen pupae	41.6	—	28[b]	0.35[c]	0.79	6.93	Schmidt, 1968
Male pupae	46.7	—	28[b]	0.36[c]	0.84	7.22	Schmidt, 1968
Queens, newly flown	29.4	—	24	0.58[c]	0.89	11.94	Gösswald & Schmidt, 1960
Queens, newly flown	—	—	25	1.06[c]	—	—	Kneitz, 1967
Queens, mature	—	—	25	0.61[c]	—	—	Kneitz, 1967
Males, newly flown	—	—	25	1.24[c]	—	—	Kneitz, 1967
Workers, winter[d]	15.20	—	25	0.74[c]	1.20	16.43	Kneitz, 1967
Workers, winter[e]	10.52	—	25	0.90[c]	—	—	Kneitz, 1967
Workers, winter[f]	12.81	—	25	0.81[c]	—	—	Kneitz, 1967
Workers, summer[g]	7.36	—	25	1.12[c]	1.00	23.67	Kneitz, 1967
Workers, summer[h]	8.60	—	25	1.08[c]	1.00	22.85	Kneitz, 1967
Formica ulkei							
Workers	9.5[i]	—	3–6	0.04	1.05	0.92	Holmquist, 1928
Workers	9.5[i]	—	20–25	0.16	0.88	3.24	Holmquist, 1928
Workers	7.7[j]	—	4	0.31	0.50	5.75	Dreyer, 1932
Workers	7.7[j]	—	7	0.31	0.60	5.85	Dreyer, 1932
Workers	7.7[j]	—	10.5	0.48	0.64	9.29	Dreyer, 1932
Workers	7.7[j]	—	15.5	0.57	0.71	11.15	Dreyer, 1932
Workers	7.7[j]	—	18.5	0.65	0.79	12.94	Dreyer, 1932
Workers	7.7[j]	—	22	0.85	0.87	17.34	Dreyer, 1932
Formica perpilosa							
Workers	—	1.01	15	0.86–1.14[k, l]	—	—	Kay & Whitford, 1975
Workers	—	1.01	25	2.00–2.36[k, l]	—	—	Kay & Whitford, 1975
Workers	—	1.01	35	3.43–4.50[k, l]	—	—	Kay & Whitford, 1975
Formica exsecta							
Workers	—	1.35	25	2.89[k]	—	—	Jensen & Nielsen, 1975
Formica pratensis							
Workers	—	2.23	5	0.29[k]	—	—	Jensen & Nielsen, 1975
Workers	—	2.23	10	0.32[k]	—	—	Jensen & Nielsen, 1975
Workers	—	1.93	15	1.10[k]	—	—	Jensen & Nielsen, 1975
Workers	—	1.84	20	1.41[k]	—	—	Jensen & Nielsen, 1975
Workers	—	2.39	25	2.33[k]	—	—	Jensen & Nielsen, 1975
Workers	—	1.54	30	4.16[k]	—	—	Jensen & Nielsen, 1975
Formica fusca							
Workers	—	1.71	5	0.20[k]	—	—	Jensen & Nielsen, 1975
Workers	—	1.27	15	0.33[k]	—	—	Jensen & Nielsen, 1975
Workers	—	1.47	25	1.17[k]	—	—	Jensen & Nielsen, 1975
Lasius alienus							
Larvae	—	0.24	25	2.80[k]	0.80	56.23	Nielsen, 1972*a*
Workers	—	0.18	20	0.76[c]	0.80	15.26	Nielsen, 1972*a*
Workers	—	0.46	25	2.35[k]	—	—	Jensen & Nielsen, 1975
Lasius niger							
Workers	—	0.65	5	0.20[k]	—	—	Jensen & Nielsen, 1975
Workers	—	0.61	15	0.84[k]	—	—	Jensen & Nielsen, 1975
Workers	—	0.58	25	1.85[k]	—	—	Jensen & Nielsen, 1975
Lasius flavus							
Workers	—	0.86	25	2.37[k]	—	—	Jensen & Nielsen, 1975
Camponotini							
Camponotus modoc							
Workers	—	—	10	0.58–0.69[c]	—	—	Kennington, 1957
Workers	—	—	25	0.60–2.79[c]	—	—	Kennington, 1957
Camponotus herculeanus							
Workers	—	6.71	25	2.02[k]	—	—	Jensen & Nielsen, 1975

Table 6.12 (*cont.*)

Species	Fresh weight (mg)	Dry weight (mg)	Temperature at measurement °C	Oxygen consumption (mm³ per mg per hour)	RQ	Energy equivalent[a] (mJ per mg per hour)	Author
Myrmicinae							
Myrmicini							
Myrmica rubra							
Queens	4.60	—	19	1.40[c]	1.00	8.45	Brian, 1973*b*
Workers	2.00	—	19	0.50[c]	1.00	10.56	Brian, 1973*b*
Workers	—	—	10	0.10[c]	—	—	Stradling, 1968
Workers	—	—	15	0.25[c]	—	—	Stradling, 1968
Workers	—	—	20	0.55[c]	—	—	Stradling, 1968
Workers	—	—	25	0.60[c]	—	—	Stradling, 1968
Workers	—	0.69	5	0.20[k]	—	—	Jensen & Nielsen, 1975
Workers	—	0.79	15	0.89[k]	—	—	Jensen & Nielsen, 1975
Workers	—	0.92	25	1.31[k]	—	—	Jensen & Nielsen, 1975
Novomessor cockerelli							
Workers	—	3.91	15	0.43[k]	—	—	Kay & Whitford, 1975
Workers	—	3.91	25	0.81[k]	—	—	Kay & Whitford, 1975
Workers	—	3.91	25	2.21[k]	—	—	Kay & Whitford, 1975
Tetramorium caespitum							
Workers	—	0.30	5	0.15[k]	—	—	Jensen & Nielsen, 1975
Workers	—	0.23	15	1.13[k]	—	—	Jensen & Nielsen, 1975
Workers	—	0.23	25	1.87[k]	—	—	Jensen & Nielsen, 1975
Pogonomyrmex badius							
Workers	6.6	1.10	20	40.8[k]	—	—	Golley & Gentry, 1964
Soldiers	31.4	9.9	20	3.8[k]	—	—	Golley & Gentry, 1964
Pogonomyrmex rugosus							
Workers	—	4.95	5	0.17–0.25[k, l]	—	—	Ettershank & Whitford, 1973
Workers	—	4.95	15	0.35–0.48[k, l]	—	—	Ettershank & Whitford, 1973
Workers	—	4.95	25	0.63–0.83[k, l]	—	—	Ettershank & Whitford, 1973
Workers	—	4.95	35	1.00–2.04[k, l]	—	—	Ettershank & Whitford, 1973
Workers	—	4.95	45	2.95–5.13[k, l]	—	—	Ettershank & Whitford, 1973
Pogonomyrmex maricopa							
Workers	—	3.69	5	0.30[k]	—	—	Ettershank & Whitford, 1973
Workers	—	3.69	15	0.71[k]	—	—	Ettershank & Whitford, 1973
Workers	—	3.69	25	1.39[k]	—	—	Ettershank & Whitford, 1973
Workers	—	3.69	35	2.42[k]	—	—	Ettershank & Whitford, 1973
Workers	—	3.69	45	5.48[k]	—	—	Ettershank & Whitford, 1973
Pogonomyrmex californicus							
Workers	—	1.39	15	0.36[k]	—	—	Kay & Whitford, 1975
Workers	—	1.39	25	1.81[k]	—	—	Kay & Whitford, 1975
Workers	—	1.39	35	3.05[k]	—	—	Kay & Whitford, 1975
Pogonomyrmex desertorum							
Workers	—	2.21	15	0.29–0.50[k, l]	—	—	Kay & Whitford, 1975
Workers	—	2.21	25	0.79–1.14[k, l]	—	—	Kay & Whitford, 1975
Workers	—	2.21	35	1.71–2.00[k, l]	—	—	Kay & Whitford, 1975
Attini							
Trachymyrmex smithi neomexicanus							
Workers	—	1.07	15	0.36–0.57[k, l]	—	—	Kay & Whitford, 1975
Workers	—	1.07	25	0.57–0.79[k, l]	—	—	Kay & Whitford, 1975

Table 6.12 (*cont.*)

Species	Fresh weight (mg)	Dry weight (mg)	Temperature at measurement °C	Oxygen consumption (mm³ per mg per hour)	RQ	Energy equivalent[a] (mJ per mg per hour)	Author
Workers	—	1.07	35	1.29–1.64[k, l]	—	—	Kay & Whitford, 1975
Incertae sedis							
'Ants'	—	—	0	0.11[c]	1.00	0.47	Slowtzoff, 1909
'Ants'	—	—	10	0.24[c]	1.00	1.07	Slowtzoff, 1909
'Ants'	—	—	20	2.02[c]	1.00	9.00	Slowtzoff, 1909
'Ants'	—	—	24	2.11[c]	1.00	9.43	Slowtzoff, 1909
'Ants'	—	—	27	2.11[c]	1.00	9.43	Slowtzoff, 1909
'Ants'	—	—	34	2.11	1.00	9.43	Slowtzoff, 1909
'Ants'	—	—	44	4.22[c]	1.00	18.85	Slowtzoff, 1909
'Ants'	—	—	48	4.71[c]	1.00	21.01	Slowtzoff, 1909
'Ants'	—	—	52	0.11[c]	1.00	0.47	Slowtzoff, 1909

RQ = respiratory quotient.
[a] Calculated from the regression line, relating calorific equivalents to various RQs, given by Brody (1945) and quoted by Southwood (1966).
[b] Temperature at which energy consumption is at a minimum.
[c] mm³ per mg fresh weight per hour.
[d] Winter workers with well-developed fat bodies.
[e] Winter workers from the outer part of the nest.
[f] Winter workers from the inner part of the nest, but without well-developed fat bodies.
[g] Summer workers from the outer part of the nest.
[h] Summer workers from the inner part of the nest.
[i] Holmquist (1928) used workers ranging from 7–12 mg in weight.
[j] Dreyer (1932) used 25 workers with a total weight of 0.192 g.
[k] mm³ per mg dry weight per hour.
[l] Depending on the vapour pressure deficit.

findings. The great majority of investigations has been on workers and they have been carried out for diverse reasons. The results, as a consequence, have been expressed in many different ways so that it is difficult to compare them and, in any case, they leave the abiding impression of great diversity amongst the species. It is nonetheless instructive to examine them more closely and as an aid to this end they have been illustrated in Fig. 6.8, where the range of recorded respiratory rates at various temperatures is shown. The greatest number of readings for any temperature have been made at 25 °C and it will be noted straight away and, perhaps surprisingly, that there is little difference in the range covered by the values, whether they are expressed in terms of unit fresh weight or unit dry weight. The range extends from about 0.6–2.9 mm³ oxygen per mg per hour, that is, the greatest is about five times that of the smallest. Now in preliminary estimates of energy flow through the components of an ecosystem it is, perhaps, acceptable to work within an order of magnitude. If this is judged reasonable, then on the basis of the information on respiration in ants it can be argued that an approximation within these

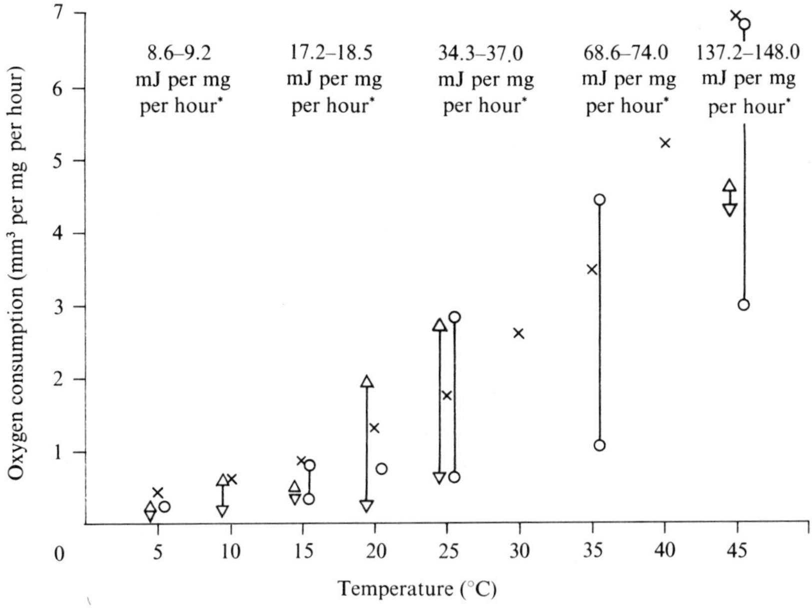

Fig. 6.8. Respiration in worker ants: the range of determined rates and predicted rates, assuming a Q_{10} of 2 and a mean rate of 1.75 mm³ per mg per hour at 25 °C, together with the equivalent energy expenditure for a range of respiratory quotients (RQs) △——△, range of fresh weight determinations; ○——○, range of dry weight determinations; '×', respiratory rate predicted by assuming a mean rate of 1.75 mm³ per mg per hour at 25 °C and a Q_{10} of 2.0; '*', energy equivalent of predicted respiratory rates for RQs ranging from 0.70–1.00.

limits can be made by assuming respiratory rates anywhere within the limits illustrated in Fig. 6.8. Extending the idea further, if a median value of 1.75 mm³ oxygen per mg per hour for respiration at 25 °C is taken and a Q_{10} of 2.0 is quite arbitrarily assumed, a series of values (marked with crosses in Fig. 6.8) not too far removed from the ranges at each temperature are obtained, and certainly within an order of magnitude. To complete the argument, the energy dissipated at 10 °C intervals about 25 °C has been calculated for an RQ range of 1.00–0.70 and incorporated into Fig. 6.8.

Returning to reality, the only tentative generalisation (that the results so far obtained allow) based on 16 species is that formicine workers respire at a higher rate than myrmicine workers.

However, as the ultimate aim is to build up an accurate picture of metabolic costs in ants, an important function of this review of existing work is to suggest areas in which future work should concentrate in order to increase our knowledge in the field profitably. Firstly, there is no doubt that the general physiological condition of the workers is an important influence in establishing the level of respiration, as the work of Dreyer (1932) and Kneitz (1967) indicates. It is important, therefore, that research should take account of the functions of the workers in the colony at the time their respiration is measured; in addition the time of the year and the stage in any cyclical events, particularly such things as brood development should be noted as all these factors will play a part in their general physiological condition. Factors such as these will not only determine the level of respiration but, as Dreyer (1932) has further shown, will also alter the nature of the respiratory substrate. As long as indirect measurement of the energy costs of maintenance is made it is important to know what respiratory substrate is utilised, in order to ascribe an energy equivalent to respiratory rates. The simplest approximation is to determine the RQ in all conditions, yet despite its importance it is a sorely neglected feature of respiratory studies. We have argued at length that the ants and termites are particularly suitable for direct measurement of costs of maintenance which would, amongst other things, obviate the need for this additional piece of information.

The influence of physical factors, besides the obvious one of temperature, is perhaps of importance only in those species living in extreme conditions, such as the desert ants investigated by Ettershank & Whitford (1973) and Kay & Whitford (1975). As far as temperature goes, its unpredictable effects have certainly been demonstrated in the work carried out so far. The range over which laboratory measurements are made, clearly have to be as wide as those experienced in the field, but in extrapolating the laboratory responses of the workers to the field greater attention has to be paid to their behavioural responses, because of their well-known ability to choose, particularly within the nest, those conditions best suited to their needs of the moment.

All these remarks, of course, apply equally to the other members of the colony. There is no doubt that there is an absolute dearth of measurements on them and much future effort must be directed towards elucidating the respiratory physiology of all components of the society. The importance of the developmental stages in contributing to respiratory costs is such that it is reasonable to make them the first target of such efforts. The study of metamorphosis in *Formica polyctena* by Schmidt (1966, 1967, 1968) sets an admirable example of the kind of work that is required and one looks forward to the time when similar studies of larval physiology have been made. Furthermore, because the rate and kind of development the larvae exhibit depends entirely on other members of the colony it will be particularly important to choose with care the experimental conditions in which they are investigated.

Finally, as the ultimate object of these studies is to determine the role of ants in the general energy flow in ecosystems, no study of respiration will be complete without an attempt to extrapolate the findings to the field. Undoubtedly, it is exceedingly difficult to do this in most cases. In the two instances where it has been attempted (Golley & Gentry, 1964 and Nielsen, 1972*a*) substantial approximations have been necessary. To avoid these in future, it will be necessary to know much more about the physical and biological conditions within the nest, so that the relevance of laboratory measurements is fully established before extrapolation is attempted. This is perhaps best tried with species that isolate themselves more completely from the general environment, as is the case in the aerial and arboreal forms, than soil-dwelling species that are more or less inextricably mixed with the soil ecosystems.

It is clear that much work remains before it is possible to quantify the important and unique role of ants in the environment.

Energy flow measurements in termites

The studies which have been done on termite respiration are rather scattered and only a few of them can be extrapolated to energy flow estimates. Work has been done with the aim of measuring the influence of several factors on termite health or tolerance or of studying the role of their symbionts in digesting cellulose. Nevertheless, 'in considering the matter from the ecological point of view, one is confronted by the difficulty that very little is known about the normal respiration of the termite itself'; this was written by Cook (1932) more than 40 years ago but is still almost true today. Let us examine the available data, and first of all the problem involving the anaerobic processes occurring in the gut of the termites which are characteristic of their metabolism.

The RQ and the anaerobic processes problem

The first known paper which refers to termite respiration is that of Cleveland (1925*a*) who maintained *Zootermopsis* and *Reticulitermes* on a diet of pure cellulose for more than a year. These experimental societies developed as well as the control ones fed upon wood. Hence this author suspected a fixation of atmospheric nitrogen; he measured the RQ and 'as would naturally be expected on a carbohydrate diet (the RQ) has been found to be practically 1/1'. He confined termites in air at constant temperature and noted that 'a negative pressure had very soon developed' which he interpreted as the sign of atmospheric nitrogen fixation.

Cook's (1932) contribution was very interesting; thanks to his experiments carried out on *Zootermopsis nevadensis* in low oxygen tension he was able to perceive the evolution of an undetermined gas and there was 'a strong possibility that it might be hydrogen or methane, or a mixture of both'. A comparison of normal and defaunated numphs showed that there was no longer production of the gas when the symbiotic protozoa had been removed. This same experiment enabled Cook to estimate the production of the gas (probably at 20 °C) as 0.0199 μl per mg per hour. He determined that the RQ of normal termites was between 0.955 and 1.010 and that of defaunated nymphs was 0.82, two weeks after removing of the symbiotic protozoa.

Hungate's (1938) work on *Zootermopsis* confirmed an 'RQ of approximately 1.0 which suggests that the metabolism is one in which carbohydrates are completely oxidised'. He identified, however, the gas as being hydrogen, by means of platinised asbestos and a solution of methylene blue (Hungate, 1939). He confirmed the order of magnitude of hydrogen production with a figure of 0.027 μl per mg per hour.

Again, *Zootermopsis nevadensis* was the material chosen by Gilmour (1940) for his experiments which were run mostly at 3 °C: 'at this temperature, the insects were motionless and their rate of respiration in air could be considered to approximate a "basal" level'. The RQs he obtained from nine measurements varied between 0.80 and 1.37 and their average (1.09) did not differ significantly from that (1.06) he obtained from three measurements at 20 °C. He also encountered an evolution of gas (hydrogen) of 0.003 μl per mg per hour (at 3 °C), which was about 10% of the rate of oxygen consumption.

Almost nothing is known about hydrogen evolution in higher termites: Grassé & Noirot (1958*b*) reported a concentration of 0.6% hydrogen in gas samples from the central part of a nest of *Macrotermes bellicosus*, but no hydrogen was present in the surrounding soil. Using palladium, Hébrant (1970*a*, *b*) failed to detect any measurable evolution of hydrogen in *Cubitermes* (humivorous termites) respiration. During recent experiments at the IBP Sonoran Desert Project (University of Tucson), production of hydrogen was detected by means of a gas chromatograph in respiration measurements of *Marginitermes hubbardi.* An evolution of methane was also mentioned in rearings of *Gnathamitermes perplexus*; in one experimental vessel, this last species produced both hydrogen and methane (CH_4) (La Fage, Josens, unpublished data).

If such evolution of incompletely oxidised gases – hydrogen, methane (and possibly carbon monoxide, Bouillon, 1970) attained relatively high levels, one might think that corrections should be made on both the expected RQ and the oxycalorific equivalent of oxygen consumption. If the evolution of hydrogen is about 10% of the oxygen consumption (according to Gilmour, 1940), that means that the oxidation of carbohydrates would not follow the classical scheme:

$$100\,(CH_2O) + 100\,O_2 \rightarrow 100\,CO_2 + 100\,H_2O \qquad (6.10)$$

but this one:

$$100\ (CH_2O)+95\ O_2 \rightarrow 100\ CO_2+90\ H_2O+10\ H_2 \qquad (6.11)$$

In this case, the RQ is close to 1.05 instead of 1.00. The oxycalorific equivalent of oxygen consumption shifts from 5.05 to 5.05/0.95−EH (in mcal per μl oxygen), EH being the energy of combusion of hydrogen, that is, 56.7 kcal per mole* or 2.53 mcal per μl hydrogen.

With, as in this example, an evolution of hydrogen equal to 10.5% of oxygen consumption, the oxycalorific equivalent of oxygen becomes $5.05/0.95-(2.53\times0.105) = 5.05$ (mcal per μl oxygen). There is no difference in the case of hydrogen, but that would not be true with methane, which is much richer in energy of combustion (218.8 kcal per mole, or 9.77 mcal per μl methane). Therefore, in energy flow studies on termites, the measurement of the amount of hydrogen production is important in order to correct the amount of oxygen consumed. The RQ may be expected to be slightly higher than 1.00 and no substantial correction need be applied to the oxycalorific equivalent of oxygen consumption. If it is confirmed that some termite species produce methane, a correction on the oxycalorific equivalent would be essential.

Meeting the energy budget in termite societies

The first attempt at calculating the energy flow through a termite population was made by Maldague (1967). The population was estimated at about 10 million individuals per hectare, a great majority of them being the large, humivorous *Cubitermes fungifaber* (Sjöstedt) (12.4 mg per worker, fresh weight). Their biomass was estimated at more than 100 kg (fresh weight) per hectare. Subsequently, Maldague used Zeuthen's (1953) figures – which had been obtained from animals as different as Coelenterates, Vermes, Molluscs, Crustacea and fishes – and extrapolated them to the respiration of his termites. Using an oxygen consumption of 8 μl per individual per hour (that is to say about 0.70 μl per mg fresh weight per hour with an average weight of 11.45 mg per individual) and an oxycalorific equivalent of 4.7 kcal per l oxygen he estimated the energy in respiration at 329 kcal per m^2 per year (1377 kJ per m^2 per year) or at the combustion of 170 g of cellulose per m^2 per year; that is to say that he used an energy equivalent of 1936 cal per g (or 8104 J per g) of cellulose! Assuming that only cellulose is assimilated by termites and that cellulose represents 30% of the soil organic matter, Maldague concluded that his termite population must consume between five and six metric tonnes of organic matter per hectare per year. It appeared later, in Hébrant's work, that humivorous

* Free energy of water formation at 25 °C; *Handbook of chemistry and physics*, 48th edn. Cleveland, Ohio: Chemical Rubber Company (1967).

termites have very low respiratory rates and therefore Zeuthen's figures are not applicable.

Hébrant's (1970*a*) thesis is at present the only good work on energy flow through a termite population. After having studied the density and structure of the population, he first approached the problem in an analytical way: each caste was examined separately and the influence of each of several factors was analysed. Thus he showed that the nature of the substrate the termites were kept on, as well as the volume they were put into during the measurements, influenced their respiration. The termites consumed more oxygen when placed upon a sheet of filter paper than they did within a sieve tube of copper; Hébrant considered that this last container reproduced the environmental conditions of the galleries the termites normally live in, and that their lower respiration was an indication of lesser excitement and more normal conditions. So his subsequent measurements were all made with termites in a sieve tube of copper at 30 °C, and with a 5% solution of potassium hydroxide which maintained an RH of about 95%. The influence of individual weight, temperature, RH, the carbon dioxide concentration within the atmosphere, as well as the society the termites came from (that is, genetical or 'historical' factors) were all analysed by Hébrant (1970*a*) and have already been referred to.

The RQ of both *Cubitermes exiguus* and *C. sankurensis* workers was in the neighbourhood of 0.7, characteristic of lipid oxidation; during the first hour of measurements, however, the RQ was higher and close to 0.9. Considering the results of his experiments in different relative humidities, Hébrant thought that the low RQ was not related to a starvation state but was a reaction to the change of environment and RH, the oxidation of lipids producing more water than that of cellulose.

Hébrant's data on oxygen consumption in all stages from eggs to queen are quoted in Table 6.13. As one will see, the lowest respiratory rates are those of the workers and soldiers, the two castes which have appreciable amounts of soil in their gut: the inorganic fraction represents almost 20% of the workers' fresh weight, or 35% of their dry weight (calculated after Hébrant's (1970*a*) Table 13). Late nymphs also exhibit low respiratory rates; here also an appreciable amount of the fresh body weight (15 to 20% lipids) does not respire. Some RQ figures determined in non-worker castes were low as well: 0.9–0.6 in soldiers and 0.7–0.4 in larvae and nymphs. These last figures are explicable by the way the larvae and nymphs are fed: they are provided by the workers with an elaborated saliva, the composition of which is still unknown.

After this analytical approach, Hébrant tackled the problem of the energy flow through a termite society by measuring the respiration of whole nests. This led him to the discovery of the strong circadian rhythm already referred to. The RQ of such whole societies was close to 1.0, in

Table 6.13. *Respiratory data on termites*

Species	Stage investigated	Fresh weight mg per individual	Temperature °C	O_2 consumption μl per mg per h	RQ	Energy equivalent mcal per mg per h (mJ per mg per h)	Author
Calotermes flavicollis	?, 323 individuals in three series	6.13	26	0.540	0.90	—	Lüscher, 1955*b*
Cryptotermes havilandi	Nymphs and pseudergates	—	—	—	*ca.* 1.0	—	Hébrant, 1970*a*
Zootermopsis nevadensis	Nymphs	—	32	0.805[a]	—	4.025 (16.85)	Cook, 1932
	Nymphs	—	20	0.423[b]	—	2.115 (8.85)	Cook, 1932
	Nymphs	—	—[c]	0.310[d]	0.986[e]	1.55 (6.49)	Cook, 1932
	Medium-sized nymphs	—	3	0.0366[f]	1.09[f]	0.183 (0.77)	Gilmour, 1940
	Medium-sized nymphs	—	3	—	1.05[g]	—	Gilmour, 1940
	Medium-sized nymphs	—	20	—	1.06	—	Gilmour, 1940
	Defaunated, medium-sized nymphs	—	3	0.0361	0.76[h]	3.56 (14.89)	Gilmour, 1940
	?, 193 individuals in three series	16.7	18.7	0.512	0.895	9.57 (40.06)	Lüscher, 1955*b*
		—	26	*ca.* 1.3 to *ca.* 2.0[i]	—	—	Damaschke & Becker, 1964
Zootermopsis angusticollis	?, Normal termites on cellulose	—	24	0.2595[j]	0.995	1.30 (5.43)	Cook, 1943
	?, Defaunated termites on cellulose	—	24	0.2595[k]	0.757	1.21 (5.08)	Cook, 1943
	?, Starving, normal termites	—	24	0.2426[l]	0.721	1.13 (4.72)	Cook, 1943
	?, Starving, defaunated termites	—	24	0.2585[k]	0.735	1.20 (5.04)	Cook, 1943
Reticulitermes lucifugus	Workers	0.603	19–20	0.172[m]	0.970[m]	0.86 (3.61)	Ghidini, 1939
	Soldiers	1.355	19–20	0.0565[m]	0.925[m]	0.28 (1.17)	Ghidini, 1939
	Nymphs of first stage	1.967	19–20	0.077[m]	0.783[m]	0.36 (1.51)	Ghidini, 1939
	Non-pigmented alates	1.296	19–20	0.173[m]	0.825[m]	0.83 (3.47)	Ghidini, 1939
Reticulitermes flavipes	Groups of 100 workers	—	24	0.78[n]	—	3.90 (16.33)	Damaschke & Becker, 1964
		—					
		—	26	1.39[n]	—[o]	6.95 (29.09)	Damaschke & Becker, 1964
		—	28	1.02[n]	—[o]	5.10 (21.35)	Damaschke & Becker, 1964
		—	30	1.86[n]	—	9.30 (38.93)	Damaschke & Becker, 1964
		—	32	1.76[n]	—	8.80 (36.84)	Damaschke & Becker, 1964
		—	26	*ca.* 0.5 to *ca.* 1.5	—	—	Damaschke & Becker, 1964
		—	28	*ca.* 0.5 to *ca.* 1.25	—	—	Damaschke & Becker, 1964
Reticulitermes santonensis	Groups of 100 workers	—	24	0.88[n]	—	4.40 (18.42)	Damaschke & Becker, 1964
		—	26	0.96[n]	—	4.80 (20.09)	Damaschke & Becker, 1964
		—	28	1.62[n]	—	8.10 (33.91)	Damaschke & Becker, 1964
		—	30	1.41[n]	—	7.05 (29.51)	Damaschke & Becker, 1964
Undetermined: small colony from the eastern USA		2.338	25	0.391	—	1.96 (8.18)	Maldague, 1967

Table 6.13 (*cont.*)

Species	Stage investigated	Fresh weight mg per individual	Temperature °C	O_2 consumption μl per mg per h	RQ	Energy equivalent mcal per mg per h (mJ per mg per h)	Author
Heterotermes indicola	Groups of 100 workers	—	24	0.74[n]	—	3.70 (15.49)	Damaschke & Becker, 1964
		—	24	0.92[n]	—[o]	4.60 (19.26)	Damaschke & Becker, 1964
		—	26	0.46[n]	—	2.30 (9.63)	Damaschke & Becker, 1964
		—	26	0.89[n]	—[o]	4.45 (18.63)	Damaschke & Becker, 1964
		—	28	0.95[n]	—	4.75 (19.88)	Damaschke & Becker, 1964
		—	28	1.06[n]	—[o]	5.30 (22.19)	Damaschke & Becker, 1964
		—	30	0.77[n]	—	3.85 (16.12)	Damaschke & Becker, 1964
		—	30	1.40[n]	—[o]	7.00 (22.30)	Damaschke & Becker, 1964
		—	32	1.28[n]	—	6.40 (26.79)	Damaschke & Becker, 1964
		—	24	*ca.* 0.75[n]	—	—	Damaschke & Becker, 1964
		—	26	*ca.* 0.45[n]	—	—	Damaschke & Becker, 1964
Cubitermes sankurensis	Queen	(50)[p]	30	0.600	—	—	Hébrant, 1970*a*
	King	(8.50)[p]	30	0.329	—	—	Hébrant, 1970*a*
	Eggs	0.06	30	0.283	—	—	Hébrant, 1970*a*
	Larvae 1	0.26	30	0.362	—[s]	—	Hébrant, 1970*a*
	Larvae 2	1.25	30	0.253	—[s]	—	Hébrant, 1970*a*
	Workers	4.85	30	0.122	—	—	Hébrant, 1970*a*
	White workers[q]	1.68	30	0.249	—	—	Hébrant, 1970*a*
	Soldiers	5.18	30	0.135	—[s]	—	Hébrant, 1970*a*
	White soldiers[q]	3.69	30	0.283	—	—	Hébrant, 1970*a*
	Nymphs 1	(1.25)[t]	30	(0.253)[t]	—[s]	—	Hébrant, 1970*a*
	Nymphs 2	1.22	30	0.385	—[s]	—	Hébrant, 1970*a*
	Nymphs 3	2.53	30	0.301	—[s]	—	Hébrant, 1970*a*
	Nymphs 4	7.48	30	0.171	—[s]	—	Hébrant, 1970*a*
	Nymphs 5	9.66	30	0.156	—[s]	—	Hébrant, 1970*a*
	♀ imagoes[r]	9.18	30	0.339	—	—	Hébrant, 1970*a*
	♂ imagoes[r]	8.69	30	0.330	—	—	Hébrant, 1970*a*
Cubitermes exiguus	Queen	(40)[p]	30	0.625	—	—	Hébrant, 1970*a*
	King	(5.0)[p]	30	0.540	—	—	Hébrant, 1970*a*
	Eggs	(0.04)[u]	30	(0.375)[u]	—	—	Hébrant, 1970*a*
	Larvae 1	0.38	30	0.371	—	—	Hébrant, 1970*a*
	Larvae 2	0.94	30	0.268	—	—	Hébrant, 1970*a*
	Workers	3.87	30	0.131	—	—	Hébrant, 1970*a*
	White workers[q]	(1.25)[u]	30	0.320	—	—	Hébrant, 1970*a*
	Soldiers	3.89	30	0.179	—	—	Hébrant, 1970*a*
	White soldiers[q]	(2.90)	30	0.345	—	—	Hébrant, 1970*a*
	Nymphs 1	(0.94)[t]	30	(0.268)[u]	—	—	Hébrant, 1970*a*
	Nymphs 2	(0.91)[u]	30	(0.330)[u]	—	—	Hébrant, 1970*a*
	Nymphs 3	(1.82)[u]	30	(0.275)[u]	—	—	Hébrant, 1970*a*
	Nymphs 4	5.47	30	0.136	—	—	Hébrant, 1970*a*
	Nymphs 5	8.12	30	0.225	—	—	Hébrant, 1970*a*
	♀ imagoes[r]	5.72	30	0.449	—	—	Hébrant, 1970*a*
	♂ imagoes[r]	5.13	30	0.542	—	—	Hébrant, 1970*a*
Macrotermes bellicosus	Small and large workers	—	—	—	0.84	—	Hébrant, 1970*a*
Macrotermes bellicosus	Large workers	—	—	0.289	—	—	Ruelle, 1962
	Small soldiers	—	—	0.276	—	—	Ruelle, 1962
	Large soldiers	—	—	0.169	—	—	Ruelle, 1962

Table 6.13 (*cont.*)

Species	Stage investigated	Fresh weight mg per individual	Temperature °C	O_2 consumption µl per mg per h	RQ	Energy equivalent mcal per mg per h (mJ per mg per h)	Author
Nasutitermes exitiosus	Workers and soldiers in normal proportions	—	27	—	0.89	—	Day, 1938
	As above, about 5000 individuals with 225 g mound material	—	27	—	0.85–0.90	—	Day, 1938
Nasutitermes ephratae	Groups of 100 workers	—	22	0.31[n]	—[o]	1.55 (6.49)	Damaschke & Becker, 1964
		—	25	0.30[n]	—[o]	1.50 (6.28)	Damaschke & Becker, 1964
		—	28	0.67[n]	—	3.35 (14.02)	Damaschke & Becker, 1964
		—	30	0.82[n]	—	4.10 (17.16)	Damaschke & Becker, 1964
Nasutitermes costalis	Workers	0.49–1.19[n]	21–25	0.56–2.70[n]	0.91	55.61–11.53	Wiegert & Coleman, 1970[w]
	Soldiers	0.30–0.44[n]		0.47–2.81[n]	1.09	?	
	Nymphs	0.06–0.23[n]	21–25	1.25–2.84[n]	—	—	Wiegert & Coleman, 1970[w]
	Alates	4.78–6.05[n]	21–25	0.264–0.385[n]	1.67	?	Wiegert & Coleman, 1970[w]
	Queen	35.6[n]	21–25	0.80[n]	—	—	Wiegert & Coleman, 1970[w]
	Average (all castes)	—	21–25	1.07[v]	—	—	Wiegert & Coleman, 1970[w]

[a] Mean calculated from Cook's (1932) experiment (10 measurements) without correction for the evolution of the undetermined gas.
[b] Mean of two measurements not corrected for the evolution of the undetermined gas.
[c] Temperature not quoted: probably 20 °C.
[d] Mean of four measurements corrected for the evolution of the undetermined gas.
[e] Production of 0.0199 µl per mg fresh weight per hour of the undetermined gas.
[f] Mean of six measurements.
[g] Mean of 30 other measurements.
[h] The RQ reached 0.59 after three days and 0.57 after 14–17 days.
[i] Variations occurring while the 'atmospherics' varied; data deduced from authors' graph.
[j] Production of 0.007 µl H_2 per mg per hour.
[k] No hydrogen production.
[l] Production of 0.00125 µl H_2 per mg per hour.
[m] Mean calculated from author's table.
[n] Deduced from authors' graphs.
[o] Anomalies depending on the date of measurements (see discussion in text about Q_{10}).
[p] Arbitrary possible values.
[q] Stages between larvae and either workers or soldiers.
[r] Without their wings.
[s] Low or very low values (see text).
[t] Same values as those for larvae 1.
[u] Values extrapolated by Hébrant (1970*a*) from his results concerning *Cubitermes sankurensis* and the stage evolution in *C. exiguus*.
[v] All data of Wiegert & Coleman (1970) refer to dry weight.
[w] Deduced from author's equation; in order to compare with other termites. this respiratory rate would be 0.16–0.21 µl per mg fresh weight per hour at a temperature between 21 and 25 °C.

flagrant contradiction with the former measurements on isolated termites, and suggested that the oxidation of carbohydrates is also normal for humivorous species such as *Cubitermes*. The Q_{10} of whole societies (between 1.6 and 2.0) has already been referred to.

The next step in Hébrant's work was an interesting comparison. Knowing the mean composition of the society of *Cubitermes exiguus*, the

Table 6.14. *Calculation of energy flow in two* Cubitermes *spp. (after Hébrant, 1970a)*

Density (nests per hectare)	Biomass (kg per hectare, fresh weight)	Respiration (l oxygen per hectare per year)	Energy equivalent (MJ per hectare per year)
	Cubitermes exiguus		
213	4.8	12343	260
510	11.5	25902	550
652	14.6	37782	800
	Cubitermes sankurensis (extrapolated)[a]		
550	21.45	42559	900

[a] Possibly 10–12% overestimated.

relation of oxygen consumption (y) (in ml per day) with the fresh weight (in mg) of whole societies (W) was predicted by

$$\log y = 0.435 + 1.118 \log W \qquad (6.12)$$

After having measured the respiration of 21 whole societies (within their nest), a new relation was established:

$$\log y = 1.577 + 0.468 \log W \qquad (6.13)$$

The comparison between expected (computed by equation (6.12)) and measured values showed that in all cases but one the oxygen consumed by the whole society was higher than expected (61.6% higher on average). Hébrant concluded that the termites respired more when they were within their nest because they were in more natural conditions, with undisturbed social contacts and RH. It is also possible that the nest material and its microflora participated in the respiration when whole nests were under measurement and it may be that the collection and carrying of the nests induced the termites to recast the inner geometry. Anyway, as Hébrant pointed it out, the nest does not seem to constitute a barrier for gaseous exchange.

Finally Hébrant (1970*a*, *b*) calculated the energy flow due to the respiration of *Cubitermes exiguus*. The composition of 92 societies collected and counted by A. Bouillon over a year was used to calculate a mean annual biomass. The lowest, modal and highest classes of biomass encountered were 2, 22 and 52 g (fresh weight) and their respective calculated respiration rates were 19, 58.6 and 87.6 l oxygen per year. Table 6.14 gives Hébrant's results using an RQ of 1.0 as determined on whole societies and respiratory rates measured at 30 °C. In estimating the density, the young societies, which are still subterranean, have of course been discounted.

Respiration is, however, only a part of the energy flow. So Hébrant tried to attain an estimate of the production: that of alates was deduced from field observations with an assumed energy equivalent of the biomass of 17.5 kJ per g dry weight, and that of neutral individuals was assumed to be twice that of alates. Thus the energy flow in the case of *Cubitermes exiguus*, with a density of 510 nests per hectare, is 36.9 MJ per hectare per year (production) plus 548 MJ per hectare per year (respiration) which equals about 585 MJ per hectare per year. That is to say that they would consume 33.5 kg of carbohydrates per hectare each year with an energy equivalent of 17.5 kJ per g carbohydrate.

In the very poor, tropical soils, with 1% organic matter, 11.4% of it being cellulose and derived products, a *Cubitermes exiguus* population of 510 nests per hectare have to swallow 17000 kg of soil per hectare each year in order to get their 33.5 kg of cellulose. This could be overestimated: perhaps the termites are able to digest non-cellulose organic material and thus can select organic-rich soils to feed on. Nevertheless, it is more probably underestimated since the termites must have a rather low assimilation rate. If it was, for example, 15%, then Hébrant concludes that they would swallow 113000 kg of soil per hectare per year.

Elements of an energy flow were presented by Wiegert & Coleman (1970) working on *Nasutitermes costalis*, a nest-building species inhabiting the Puerto Rican rainforest. A relation was found between body weight and oxygen requirements in different castes (already referred to); some data deduced from the authors' graph are quoted in Table 6.13. Another relation was established between the total oxygen consumption and a volume index of the nest. The authors did not, however, detail the proportions of the castes; nor did they estimate the nest density. The RQ was determined in three castes and was particularly high (1.67) in the case of winged reproductives; this suggests that 'these nymphs were actively producing and storing fat at this stage of their development'.

The difficulty of measuring actual respiration of termites led Josens (1973) to tackle the problem another way. The consumption of a fungus-growing termite *Ancistrotermes cavithorax* (Sjöstedt) was measured in the field; that of a grass-cutting species *Trinervitermes geminatus* was measured in the laboratory. The production of neutral castes was estimated from both field and laboratory observations. It was emphasised that termites consumed their dead and therefore that the biomass produced in the neutral castes had to be added into the consumption (assuming in this preliminary calculation that predation was negligible). It was also emphasised that, from an ecological viewpoint, termite larvae must be considered as parasites since they feed upon an actual production of the

Table 6.15. *Preliminary data of the energy budget in two species of termites at the IBP Lamto Project (Ivory Coast), in an open* Hyparrhenia *savannah with* Borassus aethiopum (*after Josens, 1973*)

	Trinervitermes geminatus	*Ancistrotermes cavithorax*
Nest density (number per hectare)	57[a]	140
Termite density (million of individuals per hectare)	1.6[a]	1.9
Percentage of larvae (in terms of individual density)	7	47
Biomass (kJ per hectare)	30.2	9.75[a]
Percentage of larvae (in terms of biomass)	0.8	17
Production (kJ per hectare per year)	31.4[b]	134[b]
Ingestion		
Plant material (kJ per hectare per year)	314	7325
Dead (kJ per hectare per year)	31.4	134
Total (kJ per hectare per year)	345	7460

[a] These data are different from those of the author's original paper (which were erroneous).
[b] Underestimated (largely in the case of *A. cavithorax*) because it does not include the 'elaborated saliva' production.

workers. However, estimation of the secretion of the 'elaborated saliva' which is the larval food is not yet possible. Table 6.15 quotes these preliminary data on the energy budget in *T. geminatus* and *A. cavithorax*. In the latter species, the plant material ingested by the termites may seem enormous: this is the actual amount of wood, litter, etc. collected by the workers but it is the food of both the termite society and its symbiotic fungi. Otherwise the production is underestimated, since the production of 'elaborated saliva' has not been taken into account. Since the life duration of larval stages is about 26 days (Sands, 1960; Josens, 1973), every 26 days, at least* 17% of the biomass must be formed from food which is already produced. Let us suppose that the assimilation rate of 'elaborated saliva' by the larvae is 20%, then a substantial supplementary production of $(9.75 \times 0.17 \times 365)/(0.20 \times 26) = 116$ kJ per hectare per year must be added to the 134 kJ per hectare per year already mentioned in Table 6.15.

Concluding remarks on termites

It appears from recent investigations carried out in Africa that at least two – probably three – groups of termites play an important role, through their energetic requirements, in tropical ecosystems.

(*a*) Humivorous species when they are abundant may dig considerable

* 'At least', because this proportion refers to both younger and older larvae.

amounts of soil. Since their actual diet is in fact still unknown, their role cannot be exactly known either. Hébrant's (1970*a*) work has made possible a first quantitative estimate of their energy flow and of the soil-dwelling activities it involves. It should be borne in mind, however, that the magnitude of these activities is based on the assumption that *Cubitermes exiguus* feeds upon cellulose. This requires two comments. Firstly, it seems surprising that a population of these termites can be so successful (500–600 nests per hectare) while their food is diluted in the soil to a level of only 0.114%. Secondly, it should not be deduced from the example of *C. exiguus* that all the humivorous termites feed upon cellulose. Humivorous (or better, geophagous) species have proved very successful in Africa since there are at least 280 species on this continent. Often several species exploit the same soil, which suggests that their diet must be diverse.

(*b*) Fungus-growing termites consume enormous amounts of plant material, due to their symbiotic fungi and their high production. This in its turn is partially due to the parasite-like feeding of their larvae. These termites are common in Africa and can easily devour 8.8% of the primary production, as was shown in a humid savannah in the Ivory Coast (Josens, 1972), or up to 27–33% in cultures of maize and soya bean as in Tanzania (Bigger, 1966). In his recent thesis, Lepage (1974*b*)* mentions a consumption of 5.8% of primary production in a dry Sahel savannah (in northern Senegal) by fungus-growing termites.

(*c*) Finally, grass-cutting termites, such as *Trinervitermes geminatus*, which seems to be very frugal in comparison with fungus-growing termites, are probably species well fitted to drier savannahs. Their lower energy flow (low production, no greedy symbionts) is better adapted to lower primary production. These termites may, however, be very abundant in some dry savannahs and then they play a dramatic role as competitors with cattle.

Most interesting is the difference of RQ observed by Hébrant (1970*a*) between termites within their nest and those isolated from it. The circadian rhythm of respiration which occurs only when the termites are in their nest is also noteworthy. This social fact must probably be one of the main factors which has influenced these results, and the importance of working with whole societies in their nests whenever possible is obvious. Of course, the microflora of the nest may introduce some bias into the measurements but the nests of the termites are realities, and studies of the role the termites play in nature must take their nests and symbionts into account.

We are grateful to Dr F. Hébrant who permitted us to quote extensively from his thesis and we wish to thank Professor W. G. Whitford and Dr M. G. Nielsen for sending results in advance of publication.

* Many other interesting pieces of information are available in Lepage's thesis.

7. Nutrient dynamics of termites

J. P. LA FAGE & W. L. NUTTING

As a result of their peculiar nutritional dynamics, termites play a significant role in the biological turnover of cellulosic materials and in the physical and chemical alteration of soils. Termites are herbivores and detritivores variously involved in the comminution and decomposition of vegetable matter, through most of the warm temperate and tropical zones. The various categories of plant material which they use as food have been classified and discussed already.

Termites rely largely on symbiotic micro-organisms for the digestion of this diet, a situation similar to that in ruminants. Consequently, their nutrition is extremely complex, involving as it does, the interrelated biologies of the insect host, its intestinal protozoans and bacteria, and the chemistry of the plant material – the food itself – which has usually been altered, even before it is eaten, by both abiotic and biotic factors in the external environment.

The reviews by Becker (1971), Honigberg (1970), Lee & Wood (1971*b*), Noirot & Noirot-Timothée (1969), and Sands (1969) consider termite nutrition from rather different viewpoints. While maintaining some semblance of balance, we have chosen to emphasize the role of termites as processors of plant material in the detritus cycle. Since entomologists and ecologists are generally unfamiliar with the biochemistry of plant materials, we have begun by giving a brief account of the composition and structure of wood and what may happen to it as a result of weathering and the action of biological agents. This food is then followed in the individual termite through the conventional stages of digestion, absorption, metabolism and excretion, with special attention to the function of symbiotic micro-organisms. We also consider the utilization of nutrients at the colony level, particularly the special social functions of food exchange, the extent and importance of which are certainly not yet known or fully appreciated. Finally, we have taken the liberty not only to point out the gaps in our knowledge, but also to speculate freely on the challenging subject of termite nutrition, with the hope of inspiring others.

Chemical constituents of plant material

The sole nourishment for many species of termites, notably among the Calotermitidae, is sound, decay-free, deadwood (Hendee, 1933; Pence, 1956) which may contain as little as 2.5–3.0% moisture. It therefore seems appropriate to introduce the complex subject of nutrient dynamics in

Table 7.1. *Chemical elements and organic components in sound wood*

Element or component	% oven-dry weight	Author
Carbon	49–50	Tsoumis, 1968, modified
Hydrogen	6	Tsoumis, 1968, modified
Oxygen	44–45	Tsoumis, 1968, modified
Nitrogen	0.03–0.10	Cowling & Merrill, 1966
Ash	0.2–1.0	Tsoumis, 1968
Polysaccharides		
α-cellulose	40–50	Côté, 1968*a*; Tsoumis, 1968
Hemicellulose(= xylans mannans, etc.)	15–30	Tsoumis, 1968: p. 61
Pectic substances	< 1.0	Browning, 1952: p. 1189
Starch	0.5–5.0	Wise, 1952: p. 644
Gums and mucilages	—	
Lignin	15–35	Stecher, 1968: p. 619; Côté, 1968*a*
Protein	0.01–0.20	Cowling & Merrill, 1966
Extractives	1.0–10.0	Tsoumis, 1968

termite societies with a discussion of the chemical composition of this relatively simple food. Although the composition of wood varies widely according to species, age, tissue and growing conditions, we have listed in Table 7.1 the major elements and organic components as they might occur in typical, decay-free wood.

Wood carbohydrates, isolation and characterization

The term wood has been used rather loosely by many of the authorities cited in this book. The reader should be aware that technically it is only the xylary portion of the vascular tissue of trees, shrubs and, to a limited extent, herbaceous plants. It is well known that cellulose, the basic energy source for all termites, forms a large part of sound woody tissues. Curiously enough there is no general agreement in the literature as to its exact chemical nature. Green's (1963) definition has gained wide acceptance: 'Cellulose is a naturally occurring polymeric fiber found in woody tissues composed primarily of β-(1$\rightarrow$4)-linked D-glucopyranose units.' Although cotton fibers are almost 'pure' cellulose (Côté, 1968*a*), a small number of non-glucose residues occur regularly in this native (naturally-occurring) form. The existence of a naturally-occurring 'true', or chemically pure cellulose composed of only glucose units is thus doubtful.

Unfortunately, chemically pure cellulose is rarely produced in the laboratory fractionation of woody tissues, and those cellulosic compounds which are obtained vary considerably according to the methodology and raw materials employed. Jermyn's (1955) admonition is worth repeating:

'It cannot be too strongly emphasized that *cellulose* and *hemicellulose* are normally determined as the resultants of certain sets of operations, rather than as chemically defined species.' A brief survey of the 'operations' employed in analyzing woody tissue carbohydrates is presented here to clarify certain commonly encountered terms and to emphasize the necessity for accurately describing methods employed. As Jermyn noted, no single set of procedures will suffice for every situation and the analyst will often be required to develop or modify existing techniques.

The preparation of wood tissues for analysis involves a careful selection of samples which will reflect the overall composition of the material under study. The optimum particle size to begin with has been debated (Browning, 1952). While some investigators prefer rather large shavings, others utilize only those particles which will pass an 80–100-mesh sieve. The first chemical step normally involves extraction of the raw material in organic solvents such as ether, alcohol, or alcohol–benzene. Materials soluble in these solvents include fats, fatty acids, resins, resin acids, phytosterols, waxes and hydrocarbons (Browning, 1952). Succeeding extractions in hot and cold water remove inorganic salts, sugars, cyclitols, and various polysaccharides including gums, mucilages, starch, pectin and galactans (Browning, 1952). The insoluble residues are subsequently delignified by the chlorite method (Jermyn, 1955). The resultant material, 'holocellulose', contains the bulk of the remaining cell wall polysaccharides, with small amounts of lignin invariably included. In the following step, holocellulose is separated by mild alkali extraction into insoluble α-cellulose, and soluble hemicellulose components. As in the previous step this separation is often not distinct and is greatly influenced by alkali concentration. The α-cellulose may contain other sugar residues such as xylose and mannose. Both the α-cellulose and hemicellulose fractions may then be hydrolysed and subjected to qualitative and quantitative analyses. Finally, Jermyn (1955) cautions that, although the terms, 'hemicellulose', 'xylan' and 'pentosan' are sometimes used almost interchangeably, they are indeed different entities. The xylans are simply those polysaccharides which consist primarily of xylose; the pentosans are likewise those made up of 5-carbon pentose sugars. The hemicelluloses, as we have seen, are defined by a set of extraction procedures rather than the characterization of chemical species.

The basic approach to woody tissue analysis has progressed little during the last 20 years and, although much of the methodology has become somewhat standardized, confusion remains because some investigators have been negligent in reporting their laboratory procedures and careless with terminology. The necessity for paying strict attention to details should be especially obvious to those who study the organisms which habitually consume woody tissues.

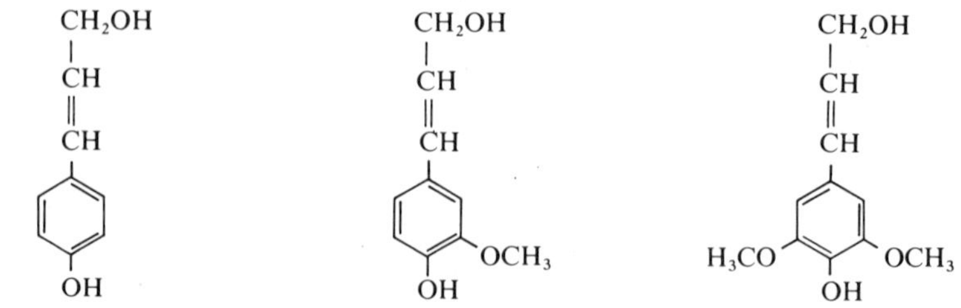

p-Coumaryl alcohol — Coniferyl alcohol — Sinapyl alcohol

Fig. 7.1. Lignin is composed of phenylpropane sub-units derived from hydroxycinnamyl alcohols such as these.

Although difficult to characterize precisely because of structural degradation during chlorite de-lignification, native cellulose molecules probably contain about 10000 sugar residues (Côté, 1968*a*). Hemicelluloses, though naturally variable, are much smaller molecules ranging from 50 to 200 residues in length. Degree of polymerization (DP) is a term often used to indicate the number of repeating units (sugar residues here) in a polymeric molecule.

Lignin

After cellulose, the lignins are normally the most abundant group of compounds in wood and the most abundant of the continuously cycled organic compounds in the biosphere (Kirk, 1971). They comprise an entire family of phenolic compounds which are structurally related to the polysaccharides. The lignins account for most of the methoxyl content of wood; they are resistant to acids, readily oxidized, soluble in hot alkali and bisulfate, and condense with phenols and thio compounds (Schubert, 1965). Kirk & Harkin (1972) describe lignin as 'an amorphous, three-dimensional, highly branched aromatic polymer', and suggest that one can best understand its structure from a 'biosynthetic viewpoint'. As such, it is composed of many phenylpropane (guaiacyl) sub-units derived from hydroxycinnamyl alcohols (Fig. 7.1). These undergo enzymatic oxidation to form free radicals which subsequently couple with one another at the site of lignification (Kirk, 1971).

The resultant lignin is deposited in close association with the cellulose of the plant cell wall to form an intractable complex which has generally been thought to consist of mutually interpenetrating polymers held together only by physical attraction (Cowling, 1961). Kirk & Harkin (1972) have suggested, however, that the association also involves a degree of chemical bonding between lignin and the wood polysaccharides of the cell wall. Regardless of the true nature of this association, it is clear that lignin strengthens the cell wall and protects the cellulose from attack by cellulolytic enzymes. For example, there are micro-organisms, such as *Bacillus polymyxa* (Prazmowski) Macé, that readily attack isolated cellulose, but cannot utilize it in wood at all.

Owing to the many possible linkages between the guaiacyl monomers and the variability among species, a definite structural formula for lignin cannot be given. Fig. 7.2 is a compilation of most of the possible functional groups and inter-unit linkages, but it is not a structural formula. Approximately 30% of the guaiacyl monomers of softwood lignin appear to contain a free phenolic group while 70% of the monomers possess an etherified phenolic group. About 20% of the monomers possess carbonyl groups on one of the carbons of the propyl group. Benzyl ether or benzyl alcohol groups are found on 43% of the units. Linkages commonly found between lignin monomers include carbon–oxygen–carbon and carbon–carbon. We suggest the texts by Schubert (1965) and Marton (1966) for a more complete discussion of the chemistry of lignin. As a group, the lignins are digested less efficiently by the termites and their symbiotes than the hemicelluloses or cellulose (Lee & Wood, 1971*b*; Wolcott, 1946). The remaining wood components, although quantitatively insignificant, may be critically important in determining the suitability of a particular wood as termite food.

Nitrogen

All living organisms require nitrogen for protein and nucleic acid synthesis. Since the nitrogenous compounds in wood are quantitatively minimal, they are of more than casual interest in any consideration of the nutrition of xylophagous organisms.

Occurrence and significance

The nitrogen content of plant materials varies tremendously. Herbaceous tissues normally contain 1–5% nitrogen by weight, whereas woody tissues contain only 0.03–0.10% (Table 7.2) (Cowling & Merrill, 1966). The carbon/nitrogen ratio in woody tissues is normally about 350–500:1. It may, however, be as high as 1250:1 in heartwood of Sitka spruce which is highly susceptible to fungus attack (Cowling & Merrill 1966). Levi, Merrill & Cowling (1968) report comparative ratios for several other plant materials: tomato foliage (10:1), tobacco stems (55:1), cotton-seed hairs (200:1), microbiological agar medium (200:1).

It should be noted that three factors contribute to the highly variable determinations which have been published on total nitrogen in wood tissues. These are the low nitrogen content, high carbon content, and

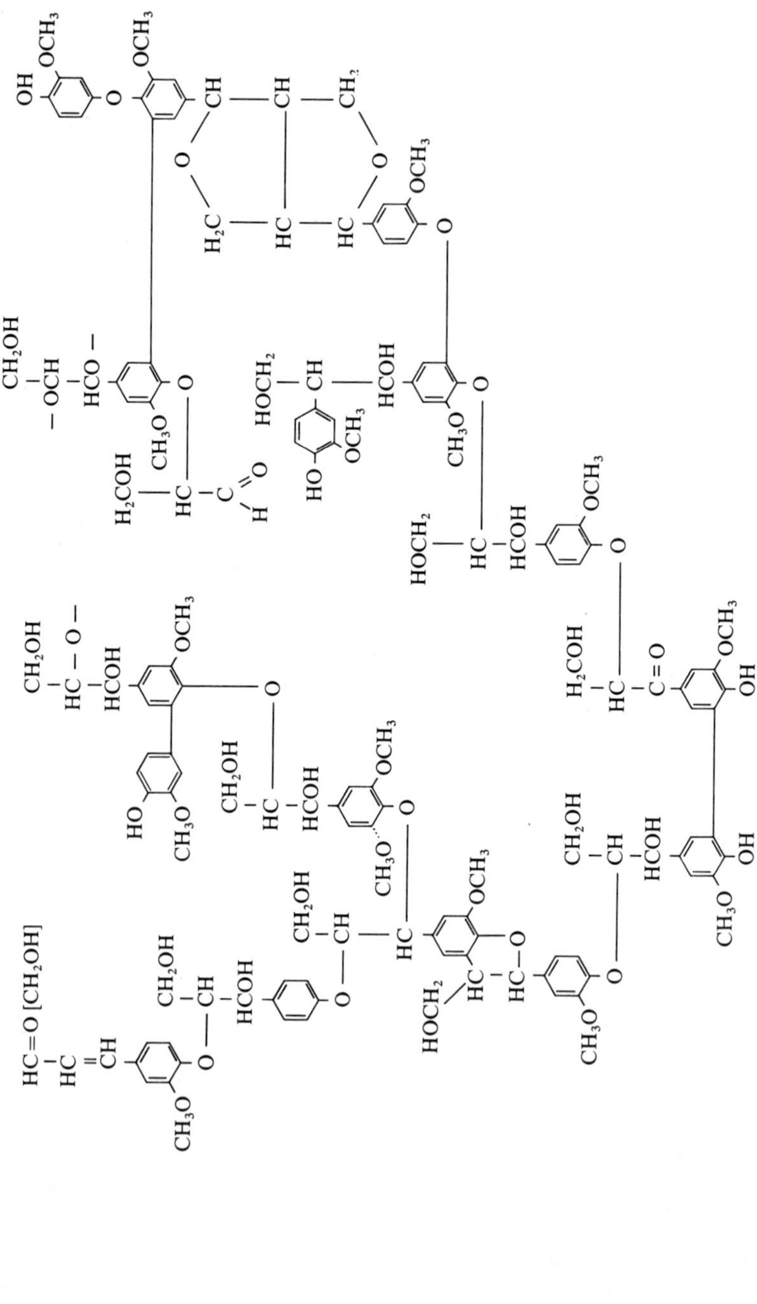

Fig. 7.2. Summary of lignin structure (Kirk, 1971).

Table 7.2. *Nitrogen content of stem wood of various gymnosperms and angiosperms. The data are for individual samples from a given tree (Cowling & Merrill, 1966)*

Tree species	Nitrogen content (% by weight)	Tree species	Nitrogen content (% by weight)
Gymnosperms		*Pseudotsuga taxifolia*	0.051
Abies concolor	0.045	*Sequoia sempervirens*	0.067
Abies magnifica	0.227	*Taxodium distichum*	0.057
Juniperus virginiana	0.139	*Tsuga canadensis*	0.106
Larix occidentalis	0.180	Angiosperms	
Libocedrus decurrens	0.097	*Carya ovata*	0.100
Picea engelmanni	0.118	*Castanea dentata*	0.072
Pinus contorta	0.071	*Juglans nigra*	0.100
Pinus echinata	0.130	*Liquidambar styraciflua*	0.057
Pinus elliottii	0.050	*Liriodendron tulipifera*	0.088
Pinus lambertiana	0.124	*Quercus alba*	0.104
Pinus monticola	0.113	*Quercus rubra*	0.099
Pinus palustris	0.038	*Quercus stellata*	0.096
Pinus ponderosa	0.052	*Quercus velutina*	0.070
Pinus strobus	0.087		
Pinus taeda	0.068		

extreme heterogeneity of woody materials. Additional variation stems from the many modifications of the basic Kjeldahl process which have been used almost exclusively for these determinations. In general all Kjeldahl procedures suffer the similar short-coming of not fully recovering nitrogen bound in nitrogen–nitrogen (N–N) or nitrogen–oxygen (N–O) linkages (Rennie, 1965). Much of the previous work on nitrogen budgeting among termite societies should probably be re-examined owing to the general lack of uniform Kjeldahl methodology and the current knowledge that the nitrogen–oxygen linkage is common in several woody tissues, notably among the Leguminosae.

Early biologists often speculated about the ability of termites to survive and reproduce on diets apparently containing very little or no nitrogen (Cleveland, 1925*a*). The specific quantity of nitrogen in wood is most critical for the dry-wood species. After swarming, the reproductives establish their nest in a dead branch or log of limited size which serves not only for shelter but also as the sole source of nutrients. The composition of the nest material undoubtedly changes qualitatively as well as quantitatively, and eventually must limit colony growth (Nagin, 1972). Although the relationship between quality and quantity of available nitrogen and termite growth is not well understood, we do have data on the

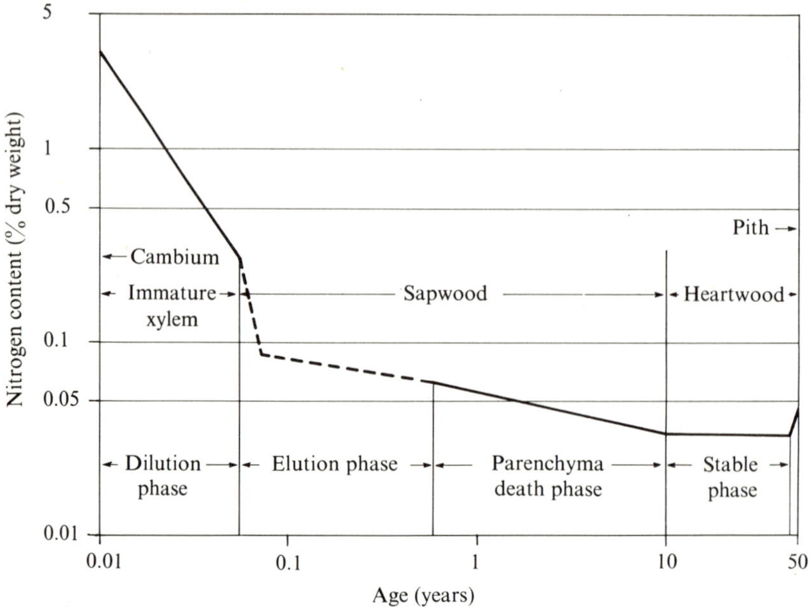

Fig. 7.3. Hypothetical graph of changes in nitrogen content during various stages in the maturation and aging of the xylary tissues in a typical angiosperm or gymnosperm. The major phases of change in nitrogen content have been named according to the predominant mechanism believed to be involved at various stages after cell division in the maturation of wood cells (Cowling & Merrill, 1966. Reproduced by permission of the National Research Council of Canada from the *Canadian Journal of Botany*, **44**, 1539–54, 1966).

distribution of this element in wood (Cowling & Merrill, 1966) and some knowledge about the effects of varying nitrogen levels on the development of two xylophagous beetles, *Hylotrupes bajulus* (L.) (Becker, 1971) and *Anobium punctatum* De Geer (Bletchley, 1969).

Origin and distribution

Nitrogen compounds are synthesized in the cambial cells as structural proteins, enzymes, peptides, amino acids, lipoprotein membranes (Cowling & Merrill, 1966) and certain, special hydroxyproline-rich proteins deposited in the primary walls of differentiating cambial cells (Scurfield & Nicholls, 1970). These cells contain 1–5% nitrogen dry weight basis (dwb) (Cowling & Merrill, 1966). Protein found in old or dead wood has often been attributed to dried cytoplasmic remnants derived from differentiating cambial cells (Wise, 1952), but Cowling & Merrill (1966) suggest that little or none of the cytoplasmic nitrogen of differentiating cambial cells remains after lignification.

Fig. 7.3 is a double logarithmic graph of the hypothetical changes in the nitrogen content of woody tissues during various stages in the maturation of a typical angiosperm or gymnosperm. The graph shows that tissue nitrogen drops rapidly as wood matures. Shortly after cambial cell formation the gradual deposition of cellulose, lignin and other secondary cell wall substances dilutes the nitrogen content of the immature xylem from 5.0 to 0.5%. Lignification is normally completed by the beginning of the second stage of maturation (elution phase), 10 to 30 days after cell division at the cambium. It is during this period that the vascular elements (wood fiber cells) die and lose most of their nitrogen-rich cytoplasmic constituents to more active tissues. Although experimental evidence is lacking, Cowling & Merrill (1966) have theorized that the xylary cell contents undergo an autolytic depolymerization and are subsequently eluted upward in the transpiration stream. This process presumably functions as an internal recycling mechanism whereby limited supplies of nitrogen and other nutrients are removed from essentially dead tissues and re-distributed among the living tissues of the stems and crown. By the end of the first growing season, and following cell division at the cambium, the nitrogen content of most woody tissues has dropped to 0.06–0.05% (Cowling & Merrill, 1966).

The longitudinal and ray-parenchyma cells of the sapwood live considerably longer and maintain a somewhat higher nitrogen concentration than other vascular elements of the xylem. The death and loss of nitrogen from these cells to the transpiration stream proceed slowly through what is known as the 'parenchyma death phase'. The nitrogen content of most wood species has stabilized near 0.03% by the time the sapwood has transformed into heartwood (stable phase). In conifers a slightly higher nitrogen concentration is often found near the pith (Fig. 7.3). Table 7.3 lists the nitrogen content of various parts of several species of stem wood. Despite considerable variation in the literature, Cowling & Merrill (1966) conclude that a higher percentage of nitrogen exists: (1) in angiosperms than in gymnosperms; (2) within the crown of the tree than below it; (3) in sapwood than in heartwood; (4) nearer the cambium than close to the sapwood–heartwood interface; and (5) nearer the pith (in conifers) than in the more recently formed heartwood.

Forms of nitrogen

Scurfield & Nicholls (1970) have isolated and characterized the protein-bound amino acids (PBAAs) in the wood of *Eucalyptus* sp. and *Pinus radiata* D. Don. Their study provides insight into the problems involved in nitrogen analysis and their data help to explain why nitrogen recoveries are often low and inconsistent. Prior to hydrolysis in 6 N HCl, samples of each wood were extracted first with water then with methanol and ethanol: benzene (1:2 v/v). Amino acids were subsequently identified and measured with an amino acid analyzer. Additional samples of raw wood,

Table 7.3. *Nitrogen content of portions of the stem wood of various tree species (after Cowling & Merrill, 1966)*

Tree species	Nitrogen content (% dry weight)			
	Cambium	Immature sapwood	Sapwood	Heartwood
Gymnosperms				
Picea mariana	1.10	—	0.056	0.059
P. mariana	1.11	0.27	0.047	0.062
Pinus sylvestris	3.25	—	0.012	—
Pinus	3.33	—	0.130	—
Corsican pine, butt	—	—	0.108	0.067
Corsican pine, crown	—	—	0.085	0.063
P. sylvestris, butt	—	—	0.086	0.079
P. sylvestris, crown	—	—	0.052	—
P. sylvestris	—	0.115–0.078	0.047–0.039	0.031–0.047
Angiosperms				
Eucalyptus regnans	2.03	0.52	0.062	0.031
Fraxinus elatior	4.59	0.88	0.22	—
Fraxinus sp.	4.70	0.89	0.22	—
Ulmus sativa	4.69	0.81	0.27	—
Ulmus sp.	4.80	0.83	0.28	—

solvent-extracted wood, and hydrolysis residues were subjected to Kjeldahl and/or Dumas analysis for total nitrogen content. Results of these analyses showed that approximately 50% of the total nitrogen in *P. radiata* was recovered as amino or ammonia nitrogen and included all of the commonly encountered amino acids. Lincoln & Mulay (1929) found a similar situation in pear wood. There were, however, both quantitative and qualitative differences between the wood species tested. While certain nitrogen-containing compounds were removed during the extractions, it is important to note that hydrolysis residues held an estimated 15% of the total (Kjeldahl-determined) wood nitrogen. This suggests that some of the nitrogen must have combined with certain cell wall constituents, presumably carbohydrates or lignin. In addition, many non-protein amino acids have been isolated from plant tissues, including γ-amino butyric acid and β-alanine (Robinson, 1963).

Wood species which contain unusually high amounts of nitrogen may immediately be suspected of possessing large quantities of alkaloids (Wise, 1952). The alkaloids comprise a fairly large, heterogeneous group of compounds, all of which contain nitrogen, usually in a heterocyclic ring. Some of the more common alkaloids are strychnine, quinine, morphine, cocaine, mescaline, ephedrine and nicotine. They are found primarily outside the xylem, and have received intensive chemical study because of their toxic properties (Wise, 1952). For unknown reasons, many of the alkaloids poisonous to man have little effect on xylophagous organisms (Robinson, 1963).

Lipids

The lipid component of wood (ether extract) is complex as well as variable. It contains numerous fats, fatty acids (FAs), phytosterols and waxes, all of which may influence wood–termite and wood–microbe relationships. The FAs in tall oil, a resinous by-product from the manufacture of wood pulp, include oleic, linoleic, linolenic and palmitic acids (Harris, 1952: p. 611). Carter, Dinus & Smythe (1972*b*) have reported the FA contents of 2:1 chloroform: methanol extractions of sound *Pinus taeda* L. (loblolly pine), *Pinus elliottii* Engelm. var. *elliottii* (slash pine) and *Acer saccharum* (Marsh.) (sugar maple). Oleic acid was the most abundant FA from all three types of wood, with concentrations of total FA present varying from 12.5 in maple to 45.4% in slash pine. The other FAs found in descending quantitative order were: linoleic, palmitic, palmitoleic, stearic and myristic. Many other minor and trace compounds have been isolated from ether extracts of wood. Among those of probable significance in termite nutrition are vitamin A precursors such as β-carotene. Tsoumis (1968) states that total wood extractives normally vary between 1 and 10% of the oven-dry weight of wood. Many tropical species, however, contain extractives in excess of 30%.

Minerals

Inorganic minerals occur in varying amounts in plant tissues, with quantitative and qualitative differences most closely related to soil type. Wise (1952: p. 658) reports that 27 different minerals have been isolated from *Pinus strobus* L. Although the mineral requirements of insects are poorly known, it would appear that termites have an ample supply of these nutrients at their disposal.

Minor constituents and their significance

Many constituents of plant tissue have been omitted from our presentation thus far. This is not to suggest that they are unimportant, but rather than their relevance to termite nutrition is as yet unknown or only suspected. Many of these compounds are probably involved in modifying the attractiveness or resistance of wood to termites, and thus its suitability as food. Although specific termite attractants, such as vanillic acid, *p*-hydroxybenzoic acid, *p*-coumaric acid and protocatechuic acid, have been

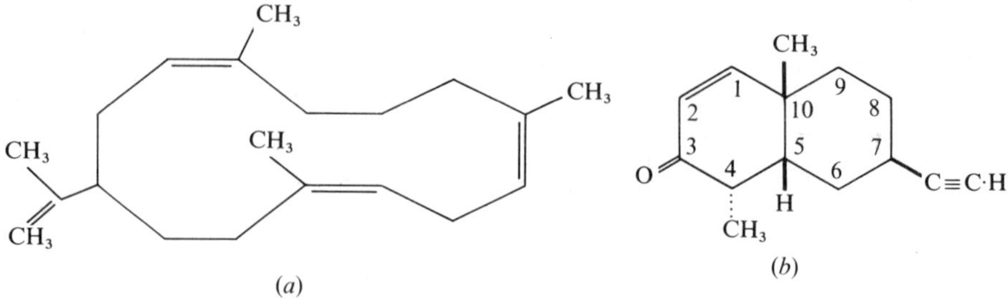

Fig. 7.4. Miscellaneous plant compounds. (*a*) nasutene (Moore & Brown, 1971–72). (*b*) chamaecynone (Stevens, 1967).

isolated from fungi (Becker, 1964), none has yet been reported from the higher plants. Becker (1971) has reviewed the studies on substances found in wood which attract xylophagous insects. For example, females of *Hylotrupes bajulus* and another large cerambycid, *Ergates faber* (L.), are attracted by the $C_{10}H_{16}$ hydrocarbons (α- and β-pinenes), carene, and sabinene (Becker, 1971). Certain wood-borers and their predators are drawn to turpentine (Becker, 1971) while wood-inhabiting ambrosia beetles may be attracted by low concentrations of ethanol (Becker, 1971). It seems reasonable to expect that similar examples will be found, particularly of the sort which would lead to the establishment of dry-wood termites in suitable deadwood nesting sites. Another example of a minor plant constituent which may affect termites involves cinnamyl alcohol. This compound already mentioned as a basic building block of lignin (Fig. 7.1), is a strong releaser of trail-following behavior in *Reticulitermes flavipes* (Kollar) (Blum & Brand, 1972). Moore & Brown (1971–72) found that the trail pheromone, nasutene (Fig. 7.4*a*), of an Australian *Nasutitermes* spp. was a terpenoid compound, apparently identical with cembrene-*A*, which has been isolated from a plant *Commiphora mukul* (Burseraceae). Although many compounds of plant origin possess strong pheromonal activity, it is not known whether they are transported directly to the glands associated with certain modes of behavior or metabolized and reconstructed at these sites if, indeed, they are utilized at all (Blum & Brand, 1972).

The resistance of wood to termite attack has been considered in Chapter 4 of this volume, and Becker's (1971) review should provide an excellent literature base for further reading. In general, natural termite resistance is a function of many variables, including toxic or repellent chemicals, nutrient imbalance or lack of specific growth factors, adverse physical characteristics, and pre-conditioning. Termites are repelled or poisoned by a number of plant compounds, among them a steam-volatile oil isolated from *Chrysocoma tenuifolia* (Berg) (Compositae) which possesses both contact and vapor toxicity for the harvester *Hodotermes mossambicus*

(Hagen) (Hewitt & Nel, 1969*a*). Saeki, Sumimoto & Kondo (1973) have isolated chamaecynone (Fig. 7.4*b*), a termiticidal agent, from the essential oil of Japanese sawara wood, *Chamaecyparis pisifera* D. Don. Becker (1971) reports that stilbenes, quinones, pyran derivatives and furfural are among the many other compounds found in plant tissues which are toxic or repellent to termites.

In addition to the B vitamins, plants contain certain water-soluble growth factors which are now distinguished as lipogenic factors because they function primarily in FA metabolism rather than as enzyme co-factors (Dadd, 1973). As such, the exogenous requirements for choline and inositol are far greater than for the B vitamins. While only phytophagous insects have a demonstrated need for inositol, Dadd speculates that dietary choline is probably essential for all insects. Fortunately for termites, both myo-inositol (= mesoinositol) and its derivatives, and choline, are common in many types of plants. Choline occurs in many phospholipids and functions as a phosphate carrier in plant sap (Robinson, 1963). It is possible that Goetsch's 'vitamin T', found in termites and other organisms, is actually a mixture of these growth substances with additional, known vitamins (Stecher, 1968).

Behr, Behr & Wilson (1972) found an inverse relationship between the quantity of wood consumed by *Reticulitermes flavipes* and wood hardness and specific gravity. Buchner (1965: p. 819) has remarked that no attempt has been made to compare the vitamin content of different species of wood or areas within wood, and that such studies are critical to understanding the symbioses of certain wood-boring beetle larvae. In addition, McMahan (1966*a*) found that young pairs of *Cryptotermes brevis* (Walker) maintained on poplar for one year showed a preference for it when given a choice of poplar or maple, the latter a normally 'more preferred' wood.

External modification of plant material

The flow of nutrients from sound deadwood to termite tissue is a lengthy and poorly understood process involving numerous degradative and synthetic steps, both abiotic and biotic. A series of external events often involving physical and meteorological factors as well as the degradative actions of fungi and other micro-organisms, combine to make wood acceptable to termites. Once the food is ingested symbiotic organisms provide still further nutrient degradation before absorption.

Abiotic factors

Under natural conditions it is difficult if not impossible to distinguish wood decay which results from the effects of light, oxygen and water from that caused by living organisms. Tiemann (1951) held that no wood decay could occur abiotically. While living organisms account for a great percentage of wood decay, other data suggest that substantial degradation may result from strictly abiotic processes. Campbell (1952) summarized the findings of several authors and concluded that both oxygen and moisture were essential for the formation in wood of oxycellulose, a brittle substance which becomes reduced to a loose powder by the expansion and contraction of the wood. In one study, poplar wood, presumably free from living organisms, was exposed to sunlight for 100 hours in Florida. Considerable oxidation of cellulose was noted, in addition to a 30% reduction in lignin content. The end-products of this decay were water soluble in all cases. Weathering appears to proceed most intensely at the wood surface and is most pronounced when the ratio of surface area to volume is greatest. Although the effects of wood-destroying fungi on the woody materials attacked by *Gnathamitermes perplexus* (Banks) have not been investigated, its habit of consuming only the outer layer of weathered fibers suggests that abiotic processes may influence food selection by this Arizona termite (La Fage & Nutting, unpublished).

Air pollutants, notably sulfur dioxide, also contribute to accelerated wood decay by causing a partial hydrolysis of cellulose as well as some loss of lignin. Additional abiotic factors which promote wood decay include prolonged heating, ultra-violet radiation, leaching of water-soluble extractives by rainfall, and degradative oxidations resulting from prolonged contact with ferrous metals (Campbell, 1952). There is general agreement, however, that no abiotic agent promotes wood decay as quickly or efficiently as living organisms, especially fungi.

Biotic factors

Fungi, micro-organisms and even the larger herbivores transform plant materials into forms which are used as food by a variety of termites.

Wood-decaying fungi

The associations of termites and fungi have been studied by many investigators: Cook & Scott, 1933; Hendee, 1933, 1934, 1935; Hungate, 1940, 1941, 1944; Becker & Kerner-Gang, 1964; and recently reviewed by Sands (1969). The significance of such relationships remains somewhat obscure and debatable. Lee & Wood (1971*b*) have reported that many termites, including all of the Hodotermitidae except the sub-family Hodotermitinae, many of the Rhinotermitidae and Termitidae and a few of the Calotermitidae, consume wood only after it has undergone extensive decay by wood-destroying fungi. Hendee (1933) found 17 genera of fungi associated with *Incisitermes minor* (Hagen) (as *Calotermes minor*) which normally feeds on sound deadwood. Although she was unable to establish a specific

Table 7.4. *Comparison of the characteristics of wood decayed by brown-rot versus white-rot fungi (after Côté, 1968*b)

Characteristic	White-rotted wood	Brown-rotted wood
Color	White, bleached appearance	Reddish-brown
Constituents removed	Holocellulose and lignin	Holocellulose
Shrinkage	Nearly normal	Abnormally high, especially longitudinal
Static strength	Reduced slightly	Reduced greatly
Toughness	Rapidly reduced	Rapidly reduced
Effect on degree of polymerization (DP)	Gradual decrease	Rapid decrease
Preferred hosts	Hardwoods	Softwoods

relationship between the termite and any given genus of fungus, she speculated that fungi may cause chemical modifications in wood which render it more attractive to termites. In all of her experimental colonies fungi were most abundant on the inner surfaces of termite galleries, suggesting that the termites consumed fungi as well as wood. She noted that macroscopic examination of wood for the presence of early fungal attack was often inadequate, and described a method for staining thin sections cut from the walls of termite galleries which would demonstrate the presence of fungi.

Termite–fungal associations apparently span the entire range of two-species interactions described by Odum (1959). Examples include neutralism (Becker & Kerner-Gang, 1964), proto-cooperation (Sands, 1969), amensalism (Kovoor, 1964*b*), parasitism (Sannasi, 1969), mutualism (Grassé, 1959; Sands, 1969), and commensalism (Williams, 1965).

Kinds of fungi

The three classes of fungi responsible for most wood decay are the Ascomycetes, Fungi Imperfecti and Basidiomycetes. Some members of each class are associated with termites. The latter group, which includes approximately 2000 wood-destroying species (Schubert, 1965), is divided into two major groups, the brown-rots and the white-rots, on the basis of gross structural changes which they cause in wood. The brown-rot species attack primarily the carbohydrate component of wood while the white-rot species derive nourishment from both carbohydrate and lignin (Schubert, 1965; Côté, 1968*b*). Additional differences between these two groups are listed in Table 7.4. The Fungi Imperfecti and Ascomycetes are known collectively as the soft-rot fungi. They are less important than the Basidiomycetes since they normally attack wood surfaces exposed to very wet conditions and consume only the cellulosic components.

Relationship to termites

Several species of termites are nutritionally dependent on fungi (Sands, 1969). It now appears that the fungi affect termite nutrition basically in two ways. First, fungi modify wood by adding or removing compounds which attract, repel, or poison termites. A number of studies have dealt with the attractive substances produced by the fungus *Lenzites trabea* Pers. ex Fr., and other Basidiomycetes: Watanabe & Casida (1963); Allen, Smythe & Coppel (1964); Becker (1964); Smythe, Allen & Coppel (1965); Matsumura, Coppel & Tai (1968); Ritter & Coenen-Saraber (1969). Becker & Kerner-Gang (1964) tested more than 60 species of Fungi Imperfecti for their effects on termite survival and found that attractive, repellent and toxic substances were produced by the fungi. They noted, moreover, that different strains of a single fungus had varying effects on the same termite species. Further, different termite species displayed variable responses to the same strain of fungus. Sands (1969) concluded that termites reared on a particular fungus-infested medium might respond favorably to that fungus simply because of habituation. Kovoor (1964*b*) found that the white-rot fungus, *Ganoderma applantum* (Pers.) Pat., produced substances which were toxic to *Microcerotermes edentatus* Wasmann. Williams (1965) reported that, while the heartwood of British Honduras pitch pine, *Pinus caribaea* Morelet, is highly resinous, in some areas it is regularly attacked by *Coptotermes niger* Snyder (80% of the trees). He also noted that all termite infestations in this tree were accompanied by a brown-rot identified as *Lentinus pallidus* Berk. & Curt., although the rot was often found in the absence of termites. From a series of laboratory tests he concluded that the heartwood was susceptible to termite attack only after the fungus had removed repellent materials such as turpentine and rosin.

Second, fungi modify wood by consuming and digesting it. The usual end-products consist primarily of decayed wood, fungal tissue and respiration products, although decay may be so complete that only carbon dioxide, water and minerals remain. Termites are attracted to, and thrive on, wood which has been partially degraded, principally to cellulose and pentosans, but generally show little interest in wood in advanced stages of decay. Termites cultured on slightly decayed wood show significant increases in survival and biomass over those on sound wood (Becker, 1971). Cowling (1961) observed a rapid depolymerization of carbohydrates and significant alteration of lignin during the early stages of wood decay by the brown-rot *Poria monticola* Murr. The resulting production of soluble degradation products greatly exceeded their metablism by the fungus. Consequently, abundant soluble carbohydrates as well as fungal tissue are available to termites during early wood decay. During the final stages of decay, however, he found that previously formed degradation products were metabolized much more quickly than they were formed. While thoroughly rotted (punky) wood is attractive to certain termites such as *Coptotermes niger* (Williams, 1965), it may be no more than acceptable to some (Hendee, 1933), or completely avoided by other species, possibly because of its low carbohydrate content. Becker (1971) recognized that increasing the nitrogen content of wood had a growth-accelerating effect on larvae of the beetle *Hylotrupes bajulus* only when carbohydrate content of the decayed wood was still high. He also suggested that fungi may effect changes in the amino acid balance of decaying wood and noted that these larvae require a specific amino acid balance for normal development. Fungi probably add nitrogen and vitamins to decaying wood which accelerate termite growth but Becker did not give any details as to how this might be accomplished.

Effects on nitrogen compounds in wood

Hungate (1941) observed substantial increases in the nitrogen content of fungus-infested wood when it was kept on soil for 13 months. The fungus presumably assimilated soil nitrogen which remained in the wood when hyphae died. Since protein probably accounts for less than 50% of the total nitrogen content of wood, it is logical to question whether fungi are capable of metabolizing and transforming non-protein nitrogen into products which are more easily utilized by termites and their symbiotes. Levi *et al.* (1968) screened 26 nitrogenous compounds known or presumed to occur in wood for their ability to support growth of the white-rot *Polyporus versicolor* L. ex Fr. The most rapid growth occurred on proteins, amino acids, a peptide, nucleic acids (ribonucleic acid and desoxyribonucleic acid) and a purine (guanine), while intermediate growth was observed with monomers of chitin, inorganic ammonium compounds, nucleotides, a nucleoside (adenosine triphosphate), purines and a pyrimidine (cytosine). Uracil, chitin (the polymer), sodium nitrate and sodium nitrite produced little or no measurable growth. Although no alkaloids were tested, they concluded that *P. versicolor* was capable of utilizing most nitrogenous compounds known to occur in wood. The metabolic fate of the tested compounds in the fungus or their availability to termites was not examined. In an earlier study, however, Roessler (1932) tested four nitrates (K^+, Ca^{2+}, Na^+, NH_4^+) for their effects on the growth rate of *Zootermopsis nevadensis* (Hagen) (as *Termopsis nevadensis*). Three of these compounds (K^+, Ca^{2+}, NH_4^+) appeared to stimulate development when filter paper moistened with 0.001 M salt solutions was fed to groups of 100 termites. Experimental groups were compared with control groups on filter paper soaked with distilled water but no statistical analysis was presented.

Table 7.5. *Chemical constituents of sweetgum sapwood in progressive stages of fungal decay expressed as percentages of the dry weight of the original sound wood (after Cowling, 1961)*

Type of decay and causal organism	Average weight loss (%)	Total carbo-hydrate	Glucan	Hemicelluloses: Galac-tan	Man-nan	Xylan	Araban	Lig-nin	Total extrac-tives	α-cellu-lose
Sound wood	0	77.1	52.3	1.1	2.7	20.1	0.9	22.9	4.3	46.4
White-rot	5.2	74.0	—	—	—	—	—	20.9	4.1	43.1
caused by	14.5	67.0	—	—	—	—	—	18.5	3.6	37.9
Polyporus	55.2	33.9	22.8	0.3	1.7	8.9	0.3	10.9	3.7	16.3
versicolor	96.7	2.5	2.0	0.03	0.1	0.3	0.03	0.8	—	—
Brown-rot	4.5	72.1	—	—	—	—	—	23.4	7.8	32.8
caused by	15.3	61.6	—	—	—	—	—	23.1	14.8	12.1
Poria	55.0	23.5	—	—	—	—	—	21.5	14.7	0.0
monticola	69.5	9.2	5.4	0.1	0.2	3.4	0.1	21.3	13.9	0.0

Effects on lignin and cellulose

Although the effects of fungi on the nitrogen-containing compounds of wood remain poorly understood, the mechanism, and end-products of lignin and carbohydrate decay are somewhat clearer. Cowling (1961) performed an extensive series of experiments to study the effects of decay by white- and brown-rot fungi on sweetgum, *Liquidambar styraciflua* L. Table 7.5 compares the chemical composition of sweetgum sapwood at various stages of decay with that of sound wood. With the brown-rot, 72.1% of the total carbohydrate complement remained after a 4.5% weight loss and 61.6% remained after a 15.3% weight loss. When 69.5% of the wood weight was lost, however, a mere 9.2% of the total carbohydrate remained. Equally dramatic are the decreases in the xylan and araban content. During brown-rot decay the lignin content remained fairly constant; however, with the white-rot fungus, both carbohydrate and lignin were consumed as decay progressed. Similar results were reported by Kovoor (1964*b*) for the white-rot *Ganoderma applantum* and the brown-rot *Trametes trabea* (Pers.) Bres., grown on *Populus virginiana* Foug. In the light of this information it is interesting to consider Becker's (1971) finding that termites prefer wood which has lost from 5 to 15% of its mass as a result of attack by brown-rot fungus. Depolymerization of the long cellulose molecules found in sweetgum sapwood (DP > 1600) and other woody tissues apparently occurs as a result of enzymatic hydrolysis similar to that by which other polysaccharides are degraded. Points of attack within the molecule have been theorized to occur randomly within the polymer such that shorter but primarily polymeric structures result from degradation (Cowling, 1961). An alternative hypothesis suggests that very short chains or single glucose monomers are hydrolyzed from the ends of the cellulose molecule.

The mechanisms and end-products of lignin degradation by fungi have been studied most extensively with the white-rots which not only degrade lignin but are capable of assimilating and metabolizing it. The brown-rots, although apparently capable of partially degrading lignin by demethoxylation, neither assimilate nor metabolize it (Kirk, 1971). According to Kirk (1971) the exact mechanism by which white-rot fungi degrade lignin is still unknown. Two fungal enzymes, laccase and phenolase, have often been implicated in the degradation of lignin; however, their role is not yet fully understood and Kirk has concluded that additional enzyme systems must be required for complete lignin destruction by white-rot fungi. It is not known if the enzymes necessary for these further conversions are present in termites; however, if they are, at least some lignin might be metabolized via the Krebs cycle to provide them with usable energy.

Other effects

In addition to their primary effects on biopolymers and nitrogen compounds, fungi may cause other changes in the chemistry of wood which affect termites. Among these, the possibilities that they synthesize vitamins required by the termites has been mentioned by a number of authors: Grassé, 1959; Sands, 1969; Lee & Wood, 1971*b*; and Wood, Chapter 4, this volume. Although we have never seen it mentioned, fungal metabolism of wood may provide termites with substantial quantities of free water. Cartwright & Findlay (1958) estimated that an impressive 139 l of water would be produced from the respiration of 50% of the cellulose in a cubic meter of wood by the brown-rot fungus *Coniophora cerebella* Pers.! The external biotic modification of living and dead plant material is also affected symbiotically. Although many termites are partially dependent on the digestive capabilities of fungi, the Macrotermitinae and wood-destroying Basidiomycetes of the genus *Termitomyces* participate in a more binding relationship. Details of their symbiosis have been studied by numerous authors whose work has been discussed by Sands (1969) and reviewed more briefly by Lee & Wood (1971*b*), and Wilson (1971). The role of the fungi in the termite economy is explained below under *Social recycling of nutrients*.

Other micro-organisms

According to Lee & Wood (1971*b*) numerous fungi and bacteria are always present in the walls of termite mounds. Pathak & Lehri (1959) reported that soil from termitaria of *Hypotermes obscuripes* (Wasmann) contained significantly greater carbon and nitrogen concentrations than the surrounding soil. They also noted that nitrogen-fixation rates were greater in mound materials than in adjacent soil. Boyer (1955) found elevated levels of nitrogen-fixing bacteria (*Azotobacter chroococcum* Beijerinck, *Azotobacter indica* (Starkey & De) Derx, *Azotobacter lacticogenes* (= *A. indica*), *Beijerinckia* spp.) and cellulolytic bacteria in the walls of termite mounds. Meiklejohn (1965) found that termite mounds in Rhodesia contained more cellulose decomposers, denitrifiers, ammonifiers and nitrifiers than surrounding soils but fewer nitrogen-fixing bacteria of the genera *Beijerinckia* (aerobic) and *Clostridium* (anaerobic). There is little doubt then that the presence of these micro-organisms in mound walls affects the quality of materials ingested by the termites during nest enlargement. Many of these modifications probably resemble those caused by the previously discussed fungi.

Certain soil bacteria also modify the constituents of wood, particularly that on or in the ground, prior to its attack by termites. Recent data suggest that the following genera of bacteria at least partially degrade lignin: *Flavobacterium*, *Micrococcus*, *Mycobacterium*, *Pseudomonas* and *Xanthomonas* (Kirk, 1971).

Herbivores and wood-borers

To this point the pre-modification of termite nutrients has been viewed only in terms of abiotic phenomena and microbial intervention. Other alterations, however, are possible, such as those caused by the digestive processes of large herbivores and wood-eating arthropods. Ferrar & Watson (1970) listed 46 termite species which feed on herbivore (primarily bovine) droppings in Australia, of which 32 belong to the genus *Amitermes*. They further noted that six rare species were found only in dung. They suggested that termites are attracted to dung because it contains considerable quantities of undigested cellulose and, when fresh, moisture. In the Sonoran desert cattle consume range forage which contains 7–15% crude protein (dwb); their droppings contain about 7–10% crude protein. *Gnathamitermes perplexus* commonly feeds on the dung, perhaps somehow attracted to this easily accessible, high-nitrogen diet (La Fage, unpublished). A primitive dry-wood calotermitid *Pterotermes occidentis* (Walker), often drives its galleries through the tunnels of wood-boring beetle larvae which are tightly packed with coarse, sawdust-like frass. This material has not been analyzed but it appears to contain a great deal of undigested wood (Nutting, unpublished).

Utilization of nutrients at the individual level

Whether they be wood- or soil-dwelling termites, varying numbers of foragers gather food as environmental conditions (temperature, moisture,

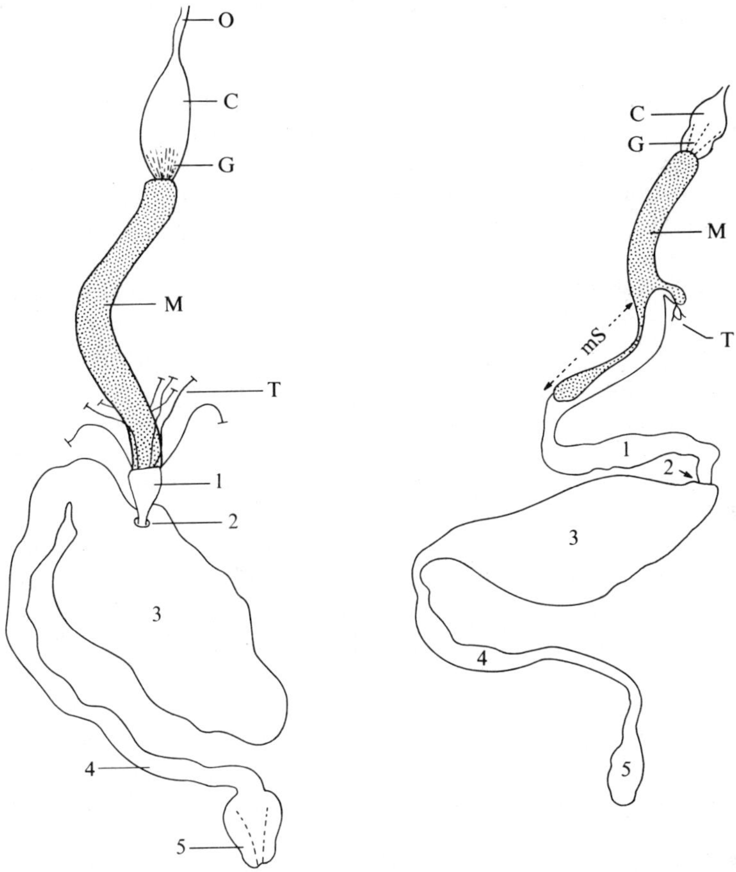

Fig. 7.5. Digestive tube of (*a*) *Calotermes flavicollis*; (*b*) *Termes hospes*. C, crop; G, gizzard; M, midgut; mS, mixed segment; O, oesophagus; T, Malpighian tubules; 1–5, segments of the hindgut; 1, first segment; 2, enteric valve; 3, paunch; 4, colon; 5, rectum (Noirot & Noirot-Timothée, 1969).

etc.) permit (La Fage, Nutting & Haverty, 1973). Raw food, variously modified by external factors, is processed primarily by individuals of certain forms or castes before being distributed among the various dependent castes.

The digestive system, morphology and physiology

The morphology of the alimentary system of worker termites as it relates to feeding habits has been described in Chapter 4 of this volume. Somewhat more detail is needed to provide sufficient background for the discussion of digestion and absorption which follows. Further information and references on most of the points covered here are given in the excellent review by Noirot & Noirot-Timothée (1969). The digestive tube of the termite (Fig. 7.5*a*), divided into a foregut (stomodeum), midgut (mesenteron) and hindgut (proctodeum), differs little from that of a cockroach. The fore- and hindguts, of ectodermal origin, possess cuticular linings. The midgut, of endodermal origin, is lined with a glandular epithelium derived from specialized cells in regenerative crypts. These linings are shed and renewed during moulting.

A slight posterior dilation of the foregut serves as a crop where small amounts of masticated food can be held before further comminution by the hardened cuticular teeth of the gizzard. At the anterior end of the midgut a ring of cells secretes the so-called peritrophic membrane, which resembles a sausage casing lying within the midgut. It has been studied in only two species of termites and whether it protects the delicate secretory epithelium as presumed in many other insects is not known. Further on, the excretory organs (Malpighian tubules), varying from twelve or fifteen to four according to family, empty into the gut at or near the junction of mesenteron and proctodeum. A curious 'mixed segment' (Fig. 7.5*b*) occurs in many of the termitids where the wall of the mudgut extends somewhat beyond the origin of the Malpighian tubules. The gut wall here is thus mixed – composed partly of mesenteron (sometimes dilated) and partly of proctodeum. An apparently pure culture of bacteria typically resides here between the gut wall and the peritrophic membrane. The excretory tubules are closely associated with the thickened mesenteral epithelium of the mixed segment; the possible function of this complex in recycling nitrogenous excretions is discussed below under *Nitrogen.*

The hindgut includes five distinguishable segments (Fig. 7.5*a*), although considerable variation occurs especially among the Termitidae. The short first segment ends in an enteric valve which prevents the contents of the hindgut from coming forward into the midgut. The most prominent in all termites is the enlarged second segment or paunch which contains most of the symbiotic fauna or flora. Katzin & Kirby (1939) estimate that the micro-organisms in the hindgut of *Zootermopsis angusticollis* (Hagen) constitute about one-third its wet weight. In some termitids the first segment is also enlarged and may then be referred to as the first paunch. Most of the volatile FAs produced by the symbiotes are probably absorbed in the hindgut, across the cuticle and through the thin epithelium of the paunch(es). The specialized epithelium of the rectal pads is apparently important in water and ion resorption, particularly in the Calotermitids which typically feed on dry wood (Collins, 1969).

At least two sets of accessory glands are associated with the digestive system. The mandibular glands have their outlets in the pre-oral cavity

Table 7.6. *Hydrogen ion concentration in major segments of the digestive tract (after Noirot & Noirot-Timothée, 1969)*

Species	Hydrogen ion concentration of			
	Foregut	Midgut	Hindgut	Paunch
Calotermes flavicollis[a]	5.2–5.4	6.8–7.5	5.0–7.5	—
Anacanthotermes ahngerianus[b]	7.7	6.9	7.9	—
Anacanthotermes turkestanicus[b]	—	—	7.7	—
Zootermopsis angusticollis[c]	5.2–6.8	5.2	3.0–6.8	—
Reticulitermes lucifugus	5.6[g]	6.8–7.2[g]	7.8[b]	7.4[g]
Reticulitermes sp.[d]	5.5	—	5.0	—
Microcerotermes edentatus[e]	8.8–9.6	8.8–9.6	6.0–9.6	7.2–7.6
Microcerotermes sp.[b]	—	—	8.3	—
Trinervitermes trinervoides[f]	—	5.6	—	—

[a] Noirot & Noirot-Timothée, 1969.
[b] Korovkina, 1971.
[c] Randall & Doody, 1934.
[d] Brown & Smith, 1954.
[e] Kovoor, 1967*b*.
[f] Potts & Hewitt, 1974*a*.
[g] Grassé & Noirot, 1945.

between the mandibles and the maxillae. Although nothing definite is known of their function, Lebrun (1973) has suggested that they might secrete one of the caste-controlling pheromones. (See *Caste determination* below.) Salivary glands and reservoirs have their ducts opening at the base of the hypopharynx ('tongue') and labium. They secrete certain digestive enzymes whose functions are discussed below under *Non-symbiotic digestion*. The glands of the workers also elaborate a 'saliva' which is an important and, sometimes, the exclusive food of the dependent castes. (See *Stomodeal food* below.)

The pH of various segments of the digestive system has been determined in about nine species of termites. We have added the results from a few additional studies to the summary given by Noirot & Noirot-Timothée (1969) (Table 7.6). Although no clear trends are evident, Kovoor (1967*b*) suggests that pH might vary according to the food consumed and, in the hindgut, the functional state of the Malpighian tubules.

Passage of food through the digestive tract

In the only comprehensive study to date Kovoor (1967*c*) allowed workers of *Microcerotermes edentatus* to consume barium sulfate for about two hours to provide a bolus opaque to X-rays which could be subsequently observed at intervals as it passed through the gut. She found that the rate of movement was slowed by low temperature, low relative humidity and isolation of the termites. Under culture conditions of 25–30 °C and about 90% relative humidity the mean time for such a bolus to pass through the entire gut was about 24 hours. Movement through the foregut was rapid. Noirot & Noirot-Timothée (1969) found that passage through the foregut of *Calotermes flavicollis* (Fabr.) takes 30 min to one hour. In Kovoor's (1967*c*) study the midgut was traversed in one to one and a half hours but movement was slower in the mixed segment. The bolus moved into the hindgut on an average four hours after ingestion; Krishna & Singh (1968), using chromatographic methods, found that various sugars reached the hindgut in *Odontotermes obesus* (Rambur) about 10 min after ingestion. In *Microcerotermes edentatus* the first proctodeal segment is considerably dilated and retains the bolus for four to five hours. Then, after passage through the enteric valve, the bolus moves into the second paunch where most digestion takes place. Some 8–12 hours are passed here with an additional two to three hours required for transit through the colon to the anus. Kovoor's study clearly demonstrates that nutrients do not pass from the paunch back into the midgut through the enteric valve. Thus, absorption must occur through the proctodeal wall or else in the fore- and midguts of other individuals who have solicited 'proctodeal food'. (See *Trophallaxis* below.)

Digestion, absorption, metabolism, and excretion

Non-symbiotic digestion

While it is widely known that the lower termites derive nutrients from the activities of their symbiotic micro-organisms, it is rarely recognized that the termites themselves are capable of extracting nutrients from dietary food sources. Cleveland (1924) noted that, following defaunation, workers of *Reticulitermes flavipes* died within 10–20 days when fed sound wood. On a diet of somewhat decayed wood, however, their survival was prolonged and when fed dextrose (glucose), peptone and starch, either separately or together, defaunated individuals survived considerably longer. These observations suggested that the defaunated termites were indeed capable of digesting and absorbing certain nutrients and, thus, must possess digestive enzymes of their own.

Hungate (1938) quantified the relative importance of protozoans and host in digestion and utilization of wood by *Zootermopsis angusticollis*. Extracts from fore-, mid- and hindguts were incubated with a substrate prepared from precipitated filter paper which contained cellulose and small amounts of other carbohydrates. Reducing substances were found in these preparations after 16 hours of incubation at room temperature: slight accumulations in the fore- and midgut, relatively large amounts in the hindgut preparations. Similar experiments using more highly purified cellulose, however, revealed no reducing substances in either the fore- or midgut regions, and considerably less in the hindgut. These results suggested that, although cellulose was attacked only by microbes, enzymes other than cellulase were present throughout the gut which could hydrolyze soluble carbohydrates. From additional experiments involving defaunated termites Hungate concluded that enzymes produced in the wall of the gut itself could account for about one-third of the digestion of normal termites. *Zootermopsis* could potentially oxidize about one-third of its total wood intake independently of its symbiotes, yet must oxidize about seven-eighths of its intake to satisfy its energy requirement (Hungate, 1938). These data were based on sawdust of 'fairly sound' Monterey pine as the nutrient source and it is probable that assimilation efficiency would vary according to level of decay in wood.

Cook (1943) fed several different carbohydrates to normally faunated and defaunated groups of *Z. angusticollis*. Those fed wood or cellulose showed respiration quotients (RQs) between 0.9 and 1.0, typical of carbohydrate utilization; however, post-absorptive RQs fell to between 0.70–0.75, suggesting that lipid reserves were being metabolized. The participation of protozoans in metabolism, signaled by the evolution of hydrogen, was noted on all diets except starch. Defaunated termites showed RQs near 1.0 when fed starch, sucrose, maltose, lactose, glucose, fructose, or galactose. Since no hydrogen was evolved, Cook concluded that this termite must possess enzymes to digest, absorb and oxidize several carbohydrates independently of its symbiotic microfauna.

Among other attempts to feed termites non-cellulosic diets are those of Montalenti (1927) and Lund (1930). Montalenti was able to maintain *Calotermes flavicollis* for several months on soluble starch, either alone or mixed with glucose. During this period the number of gut symbiotes declined rapidly. Lund, on the other hand, found that when carbohydrates other than cellulose were fed to *Zootermopsis* (as *Termopsis*) sp. the termites generally died shortly after the disappearance of the last protozoans. In general, his experimental diets enabled termites to live no longer than starved controls. Dextrose prolonged the life of the defaunated termites for approximately 40 days although bacteria became numerous during this period.

Digestive enzymes and their origins

Considerable effort has been directed toward the identification of specific digestive enzymes. Montalenti (1932) found an amylase in the foregut of *Calotermes flavicollis*, an amylase, invertase and a protease in the midgut, and an amylase and invertase in the hindgut; Noirot (1969*a*) also noted an amylase in its salivary secretion. Hungate (1938) recorded an amylase from the foregut but not the hindgut of *Zootermopsis angusticollis*. Using Child's (1934) histological studies of this termite he concluded that the amylase was most likely secreted by the salivary glands. While he found proteolytic activity only within the midgut, he did not discount the possibility that enzymes produced by symbiotes in the hindgut might have diffused forward or been obtained by trophallaxis. To test the latter possibility he measured the enzyme activity of freshly collected fecal pellets from normal and defaunated individuals. Some activity was noted in both groups; however, it is now known that materials passed via proctodeal feeding are quite different from fecal material (Noirot & Noirot-Timothée, 1969). Since the enteric valve apparently prevents the hindgut contents from escaping forward into the midgut, it is probable that any midgut proteases come from feeding on proctodeal material from other individuals.

Uttangi & Joseph (1962) concluded that amylases are generally present in the foregut of most termites, while proteases are formed in the midgut. Rao (1962) studied the distribution of proteolytic enzymes in the gut of *Heterotermes indicola* (Wasmann) and found cathepsin, chymosin (a peptic enzyme) and rennin. All but cathepsin were found in defaunated as well as faunated individuals. In defaunated individuals of the same species Krishnamoorthy (1960) also found amylase, invertase, sucrase, maltase, lactase and lipase. In no study mentioned, however, have bacteria-free termites been employed. The bacteria found in termitids appear to be especially resistant to removal (Speck, Becker & Lenz, 1971; Potts & Hewitt, 1973). As Lee & Wood (1971*b*) suggest, the non-cellulosic materials may very well be digested in part by these micro-organisms.

The most definitive studies on termite enzymes are those by the South African group led by P. H. Hewitt (Potts & Hewitt, 1972, 1973, 1974*a*, *b*; Retief & Hewitt, 1973*a*, *b*, *c*; Hewitt, Retief & Nel, 1974). Retief & Hewitt (1973*a*) found that enzyme preparations from whole guts of soldiers, workers, or larvae of a harvester *Hodotermes mossambicus* hydrolyzed sucrose, maltose and trehalose. Soluble starch was rapidly hydrolyzed by preparations from all three castes. A specific α-glucosidase, responsible for the hydrolysis of both maltose and sucrose, was most active in the midgut of larvae and workers, the first paunch of larvae and the second paunch of soldiers. No activity was found in the foregut of soldiers, first paunch of workers, or second paunch of larvae. An α-galactosidase was noted in workers and larvae but not soldiers. A gut trehalase was also found. Trehalose occurs in the spores of many fungi (Nielsen, 1962); is it possible that this termite is equipped to utilize such a source of energy? All individuals studied were normally faunated so that no conclusions could be drawn concerning the origins of these enzymes.

Subsequently, Retief & Hewitt (1973*b*) studied the β-glycosidases of *H. mossambicus*. During the period shortly before to slightly after ecdysis, protozoans and food particles are largely absent from the termite gut, a condition easily recognized because the empty gut appears light colored or white through the transparent cuticle. Total body extracts prepared from faunated larvae, workers and soldiers rapidly hydrolyzed salicin, cellobiose, ONPG, lactose, niphegal, and CM-cellulose, which suggests the presence of β-glucosidase, β-galactosidase, and cellulase, an enzyme which will be discussed under *Cellulose digestion.* β-galactosidase was absent from the foregut of all castes and generally most active in the midgut or in the first paunch of larvae. Its presence was noted in both faunated and defaunated termites. β-glucosidase was active throughout most of the gut in all castes but especially active in the first paunch of larvae. Unlike the β-galactosidase this enzyme was more active in faunated than in defaunated individuals. Although specific sites of secretion remain unknown, non-cellulolytic enzymes are found throughout the mid- and hindgut of all castes studied, even in segments where symbiotes are absent. Here is additional evidence that termites elaborate at least some of these enzymes independently of their symbiotic fauna and flora.

Potts & Hewitt (1974*b*) discovered laminaranase in whole abdomen extracts of *Trinervitermes trinervoides* (Sjöstedt) workers. The β-(1→3)-glucans, which are the substrates for this enzyme, are found in marine algae, callose of pollen and woody tissues of higher plants, cereal plants and the endosperm of some seeds. Only two other animals have been reported to possess laminaranase, a marine mollusk and a terrestrial snail, although the enzyme is common in bacteria, fungi and higher plants. It is not known to have any digestive function in termites. An aryl-β-glucosidase, isolated from heads and guts of the harvester termite *T. trinervoides* (Potts & Hewitt, 1972; Hewitt *et al.*, 1974) appears also to have no digestive function. A chitinase has been found in whole body extracts of *Nasutitermes exitiosus* (Hill) and *Coptotermes lacteus* (Froggatt) (Tracey & Youatt, 1958); the possible significance of this is discussed below under *Mutilation and cannibalism.*

Cellulose digestion

Cellulose is probably the most abundant of the continuously cycled organic materials (Kirk, 1971) and the soil bacteria and fungi are the primary effectors of its degradation. Nearly all plant materials ultimately pass through this 'microbial bottleneck' before recirculation can proceed (Nielsen, 1962). Although cellulose is metabolized by relatively few animals, there are two groups that employ basically different digestive strategies. The first group includes some protozoans, mollusks and arthropods which actually produce their own cellulolytic enzymes. Members of the second group lack this ability themselves but possess a rich gut fauna and/or flora which partially metabolize the cellulose. There is mounting evidence that termites belong to both groups. Among the cellulolytic animals few have been studied more intensively than the termites and their symbiotes, yet we still know very little indeed about their digestive physiology. The lower termites have symbiotic protozoans in the hindgut; the higher termites do not, but contain symbiotic bacteria instead. Because of this difference the problem of cellulose digestion must be considered separately for the two. Moore (1969), Noirot & Noirot-Timothée (1969), Honigberg (1970) and Lee & Wood (1971*b*) have written general reviews of cellulose digestion by termites. Lasker (1959) presents a good discussion of the problems of defining the relative importance of host and micro-organism in cellulose digestion.

The lower termites

Although early investigators had recognized the presence of protozoans in the termite gut, Cleveland (1923, 1926) was the first to present experimental evidence that the protozoans actually contributed to the digestion of wood. He observed that, when the protozoans were killed, either by starvation, incubation at elevated temperatures, or by increased oxygen tension, the termites lost their ability to utilize wood as a nutrient. When, however, the same termites were refaunated by proctodeal feeding they regained their former capacity. Trager (1932) was the first to demonstrate a cellulase from the intestinal flagellates of the wood-feeding cockroach *Cryptocercus punctulatus* Scudder, *Reticulitermes flavipes* and *Zootermopsis angusticollis*. He maintained one of the cellulolytic flagellates, *Trichomonas termopsidis* Cleveland, for three years, a feat yet to be equaled. Honigberg (1970) discusses Trager's work in detail along with several other less successful culture attempts. Lasker & Giese (1956) point out that the mere presence of cellulolytic organisms in the gut of cellulose-consuming animals does not necessarily signify that they benefit the host. It is possible that other microbes are contributing to the host's well being and/or that the host is partially self sufficient.

Few efforts have been made to assess the cellulolytic capability of the microflora which also occurs in the hindgut, or of completely defaunated and deflorated termites. Brief reviews of these attempts have been given by McBee (1959) and Mannesmann (1972). Work which suggests that gut bacteria of the lower termites contribute to their host's digestive capacity includes that of Beckwith & Rose (1929), Tetrault & Weis (1937), Ritter (1955) and Mannesmann (1969). Cleveland (1928), Dickman (1931), Hungate (1936, 1955) and Honigberg (1970) all argued that, although cellulolytic gut bacteria may in fact be present, they are insignificant in the overall process of providing energy for their host. Mannesmann (1972) demonstrated cellulolytic bacteria in the guts of two rhinotermitids *Reticulitermes virginicus* (Banks) and *Coptotermes formosanus* Shiraki. The slow rate at which the bacteria attacked filter paper, however, suggested that their importance was secondary. Tetrault & Weis (1937) demonstrated that bacteria from the gut of *R. flavipes* are also capable of degrading cellulose.

An objection to most searches for cellulolytic bacteria in the lower termites might lie in the frequent use of cotton or filter paper as a cellulose source in the experiments. Recall the fact that wood, the normal food of many termites, is a far more complex material with its cellulose partially protected from enzyme attack by lignin and other binding compounds. No attempts have yet been made in any of these studies to determine the relative importance of the bacteria versus the protozoans in the overall digestive regime of these termites. However, as several authors have already concluded, it appears that the lower termites depend primarily on their symbiotic flagellates for the digestion of cellulose.

The termite gut has generally been considered to be an anaerobic milieu although no direct measurements of oxygen tension have been attempted. This statement is based on the observation that efforts to culture gut fauna and flora have generally succeeded only under anaerobic conditions (Trager, 1934). The conclusion has followed that all symbiotes in the termite gut are obligate anaerobes (Hungate, 1939; Honigberg, 1970). The bacteria of *Coptotermes formosanus* and *Reticulitermes virginicus*, however, have been successfully cultured under aerobic conditions (Mannesmann, 1972). Further, succinate dehydrogenase, an enzyme normally associated with the (aerobic) Krebs cycle, has been demonstrated in atypical mitochondria of flagellates from the gut of *Reticulitermes lucifugus* (Rossi) (Lavette, 1970). Tricarboxylic acid cycle enzymes have also been isolated from strictly anaerobic ruminal bacteria where they may be involved in the synthesis of oxaloacetate and α-ketoglutarate (Allison, 1969).

Lee & Wood (1971*b*: p. 137) summarize Seifert & Becker's (1965) experiments on the lignin-degrading capabilities of various termites. From these data it is apparent that termites have a real facility for lignin degradation; however, the enzymes which attack lignin require oxygen, a fact inconsistent with the concept of an anaerobic hindgut. Clearly this is a problem which must be resolved.

The higher termites

As a result of the early work of Cleveland (1923) and others, most students of termites know that there are no protozoans in the hindgut of many species of higher termites and that, when they do occur, they are apparently neither numerous nor mutualistic (Honigberg, 1970). Cleveland (1926) wondered whether this situation signaled the beginning or the end of a symbiotic relationship. With the rather meager knowledge of food habits of termites then available, he believed that the Termitidae rarely fed on wood or cellulosic material but that when they did it was more decayed (partly digested) than it was in the diets of the lower termites. Although he did not deny that they might be able to digest cellulose, he felt that there was some cause and effect relationship between their general lack of protozoans and their choice of decayed plant material. If they were able to utilize cellulose, he thought they were probably assisted by micro-organisms. It has since become better known that many of the termitids do feed on sound deadwood, particularly members of the Macrotermitinae and Nasutitermitinae (Noirot & Noirot-Timothée, 1969; Lee & Wood, 1971*b*). However, whether or not their complex intestinal flora play any role in cellulose digestion remains unsettled.

Although Misra & Ranaganathan (1954) did not conclusively prove that bacteria were producing them, the mere presence of cellulolytic enzymes in whole gut preparations of the termitid *Odontotermes obesus* (as *Termes*) was a significant discovery. Pochon, De Barjac & Roche (1959) found a cellulolytic flora in *Sphaerotermes sphaerothorax* (Sjöstedt) (Macrotermitinae) similar to that in the rumen of cattle. Tracey & Youatt (1958) confirmed the presence of a cellulase in *Nasutitermes exitiosus* but did not determine its source. An anaerobic, cellulose-decomposing actinomycete *Micromonospora propionici* Hungate was isolated from the crushed alimentary tracts of *Amitermes minimus* Light by Hungate (1946*a*). The slow growth rate in culture plus the low number of colony-producing units per termite suggest that this bacterium is not very important in the termite's overall energy budget.

It now appears that a non-microbial cellulase is produced by at least two termitids, *Trinervitermes trinervoides* and *Microcerotermes edentatus*. In *T. trinervoides*, Potts & Hewitt (1973) found approximately 70% of all cellulase activity and 50% of the cellobiase activity in the midgut which contains no bacteria. The paunch, which contains abundant bacteria, possessed very little cellulase activity but 43% of the cellobiase activity. In addition, they found about 40% of the cellulase activity of the midgut localized in the gut wall. The authors concluded that this cellulase was of termite origin as no intracellular bacteria are known in any termite except the primitive *Mastotermes darwiniensis* Froggatt. The cellobiase, on the other hand, could well have a bacterial origin. After partial purification, including the removal of cellobiase, the cellulase from *T. trinervoides* was tested for temperature and pH effects (Potts & Hewitt, 1974*a*) and for its hydrolytic mechanism (Potts & Hewitt, 1974*b*). The temperature/activity peak of the enzyme was found to be 60 °C, above which denaturation was rapid. The pH optimum was 5.8 ± 0.2. When tested with CM-cellulose, the primary hydrolysis product was cellotriose with some cellobiose and very little glucose. Potts & Hewitt (1974*b*) concluded that the enzyme was probably an endo- rather than an exo-cellulase, since the latter would produce nearly all glucose residues. Although they did not test it, cellobiase would subsequently hydrolyze cellotriose units to cellobiose and glucose. CM-cellulase and cellobiase activity were also studied in *Microcerotermes edentatus* by Kovoor (1970). She found as much cellulolytic activity in the mixed segment, mid-, and foregut as in the hindgut. These results are somewhat less convincing than those of Potts & Hewitt (1973) because of the small number of samples and variation in the results.

Lasker & Giese (1956) have effectively demonstrated the aposymbiotic production of cellulase by an insect. They fed uniformly labelled ^{14}C-cellulose to bacteria-free silverfish *Ctenolepisma lineata* (Fabricius). Because $^{14}CO_2$ was respired, they concluded that the cellulose was metabolized and therefore must have been digested. A similar investigation of the capacity of symbiote-free termites to metabolize ^{14}C-cellulose would go far toward clarifying the question of whether or not termites produce their own cellulase. In summary we can do no better than to repeat Honigberg's (1970: p. 25) challenge that still 'there is... no incontestible evidence in support of any particular mechanisms' for cellulose digestion among the higher termites. We can only agree with Cleveland's (1926) early suspicions and those of recent reviewers (Noirot & Noirot-Timothée, 1969) that the intestinal bacteria are probably involved; however, the possibility that the termitids themselves possess some capability – as do the lower termites – cannot be ruled out.

End-products and metabolic pathways

In his studies with a flagellate *Trichomonas termopsidis* from *Zootermopsis angusticollis*, Trager (1932) demonstrated a protozoan enzyme complex which hydrolyzed cellulose and cellobiose to glucose, which was in turn metabolized by the termite host. Although he presented some indirect evidence to substantiate this, he was unable to demonstrate reducing substances in the fluid of any active flagellate culture. This inconsistency was reconciled, in his opinion, by the presence in all cultures of an active, glucose-fermenting bacterium.

Randall & Doody (1934) noted of *Z. angusticollis* that 'when the gut contents were acidified, there was always a strong odor of acetic or similar organic acid of low molecular weight. This hints strongly at the production of acetates or butyrates in the digestion of cellulose.' Hungate (1939) subsequently reported that acetic acid, hydrogen and carbon dioxide were the primary end-products of the protozoan digestion of cellulose. The argument for volatile fatty acid (VFA) production was reinforced by Hungate's (1943) observation that the termite hindgut was permeable to acetate and that, although acetate could be detected in the hindgut, it was never found in fecal material. Glucose, on the other hand, was present only in trace quantities in the hindgut. The primarily argument against the 'acetate' theory stems from Cook's (1943) failure to maintain termites on a diet of sodium acetate in agar. Honigberg (1970) concluded that the pH (5–6) of Cook's medium might have been lethal. Additional work by Brown & Smith (1954) showed that the hindgut contents of an unidentified termite (probably *Z. angusticollis*), contained large quantities of acetic acid and smaller amounts of propionic acid. Hungate (1943) showed that, in *Zootermopsis* sp., the ratio of carbon dioxide/hydrogen/acetate varied according to differences in the protozoan faunules of individual colonies. He was able to account for 70–75% of the ingested carbon in the form of carbon dioxide and short-chain organic acids, primarily acetic. Bacteria isolated from the gut of *Reticulitermes flavipes* fermented cellulose powder to acetic, lactic and butyric acids *in vitro* (Tetrault & Weis, 1937).

Kovoor (1967*a*) found at least three VFAs in the paunch of a higher termite *Microcerotermes edentatus* and attributed their production to its symbiotic microflora; she later reported that this termite produced its own cellulase (Kovoor, 1970). The fact that workers in the process of molting did not hydrolyze cellobiose as they did during intermolt periods suggests that cellobiase is produced by the microbes. Cellobiase activity in nymphs, however, was much greater in the midgut than in hindgut regions, suggesting a possible termite origin. An examination of the gut contents of *Trinervitermes trinervoides* for VFAs would be enlightening, since Potts & Hewitt (1973) concluded that this species produces its own cellulase. Pochon *et al.* (1959) found VFAs, particularly acetic and butyric, in the digestive tube of *Sphaerotermes sphaerothorax*.

Hungate (1939, 1946*b*) outlined a general theory for the symbiotic utilization of cellulose as it occurs in the lower termites. The symbiotic protozoans first satisfy their own energy requirements through anaerobic fermentations. The end-products of these fermentations, primarily VFAs, are then oxidized by the host for its energy needs. Such a scheme seems far more plausible to us than that presented by Trager (1934), particularly with regard to energy utilization. The hydrolysis of cellulose alone simply cannot provide adequate energy for the symbiotes. The standard free-energy change resulting from the phosphorolysis of glycogen is 3.05 kJ (Lehninger, 1970). It is probable that the removal of the terminal glucose residue from cellulose is likewise endergonic, i.e. energy requiring. It is thus obvious that, with glucose as the end-product of microbial digestion, no energy would be conserved in the phosphate bond of microbial ATP (*ca* 30.5 kJ are required to generate 1 ATP from ADP+phosphate). On the other hand, if VFAs are produced during anaerobic fermentation, a considerable amount of energy could be captured by the symbiotes. For each glucose residue derived from cellulose, three molecules of ATP are generated during glycolysis. In addition, the oxidation of pyruvate to acetyl-CoA releases 33.4 kJ which are potentially available for phosphorylation. The hydrolysis of acetyl-CoA is also an exergonic reaction, releasing 31.4 kJ. Altogether, approximately one-third of the energy derived from the complete oxidation of glucose (from cellulose) is potentially available to the gut symbiotes. Other energy-conserving steps have been postulated as occurring in the rumen (Fig. 7.6) but have not been documented.

It is perhaps worth speculating on the products of cellulose degradation in a termite which produces its own cellulase and cellobiase. VFAs would probably not be found in the gut of such an animal because the high-energy phosphate bond can be produced only intracellularly. It is more likely that cellulose would be hydrolyzed in the gut (extracellularly), first to cellobiose and then to glucose which would be absorbed.

Although the metabolic pathways leading to the production of VFAs by termite symbiotes are unknown, it is likely that they will be found to follow the general pattern for rumen micro-organisms which have been studied in much more detail. Fig. 7.6 shows the hypothetical pathways for carbohydrate metabolism in ruminants as presented by Hungate (1966). Acetate appears to be the primary fermentation product in both termites and ruminants so that we might expect the pathways leading to its formation to be similar or identical in both groups of animals. The following evidence tends to support such a postulate. Pochon *et al.* (1959) found a species of cellulolytic bacterium (*Clostridium*) in the gut of a higher termite *Sphaerotermes sphaerothorax* which was very similar to species present in the rumen of cattle. Hungate (1946*a*) also noted quantitative similarities between the fermentation products of *Clostridium cellobioparum* Hungate and those which occur in *Amitermes minimus*. Finally, so far as we are aware, in all anaerobic organisms studied up to 1965, the conversion of pyruvate to acetate yields 'a high-energy bond in the form of acetyl-CoA or acetylphosphate' (Walker, 1965).

Absorption

According to Noirot & Noirot-Timothée (1969) absorption of nutrients, primarily VFAs, must proceed along two basic pathways. Because nutrients cannot pass anteriorly from the paunch through the enteric valve, absorption in all termites probably occurs in the hindgut. This idea tends to be supported by Hungate's (1943) observation that the hindgut of *Zootermopsis angusticollis* is permeable to acetic acid and Noirot & Noirot-Timothée's (1967) note that the microscopic appearance of the paunch epithelium is similar to that of other absorptive epithelia. In addition, in the lower termites, absorption might take place in the midgut of individuals who have fed proctodeally. There is, however, no direct evidence that VFAs are passed in the proctodeal food.

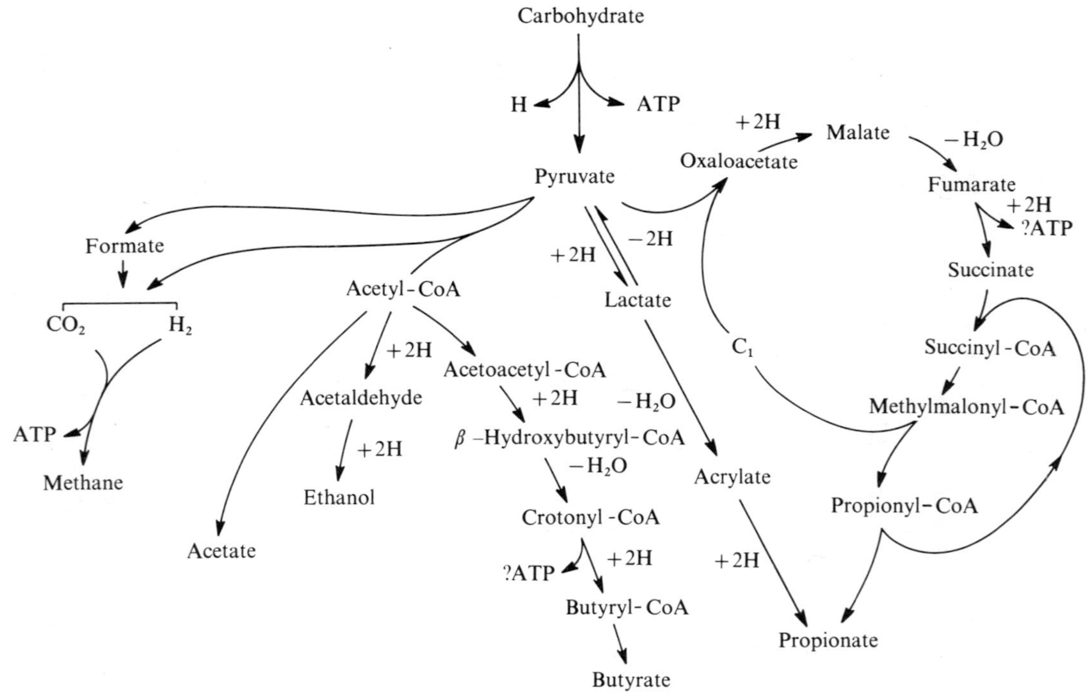

Fig. 7.6. Chart to show the pathways leading to the principal products of carbohydrate fermentation in the rumen (Hungate, 1966).

Hemicellulose

Although considerable data have accumulated on the cellulose-digesting capacities of termites, far less has been learned of their ability to process hemicellulose. A major stumbling block to understanding the phenomenon stems from the lack of precise methods for quantifying these compounds. The techniques, requiring extraction in strong alkaline solvents, have undergone little modification since Browning (1952: p. 1135) remarked that 'Many constituents of wood are severely attacked by alkalies and, consequently, the alkaline solubility value possesses no fundamental significance. Portions of the extractives, hemicelluloses, and lignin are removed to a degree depending upon the temperature, time, and concentration and kind of alkali.' Thus, researchers should take special care to adopt standard methodology, not only in dealing with hemicelluloses, but also in all aspects of wood component analysis. An extensive set of standardized procedures ('TAPPI Standards') is kept updated and offered for sale by the Technical Association of the Pulp and Paper Industry.*

Hungate (1938) employed the acid hydrolysis method for determining hemicellulose content of Monterey pine. This method, however, has been severely criticized (Browning, 1952) because of its high variability and has been replaced largely by the more dependable alkaline extraction procedure. Nevertheless, Hungate reported that *Zootermopsis* digested about 44% of the hemicellulose consumed in its food. Oshima (1919) also reported that *Coptotermes formosanus* utilized some of the hemicellulosic component in its diet.

Nel *et al.* (1970) determined that *Hodotermes mossambicus* metabolized 90% of the hexosans and 84% of the pentosans in redgrass (*Themeda triandra* Forsk.). Both of these groups of compounds include certain hemicellulosic components and it is presumed that they were digested by the termites although no quantitative data were presented. Finally, Potts & Hewitt (1974*b*) found that the cellulase produced by *Trinervitermes trinervoides* itself was capable of hydrolyzing the hemicellulose xylan. It should now be very clear that the means by which termites exploit cellulosic materials are far from being completely understood.

Lignin

Although the chemistry and occurrence of lignin have received considerable attention, the mechanisms by which termites attack it remain almost entirely unknown. There is at present no consensus as to whether termites do or do not degrade lignin. Moore (1969) stated that, among the major wood constituents that pass through the termite gut, only lignin is unaltered or only slightly modified. Hungate (1938) expressed a similar opinion as

* TAPPI Publications Order Dept., 1 Dunwoody Park, Atlanta, Georgia 30341, USA.

a result of his studies of *Zootermopsis*. Such was the view of Leopold (1952) who compared the lignin content of the douglas fir *Pseudotsuga taxifolia* (Poir.) Britt. with that of fecal pellets of *Cryptotermes brevis* fed on the same species. The termites assimilated most of the carbohydrate but almost none of the lignin, in marked contrast with fungi which caused great reduction in the lignin content. In laboratory colonies of *Hodotermes mossambicus*, Nel *et al.* (1970) noted that less than 1% of the lignin in the redgrass *Themeda triandra* was utilized.

Other authors, however, have observed that termites do attack and degrade, often vigorously, the lignin component in wood. The most impressive data are those of Seifert & Becker (1965), which have been conveniently summarized by Lee & Wood (1971*b*: p. 137). Six wood species were subjected to attack by four termite species (*Calotermes flavicollis*, *Heterotermes indicola*, *Reticulitermes lucifugus* var. *santonensis* de Feytaud, *Nasutitermes ephratae* (Holmgren)). The feces produced during the feeding trials were subsequently analyzed to determine the percentage of lignin and cellulose which had been removed from the ingested food. In one trial (*R. lucifugus* on beech) 83.1% of the lignin was removed from the original food. In all but two of the trials (excluding five in which no termites survived) more than 15% of the lignin was removed from the original food. The very high lignin utilization by *R. lucifugus* and *N. ephratae* may have resulted from the fact that these species consume their own fecal matter. Lavette (1964) found that the flagellates of *C. flavicollis* and several rhinotermitids were capable of partially degrading lignin. Kovoor (1964*a*) reported that *Microcerotermes edentatus* also degrades lignin. She determined that the primary means by which this is accomplished is via demethoxylation rather than by destruction of phenolic structure or depolymerization (Kovoor, 1964*b*). If termites did degrade lignin by one of the last two ways, the mechanisms by which they did so would be difficult to visualize. While we have already mentioned that lignin degradation appears to proceed only in an aerobic milieu, and that the termite gut is generally believed to be anaerobic, it is curious to find that lignin is at least partially degraded by both the higher and the lower termites.

Among the micro-organisms, only the white-rot fungi have the capacity for complete decomposition of lignin. Many micro-organisms appear to effect limited changes in lignin structure, primarily through demethoxylation, but the ability of any true bacteria to substantially decompose lignin has not been shown (Kirk, 1971). We known of no *in vitro* or *in vivo* evidence to suggest how termite flagellates or any higher animals accomplish this. It is thus especially perplexing that *N. ephratae* destroyed so much of the lignin (43.4–51.5%) in Seifert & Becker's (1965) feeding trials.

One group of termites does, however, derive considerable nutritive

Table 7.7. *Nitrogen content of termites*

Species	Sex	Caste	% Nitrogen (dwb)[a]	Author
Pterotermes occidentis	M	Alate	9.94	La Fage, unpublished
P. occidentis	F	Alate	10.32	La Fage, unpublished
Marginitermes hubbardi	M	Alate	10.93	La Fage, unpublished
M. hubbardi	F	Alate	10.96	La Fage, unpublished
Zootermopsis angusticollis	M+F	Larva and nymph	9.07	Hendee, 1935
Zootermopsis nevadensis	M+F	Alate	14.7	Hungate, 1941
Z. nevadensis	—	Nymph	9.1	Hungate, 1941
Undetermined	—	Alate	5.71	Leung, 1968

[a] Dry weight basis.

benefit from lignin. The Macrotermitinae, via their symbiotic fungi, derive not only energy from it but also additional cellulose which would normally remain bound in the refractory cellulose–lignin complex. This subject, as well as other benefits which termites may derive from fungi, are discussed in detail by Sands (1969). Cmelik & Douglas (1970) recently reported that lignin in certain macrotermitine fungus gardens may be converted into readily digested D-mannitol and lipids.

Nitrogen

The ability of xylophagous organisms to survive and grow on diets containing very little nitrogen has puzzled biologists for many years. Cleveland (1925*a*) reported that *Zootermopsis* was able to survive and multiply on a diet of 'pure' cellulose. After 18 months, his colonies showed a 40-fold weight increase which led him to suggest that either atmospheric nitrogen had been fixed or the termites possessed a new and unique capacity to convert carbohydrate into protein. Other investigators had less success in maintaining termites on a diet of 'pure' cellulose; for example, Roessler (1932) found that *Zootermopsis* survived only marginally on such a diet. Cook & Scott (1933) were the first to demonstrate a true dietary nitrogen requirement for termites. Those maintained for 12 days on mercerized cotton (changed daily) lost large quantities of nitrogen while others, kept on wood, gained nitrogen. Hendee (1935) reported that *Z. angusticollis* lost weight or died when fed fungus-free filter paper. Cleveland's (1925*a*) success in culturing *Zootermopsis* was partially explained by Hungate's (1941) observation that Whatman No. 43 filter paper (used by Cleveland) contained about 0.03% nitrogen.

Nitrogen analyses of whole termites (Table 7.7) suggest that their protein content is comparable with that of most other animals. Table 7.8

Table 7.8. *Analysis of total body amino acids in alates of two species of dry-wood termites from Arizona.* (*La Fage, unpublished*)

	% of dry weight, all samples	
Amino acid	*Pterotermes occidentis*[a]	*Marginitermes hubbardi*[b]
Lysine	3.540	1.805
Histidine	1.702	0.855
Arginine	3.419	1.677
Aspartic acid	4.669	2.369
Threonine	2.262	1.155
Serine	1.973	1.005
Glutamic acid	7.303	3.691
Proline	3.833	2.000
Glycine	6.955	2.664
Alanine	5.011	2.401
Valine	3.720	1.901
Methionine	0.920	0.450
Isoleucine	2.692	1.200
Leucine	4.427	2.203
Tyrosine	3.311	1.723
Phenylalanine	2.181	1.142

[a] Based on four samples ranging from 10.0–18.3 mg (dry weight).
[b] Based on six samples ranging from 5.2–85.7 mg (dry weight).

lists the amino acid composition of alates of *Pterotermes occidentis* (Walker) and *Marginitermes hubbardi* (Banks) collected during dispersal flights. These and similar data presented by Manzanilla & Ynalvez (1958), Fujii (1964), Speck *et al.* (1971), Carter & Smythe (1973) and Mauldin & Smythe (1973), indicate that the amino acid composition of termites is also generally similar to that of other animals.

Mauldin & Smythe (1973) recognized four routes by which termites could maintain or increase their metabolic nitrogen pool: (1) extraction from the diet, (2) fixation of atmospheric nitrogen, (3) digestion of symbiotes, and (4) utilization of microbial (and we would add termite) waste products. New nitrogen could enter the termite–symbiote complex only *via* (1) and (2), and the dietary source is probably the more important. Numbers (3) and (4) are concerned with the internal cycling of nitrogen. The diet contains a wide range of nitrogenous compounds which have been considered above under *Chemical constituents of plant material – Nitrogen* and under *External modification of plant material – Biotic Factors.* The important ones include proteins, free amino acids and peptides, nucleic acids and, perhaps, alkaloids. As far as our very limited knowledge goes, these compounds are attacked by appropriate enzymes, generally as follows: the proteins by the termite, protozoans and bacteria, the other compounds probably by bacteria of varying capabilities. The bulk of the nitrogen, of course, becomes incorporated into termite, protozoan and bacterial protein. The end-products of protein metabolism are probably uric acid in termites and ammonia in protozoans.

The discovery by Benemann (1973) and Breznak *et al.* (1973) that symbiotes of the lower termites, at least, are capable of fixing atmospheric nitrogen probably ranks among the most significant advances in termite biology since Cleveland's work in the 1920s. The microbes responsible for this activity have not yet been specifically identified. However, only procaryotic cells are known to fix nitrogen (Toth, 1953; Lehninger, 1970). When Breznak's group administered antibiotics to termites there was a parallel decrease in their bacterial populations and nitrogen-fixing capacity, while protozoan populations remained stable. Additional data presented by these authors prompted them to conclude with reasonable assurance that bacteria are, indeed, the responsible nitrogen-fixing agents. The importance of nitrogen fixation versus dietary nitrogen to the termite, the protozoans or to both, remains to be established. It has been suggested that termites might recover some nitrogen by digesting the dead bodies of their symbiotes (Leach & Granovsky, 1938). Although the protozoans contribute greatly to their host's energy requirements, Honigberg (1970) concluded that their turnover rate is not sufficiently high to supply the host with an important amount of nitrogen. The protozoans themselves require protein nitrogen and apparently obtain it by consuming bacteria (Wagtendonk & Soldo, 1970). They thus may actually be competing with their host for this source.

The role of the symbiotic microflora in termite nitrogen metabolism is not at all clear. After the bacteria were removed from the intestine of *Nasutitermes nigriceps* (Haldeman) no important metabolic changes were noted (Speck *et al.*, 1971). When, however, U-^{14}C-glucose was fed to bacteria-free *Reticulitermes santonensis*, incorporation of the labelled ^{14}C into the 10 normally essential amino acids was greatly reduced. Mauldin & Smythe (1973) found that *Coptotermes formosanus* maintained normal PBAA levels in their tissues three weeks after all protozoans and *most* bacteria had been removed. It may be pertinent to recall here that bacteria in the rumen are alone responsible for the biosynthesis of protein from non-protein sources such as ammonia and urea (Allison, 1969). We have already mentioned that wood may contain as much as 50% of its nitrogen in non-protein forms. These facts, coupled with Roessler's (1932) observation that termites can apparently assimilate some ammonium compounds, suggest a parallel possibility where hindgut bacteria synthesize protein which is ultimately utilized by the host.

Within the confines of the hindgut there is a possibility that the waste products of the termite and its symbiotes are mutually useful. Leach & Granovsky (1938) suggested that urates and other nitrogenous compounds, excreted by the Malpighian tubules into the anterior end of the hindgut, could be elaborated into microbial protein. This protein could then be taken up by the termite, either through the ileal wall of the same individual or through proctodeal feeding by another. Such internal nitrogen cycling would theoretically permit a colony to maintain its *status quo* as long as carbohydrate was available. There is no evidence to support this idea but Noirot & Noirot-Timothée (1969) have made a similar suggestion with regard to the bacteria in the mixed segment of some termitids.

Hungate (1955) reviewed the work on the all but forgotten intracellular bacteria which have been considered as symbiotes in some of the protozoans in *Calotermes flavicollis*. The early Italian workers who studied them thought they were responsible for cellulose digestion. Their function remains unknown and knowledge of their existence only compounds an already complex relationship. It is interesting to note that the most abundant waste product of protozoans is ammonia (Wagtendonk & Soldo, 1970), a compound readily utilized by many bacteria.

It should be obvious that we know little about the ways in which termites meet their nitrogen requirements except that they are many, complex and probably conservative. According to Hungate (1944) *Zootermopsis nevadensis* uses nitrogen very efficiently, assimilating about 50% of the nitrogen in its diet. While we have mentioned possible ways in which nitrogen could be cycled internally, other ways of conserving it are considered below under *Social recycling of nutrients*. Except for odd bits of information such as the occurrence of uric acid in feces (*Zootermopsis*, Hungate, 1941; *Nasutitermes*, Moore, 1969), we are unaware of any studies dealing directly with the absorption, transport or metabolism of nitrogen compounds in termites. The subject is, however, rich in speculation and many ideas are ripe for cooperative investigations by insect physiologists, microbiologists and biochemists.

Lipids

Moore (1969) remarked that lipids make up only a small portion of the termite diet, but play a most important role physiologically. Some functions served by these compounds include energy storage (Moore, 1969), protection from desiccation (Collins, 1969), incorporation into defensive secretions (Moore, 1969), as vitamins (Sannasi, 1968), as pheromones (Sannasi & George, 1972) and incorporation into cell membranes (Moore, 1969). Lipids may be obtained not only directly from plant materials but also from the fungi consumed in many termite diets. The lipid content of woody tissues is poorly known and, while generally present in a concentration of less than 1% (as ether extract), some tropical plant species contain considerably more. Carter *et al.* (1972*b*) analyzed the lipid components of a fungus *Lenzites trabea* and three different woods which it had attacked. When cultured on loblolly pine, the fungus contained 65.2% of its FA complement as linoleic acid, 17.0% as oleic acid, 6.0% as palmitic acid and 1.4% as stearic acid. The FAs of the decayed woods reflected the utilization of specific FAs by the fungus as well as the composition of the fungus itself. The fungus consumed much of the palmitic acid in two of the woods and much of the linoleic in the third, which was low in palmitic.

The total lipid content of the fungus cultured by the ant *Atta colombica tonsipes* Santschi was found to be about 0.2% (Martin, Carman & MacConnell, 1969: Chapter 4, Table 1). Cmelik & Douglas (1970) investigated the chemical composition of fungus combs constructed by two fungus-growing termites *Macrotermes falciger* (Gerstächer) (as *M. goliath*) and *Odontotermes badius* (Haviland) and concluded that their lipid content, approximately 2.5% of the dry weight of the comb, was probably of fungal origin. They also showed that fungal tissue made up 15% of the dry weight of the combs, which suggests that the fungus contained about 17% lipid. In a later paper, Cmelik (1971) reported the lipid content of the gut and hemolymph from an undetermined *Macrotermes* queen. The gut lipids contained 17% free fatty acids (FFA), 46% neutral lipids and 37% phospholipids. The hemolymph composition did not closely reflect the gut contents, as it consisted of 64% neutral lipids, 28% phospholipid and 8% FFAs.

Gilbert (1967) has reviewed the subject of lipid digestion and absorption for insects in general but all we have are a few pieces of the puzzle for the termites. Its solution seems to be confounded by the variability found among the different castes and by the unknown role of the symbiotes. In *Macrotermes falciger*, the guts of alates contained nearly all triglycerides (TGLs) with minor amounts of FFAs and cholesterol (Cmelik, 1969*a*). The guts of queens contained TGLs which are possibly degraded to mono- and diglycerides (MGLs and DGLs) (Cmelik, 1971). The presence of these compounds in the hemolymph, however, does not necessarily signify that they were absorbed and released from the gut wall in these forms; they might have been released from the fat body after considerable structural modification. In humans an intestinal lipase splits TGLs primarily into MGLs and FFAs which are absorbed intact by the mucosal cells. Here the TGLs are reconstituted and then passed into the lymphatics. Is it possible that TGLs are similarly reconstituted in the insect gut epithelium?

While the sterol content of the diet of *M. falciger* is very low – not more than 1–2 mg per 100 g of dry fungus comb (Cmelik & Douglas, 1970) – Cmelik (1969*a*) showed that 10% of the body content of workers of this species is composed of sterols. It seems probable that some microbial synthesis of sterols is occurring here. Dadd (1973) noted that sterols are synthesized by the intracellular symbiotes of some aphids.

Owing to their very high energy/mass ratio but unlike carbohydrates, lipids can be stored in nearly an anhydrous form. They are the most commonly stored metabolic fuel and, for fasting and hibernating animals, the only energy source (Lehninger, 1970). When oxidized they generally yield twice as much energy per unit mass as either carbohydrates or proteins. As with most other animals, it appears that lipids, as TGLs, constitute the most important long-term energy reserve for termites. Because of their small size and the diffuse nature of the fat body, few analyses have measured the lipid content of specific insect organs or tissues. Cmelik (1969*a*) used 'semi-quantitative' methods to estimate the percentage composition of total lipid found in various organs of unflown *Macrotermes falciger* (as *M. goliath*) alates. He found approximately 60% in fat body, 6% in thoracic muscles, 5% in gut, 5% in reproductive organs, 2% in heads and 22% in other body parts. Of the total dry body weight 47% was fat. In physogastric queens of *Macrotermes natalensis* (Haviland) the same author (Cmelik, 1969*b*) found that fat body contained only 22% of the total body fat, whereas the reproductive organs (including eggs) contained 72%. Furthermore, in *M. falciger* queens, only 11% of the total fat was located in the fat body while 83% was found in the reproductive organs. Combined, these two organ systems accounted for 94% of the total fat in both species. The fat content as a percentage of total dry body weight was not reported for either species; however, it would likely be considerable.

It should be noted that the fat body of mature, egg-laying queens bears little resemblance to that of other castes. Grassé & Gharagozlou (1963, 1964) described the fat body of *Calotermes flavicollis* queens as containing very little fat or glycogen. It does contain an extensive endoplasmic reticulum and associated ribosomes and is most likely specialized for protein synthesis. The queens are apparently fed large quantities of protein in salivary secretions from the larvae. Also, special cells, 'endolophocytes' are present that may perform an excretory function. Similar observations have been made on mature queens of five species of termitids by Gabe & Noirot (1960). Cmelik (1969*b*, 1971) stated that food received by *Macrotermes falciger* and *M. natalensis* queens might travel directly to reproductive organs, by-passing the fat body completely. He cites a study by Martin (1969) with *Pyrrhocoris apterus* L. which suggests that such a scheme is possible.

The lipid content of the worker caste has been determined for several species: *Reticulitermes flavipes*, 5.0% (Carter, Dinus & Smythe, 1972*a*); *Coptotermes formosanus*, 4.2% (Mauldin, Smythe & Baxter, 1972); *C.*

lacteus, 0.4–0.6% (Moore, 1969); *Macrotermes falciger*, 2.2% (average) (Cmelik, 1972); and *Nasutitermes exitiosus*, 1–3% (Moore, 1969). Major soldiers of *M. falciger* contained 1.8% lipids (Cmelik, 1972). The value of these figures, especially for comparison, is questionable. It might be assumed that they were figured on a dry weight basis, but this was not clearly stated in any case. Colonies of *C. lacteus* maintain larger cellulose stores than *N. exitiosus* and are thus apparently less dependent on body lipid reserves (Moore, 1969). No attempts have been made to extend any of these analyses to the organ or tissue level; however, Carter *et al.* (1972*a*) claim that total body extracts of lipids tend to approximate fat body composition. While TGLs are the primary lipids stored in insect fat body (Gilbert, 1967), Cmelik (1972) found phospholipid to be the predominant class in workers of *M. falciger* (57% of total lipid). It seems likely that this high phospholipid content represents membrane components rather than fat body lipids since the fat body is relatively small in workers.

Basalingappa (1970*a*) studied the quantitative variation in lipids among the castes of *Odontotermes assmuthi* Holmgren. On a dry weight basis lipids constituted 9.7% of workers, 17.8% of soldiers, 26.4% of royal pairs, 17.8% of the undifferentiated instars and 48.4% of alates. In *Paraneotermes simplicicornis* (Banks), a North American subterranean calotermitid, older nymphs contained 45.5% lipid, younger nymphs, 39.3%; larvae, 22.0%; pseudergates, 28.4%; and soldiers, 16.6% (dwb) (La Fage, unpublished). In this species and probably in many others there appears to be a progressive lipid build up in the reproductive line. Collins (1951) also recognized the variation in lipid content among the castes in *Reticulitermes flavipes*. She found that lipids from nymphs and young imagoes (up to six days after metamorphosis) were histochemically much different from those in workers and imagoes 11 days old or more. She concluded that products stored during nymphal development were being used during alate development. From data given by Hewitt, Nel & Schoeman (1972) it can be calculated that unflown alates of the termitid *Trinervitermes trinervoides*, contain approximately 42% (dwb) lipid in the form of neutral- and phospholipids, with the former predominating. Alates of another termitid *Gnathamitermes perplexus* caught in flight, are also lipid-rich, containing 56.3% (dwb) (M. Dimmitt, personal communication). Unspecified winged termites offered for sale in a Leopoldville (Kinshasa) market contained 44.4% fat (Tihon, 1946).

Among these admittedly random observations only one trend seems to surface: the sub-imaginal instars and alates contain substantially more lipids than other castes. These stores are undoubtedly used both for energy and for egg production during the critical period of colony foundation and early development, by many of the Termitidae and perhaps, on occasion, by the lower termites. The belief that thoracic flight musculature is rapidly mobilized for nutrient requirements during this period, as in the ants, requires further investigation, since Noirot (1969*b*) has noted that this process continues over a period of years and probably contributes very little to the nourishment of the royal pair. Hewitt & Nel (1969*b*) also noted that primary reproductives of *Hodotermes mossambicus* had lost only one-third to one-half the mass of their dorso–ventral thoracic musculature after 18 months. From the data available (Cmelik, 1969*a*; Carter *et al.*, 1972*a*; Mauldin *et al.*, 1972) it appears that TGLs are quantitatively the most important lipid class stored by termites and C^{16} and C^{18} FAs the most important TGL constituents.

Insects apparently show no generalized pattern of lipid mobilization from fat body reserves. Gilbert (1967) concluded from studies of *Periplaneta americana* (L.), *Melanoplus differentialis* Uhler, and of larvae and pupae of *Hyalophora cecropia* (L.), that DGLs were the major lipids released from the fat body. It appeared likely that these compounds were subsequently conjugated with proteins to enhance their solubility and thus transport in the aqueous hemolymph. The actual site and means of conjugation remains unknown. Cook & Eddington (1967) found, however, that TGLs and FFAs were the primary lipids released from *P. americana* fat body *in vitro*. Using isotopically labelled compounds, Chang & Friedman (1971) observed a slow release of FFAs, DGLs, and TGLs from the fat body into the hemolymph of *H. cecropia*. In *Schistocerca gregaria* Forskål) DGLs were the major compounds released, while in *Manduca sexta* Johannson FFAs predominated. Wlodawer, Lągwińska & Barańska (1966) found that ^{14}C-labelled sodium palmitate added to the hemolymph of *Galleria mellonella* (L.) *in vitro* became esterified to glycerol and formed TGL. FFAs are the primary lipids released from the fat body of *G. mellonella* while protein–TGL conjugates are the major transport structures. Although there are no comparable data for termites, Cmelik (1969*a*) noted that absence of DGLs and MGLs from all organs of unflown alates of *Macrotermes falciger*. He did find a substantial concentration of FFAs in the flight muscles and suggested that these might be utilized for fuel during dispersal. Hewitt *et al.* (1972) rejected this idea since they knew that sarcosomal preparations from unflown alates of *Trinervitermes trinervoides* oxidized carbohydrates much more rapidly than palmitic acid. While carbohydrates are probably utilized during dispersal flights, lipids are likely to be the substrate metabolized during the first 10 weeks in the development of the young colony. Hewitt *et al.* further reported that during this period the lipid reserve of *Hodotermes mossambicus* declines rapidly; their failure to demonstrate a strong β-oxidation activity may have stemmed from the extremely labile nature of this system.

Vitamins

Moore (1969) discussed the few pieces of information available and admitted that the vitamin requirements of termites were unknown. The statement is still true. Most of the compounds in the B complex have been found in a 'termite' (Spector, 1956) and Moore suggested that some may be synthesized by symbiotes. The requirements of insects for the fat-soluble vitamins (A, D, E, K) have only recently become appreciated. It is believed that vitamin E is important for reproduction in at least some species and that the carotinoids, vitamin A precursors, are probably essential for vision in most if not all species (Dadd, 1973). Sannasi (1968) has found the ratio of retinene to Vitamin A to be greater than 4:1 in the pre-flight alates (negatively phototactic) of a fungus-grower *Odontotermes redemanni* (Wasmann) (as *Termes*). The situation is approximately reversed in swarming alates, which are positively phototactic. These compounds were restricted to the head and prompted the author to conclude that they were involved with vision. Neither the source, identity nor specific function of Goetsch's (1947) growth-promoting 'vitamin T' is known. (See *Minor constituents and their significance.*)

Effects of diet on body composition

Lipids

During transport, lipids in mammals undergo cycles of lipolysis and re-esterification at various sites, which include the intestinal lumen and wall, the liver and adipose tissue. Dole (1964) supposes that during these cycles FAs tend to become mixed and the body TGL composition standardized. Wlodawer *et al.* (1966) theorize that similar mixing of FAs occurs in the larva of the wax moth *Galleria mellonella* and possibly other insects. Their data suggest that dietary lipids have little effect on the overall lipid composition of *G. mellonella*. Various investigators have found that the body lipid content closely reflects the dietary lipids in some insects, while in others fats not only undergo considerable structural modification but also are readily synthesized from other FAs and non-lipid sources such as carbohydrates. The FA requirements of insects are poorly understood and by no means uniform; to date, only the di- and trienoic C18 FAs are known to be essential for some insects (Dadd, 1973). Carter *et al.* (1972*b*) found that C18:2 was one of the major FAs in several species of sound and fungus-rotted wood. The general body lipids of *Reticulitermes flavipes* reared on these substrates, however, did not reflect the lipid composition of the food. Mauldin *et al.* (1972) found that partially and completely defaunated larva–workers of *Coptotermes formosanus* still possessed the capability of synthesizing unsaturated FAs, though at a significantly lower rate than normally faunated individuals. The lowered rate was attributed to a reduction in the availability of acetate in the defaunated individuals. Cmelik (1972) compared the lipids from workers of *Macrotermes falciger* with those of the fungus comb and found very little similarity between the two. Based on these data and the observation *a priori* that FAs are normally deficient in the termite diet, it is probable that they synthesize the lipids which they require, and that dietary lipids have little effect on body lipids.

Proteins

Carter & Smythe (1973) compared the PBAAs of *Reticulitermes flavipes* maintained for four weeks on sound wood with those from termites fed wood which had been partially decayed by fungi. The three species of sound wood tested contained an average of 19.3 μM PBAA per g of lipid-extracted residue versus 31.7 μM per g for the decayed material. Furthermore, there were differences in the molar ratios of PBAAs in sound wood and decayed wood. These differences appeared to have little effect on the PBAAs of the termites although the free amino acid (FAA) composition (molar ratios) of the termites was somewhat affected. The authors noted, however, that FAA composition normally varies with changes in metabolism as well as nutritional state.

Assimilation efficiency

Because they produce little or no cellulase, most metazoan soil organisms are inefficient processors of woody detritus. This is true also of many herbivorous insects such as the larvae of Lepidoptera. In contrast, a few animals, including the silverfish *Ctenolepisma lineata*, land snails of the genus *Helix*, ruminants and termites, are surprisingly efficient cellulose assimilators. Lasker & Giese (1956) reported that *C. lineata* absorbed 71.7–87.0% of a pure cellulose diet. The assimilation efficiency of laboratory colonies of termites has been summarized in Chapter 4 of this volume (Table 4.3). For the termite species and four wood species listed, efficiencies vary from 54–61% (dwb). This may be compared with a dairy cow (a ruminant) which digested 72% of a ration of dry grass (Maynard, 1937).

No direct comparisons or generalizations can be made using the above figures because a standard terminology for gauging the efficiency of nutrient processing has not been uniformly adhered to. The Committee on Animal Nutrition of the National Academy of Sciences – National Research Council has recently published a glossary of energy terms including a consideration of various energy distribution schemes (Harris, 1966). Although originally intended for use with domestic farm animals, these standards should apply equally well to insects and should, if used universally in the future, provide a basis for meaningful comparisons between diverse groups of animals.

Utilization of nutrients at the colony level

While nutrients are being processed by individual members of all instars and castes, they are routinely exchanged among colony members so that digestion becomes a colonial or social affair as well. Such an exchange serves both nutritional and social functions – a fairly equitable distribution of nutrients, some control over reproduction and caste determination and, most importantly, communication by chemical messengers. Beyond this, the assimilation efficiency of available nutrients is enhanced at the colony level through fungus cultivation by the Macrotermitinae, the recycling of carton and fecal material, and occasional cannibalism.

Trophallaxis

Although trophallaxis undoubtedly plays an important role in the organization and regulation of most insect societies, its significance has been overstated, occasionally to the point of absurdity, by the more philosophically inclined writers. Current usage seems to favor its original meaning: the exchange of nutrients, either reciprocally or unilaterally, among members and guests of social insect colonies. As with most insect behavior, it is probably instinctive, with some learning entering into its more elaborate manifestations as, for example, between older members of the various castes. Its adaptive value has been further increased where it serves to circulate the semiotic ('message carrying') chemicals (Law & Regnier, 1971) which form the basis for a highly versatile and sophisticated communication system. Wilson (1971) has provided an enlightening historical review of this phenomenon for all social groups, including a critique of the varied thinking by the memorable students of the subject. Harris & Sands (1965) have reviewed the role of trophallaxis in the social organization of termite colonies. More recently McMahan (1969) has focussed on the complexities of feeding relationships among the termites and described the fruitful application of radioisotope techniques to this area.

Nutrients are transferred by donors to more or less passive recipients, such as young larvae, or to active solicitors, such as recently molted young nymphs. The type of material passed depends upon the recipient, or the type of solicitation. A donor may dispense either 'saliva' or regurgitated (stomodeal) food from the mouth (Fig. 7.7), or liquid contents of the hindgut (proctodeal food) from the anus (Fig. 7.8). Saliva and proctodeal food are often called 'elaborated' foods as opposed to the relatively undigested 'raw' food from the crop.

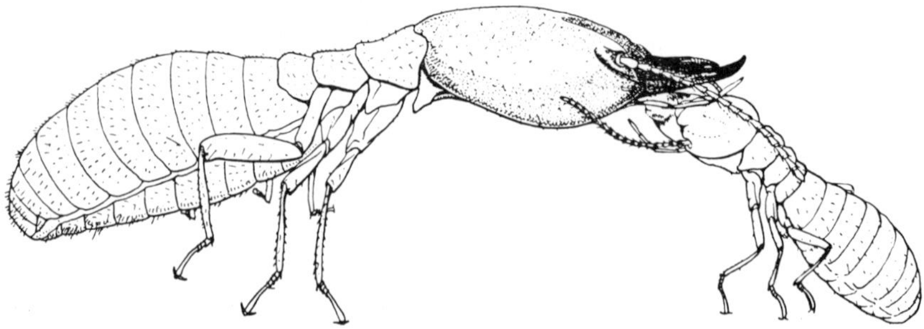

Fig. 7.7. Small worker of *Macrotermes natalensis* (as *Bellicositermes*) giving stomodeal food to a large soldier (Grassé, 1949).

Fig. 7.8. Large nymph of *Cryptotermes brevis* receiving proctodeal food from a soldier (from a color transparency by E. A. McMahan).

Saliva and stomodeal food

The saliva is a clear or slightly opalescent liquid which is apparently secreted by the salivary glands and held in the salivary reservoirs (Grassé, 1949). Little is definitely known of its composition, beyond the observation that the salivary gland foam cells secrete lipids and perhaps mucopolysaccharides, and the vacuole cells a material with glycoprotein characteristics. The saliva evidently sustains larvae, functional reproductives and soldiers of some species. It appears to be digested and absorbed by the time it reaches the hindgut since the bacterial flora of these forms is absent or nearly so (Noirot & Noirot–Timothée, 1969). Grassé & Gharagozlou (1963) considered the saliva fed by the larvae to queens of *Calotermes flavicollis* to be very rich in proteins. As a result of his detailed analyses of lipids from the guts of queens and from whole workers of *Macrotermes falciger*, Cmelik (1971, 1972) suggested that certain classes could be considered as possible constituents of stomodeal food (probably saliva as understood here). On the basis of their relative abundance, these are FFAs and easily emulsifiable phospholipids, but probably not TGLs. Such high-energy compounds would seem ideal ingredients of a diet for the dependent castes. A hint of the possible importance of saliva as a brood nutrient comes from Pasteels' (1965) observation that stage I workers of *Nasutitermes lujae* (Wasmann) have better developed salivary glands than stage II or III. The fact that stage I workers remain in the nest further suggests that they are specialized for tending and feeding the brood. A number of other functions have been attributed to salivary secretions. After pairing, the alates of *Hodotermes mossambicus* drink large quantities of water, which is rapidly transported via the hemolymph to the water sacs (salivary reservoirs). Before their eggs hatch, this liquid is very low in solutes and probably at least serves the incipient colony during the drought which usually follows their dispersal flight (Hewitt *et al.*, 1971). Watson, Hewitt & Nel (1971) speculate that if the reproductives do feed their brood this liquid, its nutrient level may increase later or the salivary acini may add nutrients to the fluid as it is fed.

The presence of amylase in the saliva of *Calotermes flavicollis* suggests at least a minor role in digestion; the soldier secretions of many Macrotermitinae and certain other termitids have an obvious defensive function (Noirot & Noirot-Timothée, 1969); the saliva of workers of the Macrotermitinae, and probably many other termites, is evidently used as a cement in construction (Noirot, 1970); and that of *Odontotermes* workers exerts a fungistatic effect on unwanted fungi in their gardens (Batra & Batra, 1966).

Stomodeal food varies from a slightly viscous liquid to a paste of wood fragments or humus regurgitated from the crop. Noirot & Noirot-Timothée

(1969) ventured that soldiers generally, and older nymphs of certain termitid species, are fed stomodeal food. However, on the basis of her careful and detailed studies of *C. flavicollis*, Alibert-Bertho (1969) has determined that regurgitated food is not fed to the dependent castes and may, in fact, be exchanged rather infrequently between the other members of the colony. McMahan (1963) came to a similar conclusion with *Cryptotermes brevis*, as did Andrew (1930) with *Zootermopsis angusticollis* (as *Termopsis*).

Proctodeal food

Proctodeal food consists of the viscous hindgut liquid with its suspended fine particles of wood and symbiotic protozoans. It thus differs fundamentally from fecal material and probably from other anal 'exudates' (see below). The concept of proctodeal trophallaxis is generally restricted to the termites and, further, to those harboring symbiotic flagellates; that is, to all but the Termitidae. By all accounts the xylophagous protozoans of these lower termites degenerate during the molting period and their remains are eliminated at about the time of ecdysis (Cleveland, 1923; Andrew & Light, 1929; Nutting, 1956). It is generally known, however, that young larvae become inoculated and that recently molted and defaunated individuals regain their infections by the solicitation of proctodeal food from their non-molting associates. It may be presumed that this liquid also contains nutrients since it is widely exchanged among most ages and castes and is part of the diet of all dependent castes of *Calotermes flavicollis* (Alibert-Bertho, 1969). Although droplets of 'pure' proctodeal food itself have not been analyzed, its liquid component is probably representative of the hindgut fluid which contains a complex mixture of the end-products of digestion, particularly easily metabolizable VFAs. (See *End-products and metabolic pathways* above.)

Exceptions to our statements above involve proctodeal liquid exchanges among certain ants: from larvae to workers (Le Masne, 1953), and from army ant (*Eciton*) queens to attending workers (Rettenmeyer, 1963). Similarly, queens in three genera of the Termitidae (*Microtermes, Microcerotermes* and *Odontotermes*) produce an 'anal' exudate which is eagerly taken by workers and soldiers. This liquid is reported to be a mixture of waste products from the hindgut and a secretion from the colleterial glands (which open into the genital chamber below the rectum) (Sannasi, Sen-Sarma, George & Basalingappa, 1972). Implications that components of this material may affect caste determination, reproduction and the general economy of the colony are considered below under *Special social functions of trophallaxis*.

Such intriguing discoveries should spur attempts to survey the other termite families for similar exudates and to complete chemical analyses of the three types of trophallactic materials. *Calotermes flavicollis* is well known and thus the most promising candidate for such studies, particularly since the information would complement the excellent behavioral work of Alibert-Bertho (1969). Parallel studies have revealed that the saliva of vespine wasp (*Vespula*) larvae is both attractive and nutritive (trehalose, glucose, etc.) (Maschwitz, 1966), and that there is considerable biochemical division of labor between adults and larvae of the wasp *Vespa orientalis* Fabricius (Ishay & Ikan, 1969).

Nutritional functions of trophallaxis

All termites apparently engage in trophallaxis. Rather set patterns of food exchange between individuals occur within families or sub-families, although generalizations are not easily based on the behavior of a few well-studied examples. The subject has been reviewed by Grassé (1949), Noirot & Noirot-Timothée (1969) and McMahan (1969). Colony members which are capable of feeding and nourishing themselves directly, act as both donors and recipients. Others, partly or completely incapable of feeding themselves, are thus facultative or even obligate recipients of partly digested food; they may occasionally act as donors. In the lower termites the independent castes, or functional workers, are the older larvae, pseudergates (larva-like forms which undergo stationary, non-differentiative molts), and young nymphs; the dependent castes are generally the young larvae, soldiers, older nymphs, reproductives and, perhaps, unflown alates.

Caste interactions

The trophallactic behavior of *Calotermes flavicollis* has been described in a series of classical papers by Grassé & Noirot (1945), Noirot (1952), Gösswald & Kloft (1963) and more recently by Alibert-Bertho (1969). In the hope that this species is representative of the lower termites, we present the following summary to illustrate the complex interactions which take place among the castes of its colonies. The founding pairs feed directly and nourish their first brood until the larvae acquire their own protozoan infections through proctodeal feeding and thus become nutritionally independent in the third instar. Third- and fourth-instar larvae and first-instar nymphs are the most active in reciprocal food exchange and as general donors to the dependent castes. The dependent castes are also solicited but vary in their dependence. Reproductives are very good (willing?) donors, while soldiers and active last-instar nymphs are poor donors. First- and second-instar larvae are rarely solicited. Alate behavior depends on many factors. Unflown individuals in groups never feed directly. However, as de-alated pairs heading active colonies, they may continue some direct feeding for as long as two years. Pairs isolated from their colonies after six years still retain the ability to feed directly.

Pure saliva is regularly fed to first- and second-instar larvae, the royal pair in large, old colonies and to pigmented alates. Stomodeal food is not often, and probably never, passed to young larvae, soldiers, alates or the royal pair. Proctodeal food is a part of the diet of all dependent castes.

The trophallactic behavior of only two other lower termites has been studied in comparable detail: *Cryptotermes brevis* (Calotermitidae) (McMahan, 1963) and *Reticulitermes lucifugus* (Buchli, 1958). The broad patterns of food exchange described above hold generally true for them. The following observations seem to be exceptional or of unusual interest. The larvae in incipient colonies of *R. lucifugus* are fed stomodeal and proctodeal food until the third instar, when they are able to feed themselves (Buchli, 1950). The primary and neotenic reproductives of this species lose their symbiotic flagellates as their offspring begin to feed them and, if isolated, die of starvation in about a month (Buchli, 1958). A somewhat similar situation was found in *R. flavipes* by Cleveland (1925*b*). Beard (1974) concluded that the exchange of food material by *R. flavipes* was relatively slow and incomplete and, after 24 hours, went little beyond about one-third of the unlabelled members of a group (70 labelled larvae, 317 unlabelled larvae, nymphs and soldiers). Nel, Hewitt, Smith & Smith (1969) have determined that the so-called adult workers of *Hodotermes mossambicus* feed only to a limited extent and cannot survive without being fed by the large larvae. The functions of this pigmented caste are apparently to forage on the surface and to excavate the underground galleries. Perhaps the most noteworthy example of nutrient recycling occurs in young founding pairs of *Cryptotermes havilandi* (Sjöstedt), who practice not only buccal–anal exchanges, but occasionally even auto-proctodeal feeding (Wilkinson, 1962)!

Among the higher termites (Termitidae) the trophallactic behavior of *Cubitermes fungifaber* (Sjöstedt) (Termitinae) has probably been studied more intensively than any other (Alibert, 1963, 1968). The workers are the only independent caste; larvae, soldiers, nymphs, alates and reproductives are dependent to varying degrees. The immature soldiers, alates and queen are fed generously with salivary liquid, while the young brood, king and nymphs of the penultimate instar receive lesser amounts. Some of the workers and the soldiers are fed regurgitated food. However, it is doubtful that the behavior of this species can be taken as typical of this large family whose members feed upon a very wide range of plant materials. Certain exceptions or points of interest again seem worth noting. Larvae generally remain dependent until they transform into workers when they gain hardened, functional mandibles and adequate intestinal bacteria. Male reproductives in incipient colonies of *Cubitermes ugandensis* Fuller apparently do more grooming and feeding of the young larvae than their mates (Williams, 1959*b*). Active larvae and older nymphs of Amitermitinae and Nasutitermitinae take raw as well as elaborated foods (Grassé, 1949). Notes on the feeding behavior of unflown alates and the reproductives of both young and mature colonies have been summarized by Nutting (1969).

Such complex interactions obviously require a system of signalling the need for giving or receiving food – how is a nest-mate informed of a full stomach, or an empty one? Grassé & Noirot (1945) remain the basic sources for descriptions of the behavioral rituals initiating and terminating food exchange among the termites. Notes regarding obvious displays of intent by prospective donors are rare and unenlightening; for example, McMahan (1963) observed individuals of *Cryptotermes brevis* 'offering', proctodeal food (droplets?) to others. Perhaps such feeding is usually initiated by the solicitor as McMahan suggests, and the donors chosen (in the dark!) mainly on the basis of odor. This possibility is suggested by Alibert's (1968) finding that, in *Calotermes flavicollis*, individuals during the molting period are somehow more attractive than their nest-mates. During the time they do not feed, they invite increasing solicitation of proctodeal food for some 10 days prior to ecdysis, and the material released from the hindgut just before ecdysis is especially sought by the others. For one or two days thereafter the situation is reversed. Recently molted termites become eager proctodeal solicitors until their intestinal faunas have been replaced.

Solicitors of stomodeal food or saliva touch the head of a prospective donor with their antennae (soldier to worker, for example), or rub their mandibles against the worker's frons and mouthparts. This behavior is presumed to release a disgorging reflex. Proctodeal food is solicited by tactile stimulation of the posterior of the abdomen with antennae, palpi and even the forelegs. A defecatory reflex may produce fecal material, which may be accepted or not, or hindgut fluid (Grassé, 1949). From her observations on *Cryptotermes brevis*, McMahan (1963) has added that the recipient generally initiates and terminates proctodeal feeding. It may even hold the donor's abdomen with the mouthparts until it is finished, or the donor may simply walk away.

Interactions with termitophiles

We mention the termitophiles here because they are involved in the nutrient dynamics of their host's societies – either as scavengers in their nests, croppers of their fungus gardens, ectoparasites or predators on their hosts, thieves during trophallactic exchanges or, the *ne plus ultra*, as solicitors of regurgitated food on an equal basis with their hosts.

The less well-integrated forms, relatively normal in appearance or limuloid (tear-drop shaped), may go so far as to groom their hosts but otherwise tend to avoid them through a combination of speed, agility, and perhaps protective body form and lack of distinctive odors. Such a generalization may not hold, however, since behavioral observations on the limuloid forms are few and some of them contradictory. The more highly integrated forms owe their acceptance as equals in their host's societies to a few special characteristics, notably a complex system of epidermal glands (Pasteels, 1969), a host-like behavioral repertoire and, probably, in a few cases, even morphological or tactile mimicry. Some of the glands produce what Wilson (1971) has called 'appeasement substances', attractive to the termite hosts. They may eventually prove to be mimics of one or more host odors, or perhaps even simple nutrients.

At any rate, in some of the genera with bloated (physogastric) abdomens, mutual licking of exudates from body and appendages has been observed between termites and beetles (Kistner, 1968). Furthermore, appropriate begging behavior – antennal touching, then mutual contact and licking of mouthparts – in a very few genera of beetles is known to result in their receiving regurgitated fluid food (saliva?) from their hosts (Emerson, 1935; Kistner, 1973). It is not known whether there is any return of this material to other host individuals as reported by Le Masne & Torossian (1965) for a species of *Camponotus* and a brenthid beetle. However, Kistner's (1973) observations of lingering mouth-part contact between some termites and their guests have suggested some sort of mutual exchange.

Effectiveness of trophallaxis

McMahan (1969) has reviewed the few studies of the rate of food transfer within groups or colonies. Rate depends upon such parameters as species, group composition and size, time and temperature. Results vary and are not readily compared; however, rates are generally slower than those reported for the honey bee and some ants. For example, McMahan (1966*b*) reported that radioactivity could reach saturation in a 25-member group, or 50% saturation in a 200-member group in 96 hours. The effects of group composition and size were shown in tracer studies with *Calotermes flavicollis* and *Cubitermes fungifaber* by Alibert (1968). With the king and queen present, rate and intensity of trophallaxis between larvae and nymphs (or workers) increased with the number in the group while, if the royal pair were removed, there was little change with *Calotermes*, but in *Cubitermes fungifaber* stomodeal feeding declined and social feeding habits became disorganized. In similar studies with *Cryptotermes brevis*, McMahan (1963) found that large and small nymphs exchange food more or less in proportion to their sizes and that soldiers and supplementary reproductives have relatively light appetites. As might be expected, she also noted (McMahan, 1966*b*) that, as colony size increases, the average amount of radioactive material (and pheromones as well?) transmitted to each individual decreases. Finally, Gösswald & Kloft (1963) and Sen-

Sarma & Kloft (1965) have made the very simple but important point that, because of continual recycling, the radioactive food remains longer in the individuals of a group than in isolated individuals. All of this might explain to a considerable degree why individuals in groups or colonies live longer than lone individuals. Such a benefit of living in a society is a prime example of the group effect (Wilson, 1971: Chapter 15).

Even though the rate of trophallaxis in termites may be slower than in some ants, with all members of a colony participating, the primary dispensers of food are probably sensitive to the nutritional status of the colony (Wilson & Eisner, 1957; Wilson, 1971). Very much to the point here, Nel *et al.* (1969) have speculated that the large larvae of *Hodotermes mossambicus*, which function as nurses, might induce foraging by the pigmented workers by withholding food from them.

We have already mentioned the type of food generally used by each caste. Since we know essentially nothing about the composition of these materials little can be said about their equitable or preferential distribution within the colony. However, there is apparently nothing approaching the aggressive dominance system of feeding as has been described for some of the Hymenoptera. Trophallaxis probably continues as long as food is available.

Special social functions of trophallaxis

Caste determination, influence of nutrition

From whatever source, the direction and rate of transmission of a substance through a colony depends upon the strength of inter-individual attractions and the consequent levels of trophallactic exchange. The successful functioning of colonies containing many thousands of individuals attests to the remarkable balance that is apparently maintained in this way. Although pheromones are clearly involved in caste control of a few species, many investigators have felt that the quantity or quality of nutrition is also important, especially among the Rhinotermitidae (Miller, 1969). No one has provided any conclusive support for this contention; nevertheless, there are many intriguing bits of circumstantial evidence in its favor. We are not aware of any significant observations from studies on the Termitidae.

In early studies of developing colonies of *Zootermopsis angusticollis* and *Z. nevadensis*, Heath (1931) noted a relation between increasing frequency of molting and the increasing size of the laboratory group or colony. He concluded that an increasing amount of food was the stimulus here. Later, Becker (1965*a*) found that wood attacked by certain brown-rot fungi influenced feeding activity and could increase or inhibit the rate of development of *Calotermes flavicollis* toward different castes. Many factors appear to be involved, including the species and strain of fungus, the host wood and the stage of its decomposition. He believed that the effects on the termites depended upon the protein and vitamin content of the wood.

Lüscher (1960) had noted that larvae of *C. flavicollis* develop into nymphs in nature at about the time of swarming, when the pseudergates have fewer mouths to feed and nutritional conditions are probably good. Conversely, nymphs undergo regressive molts if they are transferred from large to small laboratory colonies at the wrong time, probably because of a lower level of nutrition. He has reasoned that nutrition may in some way be responsible for activation or inhibition of the corpora allata, and hence the juvenile hormone titer which, in turn, determines the path of differentiation toward the next molt. In *Hodotermes mossambicus* a reduction in the number of pigmented, foraging workers, as by predation, can be compensated for by production of individuals to replace them (Hewitt *et al.*, 1969). Since they bring in nearly all of the food, it is tempting to think that their differentiation is similarly related to a colony's nutritional status.

Most dry-wood termites nest within their food supply, the deadwood of branches and logs, which eventually is either essentially consumed or decomposed to the point that it becomes nutritionally unsuitable. Nagin (1972) has described the strategy used by *Neotermes jouteli* (Banks), which is probably typical of species whose colonies flourish in the face of such finite and changing food supplies. Immature colonies divert nutrients from their nests toward building up large numbers of larvae. During this phase colonies have a strong capability for producing replacement reproductives to insure that this goal is reached. Mature colonies are concerned with preserving and fattening up their pseudergates for the production of alates. Their capacity for producing replacement reproductives thus gradually declines. This may explain a phenomenon which has been rather casually noted by a few observers (Nutting, 1969: p. 236) – an apparent final flare of alate production in declining colonies of certain members of the Calotermitidae and Hodotermitidae.

Nutrition seems to play an important role in determining the developmental route taken by the larvae of *Reticulitermes lucifugus* (Rossi). In contrast with the striking inhibitory influence of reproductives on the immatures of *Calotermes flavicollis*, larvae may develop into neotenic reproductives even when reproductives are present, if the 'alimentary equilibrium' is favorable. The numerical regulation of the castes is a result of the group effect which acts by way of the 'alimentary potential' of a colony: the number of nurses (workers and pseudergates) determines the formation or suppression of the 'parasitic' castes (larvae, nymphs, soldiers and reproductives). In other words, balanced differentiation proceeds in a well-fed colony (Buchli, 1958). Good nutrition apparently produces the well-developed fat body which is a prerequisite for the development of workers, pseudergates or nymphs into neotenic reproductives (Buchli, 1956).

Esenther's (1969) studies of the activity and development of colonies of *Reticulitermes flavipes* near the northern limit of its distribution led him to agree with Buchli. Although neither denied that pheromones might be involved in caste control, they were convinced of the more obvious importance of nutrition. In Wisconsin the reproduction of *R. flavipes* ceases during the winter, when the workers cannot forage at or near the surface. During the rest of the year the nutritional state of its colonies appears to be good when the workers are abundant and relatively free to forage, and poor when they are occupied with new broods or overburdened with nymphs and other dependent castes. Good nutrition favors successive flares of alates, nymphs and soldiers; moderate nutrition permits growth and molting with little progressive development; poor nutrition results in little or no growth, development or reproduction, and even in mortality of partially differentiated forms.

Smythe & Mauldin's (1972) selective defaunation experiments with *Coptotermes formosanus* have raised the intriguing possibility that one or two of its three species of symbiotic protozoans may be responsible for producing the nutrients required for soldier differentiation and maintenance.

We obviously have very little idea of what constitutes good or poor nutrition for any species of termites, either quantitatively or qualitatively. The logical approach to this problem requires an artificial diet. This had seemed a rather remote possibility until Mauldin & Rich (1975) reported limited success with a diet of α-cellulose, water, a sterol such as β-sitosterol and methyl-*p*-hydroxy benzoate for *Reticulitermes flavipes*.

Control of reproduction

That reproduction should also be strongly influenced by nutrition seems obvious; yet, here again, we can scarcely relate it either to quantity or to any particular quality of food. Nutting (1969) reviewed colony foundation and outlined colony development through maturity and decline. Founding pairs of the Calotermitidae, Hodotermitidae and Rhinotermitidae which nest in wood or near to it are generally capable of feeding themselves and their brood for many months if necessary. Although the alates mature with well-developed fat bodies, these may become severely depleted. The alates of some species apparently do not feed during the interim between maturation and the time they fly from the parent colony (see below).

Weesner (1956) reported that the initiation of egg laying by founding females of *Reticulitermes hesperus* Banks is partly dependent on feeding. She concluded that the fertility of these young queens is related to their 'physiological state', i.e. those which fly after a long delay in the parent colony are less well nourished and hence are slower to oviposit and lay fewer eggs than those flying sooner, shortly after their final molt. Buchli (1960) was reasonable sure that imagoes of *R. lucifugus* awaiting flight do not feed or have any trophallactic exchanges with the workers. He also suggested that oviposition is temporarily halted in incipient colonies because food reserves of the young pair became exhausted while they are tending their first brood Buchli (1950). Oviposition resumes when this brood is able to feed its parents.

Most authors have reasoned that the founding pair must depend initially on stored reserves. Hewitt *et al.* (1972) have determined that the alates of *Trinervitermes trinervoides* contain a large quantity of TGLs which is rapidly utilized during the first 10 weeks or so of colony development, until the first workers mature and begin foraging. This material thus appears to be an important energy source during this critical period. Various members of Nel's group have contributed to the finding that, at flight time, the alates of both *T. trinervoides* and *Hodotermes mossambicus* store large amounts of water in their salivary reservoirs which must supply the incipient colony during droughts that may last as long as three months (Hewitt *et al.*, 1972). During subsequent development of the colony, the nutritional state of the king and queen, and hence their reproductive capacity, is dependent upon the food which the workers are able to bring in. The availability of food, which depends upon weather conditions, as well as the territorial and aggressive behavior of neighboring colonies, thus affects both size and density of colonies (Hewitt *et al.*, 1969).

Perhaps the deterioration and eventual exhaustion of the food supply of dry-wood (most of the Calotermitidae) and damp-wood (Termopsinae) colonies is the reason for the declining egg production and the gradual shift from larval to alate production in these species which are usually confined to the wood originally chosen by the founding pair (Nagin, 1972). This is almost certainly not the case with many species of Mastotermitidae, Hodotermitinae and Rhinotermitidae, whose workers are able to forage widely beneath, on or above the soil surface.

On the other hand, Smythe (1972) has made an interesting observation that may have a bearing on the decline of certain species, even in the Rhinotermitidae. He has found three abnormally faunated colonies of *Reticulitermes virginicus* in nature, all in extensively worked logs. Further, he has noted that normally faunated colonies held in the laboratory for 10 months or more until their food supply has greatly deteriorated have tended to lose all or nearly all of their protozoans of the genera *Pyrsonympha* and *Dinenympha*. Inadequate nutrition may well eliminate these important symbiotes, with the result that such colonies cannot recover even by exploiting other more adequate food resources that appear to be available to them.

Most of the evidence on the Termitidae indicates that oviposition begins at a relatively high rate within a few days after swarming and the establishment of a pair in their initial cell. The literature continues to promote the widely held belief that the pair take no food until the maturation of workers in their first brood and assurance of an ample food supply. The royal pair are generally assumed to subsist on, and feed their brood from, nutritional reserves derived from the fat body, flight muscles and whatever might have been in the gut at the time of flight. However, there is evidence that the royalty of some species supplement these reserves by feeding on soil, cast skins and even some of their brood (Nutting, 1969). By the time they fly from the parent colony, young females of *Tenuirostritermes tenuirostris* (Desneux) contain some well-developed eggs so that they are ready to begin laying within a few days after pairing (Weesner, 1955). This is apparently also the case with *Macrotermes natalensis* (Geyer, 1951). Wiegert & Coleman (1970) have pointed out that the primary reproductives of *Nasutitermes costalis* (Holmgren) have two physiological adaptations which carry them through the critical period of colony founding in situations where food may not be available for several months: the ability to store large amounts of high energy fat and an unusually low fasting metabolic rate. We might reasonably expect adaptations such as these to be widespread within the Termitidae.

In several species, egg production is reduced or suspended after the first clutch is laid and resumed only after these have hatched, or even later when the first workers mature (Nutting, 1969). Sands' (1965*a*) studies of established colonies of *Trinervitermes ebenerianus* (Sjöstedt) suggest a direct and continuing relation between good nutrition and oviposition. He found that its most active foraging period occurs in September and October, toward the end of the rainy season, when the grass collected for storage is at its highest nutritive content. A large increase of brood follows in November and December. Skaife's (1955) observations on the biology of *Amitermes hastatus* (Haviland) (as *atlanticus*) are probably as complete as those on any termitid. In thriving colonies the queen lays no eggs at all during the winter from May to September, apparently as a result of the workers' limiting her food supply. With the return of warm, spring weather, the workers begin to feed her more generously and she resumes laying. The queen's rate of oviposition fluctuates through the summer even in good times. According to Skaife, the workers regulate egg production by increasing or decreasing her food supply according to their abilities to care for the brood. If food is short, as in a prolonged drought, they halt oviposition altogether.

Social recycling of nutrients

While trophallaxis has led to the efficient distribution of food, termite societies also practice nutrient recycling to a degree rarely found among animals. Dietary items which they rather commonly subject to further digestion are fecal material, carton and fungus combs. Conservation of the nutrient pool of a colony also extends to the eating of dead individuals and to the practice of 'controlled' cannibalism (Moore, 1969).

Fecal material and carton

Although the fecal material of the Calotermitidae and the wood-inhabiting members of the Hodotermitidae is conspicuously packed into abandoned galleries or 'dumps', probably all termites use their excreta for constructing, modifying and lining their nests and associated structures. Carton is a hard, dark brown and highly adaptable construction material which is used for a wide range of purposes, from making minor modification to nests in wood to the building of entire nests and enclosing mounds. It is made from semi-liquid excreta which dry to form a tough, pasteboard-like material. It may contain varying amounts of soil particles, and bits of masticated wood or more finely comminuted plant materials which have passed, largely undigested, through the gut.

The composition and uses of fecal material and carton have been thoroughly covered by Lee & Wood (1971*b*) and are briefly reviewed in Chapter 4 of this volume. Both contain variable amounts of cellulose, hemicelluloses and lignin. Feces, and presumably carton, may also contain uric acid and related compounds (Moore, 1969). There appears to be a tendency toward higher lignin/carbohydrate ratios in the feces and carton of the Termitidae, signifying less undigested cellulose and hence, more efficient carbohydrate digestion. The evidence will not support any such generalization, however, for it is based on very few representatives of each family and has not been related to the lignin–carbon content of the food eaten by the test species.

Although it is well known that the lower families regularly engage in proctodeal feeding and that the Macrotermitinae incorporate fecal material into their fungus gardens, reports of more than incidental observations on the ingestion of fecal material are uncommon: *Zootermopsis angusticollis* (Hungate, 1938), *Hodotermes mossambicus* (Nel *et al.*, 1969), *Ancistrotermes guineensis* (Silvestri) (Sands, 1960), *Amitermes hastatus* (Skaife, 1955). During some of his experiments Hungate noted that individuals isolated on pieces of wood ingested fecal pellets. As a result of his experiments and from the resemblance in composition between the foregut contents and the pellets from wild termites, he was convinced that pellet feeding was quantitatively of very considerable importance to natural colonies of *Zootermopsis*. Nel and his colleagues have seen larvae in normal colonies of *H. mossambicus* feed avidly on soldier feces. Although the attraction or reason for this is unknown, the group has suggested that the feces may contain a pheromone inhibiting soldier determination which the larvae could easily dispense to the colony. Skaife has reported that only the workers and soldiers of *A. hastatus* partake of 'this unpleasant form of nourishment'. Leach & Granovsky (1938) theorized that feeding on fecal material, and thereby salvaging nitrogen from urates and dead micro-organisms, might increase the efficiency of the termites' nitrogen economy. However, considering our present knowledge, we are inclined to conclude that the recycling of raw fecal material, as opposed to structural carton, is neither a general nor a particularly economical practice.

Lee & Wood (1971*b*) reported that termitids that enlarge their nests from the inside outwards (e.g. *Nasutitermes exitiosus*) ingest carton during the remodeling; however, we suspect that some members of each family probably eat carton on occasion. Members of the two Australian genera *Ahamitermes* and *Incolitermes* which are obligate inquilines in the nests of *Coptotermes* deserve special mention here (Gay & Calaby, 1970). These social parasites apparently subsist entirely on the carton walls of the host's nest. Lee & Wood (1971*b*) have determined that the carton of *Coptotermes acinaciformis* (Froggatt) and *C. lacteus* contains much undigested wood. Certain species of the genus *Termes* appear to have established similar relations with dominant genera in Australia, South America and Africa (Wilson, 1971).

There is a growing body of evidence that the recycling of feces and carton does accomplish further digestion of carbohydrate, although Moore (1969) found no support for the idea that termites might be able to utilize the urates in these materials. *Reticulitermes lucifugus* var. *santonensis* and *Nasutitermes ephratae* digested higher proportions of cellulose in all woods tested than did *Calotermes flavicollis* or *Heterotermes indicola*. Seifert & Becker (1965) suspected that the greater efficiency resulted from re-ingestion of wood in their fecal material. Larvae of *Hodotermes mossambicus* isolated shortly before or after ecdysis, and which have lost most or all of their protozoans, showed 85–90% survival on water and fresh fecal material for at least two weeks. The survivors had obviously become refaunated through either trophallaxis or coprophagy. The fact that they lost weight was probably due to the low nutritive value of the fecal material (Retief & Hewitt, 1973*b*). Hegh (1922) mentioned that *Microcerotermes fuscotibialis* (Sjöstedt) ate the inner carton walls of its nest when food was scarce. *Nasutitermes exitiosus* is able to survive for at least 12 weeks on carton from the nest of *Coptotermes lacteus*, apparently because it contains a sufficient quantity of undigested plant fragments; however, the reverse is not true.

It is pertinent to mention here that the very origins of termite society may rest on a history of feeding on fecal material. Ancestral cockroaches or insects ancestral to the wood-feeding cockroaches (*Cryptocercus* spp.) and the termites may have remained together in family groups in favorable micro-climatic situations such as damp wood. Food shortages may have enforced feeding on fecal material for maximum use of nutrients. Stomodeal and proctodeal food exchange could then have followed logically and perhaps promoted a symbiotic relationship between free-living protozoans and insects (Sands, 1966). The necessity for maintaining their cultures of symbiotes might then have provided the binding force which led to the development of true sociality in these small feeding groups (Cleveland, Hall, Sanders & Collier, 1934).

Fungus combs

The fungus-growing termites of the sub-family Macrotermitinae construct within their nests sponge-like structures known as 'fungus combs' or 'fungus gardens'. Sands (1969) has reviewed early interpretations of the function or significance of these unique structures, which included food for the young, nutritional adjuvants or vitamins, homeostatic mechanisms for heat and humidity control, a characteristic of normal nest architecture and a tolerated commensal expelled when overgrown. Sands himself has concluded that the fungus combs serve primarily a nutritive role, while Lee & Wood (1971*b*) have suggested that function may vary with species of termite and time.

Although the origin of comb material is still disputed, the work of Alibert (1964), Josens (1971*a*) and Sands (1969) provides adequate evidence for at least some species, that chewed wood and other plant materials are not used directly, but are eaten and digested prior to incorporation into the fungus garden. The alternate hypothesis, promoted by Grassé & Noirot (1961) and Noirot & Noirot-Timothée (1969), maintains that the masticated but undigested plant fragments are placed directly on the combs. Lee & Wood (1971*b*) reasonably suggest that the ingredients of the combs may vary according to the species and that some termites may even use a combination of feces and plant material, as is sometimes the case with carton.

Josens (1971*a*), in work mentioned earlier in this book, studied the dynamics of comb construction in four species of Macrotermitinae from the Ivory Coast. He induced termites to feed on small cakes of sawdust and carbon black, then followed the subsequent movements of this marked food. The entire length of the gut was black one day later. The termites deposited marked fecal material on the upper surfaces of the comb and consumed older material from the lower surfaces. By noting the advance of carbon black through the comb, Josens measured the turnover rates which, for the four species studied, varied from five or six weeks to two months.

The fungus combs are composed primarily of lignin, cellulose and mineral matter. The lignin/cellulose ratios vary according to location in the comb, the outer or newer layers containing more lignin and less cellulose than the inner parts where the lignin has been degraded by the fungi. Nitrogen content in fungus combs varies from 1.1–2.0%, probably much higher than in carton, and mineral matter from 9.9–31.5% (Cmelik & Douglas, 1970; Lee & Wood, 1971*b*). Comb composition will probably be found to vary widely, not only with respect to position or age, but also according to the type of plant material incorporated into it.

There is no doubt that these termites regularly eat the fungal tissues along with the degraded fecal material; however, it is still not entirely certain whether the fungus combs are essential in their diet or simply provide useful supplements. Sands (1969) has concluded that the relationships between the termites and *Termitomyces* is truly symbiotic although, perhaps, more specific on the part of some of the fungi than for the termites. Sands (1956) and Ausat *et al.* (1962) found that, while laboratory groups of *Odontotermes* sp. could survive no longer on cellulosic diets than under starvation, the addition of fungus comb, bearing conidial spores, greatly increased survival. On the other hand, Grassé (1959) concluded that the fungus-growers must also eat other foods since he was able to maintain workers and soldiers of *Macrotermes natalensis* for 18 months on rotten wood.

It has been suggested that the fungi might supply vitamins (Grassé, 1945) and nitrogenous compounds (Sands, 1969). Certainly, the most significant finding was that of Grassé & Noirot (1958*a*), who demonstrated that the fungi degrade lignin in the particles of wood passed in the fecal material, thereby exposing additional cellulose which can be utilized by the termites when they feed on the comb. We might then consider that the Macrotermitinae are able to make extremely efficient use of their food through a combination of fungus culture and a specialized form of coprophagy. McBrayer (1973) has pointed out that a number of other animals, which consume cellulose-rich food but produce little cellulase, also practice coprophagy. These include certain wood-boring beetles, isopods, lagomorphs and rodents. He demonstrated experimentally that the eurydesmid millipede *Apheloria montana* (Bollman), deprived of its own feces but allowed access to an otherwise normal diet of leaf litter, becomes inactive or dies in about a month.

Fungus cultivation is a behavior which has evolved few times and is currently restricted to the Attini, the Old-World Macrotermitinae and a few wood-boring beetles (Wilson, 1971). Wilson suggests that these ants and termites are not true ecological equivalents because, were both present in the same environment, their different foraging habits would permit sympatry. Their success, evidenced by immense populations, is largely attributable to the fungal associations which permit exploitation of an almost inexhaustible energy source, cellulose. While the competitive advantage for ants is immediately apparent, it is less so for the Macrotermitinae because all termites subsist on cellulosic materials in various stages of decomposition. The probable advantage of external fungus cultivation to the Macrotermitinae lies in the high degree of assimilation of both lignin and cellulose made possible by the fungi. These termites are thus able to forage for a large variety of plant materials both live and dead because the fungi are far less selective than the termites in their food habits.

Cannibalism and mutilation

Emerson (1939) discussed a number of factors which influence the populations and biology of social insects, cannibalism among them. He considered that this seemingly irregular but characteristic behavior of termite societies must certainly enhance the survival of incipient colonies, probably conserves valuable nutrients such as nitrogen, and might even serve as a means of population control, particularly of the numbers and proportion of each caste at the colony level. Wilson (1971) tended to agree that cannibalism could be considered as a protein-conserving device since the termite diet is generally low in this constituent. Noting that dead individuals are found in the nests of relatively few species, Lee & Wood (1971*b*) felt that the practice probably served the primary function of nest sanitation.

A few who have worked closely with laboratory colonies have reported instances of accidental cannibalism: Buchli (1950), *Reticulitermes lucifugus*; Castle (1934), *Zootermopsis angusticollis*; Williams (1959*b*), *Cubitermes ugandensis*. The observations of Buchli and Williams involved larvae or workers that were being assisted with shedding of their integument by the young reproductives. If one of these individuals was injured during the process, the reproductives promptly ate it. Castle observed that any member of a colony unfortunate enough to incur injury during mutual grooming was consumed. Whether individuals injured at this critical point survive or not, their nutritive investments are not entirely lost.

During the development of young colonies the founding reproductives occasionally cannibalize some of their eggs and brood, apparently to supplement whatever reserves they possess. Termite workers, either functional or true, do not produce special, unfertilized 'trophic' eggs, which are routinely eaten, as do the workers of many ant and stingless bee species. Nevertheless, there are many reports of young reproductives, and occasionally the brood, feeding on their eggs (*Calotermes flavicollis* (Grassé, 1949; Alibert, 1968); *Cryptotermes havilandi* (Wilkinson, 1962), *Reticulitermes hesperus* (Weesner, 1956), *R. lucifugus* (Buchli, 1950)) or on both eggs and brood (*R. flavipes* (Snyder & Popenoe, 1932),

Amitermes hastatus (Skaife, 1955), *Cubitermes ugandensis* (Williams, 1959*b*), *Tenuirostritermes tenuirostris* (Light & Weesner, 1955) and *Nasutitermes ephratae* (Becker, 1961)). Skaife has noted that the workers of *A. hastatus* may also turn to eating the young during droughts when their food supply is deficient.

Injured, moribund and dead individuals, as well as exuviae, are routinely eaten, usually by larger larvae, nymphs, pseudergates or workers. Large numbers of dead individuals are often found in the outer galleries of the mounds of three species of Australian termitids (Grassé, 1949: p. 496; Gay & Calaby, 1970). Whether or not they are used for food, this seems an unusual habit for a termite. Perhaps it is simply a result of the highly stereotyped 'necrophoric' behavior, as described for some ants (Wilson, 1971). Decomposition products of insect corpses, commonly long-chain FAs, stimulate the workers to remove dead nest-mates to established refuse piles.

One or more individuals of a particular caste may occasionally become a burden or liability to a colony. It is apparently general practice to kill and eat these forms. Although the reasons for this behavior are not always clear, such selective cannibalism probably serves to control the proportional caste composition and, at the same time, to conserve their essential nutrients. For example, toward the time of swarming, alates of *Reticulitermes lucifugus* (Buchli, 1961) react to increasing hostility of the workers by retreating to peripheral nest chambers. Those alates which fail to do this or which are later somehow prevented from swarming are killed, but Buchli does not mention whether or not they are eaten. Alates of *Coptotermes lacteus* are killed and eaten if they do not leave the nest at the proper time (Ratcliffe *et al.*, 1952).

Occasionally three or more newly de-alated reproductives become established in the same or adjacent cells. While these joint efforts at colony founding may be temporarily advantageous, their sequel nearly always results in the destruction of all but a single pair (Nutting, 1969). The practice of eliminating supernumerary neotenic reproductives is well known from the studies of Grassé & Noirot (1946, 1960), and Lüscher (1952) on *Calotermes flavicollis*. Ruppli & Lüscher (1964) later reported that hostility between the reproductives leads to fighting and that the injured are eaten by larvae and nymphs. Further experiments by Ruppli (1969), some of them involving antennectomized reproductives, led him to conclude that all reproductives must emit odors, detectable by antennal sense organs, which permit not only sexual but also individual identification. Females are more aggressive, particularly toward males. Should this kind of pheromonal control be found to extend to other castes, cannibalism may indeed prove a highly selective regulator of the composition and size of termite colonies. Such observations as those made by Castle (1934) and a few others suggest that this may be the case. In both natural and laboratory colonies of *Zootermopsis angusticollis* he noted that an increase in the relative number of soldiers occurred after swarming. In the laboratory he was able to determine that the older and smaller soldiers were selectively killed to restore a certain soldier/non-soldier ratio. Elimination of excess neotenic reproductives and soldiers has recently been reported in *Neotermes jouteli* by Nagin (1972). In some cases the reproductives were cannibalized, in others they were forced outside the galleries, apparently to starve to death. A different approach is apparently taken with old reproductives. Skaife (1955) observed in laboratory nests that workers of *Amitermes hastatus* literally licked to death primary queens whose fertility was declining. The licking continued until nothing remained but their empty bodies which were eaten or buried.

On the other hand, there are several, well-known cases in which more than a single pair of reproductives is obviously encouraged. As many as six reproductives, of both sexes and three different forms, have been found in natural colonies of three species of calotermitids (Nutting, 1970). Castle (1934) reported that colonies of *Zootermopsis angusticollis* may be headed by large numbers of supplementary reproductives – 26 in one colony of about 2000 individuals, for example. This sort of observation suggests that selective elimination of reproductives is not necessarily the rule. One wonders not only about the extent of these opposing types of behavior in nature, but also whether there might be some combination of factors which favors selective cannibalism under laboratory conditions. Pickens (1934) reported that multiple queens in both field and laboratory colonies of *Reticulitermes hesperus* made possible the very high rates of egg production. According to Hendee (1935) the reproductive capacity of the supplementaries in natural colonies of *Z. angusticollis* probably more than offset losses due to cannibalism. More likely, perhaps, the production and tolerance of multiple reproductives is a further strategy of termites living within limited food supplies, which allows young colonies to produce large numbers of larvae for transformation into alates as the colonies mature. (See *Caste determination, nutritional influences* above.) Since it is fundamental to, but remote from, our theme here, the reader is referred to Wilson's (1971) review of the evolution and significance of monogyny and the advantages of single versus multiple queens.

However, cannibalism occurs even when nutrition appears to be adequate (Hendee, 1935; Grassé, 1949; Nutting, 1969). Hendee has presented a strong case for cannibalism being a normal occurrence among termites under natural conditions. The termites in her experiments (*Zootermopsis angusticollis*) appeared healthy and showed normal behavior while exhibiting a uniform rate of cannibalism on a diet of rotten, fungus-containing wood similar to that available to them in nature. Individuals which were attacked appeared to have been healthy and to have died from no other cause than cannibalism. Compared with termites on a natural diet, those on sound, fungus-free wood showed an increased rate of cannibalism and made smaller gains in average dry weight and in nitrogen content. Andrew (1930) and Cook & Scott (1933) have shown that *Z. angusticollis* becomes intensely cannibalistic on diets lacking nitrogen. We are thus led to ask, as did Cook & Scott, whether cannibalism would cease on an adequate diet. In short, is the practice of cannibalism a response to dietary deficiencies?

Tracey & Youatt (1958) found chitinase activity in extracts of whole workers of *Nasutitermes exitiosus* and *Coptotermes lacteus*, and Waterhouse, Hackman & McKellar (1961) later confirmed this finding in the latter termite. Moore (1969) reasoned that the chitinase might be in the digestive tracts of mature workers (which no longer molt) and therefore available for recycling the nitrogen in the exuviae and bodies of their cannibalized nest-mates. According to Retief & Hewitt (1973*b*) the generality of this possibility tends to be negated by the fact that they were unable to detect any chitinase activity in larvae, workers or soldiers of *Hodotermes mossambicus*. They suggested that it would be advantageous for the large larvae to be equipped for digesting chitin since they partially digest most of the raw food and feed the dependent castes. On the other hand, we suspect that *H. mossambicus* is probably rarely stressed for nitrogen since its main diet of dead grass is relatively high (0.75%) in this element when compared with that of many other termites which feed on wood (0.03 to 0.1% nitrogen in wood, Table 7.1). Whether or not it is cannibalistic, we submit that it may therefore not require a chitinase for this purpose.

During the establishment of incipient colonies the mouthparts, legs and, most often, the antennae of the founding pair may become mutilated to varying degrees. Williams (1959*a*) has reviewed the suggestions of various authors relating this condition to courtship behavior, mutual grooming and nutritional needs and has concluded that such mutilations were probably a manifestation of cannibalism. Examples of this restrained kind of cannibalism occasionally involve other members of older colonies, most often the nymphs. Two very significant pieces of research have recently proposed two very different functions of mutilation, each only remotely related to nutrient recycling. The mutilation of a single wing pad of an isolated nymph in the penultimate instar of *Calotermes flavicollis* is sufficient to prevent its metamorphosis to a mature, winged individual. The wing itself is rarely regenerated at later molts. Mutilations of an antenna or leg impede metamorphosis to a lesser degree. These injuries appear to act directly, or via the brain, on the thoracic glands and corpora allata. Springhetti (1969) believes that through such mutilations the members of a colony are able to effect some control over the number of alates produced, and hence the economy of the colony.

Hewitt, Watson, Nel & Schoeman (1969) have cleverly shown that this apparently trivial piece of behavior (antennal mutilation) is part of the causal mechanism of the striking change from typical pre-flight, group behavior to post-flight, pair behavior by alates of *Hodotermes mossambicus*. Alates held in large groups exhibit social behavior in that they are not aggressive, do not drink much water, mate nor lay eggs. Alates kept in pairs become anti-social; i.e. they become aggressive, drink large amounts of water, mate, lay eggs and tend their brood. In a series of carefully designed experiments these investigators determined that the major factor which maintains the social behavior is frequent body contact by the antennae of the termites within the group. Since complete removal of antennae prevents successful colony foundation, they reasoned that the partial antennal mutilation frequently observed in pairs might serve to reduce antenna to body contact and thus prevent reversion to group behavior.

In spite of Hendee's (1935) argument that cannibalism is natural, the fact that we know nothing of the incidence of cannibalism in colonies under natural conditions makes any evaluation highly subjective. While it is tempting to conclude that the primary function of cannibalism is nitrogen conservation, nearly all of our information on the subject comes from observations and experiments, many of the latter very unnatural, on laboratory groups. In cases involving accidental cannibalism, or the consumption of the moribund, the dead and exuviae, and perhaps even the selective removal of excess individuals of any caste, this behavior probably does serve a sanitary function. It incidentally conserves valuable nitrogen, a dietary commodity which is not usually abundant in the termite economy. In cases of ordinary, non-selective cannibalism of apparently healthy individuals, we are inclined to view the practice as a response to nutrient deficiencies. Extreme cases, involving the consumption of eggs and brood, are almost certainly concerned with survival. Owing to the substantial metabolic losses incurred in recycling their own eegs and brood, it is costly for a founding pair of termites to resort to this practice. Wilson (1971) has aptly said that these items serve as 'energy capital' which can be drawn upon in hard times.

The writing of our contribution and some of the unpublished research reported herein were carried out as a part of the USA/IBP Desert Biome, and were supported in part by National Science Foundation Grant no. GB-15886. Additional support was also provided under State Research Project 144 in the Agricultural Experiment Station, University of Arizona, Tucson. For reading the manuscript and contributing a number of helpful suggestions, we are very grateful to Dr James W. Berry, Department of Agricultural Biochemistry, and Dr Floyd G. Werner, Department of Entomology, University of Arizona.

8. Nutrient dynamics of ants

A. ABBOTT

It has long been known that the exchange of food among ants is important in the social organisation of the colony. The advantage of the social way of life is that it allows the division of labour, not only into reproductive and non-reproductive castes, but also into classes of workers fulfilling different functions. Some workers leave the nest and forage for food; others carry out various functions in the nest, and so must be fed by the foragers. Apart from the reserves of the founding queen, which may be sufficient to rear an initial labour force of workers, all the food given to the queens, sexuals, larvae and nest workers is ultimately derived from the material brought in by these foragers. However, there are very indirect routes by which these nutrients may reach their eventual recipients, as well as the direct one of forager donations.

Storage

The development of the proventriculus into a self-closing valve (Eisner, 1957; Eisner & Brown, 1958) has allowed the crop to be used as a reservoir for liquid food. The crop can expand greatly if required, extreme cases of this leading to the formation of a 'replete' caste whose function is entirely to act as a food store. Repletes are well known in some desert ants, such as *Myrmecocystus*, which store a carbohydrate-rich liquid collected from plant galls, and honeydew. More recently, large workers of *Solenopsis invicta* have been shown to act as repletes in storing oil (Glancey *et al.*, 1973), while the honey ant *Myrmecocystus mexicanus* has been found to store lipid as glycerol and cholesterol esters (Burgett & Young, 1974). More commonly, however, the crop acts as a short-term store, allowing foragers to bring quite large quantities of liquid food back to the nest to be shared out. Much of the space in the gaster of laying queens is taken up by the ovaries, so in some species (of the sub-families Myrmicinae and Formicinae) an alternative oesophageal 'crop' has been developed in the thorax, after the breakdown of the flight muscles and the adipose tissue replacing them (Petersen-Braun & Büschinger, 1975). This is probably used as a short-term food store to compensate for fluctuations in the supply of food secretions by young nurses.

Food exchange between workers

The exchange of regurgitated liquid food between worker ants is an accepted phenomenon occurring in all sub-families, though not in some ponerines (Haskins & Whelden, 1954), and very little in the dorylines (Wilson, 1971). The use of radioactive tracers and dyed foods has enabled this exchange to be studied. The rate of exchange has been shown to vary with season as well as temperature, more markedly in some species than in others (Khamala & Büschinger, 1971), and the pattern of exchange differs greatly with different types of food.

Carbohydrate

Carbohydrate, when given as honey or sugar solutions to worker ants, is not distributed evenly to the rest of the colony. Gösswald & Kloft (1956), working with *Formica rufopratensis minor*, found that foragers fed honey solution preferentially to large workers and that even over a period of time smaller workers received a smaller share of the food than the large or medium workers. Länge (1967) also found this differential distribution, though in *Formica polyctena* Foerst. it was independent of worker size. Workers with degenerate ovaries, that is, the foragers, received honey before workers with active ovaries, the nurses. Schneider (1972), too, found that more honey was given to foragers of *F. polyctena* and that these individuals were reluctant to pass it on when solicited. Wilson & Eisner showed that little honey reaches the larvae, but its transmission between workers varies between different species, from very little exchange in *Pogonomyrmex badius* Latreille to rapid colony saturation in two *Formica* spp. (Wilson & Eisner, 1957; Eisner & Wilson, 1958). These authors suggest that very different results would be obtained for protein distribution.

Protein

Using insect haemolymph as a protein food Länge (1967) did in fact find the opposite result to that for honey distribution. The protein was given preferentially to nurses, and the foragers were more ready to regurgitate haemolymph than the nurses. In the fire ant, very little protein is taken by the workers, most of it being given to the larvae (Vinson, 1968).

Lipids

Lipids, like carbohydrates, are quite rapidly spread among the workers (Echols, 1966), though they are also given in large amounts to the larvae (Vinson, 1968; Lofgren, Banks & Glancey, 1975).

The differences in distribution of these classes of liquid foods are understandable in the light of their different uses. Carbohydrates are primarily an energy source, and so the workers, especially the highly active foragers, take a greater share. The growing larvae, and egg-laying queens, have more need of proteins, so most of this is taken by the nurses to pass on to them, either directly, by regurgitation, or indirectly by egg-laying. Lipids are also used as a source of energy and are particularly suitable as a reserve material, so although the workers do take much of it, a lot of the lipid is given to the larvae to store. The role of the post-pharyngeal gland in this process will be discussed later. This interpretation of food exchange in ants as evidenced by laboratory experiments using single, pure foods must be a simplification of what happens in the field, where natural food sources are less homogeneous and several different foods may be brought into the nest at once. Indeed, Markin (1970) has shown that different foods do get mixed in the crop of a single ant. The same experiment demonstrated, however, that workers of the Argentine ant *Iridomyrmex humilis* (Mayr) do have separate appetities for carbohydrate and protein. Sugar-sated workers were more likely to accept protein than honey.

Food exchange between workers and larvae

Liquid food and trophic eggs

Liquid food forms an important part of the larval diet of many ant species. This may be the workers' crop contents or the contents of the post-pharyngeal gland, droplets of which are regurgitated and placed on the mouthparts of the larva. Younger larvae particularly may depend on liquid food. Those of the Argentine ant receive a high proportion of post-pharyngeal gland secretions in their diet (Markin, 1970), while the fire ant *Solenopsis invicta* Buren feeds its young larvae a liquid diet augmented with eggs (O'Neal & Markin, 1973). In several species the workers have been found to lay two types of egg; reproductive eggs which can develop into males, and trophic eggs which serve as a protein-rich food. Examples are *Plagiolepis pygmaea* Latr. (Passera, Bitsch & Bressac, 1968), *Aphaenogaster subterranea* (Bruniquel, 1972), *Temnothorax recedens* (Dejean & Passera, 1974) and *Myrmica rubra* (L.) (Brian, unpublished). The fire ant queen similarly lays two types of egg, reproductive and trophic (Glancey, Stringer & Bishop, 1973).

Solid food

Of course, in many species solid food is given directly to the larvae. Le Masne (1953) states that all the ant species that he has studied in any depth provide solid food for their larvae, in the third instar at the latest.

Fragments of food are placed on the mouthparts of the larvae which are then capable of consuming and digesting the food themselves. In at least one species, *Aphaenogaster subterranea*, which brings quite large prey insects back to the nest, workers carry the larvae to the food, and place them in a position where they can feed on the prey (Büschinger, 1973). It is probable that the larvae are sometimes essential if use is to be made of certain solid foods. Workers of *Veromessor pergandei* (Mayr) are unable to get their mandibles into the hole which they cut in one end of some of the hard-shelled seeds of the Compositae which they exploit. After enzymatic softening of the embryo, the head of a larva can be inserted into the seed, and the contents consumed (Went *et al.*, 1972). These authors suggest that this is the only way that the food value of these seeds can be made available to the ant colony. The question of larval exploitation by the rest of the colony that this implies is discussed later.

Differential feeding of solid food to different brood classes, possibly of significance for caste determination or polymorphism, may occur. Bivouacs of the army ant *Eciton* containing young worker broods have the larvae arranged with the smallest at the centre. Only the minim workers carrying the smallest food particles can reach these larvae in the middle, which consequently may receive a different diet from the larger larvae towards the outside of the bivouac, which can be reached by all workers, whatever food they are carrying. These larvae then become polymorphic workers (Schneirla, 1971).

Larval secretions

The donation of substances between workers and larvae does not occur in one direction only. Lofgren *et al.* (1975) have found a transfer of material from larvae to workers of the fire ant, though they did not study the method of transfer. Larval secretions are collected and consumed by workers in many species, and sometimes actively solicited. Apart from the cuticular secretions which may be attractive to the workers and elicit grooming behaviour, ant larvae frequently produce droplets of clear fluid at the mouthparts, and clear or sometimes milky fluid from the anus. These fluids are sometimes referred to as stomodeal and proctodeal secretions, respectively. Stomodeal secretions, produced in the sub-families Myrmicinae, Ponerinae and Formicinae, but not by the Dolichoderinae (Maschwitz, 1966), are probably formed in the salivary glands. Le Masne (1953) records that this saliva is sometimes ejected spontaneously, and sometimes solicited by the workers with similar antennal movements to those preceding feeding of the larvae. Fire ant nurses also stroke the head and mouthparts of the larvae to solicit saliva which is then quickly consumed (O'Neal & Markin, 1973). Maschwitz (1966) found that the oral secretions of *Tetramorium caespitum* Latr., *Myrmica ruginodis* Nyl. and *Formica sanguinea*, which he thought 'very probably come from the salivary gland', contained high concentrations of amino acids and proteins, but no carbohydrates. The amino acid concentration, at least in *Tetramorium* where he made the comparison, was higher than that of the haemolymph, implying active secretion of amino acids by the salivary glands, if these are in fact the source of the fluid. Wilson (1971) doubts this, but Wüst (1973) states that 'the stomodeal secretion comes from the labial gland' (i.e. the salivary gland). She also found amino acids and proteins but no carbohydrate in the stomodeal secretion of pharaoh's ant *Monomorium pharaonis* (L.) and mentions unpublished work giving similar results for other ant species.

The fire ant produces two types of anal secretion, a milky liquid which may be solicited by workers squeezing and licking the anal region and which is rapidly consumed, and a clear, watery liquid which is carried away between the mandibles of the worker and discarded, never being ingested (O'Neal & Markin, 1973). The urine may be of some food value, as the rectal fluid of two *Formica* spp. is known to contain amino acids (Maschwitz, 1966), as is that of *Monomorium* (Wüst, 1973).

Digestive enzymes

Comparatively little has been published on the digestive enzymes of ants. Delage (1968) conducted a comprehensive investigation of those of *Messor capitatus* Latr., whilst Ricks & Vinson (1972) looked at the fire ant *Solenopsis richteri*. Ayre (1967) compared the enzymes of five ant species and found that the carbohydrases, at least, could be related to the feeding habits of the species. Other authors have covered various digestive enzymes and organs, and an attempt is made here to bring this work together, summaries being given in Tables 8.1 to 8.6.

Martin and his colleagues carried out extensive investigations into the proteases excreted by the workers of attine ants (Martin, 1970; Martin, J. S. & Martin, M. M., 1970; Martin & Martin, 1971), in the course of which they found other enzymes in the rectal fluid (Martin, Gieselmann & Martin, 1973; Martin, Boyd, Gieselmann & Silver, 1975), and also looked for proteases in the head glands and labial glands (Martin, M. M. & Martin, J. S., 1970; Martin, 1974). Although it now appears probable that the rectal proteases, and possibly pectinase and enzymes for the breakdown of xylan and carboxymethyl cellulose, originate in the fungus food of these ants (Lehmann, 1974; Boyd & Martin, 1975), there still remains a considerable amount of work on the proteases of many non-attine ant species.

There have been some differences in the nomenclature of the exocrine

Table 8.1. *Summary of amylase investigations in different ant species*

Species	Author	Pharyn-geal gland	Maxil-lary gland	Mandi-bular gland	Labial gland	Crop	Midgut	Hindgut
Formica polyctena	Graf, 1964	×	×	×	×	+	+	+
	Paulsen, 1969	–	+	×	+	×	×	×
Formica integra	Ayre, 1967	–	–	×	–	×	–	×
Formica fusca	Ayre, 1967	–	–	×	–	×	–	×
Acanthomyops claviger	Ayre, 1967	–	–	×	+	×	–	×
Camponotus herculeanus	Ayre, 1967	–	+	×	+	×	–	×
Camponotus pennsylvanicus	Ayre, 1967	–	–	×	+	×	–	×
Messor capitatus	Delage, 1968	–	–	×	+	×	+	×
Solenopsis richerti	Ricks and Vinson, 1972	×	×	×	+	–	–	–
Myrmica rubra	Abbott, unpublished	–	–	–	–	+	+	–
Atta columbica tonsipes	Martin, 1974	×	×	×	×	×	×	+

+ = present; – = absent or trace only; × = not investigated.

Table 8.2. *Summary of invertase investigations in different ant species*

Species	Author	Pharyn-geal gland	Maxil-lary gland	Mandi-bular gland	Labial gland	Crop	Midgut	Hindgut
Formica polyctena	Graf, 1964	×	×	×	×	+	+	+
	Paulsen, 1969	–	+	×	–	×	×	×
Formica integra	Ayre, 1967	–	–	×	–	×	–	×
Formica fusca	Ayre, 1967	–	–	×	–	×	–	×
Acanthomyops claviger	Ayre, 1967	–	–	×	–	×	+	×
Camponotus herculeanus	Ayre, 1967	–	+	×	–	×	+	×
Camponotus pennsylvanicus	Ayre, 1967	–	+	×	–	×	+	×
Messor capitatus	Delage, 1968	–	–	×	+	×	+	×
Solenopsis richteri	Ricks & Vinson, 1972	×	×	×	–	+	+	+
Myrmica rubra	Abbott, unpublished	–	+	–	–	+	+	+

+ = present; – = absent or trace only; × = not investigated.

glands associated with the alimentary tract of the adult ants, probably because of the differences of the gland systems in the sub-families Myrmicinae and Formicinae where much of the work has been done. Emmert (1968) has reviewed this, and related the names used by different authors, but some work is still not clear, such as that in which Ricks & Vinson (1972) use the terms 'mandibular' and 'post-pharyngeal' as synonymous.

Table 8.3. *Summary of maltase investigations in different ant species*

Species	Author	Pharyngeal gland	Maxillary gland	Mandibular gland	Labial gland	Crop	Midgut	Hindgut
Formica polyctena	Graf, 1964	×	×	×	×	+	+	+
	Paulsen, 1969	−	+	×	−	×	×	×
Messor capitatus	Delage, 1968	−	−	×	+	×	+	×
Myrmica rubra	Abbott, unpublished	−	−	−	−	+	+	−

+ = present; − = absent or trace only; × = not investigated.

Table 8.4. *Summary of trehalase investigations in different ant species*

Species	Author	Pharyngeal gland	Maxillary gland	Mandibular gland	Labial gland	Crop	Midgut	Hindgut
Formica polyctena	Graf, 1964	×	×	×	×	+	+	+
	Paulsen, 1969	−	−	×	+	×	×	×
Myrmica rubra	Abbott, unpublished	−	+	−	−	+	+	+

+ = present; − = absent or trace only; × = not investigated.

The nomenclature used here is the same as that of Bausenwein (1960), based on the work of Charles Janet (1899). Looking at the enzymes produced by the different glands in various species shows that there is no rigid overall pattern in the Formicidae, or even within a sub-family, but several interesting points emerge.

Pharyngeal glands

No carbohydrases or endopeptidase have been found in the pharyngeal glands of eight ant species studied, but three exopeptidases were found here in *Messor* (Delage, 1968). Six species do have a lipase, but in two others this is absent (the myrmicine *Myrmica rubra*, Abbott, unpublished), or only faint (the formicine *Formica fusca* L., Ayre, 1967). The pharyngeal gland usually contains an oily liquid, so the presence of a lipase might suggest a digestive function for the gland itself. Other possibilities will be mentioned later.

Maxillary glands

Though the maxillary glands do have invertase, amylase, maltase and trehalase in one or two species, myrmicine and formicine, there is again no endopeptidase. *Messor* shows the same three exopeptidases here as in the pharyngeal gland, but is the only species out of the seven examined found to have a lipase in the maxillary gland. These glands have no lumen or reservoir, so the secretions that they produce are probably collected in the crop, and any enzymes will begin their action here.

Mandibular glands

Nearly all digestive enzyme studies of ants have omitted the mandibular glands. Delage (1968) states that they are totally atrophied in *Messor*, but thinks that they are probably not involved with digestion in other species. Investigation of *Myrmica rubra* shows no carbohydrases or lipase, but does show the presence of a protease. As it is released at the base of the mandible, this enzyme may be helpful for softening prey which is being malaxated by the workers. The only other work on the mandibular gland shows that another myrmicine *Atta colombica tonsipes* Santschi, a leaf-cutting ant living on cultured fungus, does not have this protease (Martin, M. M. & Martin, J. S., 1970).

Labial glands

Amylase is found in the labial glands of six of the nine species studied, the exceptions being *Myrmica rubra*, *Formica integra* and *F. fusca*. In these three species no enzymes have been found here (Ayre, 1967; Abbott, unpublished), though Ayre did not test for trehalase which may well be present in the two *Formica* spp. as it is found in the labial gland of *F. polyctena* (Paulsen, 1966, 1971). Maltase and invertase are also found here in *Messor* (Delage, 1968). No endopeptidase is found in the nine species studied, but *Messor* has three exopeptidases and a lipase. Of seven other species looked at only the fire ant *Solenopsis richteri* has a lipase (Ricks & Vinson, 1972).

The crop

The crop is not glandular, secreting no enzymes of its own, any found here having come from the glands already discussed or from other sources. *Myrmica rubra* and *Formica polyctena* have the same four carbohydrases in the crop (Graf, 1964; Abbott, unpublished), but although these could all be from the maxillary and labial glands in *F. polyctena* (Schmidt, 1974),

Table 8.5. *Summary of protease (endopolypeptidase) investigations in different ant species*

Species	Author	Pharyngeal gland	Maxillary gland	Mandibular gland	Labial gland	Crop	Midgut	Hindgut
Formica integra	Ayre, 1967	–	–	×	–	×	+	×
Formica fusca	Ayre, 1967	–	–	×	–	×	+	×
Acanthomyops claviger	Ayre, 1967	–	–	×	–	×	+	×
Camponotus herculeanus	Ayre, 1967	–	–	×	–	×	+	×
Camponotus pennsylvanicus	Ayre, 1967	–	–	×	–	×	+	×
Messor capitatus	Delage, 1968	–	–	×	–	×	+	×
Solenopsis richteri	Ricks & Vinson, 1972	×	×	×	–	–	+	–
Myrmica rubra	Abbott, unpublished	–	×	+	–	×	×	×
Atta colombica tonsipes	Martin, M. M. & Martin, J. S., 1970	–	–	–	–	–	+	+

In addition, dipeptidase, aminopolypeptidase and carboxypolypeptidase have been found in the pharyngeal, maxillary and labial glands and the midgut of *Messor capitatus* (Delage, 1968) and dipeptidase in the midgut of *Myrmica rubra* (Abbott, unpublished).
+ = present; – = absent or trace only; × = not investigated.

in *M. rubra* only invertase and trehalase are produced by the maxillary glands, none of the head glands or the labial glands producing amylase or maltase. These two enzymes therefore must either be absorbed with the food, seep forward from the midgut, or be taken up with secretions from the larvae.

The midgut

A wide range of enzymes is found in the midgut of most species studied, though few have an amylase here and some lack invertase or lipase. *Formica fusca* has only a trace of lipase and lacks amylase and invertase but does show moderate endopeptidase activity. Apart from the mandibular gland of *Myrmica rubra*, this is the first site demonstrated for an endopeptidase in any ant species, and the enzyme is found in the midgut of all species so far investigated. The midgut, with the crop and hindgut, has also been investigated for cellulase and lignase activity in two wood-attacking ant species *Camponotus herculeanus* and *C. ligniperda*, with negative results (Graf & Hölldobler, 1964).

Table 8.6. *Summary of lipase investigations in different ant species*

Species	Author	Pharyngeal gland	Maxillary gland	Mandibular gland	Labial gland	Crop	Midgut	Hindgut
Formica polyctena	Paulsen, 1969	+	×	×	×	×	×	×
Formica integra	Ayre, 1967	+	–	×	–	×	+	×
Formica fusca	Ayre, 1967	–	–	×	–	×	–	×
Acanthomyops claviger	Ayre, 1967	+	–	×	–	×	+	×
Camponotus herculeanus	Ayre, 1967	+	–	×	–	×	+	×
Camponotus pennsylvanicus	Ayre, 1967	+	–	×	–	×	+	×
Messor capitatus	Delage, 1968	+	+	+	×	×	+	×
Solenopsis richteri	Ricks & Vinson, 1972	×	×	×	+	–	+	–
Myrmica rubra	Abbott, unpublished	–	–	–	–	–	–	+
Atta colombica tonsipes	Martin, 1974	×	×	×	×	×	×	+

+ = present; – = absent or trace only; × = not investigated.

The hindgut

Some enzymatic activity continues in the hindgut, for example, in *Formica polyctena* five carbohydrases can be demonstrated (Graf, 1964). In *Myrmica rubra*, however, only two (invertase and trehalase) have been shown, amylase and maltase being absent. It is interesting to note that the latter two are the enzymes which, though present in the crop of *M. rubra*, are not produced by its labial or head glands. If their activity does not reach the hindgut from the midgut it is unlikely that it would seep forwards through the proventriculus, so an external origin of these two enzymes in the crop becomes more probable.

Implications

The most striking point about this distribution of digestive enzymes is the lack of endopeptidase before the midgut. The narrow oesophagus will allow small solid food particles to reach the crop, but the proventriculus will let only the smallest of these reach the midgut, so it seems that the adult ant is dependent to a large extent on soluble proteins and amino acids. However, we have already seen that the larvae give up a fluid rich in amino acids and proteins which the workers eagerly accept, so it appears possible that the larvae perform a lot of the digestion of protein for the whole colony. The workers gather the food, bring it into the nest, malaxate it a little and place it on the mouthparts of the larva. The larva chews the food and takes it into the large midgut, which will take quite big pieces of prey. The food, including its protein content, is then digested and enters the larval haemolymph from where much of it is used for larval growth. Excess amino acids and soluble protein are secreted by the salivary gland and stored in its reservoir until the worker performs the correct soliciting movements, when they are ejected in a drop of fluid for the worker to take up and consume.

A midgut endopeptidase is the only digestive enzyme that has been demonstrated in any quantity in *Formica fusca*. Does dependence on the brood go a bit further in this species?

Myrmica rubra workers, too, appear to use amylase and maltase externally obtained. Do these enzymes come from the larval saliva, where amylase at least is known to occur (Abbott, unpublished)?

Pharyngeal gland function

It has already been shown that a lipase is found in the pharyngeal gland of several ant species. This, with the fact that the gland is formed as an invagination of the alimentary tract, suggests a digestive function. Forbes & McFarlane (1961) thought that these glands might secrete 'a substance which aids in digestion', and that the primary function was digestive. This was contrary to the views of Bugnion (1930) who had noted the presence of secretion granules in the gland cell cytoplasm and suggested a role in the feeding of larvae by regurgitation. More recent work has thrown some light on this question.

While *Messor capitatus* was being fed on various dyed foods, it was noticed that honey was taken straight into the crop, and the pharyngeal gland did not become coloured until a few days later. However, if an oily food such as crushed nuts was given, dissection showed this to enter the pharyngeal gland first, and only to reach the crop when the former was full. This separating out of the oily food was found to occur in several other species (Delage, 1966). Sterols of plant origin were also shown to be taken up by the pharyngeal glands (Barbier & Delage, 1967), presumably in association with oils, so the possibility that the lipase is taken up attached to its substrate had to be admitted (Delage, 1968). Delage showed that some of the oily gland contents are actually absorbed into the gland cells. She also showed, however, that transmission of the gland contents to the larvae occurs. Brian (1973*c*) confirmed the selective uptake of oils by the pharyngeal gland of *Myrmica rubra* and also the absorption into the gland cells and the regurgitation to larvae of the gland contents.

Echols (1966) also demonstrated that dyed oil (soya bean oil, used in ant poison bait formulations) was taken up rapidly into the pharyngeal gland of *Atta texana*, and that it was very quickly spread to the rest of the colony. Working with another leaf-cutting ant *Atta cephalotes* (L.), Peregrine, Percy & Cherrett (1972) found a similar rapid uptake of oil into the glands. Though they could find no transfer of gland contents to the larvae, they ascribed a nutritional role to the glands on the basis of a supposed homology with the hypopharyngeal gland of the honey bee. The maxillary gland of the ant is more generally thought to be homologous with the hypopharyngeal gland of the honey bee, however (Otto, 1958*b*), the pharyngeal gland having no known homologue in the Apidae. Peregrine, Mudd & Cherrett (1973) suggest that the high energy value of the triglycerides contained in the glands supports the idea of a nutritive role, although growing larvae have more need of tissue-building protein than a rich energy source, which would be of more use to the workers. In *Myrmica rubra* large amounts of oil remain unused in the midgut of the larva, and are passed out when the meconium is ejected prior to pupation. Peregrine *et al.* (1973) also showed that the colour of the oil in the pharyngeal glands of *Acromyrmex octospinosus* workers is caused by a plant pigment absorbed with the lipid fraction of the juices produced by the process of cutting up leaves for the fungus garden.

Radioactive tracer studies have been used to clarify the function of the pharyngeal gland. Those of Naarmann (1963) showed the extent to which the gland contents were passed on to other workers. More recently, Schmidt (1974) injected labelled triolein into the haemolymph of *Formica polyctena* and found that the radioactivity gradually collected in the pharyngeal gland, from where it was spread to the queens, larvae and other workers. Markin (1970) showed the same build up of radioactivity into the pharyngeal gland, via the haemolymph, after feeding the Argentine ant *Iridomyrmex humilis* on sugar labelled with ^{32}P. As the radioactive label was not bound chemically to the carbohydrate, it is not clear exactly what was being secreted into the gland, but the pathway of at least some ingested nutrients is delineated. Most of the gland contents were fed to the queens and small larvae.

The picture which emerges of the pharyngeal gland function is still not entirely clear. It is known to take up selectively the lipid fraction of food, and any lipo-soluble compounds at the same time. Some of these, such as plant sterols, which are known to act as vitamins in insects could be of nutritional importance (Delage-Darchen, 1974). It is thought that substances are secreted into the gland lumen to mix with the oily fluid which is passed to larvae and queens and sometimes to other workers. So it seems that the pharyngeal gland collects a high-energy liquid food, possibly enriches it in some way, and uses it to feed the queens and larvae. However, absorption of the gland contents into the gland cells has been shown to occur (Delage, 1968; Brian, 1973*c*), and recent, ultra-structural studies of the pharyngeal glands of several ant species (Zylberberg, Jeantet & Delage-Darchen, 1974) show that the folding of the cuticular intima, and the structure of the cellular apex below it are typical of an absorptive epithelium. A digestive function is therefore also possible.

9. The role of termites in ecosystems

T. G. WOOD & W. A. SANDS

The role of any organism in an ecosystem can be defined in terms of three essential features. The first expresses its ability to divert a proportion of the total energy flow through the ecosystem towards increasing its own production. The second is its influence on the physical structure of its environment, and the third is its interaction with other organisms. When assessing the role of particular groups of organisms such as termites, a different aspect of populations, species diversity, must also be considered.

The first feature relates to the density and dynamics of populations and biological factors such as predation that affect their growth. It involves the various parameters of specific energetics, namely food collection, consumption and utilisation. It is apparent from earlier chapters that as far as termites are concerned, the study is in its infancy, and so little is known that the value of a synoptic and synthetic treatment of this field of research may be questionable. Recent work, only now gradually being published, has begun to remedy this deficiency. Noteworthy in this respect are the following projects within the IBP: derived savannah in the Ivory Coast (Josens, 1971*b*, 1972); Sahelian savannah in Senegal (Lepage, 1972, 1974*b*); semi-arid desert in USA (Haverty, 1974; Haverty & Nutting, 1975); tropical rainforest in Malaysia (Matsumoto, 1976). Ecological studies in natural and agricultural ecosystems in southern Guinea savannah in Nigeria by a research team (T. G. Wood, R. A. Johnson, C. E. Ohiagu, and N. M. Collins) organised jointly by the Centre for Overseas Pest Research and Ahmadu Bello University (COPR/ABU) (Wood, Johnson & Ohiagu, 1977) are also providing relevant information.

The second feature is one in which social insects in general and termites in particular have a greater impact than many more conspicuous organisms. They construct extensive and often massive nest systems in response to their need for food, environmental control and social homeostasis. As a result, they have a profound effect on redistribution of soil particles, on physical and chemical properties of soils, and consequently, on vegetation. A little is known of the extent of these modifications of the habitat, but the interactions within the termite–soil–vegetation system are complex and are nowhere fully understood.

Interaction with other organisms by termites involves a variety of direct and indirect effects, some of which are mentioned above; others include

P.C.B. LLYFRGELL U.W.B. LIBRARY

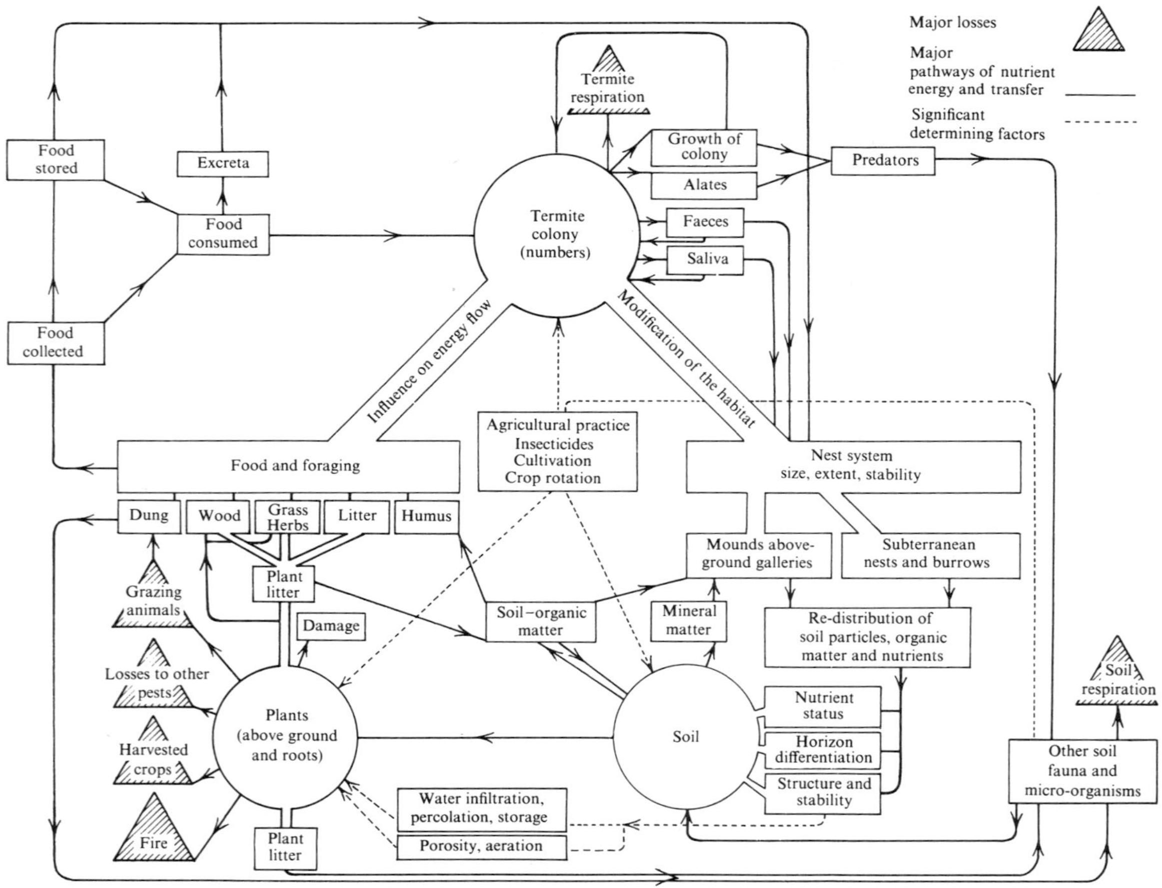

Fig. 9.1. Diagrammatic representation of role of termites in ecosystems showing the two major effects of termite communities: direct and indirect modification of the habitat by construction of nest systems and effects on energy flow and nutrient cycling by consumption and transformation of food (modified from Wood, 1975).

symbiotic and commensal relationships. These have resulted in mutual adaptations at several levels of intensity, but have never been quantified.

The role of termites in ecosystems can be usefully summarised pictorially as in Fig. 9.1. This distorts the true situation which is obviously not as termite-centred as shown in the model. However, it provides a framework for discussion and clearly illustrates the two major aspects of their role, modification of the habitat, and contribution to energy flow and the cycling of nutrients.

The order of importance, assigned above to the essential features of their ecological role, is not the most convenient arrangement for the more detailed account that follows. Here, we have first dealt with the zoogeographical aspect of species diversity, followed by the environmental effects of termites, and finally with their influence on energy flow, incorporating direct interactions with other organisms.

Diversity of species

Most termites are tropical or sub-tropical, though a few species extend into temperate regions as far as 45° N and S latitude. Roughly two-thirds of the world's land surfaces therefore, have one or more termite species. If one uses the simplest measure of species diversity, number of species per unit area, there is for most groups of animals and plants a tendency for species diversity to increase with proximity to the equator and to decrease with increasing altitude. Termites conform to this pattern, as illustrated by data from West Africa (Table 9.1); although the intensity of collecting and the area searched was not constant in the five localities there is an obvious north to south increase in the number of species. Similar trends can be discerned in the data of Harris (1961), Kemp (1955), and Sen-Sarma (1974).

The apparent exception to this general rule occurs in Australia (Gay & Calaby, 1970) where there are equal numbers of species north and south of the tropic of Capricorn, and few species in tropical rainforest. There are several peculiarities about the Australian fauna that suggest a zoogeographical rather than an ecological explanation, such as the lack of soil-feeding termites (Lee & Wood, 1971*b*), the presence of the primitive relic *Mastotermes darwiniensis* Froggatt (Mastotermitidae) and the absence of the fungus-growing Macrotermitinae.

Species lists for individual ecosystems are scarce. In addition to the data in Table 9.1, which are from localities within the tropics, there is Nutting's (unpublished) list of eight species in semi-arid shrubland (31.5° N) in Arizona and Lee & Wood's (1971*b*) list of nine species in warm temperate woodland (35° S) in south Australia. There are lists for larger (Krishna & Weesner, 1970) or more localised (Kemp, 1955; Roy-Noel, 1974) geo-

Table 9.1. *Number of species of termites and their feeding habits in different ecosystems in West Africa*

Locality	Latitude (N)	Total number of species	Living wood and fresh, woody litter	Grass and herbaceous litter	Decomposing wood	Soil
Sahel savannah, Senegal[a]	16.5°	19	11	8	0	3
Northern Guinea savannah, Nigeria[b]	11°	19	9	11	1	5
Southern Guinea savannah, Nigeria[c]	9°	23	10	10	2	8
'Derived' savannah, Ivory Coast[d]	6°	36	18	11	3	13
Rain forest, Cameroun[e]	4.5°	43	8	2	8	31

[a] Lepage (1972). Total area: 100 hectares.
[b] Sands (1965*a*). Area 1. Total area: 4 hectares.
[c] Wood, Johnson & Ohiagu (1977). Including primary and secondary savannah woodland. Total area: 7 hectares.
[d] Josens (1972). Including sparsely wooded and densely wooded savannah. Total area: 2700 hectares, but not including other vegetation types within this area.
[e] N. M. Collins (personal communication). Total area: 1.5 hectares.

graphical areas but these lists incorporate data from several different ecosystems.

It is obvious that there are relatively few species of termites, compared with the majority of invertebrates, including other social insects. Species diversity in general is often directly correlated with the biological stability of ecosystems, although opinions differ as to whether or not the relationship is causal (May, 1973; van Emden & Williams, 1974; Goodman, 1975). Diversity and biological stability may become mutually reinforcing phenomena as an end-point of evolution in physically undisturbed ecosystems; however, in the early stages of speciation, adaptive changes are more likely to be imposed by instability. Intricate habitat partitioning would appear to require a combination of rapid selection of adaptive features and mechanisms having potential for genetic isolation. These commonly take the form of courtship patterns, sexual recognition pheromones, or elaborate copulatory apparatus.

Termites have no sclerotised genitalia, and the evidence now accumulating suggests that sex pheromones may have little specific significance; courtship is minimal. Such fossils as exist are closely related to extant forms, showing little change over many millions of years. The relative longevity of most termite colonies, combined with a low success rate for the establishment of new ones, could explain their apparently slow evolution. With their internally controlled environmental homeostasis, generalised feeding habits, rather large colony biomass and efficient digestion, termites in some ways resemble vertebrate primary consumers. It is evident that their impact on ecosystems arises from their numerical abundance and elaborate behaviour patterns rather than through species diversity.

Modification of the habitat

Termite nest systems and their composition

Termite nest systems may be arboreal, either within or attached to the outside of the vegetation, epigeal (the well-known mounds), or subterranean. They consist of one or more breeding centres from which radiate a network of galleries or runways to requisites such as food, water and soil (see Chapter 4, this volume).

The nests and associated galleries and runways are constructed from soil, excreta and saliva in varying proportions. Lee & Wood (1971*b*) recognised four main types of structure based on the materials from which they are composed, as shown by microscopical examination of thin sections.

(1) Fabrics dominated by re-packed orally transported soil particles without admixture of excreta.

(2) As (1), but incorporating excreta.

  (*a*) Wood-feeders.

  (*b*) Grass-feeders.

(3) Fabrics dominated by excreta consisting of soil (i.e. from soil-feeding species) mixed with orally transported soil particles.

(4) Fabrics dominated by excreta consisting of organic matter derived from ingestion of plant material.

  (*a*) Wood-feeders.

  (*b*) Grass-feeders.

The first category, in which soil particles are cemented solely with saliva, includes the large mounds of Macrotermitinae that form a conspicuous feature of many landscapes in Africa and Indo-Malaya. The second category includes the hard, outer crusts of mounds of several other groups (Termitinae, Nasutitermitinae and also in Australia, *Coptotermes*) in which excreta is used, in addition to saliva, in varying proportions as a cementing medium, and particularly the inner regions of the nests of these species in which greater proportions of excreta are often incorporated as distinct linings of chambers and galleries. The third category includes the nests of soil-feeders and their composition therefore reflects both the feeding habits of the species concerned (about which little is known, see Chapter 4, this volume) and the extent to which orally transported soil particles are incorporated into the nest structures. The fourth category includes concentrations of faeces derived from extensively digested plant material,

Table 9.2. *Particle size analysis, organic carbon (%) and nitrogen (%) content of constructions made by termites in which soil is the dominant constituent (fabric types 1, 2 and 3, see text)*

Species no. in Fig. 9.2.	Type of fabric and species of termite	Type of construction and soil depth (cm)	Coarse sand	Fine sand	Silt	Clay	Organic carbon	Nitrogen
	Fabric type 1							
1	*Macrotermes*	Mound, external wall	46	11	8	33	0.8	—
2	*bellicosus*[a]	Mound internal	16	17	9	58	0.9	—
		Soil, 0–15	68	13	5	7	0.5	—
		30–76[g]	48	10	7	35	0.3	—
3	*M. bellicosus*[b]	Mound, external wall	10	62	2	26	0.2	—
4		Mound, internal	9	60	1	30	0.5	—
		Soil, 0–15	11	62	2	25	1.2	—
		49–95[g]	10	62	3	25	0.2	—
5	*Macrotermes*	Mound, external wall	15	11	(73)		0.6	0.06
	subhyalinus[c]	Soil, 0–10	38	30	(32)		2.6	0.26
		67–70[g]	49	31	(20)		0.5	0.25
6	*M. bellicosus*[d]	Mound, external wall	—	—	—	—	0.5	0.09
7		Runways, over tree	—	—	—	—	1.4	0.14
8	*Odontotermes* sp.	Runways, over tree	—	—	—	—	0.7	0.09
9	*Ancistrotermes* sp.	Runways, over tree	—	—	—	—	1.2	0.15
		Soil, 0–5	—	—	—	—	0.9	0.08
		75–100[g]	—	—	—	—	0.4	0.07
10	*Odontotermes*	Mound	18	19	9	44	1.0	0.13
	redemanni[e]	Topsoil	19	20	8	43	2.0	0.18
11	*O. redemanni*[e]	Mound	39	38	4	15	0.9	0.10
		Topsoil	54	39	1	6	0.7	0.08
12	*Odontotermes*	Mound	34	37	5	19	1.2	0.08
	obscuriceps[e]	Topsoil	34	34	4	23	0.7	0.05
	Fabric type 2							
	(*a*) Wood-feeders							
13	*Coptotermes*	Mound, external wall	21	41	5	31	2.7	0.08
	acinaciformis[f]	Soil, 0–25	31	46	6	16	3.4	0.17
		60–75[g]	34	43	3	21	0.2	0.01
14	*C. acinaciformis*[f]	Mound, external wall	11	39	20	25	2.4	0.10
		Soil, 0–10	18	56	21	5	0.5	0.04
		20–40[g]	10	37	24	28	0.4	0.04
15	*C. acinaciformis*[f]	Mound, external wall[1]	6	17	22	53	2.7	0.09
16		Galleries, within wood[2]	5	9	15	69	2.8	0.06
		Soil, 0–10[1,g]	8	18	24	53	2.1	0.12
		30–40[2,g]	5	10	15	71	0.6	0.05
17	*Nasutitermes*	Mound, external wall	23	37	5	33	4.0	0.10
	exitiosus[f]	Soil, 0–12	37	49	7	7	1.0	0.08
		25–30[g]	19	27	3	49	0.5	0.04
18	*N. exitiosus*[f]	Mound, external wall	26	31	9	30	2.8	0.10
19	*Coptotermes lacteus*	Mound, external wall	16	26	7	47	2.4	0.08
		Soil, 0–20	35	47	10	9	1.6	0.10
		55–70[g]	24	28	9	40	0.3	0.40
20	*Schedorhinotermes*	Galleries, over log	6	10	18	61	7.4	0.16
	intermedius	Soil, 0–10	8	18	24	53	2.1	0.12
	actuosus[f]	30–50[g]	5	10	15	71	0.6	0.05
21	*Amitermes*	Mound, external wall	24	12	(64)		4.2	0.22
	unidentatus[c]	Top soil	22	19	(59)		2.3	0.25
22	*Nasutitermes*	Epigeal, external wall	35	25	(40)		2.3	0.33
23	*latifrons*[c]	Epigeal, internal	15	30	(55)		2.3	0.33
		Topsoil	44	33	(23)		2.1	0.14

Table 9.2 (*cont.*)

Species no. in Fig. 9.2.	Type of fabric and species of termite	Type of construction and soil depth (cm)	Coarse sand	Fine sand	Silt	Clay	Organic carbon	Nitrogen
	(*b*) Grass-feeders							
24	*Nasutitermes*	Mound, external wall	18	41	12	25	2.7	0.10
25	*triodiae*[f]	Mound, internal base	13	46	12	23	2.5	0.11
		Soil, 0–6	23	24	14	9	0.5	0.04
		40–50[g]	20	45	14	18	0.2	0.02
26	*Amitermes laurensis*[f]	Mound, external wall	14	44	4	34	2.1	0.19
27		Mound, internal	19	49	9	24	1.3	0.11
		Soil, 0–20	26	65	4	4	0.5	0.04
		45–60[g]	27	40	7	24	0.3	0.03
28	*A. laurensis*[f]	Mound, external wall	7	41	27	19	3.0	0.16
29		Mound, internal	6	42	29	20	1.5	0.10
		Soil, 0–8	7	43	37	9	0.9	0.07
		16–24[g]	7	28	23	40	0.2	0.02
	Drepanotermes	Subterranean 0–5	22	52	9	14	1.4	0.12
	rubriceps[f]	nest, 5–30	24	52	8	14	0.6	0.05
		Soil, 0–30[g]	25	52	8	14	0.9	0.08
		30–60	26	54	6	14	0.2	0.03
30	*D. rubiceps*[f]	Mound, external wall	14	59	3	20	1.4	0.09
31		Mound, internal	15	60	5	20	1.1	0.07
		Soil, 0–10	19	65	4	9	0.5	0.03
		20–30[g]	17	63	2	16	0.2	0.02
32	*Trinervitermes*	Mound	—	—	—	—	1.4	0.08
	geminatus[d]	Soil, 0–5	—	—	—	—	0.9	0.08
		75–100	—	—	—	—	0.5	0.07
	Fabric type 3							
33	*Cubitermes*	Mound	6	41	14	39	3.8	—
	sankurensis[b]	Soil, 0–12[g]	13	64	3	20	2.8	—
		30–50[g]	11	66	2	21	0.7	—
34	*Cubitermes*	Mound	13	41	(46)		1.4	0.11
	fungifaber[c]	Soil, 0–5[g]	30	45	(25)		1.1	0.08
		60[g]	23	49	(27)		0.3	0.03
35	*Cubitermes* sp.[d]	Mound	—	—	—	—	1.4	0.08
		Soil, 0–5	—	—	—	—	0.9	0.08
		75–100	—	—	—	—	0.5	0.07

[a] Nye (1955).
[b] Stoops (1964).
[c] Maldague (1959).
[d] Wood (unpublished data from Nigerian savannah).
[e] Joachim & Kandiah (1940).
[f] Lee & Wood (1971*a*).
[g] Suspected or proven soil horizon used by termites for building.

collectively known as carton. These structures form the greater part of the arboreal type of nest of certain Termitinae and Nasutitermitinae and certain inner regions of the nests of *Mastotermes* and some Hodotermitidae, Rhinotermitidae, Amitermitinae and Nasutitermitinae. The nests of Macrotermitinae are unique in that all excreta are used to construct fungus combs; these structures are not regarded as carton, as the excreta consist of plant material which has undergone little chemical change.

Details of these fabrics are given in Lee & Wood (1971*b*) except for those of soil-feeders for which information is based solely on *Cubitermes* (Stoops, 1964). More detailed micro-pedological features are discussed by

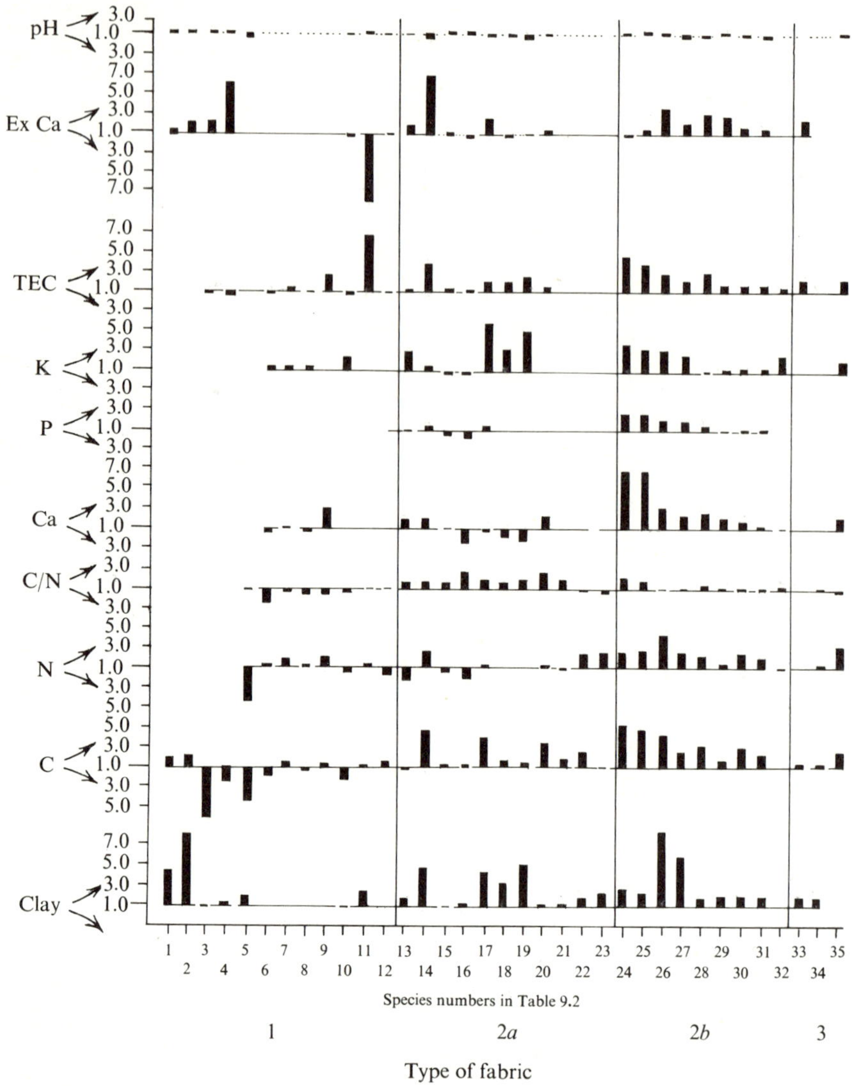

Fig. 9.2. Comparison of the clay, carbon (C), pH and nutrient content of constructions, in which soil is the dominant constituent, made by termites and topsoil of the adjacent soil. Histograms above the base line of 1.0 indicate termite construction contains 2.0, 3.0, etc. times the amount in soil; histograms below base line indicate soil contains 2.0, 3.0, etc. times the amount in termite structure. Termite structures correspond to those listed in Table 9.2 except for the subterranean nest of *Drepanotermes* which is excluded from the figure. For explanation of type of fabric see text. (P = phosphorus, K = potassium, Ca = calcium, Ex Ca = exchangeable calcium, N = nitrogen, TEC = total exchangeable cations.)

Sleeman & Brewer (1972). Physical and chemical analyses of a selection of these structures are given in Tables 9.2 and 9.3 and Fig. 9.2.

The construction of these greatly diverse nest systems results in the transport of soil from either deep or shallow horizons (often from both) to the soil surface where it is formed into mounds, covered runways or sheetings on the soil surface and over, or within, vegetation on which the termites are feeding. There is a concurrent construction of a network of subterranean galleries and the removal of food from the foraging territory of the colony with consequent incorporation of the end-products of digestion (excreta) into various structures of the nest system. The effects of these processes on modification of the habitat can be considered in relation to soils and vegetation separately, although the influence on vegetation (apart from feeding which is discussed elsewhere) is largely indirect and is brought about by modification of soils.

Modification of soils

The effects of termites on soils has been reviewed by Lee & Wood (1971*b*) and since this publication there have been few studies which either provide comparable information or illuminate the deficiencies in present knowledge indicated by this review. These authors indicated that studies had concentrated on the more obvious effects of mound-building species whereas the effects of subterranean species may be equally significant. Recent work, as yet unpublished, confirms this suggestion.

Modifications to soils can be considered under the following headings:

(i) physical disturbance of soil profiles;
(ii) changes in soil texture;
(iii) changes in the nature and distribution of organic matter;
(iv) changes in the distribution of plant nutrients;
(v) construction of subterranean galleries.

Physical disturbance of soil profiles

The modification of soil profiles as a result of termite activity depends on the fact that termites remove soil from various depths and bring it to the surface in the form of runways, sheetings or mounds from which it is re-distributed by water and wind erosion. Some of this material is re-distributed on the soil surface, some is carried away in rivers. These processes were noticed as long ago as the last century by Drummond (1888) who remarked on how Herodotus had described Egypt as the 'gift of the Nile', but that '...had he lived today he might have carried his vision farther back still and referred some of it to the labours of the humble termites in the forest slopes about Victoria Nyanza'.

The amount of soil in mounds and the area of ground covered by them was summarised by Lee & Wood (1971*b*). The weight varies from less than 10000 kg per hectare for small mounds of less abundant species to 45000 kg per hectare for small mounds of more abundant species; large mounds of *Macrotermes* in Africa may contain from $1–24 \times 10^5$ kg per hectare of soil, the maximum equivalent to a uniform layer 20 cm deep being found in one particular area in the Congo (Meyer, 1960). The percentage of land area occupied likewise varies from less than 1.0% up to 30% in the case of the *Macrotermes* mentioned above. Comparable information for the amount of soil in runways, surface sheetings and in-filled vegetation is not available except for certain unpublished information. Preliminary measurements in southern Guinea savannah woodland in Nigeria of the amount of soil in the form of sheeting constructed by Macrotermitinae (largely *Macrotermes bellicosus* (Smeathman) and species of *Odontotermes*, *Ancistrotermes* and *Microtermes*) on trunks and branches of trees was estimated at 300 kg per hectare and in an entirely different ecological situation, a maize field on the same soil type, the amount of soil carried up within the stems of maize plants attacked by *Microtermes* was estimated at 250 kg per hectare. Lepage (1974*b*) estimated that the soil-surface runways and sheetings of *Macrotermes subhyalinus* (Rambur) in Sahel savannah in Senegal amounted to 675–950 kg per hectare.

There are few measurements of the rate of erosion of soil from termite mounds and its accumulation on the soil surface. In northern Australia Williams (1968) estimated that mound-building termites (*Tumulitermes hastilis* (Froggatt), *Tumulitermes pastinator* (Hill) and *Nasutitermes triodiae* (Froggatt)) had brought 6000 m^3 per hectare of soil to the surface. Of this, 1500 m^3 per hectare remained to form a layer 15 cm deep which had formed over a period of 12000 years giving a rate of accumulation of 0.0125 mm per year; 4500 m^3 per hectare had been removed by erosion. In two other localities in northern Australia where the mound-building termites were *T. hastilis*, *N. triodiae*, *Amitermes vitiosus* Hill and *Drepanotermes rubriceps* (Froggatt), Lee & Wood (1971*b*) calculated accumulation rates of 0.02 to 0.10 mm per year. These rates are of a similar order of magnitude to that of 0.025 mm per year calculated by Nye (1955) as resulting from erosion of mounds of *Macrotermes bellicosus* in West African forest. Pomeroy (1976*a*) estimated rates of erosion from mounds of *M. bellicosus* and *Pseudacanthotermes* spp. in Uganda to be 1.3 m^3 per hectare per year, equivalent to a layer 0.13 mm deep, some of which would accumulate on the surface and some of which would be lost. A contributory factor in erosion was destruction of mounds by man to the extent that losses by erosion exceeded gains from construction by a factor of one-third which indicated that mounds would cease to exist in approximately 100 years. These rates may seem slow and over the lifetime of a single mound (which may approach 100 years in the case of large mounds of *Macrotermes* in Africa and *Nasutitermes triodiae* in Australia) eroded soil is not spread evenly but accumulates around the base of the mound. However, on ancient landscapes several thousand years old these irregularities become evened out to the extent that soil derived from termite mounds forms distinct surface horizons. Williams (1968) observed that in northern Australia, stone lines were buried under a depth of 15 cm of such material, and comparable observations by Fölster (1964), Nye (1955), Ollier (1959), de Ploey (1964) and Stoops (1968) indicate that this is a feature common to many tropical soils in both forest and savannah regions. This subject is discussed in more detail by Lee & Wood (1971*b*).

Erosion of soil sheetings on vegetation and runways and sheetings on the soil surface has not received the attention it deserves. Lepage (1974*b*) showed that of the 2000 kg per hectare of soil brought to the surface every year by *Macrotermes subhyalinus* in Senegal, between 675 and 950 kg per hectare was in the form of foraging runways and sheetings on the soil surface. Observations show that these fragile structures are rapidly eroded by rain and are just as rapidly re-built, but rates of accumulation and removal have not been measured. In mechanically cultivated agricultural ecosystems where mounds are destroyed, soil is brought to the surface by termites solely in the form of foraging galleries and runways. Measurements in southern Guinea savannah in Nigeria show that *Microtermes* populations attacking maize (Wood *et al.*, 1977) bring 250 kg per hectare of subsoil to the surface where it is re-packed within the eaten out stems of the maize plants and is re-distributed by subsequent cultivation.

In the soils of some dry regions where laterite outcrops occupy much of the ground surface, soil eroded from termite mounds and other structures may eventually bury the laterite crusts (Grassé, 1950; Tricart, 1957; Boyer, 1959). Herbaceous vegetation colonises cracks or pockets containing soil particles and is eventually followed by shrubs and trees; the laterite finally disintegrates (or under humid conditions may disappear altogether) forming a gravelly soil (Maignien, 1966). Thus termites contribute to the modification and degradation of laterite. However, after considering the evidence presented by several authors who had suggested that the abandoned galleries and nests of termites serve as sites for the formation of certain forms of laterite, Grassé & Noirot (1959*a*) and Lee & Wood (1971*b*) concluded that it is most unlikely that laterite is formed from these structures.

Abandoned subterranean galleries and nests are often filled with soil washed in from the surface. These have been observed by several authors, such as Watson (1960) who observed in-filled galleries at depths of 8.5 m, but the extent and rate of the process has never been measured.

Changes in soil texture

The significance of the physical disturbance of soil profiles depends partly on the nature of the soil brought to the surface by the termites and partly on physical and chemical changes in the soil due to the activities of the termites. Chemical changes are brought about by incorporation of organic matter and are discussed below. Physical changes appear to be due to selection and sorting of certain particles resulting in a change of structure and particle size distribution. The only instance of production of fine particles from coarse ones by termites is Boyer's (1948) claim that *Macrotermes subhyalinus* occasionally work particles of mica in their mandibles until the material has properties similar to illite and can be used in the construction of mounds as a substitute for naturally-occurring clay.

Particle size analyses of mounds and other structures made by termites and of surrounding soils have been carried out by several workers and Lee & Wood (1971*b*) summarised and discussed these along with their own data (Lee & Wood, 1971*a*) for Australian termites. A representative selection of analyses is shown in Table 9.2. By analysing clay minerals in the mounds and different soil horizons Lee & Wood (1971*a*) showed that for the 12 species they studied in 17 different localities 15 mounds were derived entirely from subsoil, six entirely from topsoil and 10 from both topsoil and subsoil.

Obviously pedological changes resulting from redistribution of mound material are greater if subsoil is used in preference to topsoil. Selection of subsoil rich in clay is a feature of the building behaviour of many Macrotermitinae (Table 9.2 and Fig. 9.2). Analyses of mounds of *Macrotermes* (Hesse, 1955; Nye, 1955; Maldague, 1959; Stoops, 1964; Pomeroy, 1976*b*) illustrate selection for clay but in soils with an extremely high clay content there may be preferential selection of sand particles although the mound is constructed primarily from clay. The present authors' observations in southern Guinea Savannah at Mokwa, Nigeria indicate that mound-building and non-mound-building Macrotermitinae often select clay subsoil for constructing runways and sheets on the outside of trees or on the soil surface (e.g. *Macrotermes, Odontotermes* and *Ancistrotermes*) or for filling-in galleries within eaten-out branches, and twigs (the above species plus *Microtermes*).

Most mound-building termites are less selective than the Macrotermitinae and build mounds with a wide range of textures (e.g. *Coptotermes acinaciformis* (Froggatt), Table 9.2). Generally there is selection for clay-sized particles in preference to sand with the result that mounds tend to have a higher content of clay than the soil from which they are constructed. Mounds are constructed entirely from materials transported by the termites whereas subterranean nests are constructed either from transported material or simply by excavating and re-packing the soil *in situ*. An example of the latter is the subterranean nest of *Drepanotermes rubriceps* (Table 9.2) whereas an outstanding example of selection are the subterranean nests of the soil-feeding *Apicotermes* which constantly consist of 81–83% fine sand (Stumper, 1923).

The changes in texture brought about by the re-distribution of mounds and other structures on the surface is likely to be accompanied by changes in physical properties, such as structure, stability, bulk density, infiltration rate, permeability and water-holding capacity. However, with the exception of Macrotermitinae, which do not use their excreta for building purposes, changes in these properties will also be influenced by excreted organic matter incorporated into the various structures. There is some evidence that aggregates from termite-modified soil are more stable than those from undisturbed soil particularly if the aggregates contain organic matter (excreta) in which case they may even resist chemical dispersion (Lee & Wood, 1971*b*). There are very few measurements of other physical properties of termite-modified soil resulting from changes in soil texture and these were reviewed by Lee & Wood (1971*b*).

Changes in the nature and distribution of organic matter

The chemical changes in food material brought about during digestion are discussed in Chapters 4 and 7 of this volume. Here we are concerned solely with effects of collection and utilisation of food on the distribution of organic matter in the ecosystem and the effects of the chemical transformations on the subsequent fate of excreted organic matter. It is apparent from Chapter 4 that plant material, in whatever form it is consumed (wood, grass, soil, etc.), once collected by foraging termites is largely utilised in the region of the central nest system housing the queen and young termites. Unlike other soil animals which deposit faeces within the soil where they are available to micro-organisms and coprophagous invertebrates, termites use their excreta to construct certain regions of the nest (e.g. nursery chamber, linings of gallery walls) or, in the case of Macrotermitinae, fungus combs which are further utilised as food. Thus organic material and the nutrients it contains are collected from a wide area and concentrated largely in the central region of the nest system.

The use of salivary secretions and faeces in building mounds and runways raises their organic content above that of the soil from which they are derived (Table 9.2 and Fig. 9.2). However, mounds are often built from subsoil, and even when modified by termites this remains lower in organic matter than the surrounding topsoils. This is generally true of mounds built by Macrotermitinae though few analyses are available (Table 9.2) for constructions other than mounds (i.e. surface runways, mud-fillings in excavated food sources, etc.).

In addition to being re-distributed, organic matter is profoundly changed during its passage through the termite's gut and ultimately one of the more important effects is a change in the carbon/nitrogen (C/N) ratio. Lee & Wood (1971*a*) showed that for 30 samples of termite-modified soil in Australia all except two samples showed an increase in the C/N ratio compared with adjacent soil. The C/N ratio expressed as a factor of the C/N ratio of corresponding soil samples was 0.8–2.0 (mean 1.4) for grass-eating termites and 1.1–2.7 (mean 1.9) for wood-eating termites. Thus erosion of these structures and re-distribution of termite-modified soil over the soil surface results in the addition of organic matter of a relatively high C/N ratio (up to 47). In contrast, mounds and other structures of Macrotermitinae in Africa (Boyer, 1955; Pomeroy, 1976*b*) and Asia (Joachim & Kandiah, 1940; Shrikhande & Pathak, 1948) appear to have C/N ratios similar to or even lower than adjacent soil (Table 9.2, Fig. 9.2), almost certainly a consequence of their behaviour in using subsoil for construction and not incorporating excreta as in the case of other species. Addition to the soil of organic matter of a high C/N ratio is in opposition to established processes of decomposition which, by the combined action of comminution of plant debris by soil-inhabiting invertebrates and microbial breakdown, reduce the C/N ratio of plant debris (25–40 for leafy tissues, 50–80 for woody tissues) to values of 10–15 found in soil humus.

This increase in the C/N values of termite-modified soil is even more obvious when analytical data for carton and other structures made entirely or almost entirely from excreta are examined (Table 9.3). In Australia, Lee & Wood (1971*a*) obtained C/N values of 32–107 for wood-feeding species and 21–25 for grass-feeding species. A further relevant property of termite excreta is the high lignin/cellulose ratio compared with that of their food. Analyses of carton by Becker & Seifert (1962) and Lee & Wood (1971*a*) and of wood and excreta produced by wood-feeding termites by Seifert & Becker (1965) show that lignin/cellulose values of 0.38–0.82 in wood are drastically changed by the termites' digestive processes. Values ranged up to 15.7 for fresh excreta (excreta of *Reticulitermes lucifugus* var. *santonensis* de Feytaud fed on *Pinus* with a lignin/cellulose value of 0.7) and up to 5.4 for carton. Returns of organic matter and nutrients to the ecosystem via the decomposition of faecal material is discussed later in this chapter and in Wood (1976).

Changes in chemical properties of soil and in the distribution of plant nutrients

The chemical properties of termite-modified soil in comparison with the soil from which it is derived can be attributed to differential selection of soil particles, incorporation of saliva, incorporation of excreta in the form

Table 9.3. *Chemical analysis of structures built by plant-feeding termites in which excreta is the dominant constituent (fabric type 4, see text). Nests as in Table 9.2. with addition of* Nasutitermes magnus *and* Tumulitermes pastinator

Termite species	Type of structure	pH	Carbon	Nitrogen	C/N[d]	Phosphorus	Calcium	Potassium
(*a*) Wood-feeders								
Macrotermes bellicosus[a]	Fungus comb, fresh	n.d.	31	0.68	46	0.07	0.70	0.10
	Fungus comb, old	n.d.	29	0.80	36	0.07	0.85	0.10
Odontotermes obscuriceps[b]	Fungus comb	n.d.	42	1.44	29	n.d.	n.d.	n.d.
Coptotermes acinaciformis[c]	Mound, carton wall	3.2	43	0.52	83	0.02	0.11	0.02
C. acinaciformis[c]	Mound, carton wall	3.2	43	0.67	64	0.02	0.17	0.11
	Mound, carton nursery	3.2	50	0.56	89	0.02	0.12	0.11
C. acinaciformis[c]	Mound, carton wall	3.1	44	0.41	107	0.02	0.30	0.05
Nasutitermes exitiosus[c]	Mound, carton wall	3.9	15	0.25	58	0.01	0.35	0.12
	Mound, carton nursery	3.9	30	0.52	58	0.02	0.52	0.07
N. exitiosus[c]	Mound, carton wall	4.0	29	0.51	57	0.03	0.11	0.23
	Mound, carton nursery	4.1	41	0.79	51	0.04	0.50	0.13
Coptotermes lacteus[c]	Mound, carton wall	3.9	29	0.45	65	0.02	0.31	0.21
	Mound, carton nursery	3.8	42	0.74	57	0.03	0.40	0.18
(*b*) Grass-feeders								
Nasutitermes triodiae[c]	Mound, carton nursery	5.6	10	0.44	23	0.03	0.22	0.28
Nasutitermes magnus[c]	Mound, carton nursery	4.9	19	0.73	26	0.03	0.18	0.23
Tumulitermes pastinator[c]	Mound, carton nursery	5.2	11	0.50	23	0.02	0.26	0.27

[a] Wood (unpublished data from Nigerian savannah).
[b] Joachim & Kandiah (1940).
[c] Lee & Wood (1971*a*).
[d] Carbon/nitrogen ratio.

of linings to gallery walls, gallery in-fillings, and in the case of the large mounds of Macrotermitinae, to pedological changes. Only a brief summary of the chemical differences between termite-modified soil and adjacent soils will be given here as this subject was discussed in detail by Lee & Wood (1971*a*, *b*).

pH

In mounds of Macrotermitinae there is generally a small increase in pH (often correlated with an increase in calcium, see below) compared with the subsoil from which the mounds are constructed but there is little difference compared with topsoil (Fig. 9.2). Structures made by other species generally have a lower pH (due to incorporation of organic-rich excreta) than subsoil but are similar to or only slightly lower than topsoil.

Carbon and nitrogen

In general there is a small increase in the abundance of these elements compared with subsoil, in mounds of Macrotermitinae, due to incorporation of saliva but the amounts involved are so low that often concentrations are lower than in topsoil (Fig. 9.2). Structures made by other species have higher concentrations due to incorporation of excreta and as excreta has a high C/N ratio (Table 9.3) the C/N ratio of the termite-modified soil is correspondingly elevated.

Calcium

All termite-modified soils have higher concentrations of calcium than adjacent soils. In Macrotermitinae the increase has been attributed to several causes (Watson, 1974), but appears to be mainly due to the large evaporating surface of the mounds, often supplemented by an internal ventilation system which results in the accumulation of salts near the base of the mound. This phenomenon has been reported by several authors (e.g. Pendleton, 1941; Watson, 1962; Weir, 1973) and was studied in some detail by Hesse (1955) who showed that the process was accelerated by the presence of calcium-rich ground-water and by impeded drainage. These concentrations may be as high as 7% (Milne, 1947) resulting in calcium carbonate concretions amounting to 2000 kg per mound. These figures are exceptional and in many localities the calcium content of mounds is similar to or only slightly greater than that of the adjacent soil and is not in concretionary form, being distributed more or less evenly throughout the mound (Hesse, 1955; Nye, 1955; Pomeroy, 1976*b*; T. G. Wood, unpublished data from Nigeria). The increased calcium concentration in structures built by other termites is due to the incorporation of excreta.

Potassium

The concentration of potassium in termite-modified soil is usually higher than in subsoil due to the incorporation of organic matter in the form of excreta and/or saliva but may be slightly higher or slightly lower than in topsoil (Fig. 9.2).

Phosphorus

The relative concentrations of phosphorus in termite-modified soil, topsoil and subsoil are associated with relative levels of organic matter (see *Potassium*, above) but Hesse (1955) found that in large mounds of Macrotermitinae on sites with impeded drainage, phosphorus tended to accumulate at the base of the mound in the manner described for calcium.

Exchangeable cations

Addition of clay and organic matter (in the form of excreta and saliva) results in an increase in the total exchangeable cations (TEC) compared with adjacent soils and a significant proportion of this increase is due to an increase in exchangeable calcium (fig. 9.2; Lee & Wood, 1971*b*).

The general effects of these localised chemical changes on the properties of soils depend on the amount of termite-modified soil available and the rate at which it is re-distributed by erosion. The only relevant studies are

Table 9.4. *Addition of organic carbon and plant nutrients to soil surface as a result of erosion of termite mounds. For each site, figures are also given for the percentage by weight, and 'standing crop' (kg per hectare) of organic carbon and nutrients in the upper 10 cm of soil (bulk density of soil = 1.2)*

Locality and termite species	Annual erosion of mounds (kg per hectare)	Organic carbon (%)	Organic carbon kg per hectare per year	Nitrogen (%)	Nitrogen kg per hectare per year	Phosphorus (%)	Phosphorus kg per hectare per year	Potassium (%)	Potassium kg per hectare per year	Calcium (%)	Calcium kg per hectare per year
North Australia (site 4)[a]											
Nasutitermes triodiae	2310	2.6	60	0.10	2.3	0.01	0.2	0.07	1.6	0.22	5.1
Soil, 0–10 cm	—	0.5	6000	0.04	480	0.004	48	0.06	720	0.01	120
North Australia (site 6)[a]											
N. triodiae	150	2.6	4	0.10	0.2	0.01	< 0.1	0.07	0.1	0.22	0.3
Tumulitermes hastilis	10000	2.4	240	0.08	8.0	0.01	1.0	0.07	7.0	0.06	6.0
Amitermes vitiosus	2000	2.2	44	0.01	0.2	0.01	0.2	0.06	1.2	0.08	1.6
Drepanotermes rubiceps	2100	1.4	29	0.09	1.9	0.01	0.2	0.17	3.6	0.13	2.7
Total	14250	—	317	—	10.3	—	1.4	—	11.9	—	10.6
Soil, 0–10 cm	—	1.1	13200	0.06	720	0.014	168	0.06	720	0.06	720
Natete, Uganda[b]											
Macrotermes bellicosus	2300	—	—	0.064	1.5	0.0004	0.01	0.022	0.5	0.05	1.1
Pseudacanthotermes spp.	230	—	—	0.056	0.1	0.0006	< 0.01	0.020	0.1	0.06	0.1
Total	2530	—	—	—	1.6	—	0.01	—	0.6	—	1.2
Soil, 0–10 cm	—	—	—	0.07	840	0.0005	6	0.027	324	0.11	1320
Muko, Uganda[b]											
Macrotermes subhyalinus	8600	—	—	0.091	7.9	0.0017	0.15	0.03	2.5	0.09	9.1
Soil, 0–10 cm	—	—	—	0.10	1200	0.001	12	0.01	120	0.06	720

[a] Calculated from data in Lee & Wood (1971*a*) assuming that rate of erosion of occupied mounds is similar to that of abandoned mounds.
[b] Calculated from data in Pomeroy (1976*a*, *b*).

those of Lee & Wood (1971*a*) in northern Australia and Pomeroy (1976*a*) in Uganda and some of these data are summarised in Table 9.4. In northern Australia the mounds were built by various grass-feeding termites which incorporate excreta into their mounds. In the two localities taken as examples, the weight of various elements eroded annually from mounds, as a percentage of their weight in the upper 10 cm of soil, amounted to: carbon (1.0–2.4%), nitrogen (0.7–1.4%), phosphorus (0.4–0.8%), potassium (0.2–1.7%) and calcium (1.5–4.3%). In Uganda the mounds were built by various Macrotermitinae and the corresponding percentages were: nitrogen (0.2–0.7%), phosphorus (0.2–1.3%), potassium (0.2–2.1%) and calcium (0.1–1.1%). However, an unknown proportion of these elements will be lost through drainage and surface run-off so that their contribution to the nutrient status of the surface soil will be less than indicated.

Little is known about the decomposition of structures composed largely of excreta (carton) but in the case of *Nasutitermes exitiosus* (Hill) these structures appear to persist until burnt by bushfires. Lee & Wood (1968) calculated that these structures contained 143 kg per hectare of nitrogen, 0.14 kg per hectare of phosphorus and 2.47 kg per hectare of calcium. On the assumption that mounds persist for 40 years and are then immediately eroded and the carton burnt the annual accession of elements is: nitrogen (0.53 kg per hectare, but much will be lost to the atmosphere by burning), phosphorus (0.004 kg per hectare) and potassium (0.062 kg per hectare).

From the above it is reasonable to conclude that erosion of termite mounds contributes very little to the nutrient status of soils; even if their contribution was doubled, to allow for erosion of runways and surface sheetings, their effect would be negligible compared with the accession of nutrients from decomposing plant material not attacked by termites, foliar leaching and rainwater. Erosion of structures built by Macrotermitinae often adds soil of a lower nitrogen content than topsoil.

Construction of subterranean galleries

The extent of foraging galleries was discussed in Chapter 4 of this volume and it was noted that they may extend up to 30 m from the central nest system. Galleries may also penetrate to great depths (presumably for the purposes of collecting water) and there are records (Lee & Wood, 1971*b*) of termites as deep as 70 m, although such depths are exceptional. The greatest concentration of galleries is found near the soil surface close to sources of food but there are no estimates of their dimensions, orientation or total extent, nor the length of time for which they persist once they have been abandoned.

These sub-surface termite galleries are commonly so numerous as to collapse under-foot, leaving deep imprints, as we and others studying the insects in the tropics have often observed. Heavy rains compact this spongy surface so that it largely disappears by the end of the wet season. Several workers (Adamson, 1943; Robinson, 1958; Maldague, 1964; Lee & Wood, 1971*b*) have commented on the fact that this dense network of galleries must affect porosity and aeration, infiltration, storage and drainage of water, and growth of plant roots, but these important effects have not yet been measured. Such measurements will be crucial to a proper understanding of the total influence of termites on soils.

Modification of vegetation

Termites can modify vegetation directly by consuming it and indirectly by influencing its growth on or around termitaria through modifications in soil properties. The direct effects of food consumption are discussed later in this chapter.

Influence of feeding activities

The removal of parts of a living plant for food often affects the growth and subsequent survival of the plant. Stimulation of growth has rarely been observed but an example is provided by Kaiser (1953). He showed that the nests of *Anoplotermes pacificus* Fr. Muller in Brazilian rainforest are invaded by a dense mat of plant roots (e.g. *Mapouria cephalantha* Muell.). The root apices are preferentially consumed by the termites, and the relationship appears to be symbiotic, as in abandoned termitaria the roots are dead. More often the effects of feeding on a particular part of a plant by termites result in loss in vigour, or in the death of the plant, but much will depend on the condition of the plant when attacked. For instance, grass is harvested by various species of termites in many tropical and semi-arid regions (Chapter 4, this volume). In normal years, grass growth may be sufficient to support all herbivores including termites, and vegetative growth may possibly be stimulated by cropping. During drought and heavy overgrazing by vertebrate consumers, the termites may kill the plants by taking or attempting to take their usual quota of food. This is because termites, unlike other primary consumers, are not limited by their accessibility to the aerial parts of the plant. There are no published measurements of these effects, but damage to plants in agriculture and forest plantations based on many observations was summarised by Harris, W. V. (1969) as follows:

(*a*) destruction of seedlings;

(*b*) ring barking, leading to death of the plant;

(*c*) destruction of root system often followed by tunnelling within stems, leading to wilting or lodging;

(*d*) removal of leaf tissue, resulting in loss of production and photosynthetic potential.

Influence through modification of soil properties

A resumé of this subject was given by Lee & Wood (1971*b*) who distinguished between growth of vegetation on termitaria and growth on termite-modified soil around termitaria.

The stimulation of plant growth on the mounds of certain Macrotermitinae in Africa and Asia, which often support a more luxuriant and woody vegetation than the surrounding scrub or savannah has been recognised for many years. The difference may be so marked in some areas that distinct terms have been coined to describe the vegetation, such as 'termitensavannen' (Troll, 1936) or 'park-formation' (Fuller, 1915). Lee & Wood (1971*b*) summarised the more important factors responsible for enhancing growth on these termitaria as:

(*a*) protection from burning, allowing the survival of certain seedlings that would otherwise be destroyed by fire;

(*b*) improvement of drainage in areas subject to seasonal waterlogging;

(*c*) accumulation of a greater depth of soil in areas with shallow, stony soils;

(*d*) provision of moist soil in otherwise dry areas;

(*e*) producing chemically altered soil, chiefly by increasing pH and raising base-status;

(*f*) resistance to termite attack by species growing on the mounds;

(*g*) possible stimulation of growth (e.g. the case of *Anoplotermes pacificus* quoted above).

The large mounds of various Macrotermitinae in Africa and Asia appear to be the exception in supporting a distinct type of vegetation. Mounds of other species of termites (e.g. *Cubitermes*, *Trinervitermes*, *Amitermes* in Africa and *Coptotermes*, *Nasutitermes*, *Amitermes* in Australia) do not appear to support plant growth, even when abandoned by the termites. However, there are reports from South America (where Macrotermitinae do not occur) that vegetation similar to that of the surrounding savannah grows on abandoned mounds (Goodland, 1965: termites not identified) or that mounds (termites not identified and condition of mounds not indicated) support a distinct woody vegetation (i.e. having the appearance of 'termitensavannen') in contrast with the surrounding savannah (Smith, 1971).

Retardation of growth on termitaria is often regarded as being a direct effect of hardness, preventing infiltration of water and establishment of seedlings (Glover *et al.*, 1964; Watson & Gay, 1970). It may also be due to the presence of toxic substances in the termites' excreta (Lee & Wood, 1971*b*).

Distinct patterns in the growth of vegetation around termitaria have also been recorded. Frequently these observations relate to the lack of plant growth on the areas of surrounding soil eroded from termite mounds (Macfadyen, 1950; Glover *et al.*, 1964; Lee & Wood, 1971*b*). When these areas are colonised, it is by plants normally associated with hardpans (Burtt, 1942; Glover *et al.*, 1964). A more detailed study of vegetation growing around mounds of *Odontotermes* sp. in Kenya was made by Glover *et al.* (1964), who suggested that the differential growth was due to response by the plants to variations in soil moisture and amounts of clay and salts washed off the mounds and moved downslope by water.

U.V.

Effects of modification of habitat on fauna

The intimate relationships of lestobiosis, termitophily and similar phenomena existing between termites and other organisms is discussed later in this chapter. There are many less intimate, often casual or even fortuitous, relationships between termite nest systems and other organisms which Berg (1900) described under the general term of 'termitariophily'. Araujo (1970) wrote 'Termite nests provide an alluring biotope for a great assemblage of organisms, yet although extensive lists could be compiled the nature of these associations is poorly understood in most cases.' This author provided many examples of mammals, birds, reptiles, amphibians and invertebrates using termitaria as vantage points, nesting sites or for shelter. None of these associations appeared to be obligatory but Lee & Wood (1971*b*) reported on the apparently obligatory nesting behaviour of certain species of Australian parrots which excavate and nest in holes in the sides of termite mounds.

The effects of termitaria on vegetation, for instance, in preserving or developing small areas of forest in derived or natural savannahs, must have an indirect effect on the distribution of many other animals. It is well known that the clumps of trees growing on termite mounds of one or more species provide an environment for forest-limited members of other termite genera in Africa, South America and the Orient. The same must be true for many other groups of invertebrates and possibly some vertebrates but there is little information and this is undoubtedly a rich field for further ecological study.

The accumulation of salts in the large mounds of *Macrotermes* are occasionally exploited as salt licks by elephants (Weir, 1972) and possibly by other animals, but this is not a general phenomenon and is probably resorted to only when more readily accessible salt is unavailable.

Population energetics and interactions with other organisms

Energetics of individual species

Energy flow through a population is described by the following equations:

$$C = A + F \tag{9.1}$$

$$A = P + R \tag{9.2}$$

where C = consumption, P = production, R = respiration, A = assimilation and F = faeces production. Measurements of some of these parameters are available only for a few individual species of termites. They can be combined with estimates of population density to assess energy flow through termite communities and the means by which nutrients are returned to the ecosystem.

Population density

Attempts to sample termite populations quantitatively encounter two main problems. Their social behaviour makes termites more discontinuously distributed than non-social insects and most species have subterranean nests that may penetrate the soil to depths of several metres.

Sampling problems and methods were reviewed by Lee & Wood (1971*b*) and Sands (1972*b*), both of whom pointed out that figures for abundance were notably confined to individual mound-building species. Until recently, there had been no quantitative estimate of the entire termite community in any ecosystem but three separate attempts have now been made in different West African savannahs. Lepage (1972, 1974*a*, *b*) studied Sahelian savannah in Senegal, Josens (1971*a*, *b*, 1972) investigated a derived savannah in the moist forest-savannah mosaic area of the Ivory Coast and Wood *et al.* (1977) worked in Guinean savannah in Nigeria. The full details of these studies are as yet unpublished in Doctoral theses and the records of the COPR/ABU Termite Research Team (Wood, Johnson, Ohiagu and Collins) and we are grateful for permission to use much of this information.

Populations of mound-building termites in all three investigations were estimated by linear regressions between some measure of mound size and mound population. Matsumoto (1976) adopted the same method for three of the four mound-building species he studied in Malayan rainforest. Lepage (1972, 1974*a*, *b*) and Josens (1971*a*, *b*, 1972) sampled subterranean termites by carefully digging pits to a depth of 50–75 cm and collecting termites as they were encountered. Wood *et al.* (1976) took cores of soil (10 cm diameter) 2 m in depth and extracted termites by a combination of hand-sorting and flotation. Hand-excavation has the disadvantage that termites will escape while the pits are being dug; hand-sorting in the field is notoriously inefficient and there is an unknown proportion of the population below 75 cm. The efficiency of hand-excavated pits has not been tested thus no correction factors can be applied to data obtained in this way. The vertical distribution of the population is indicated by Wood *et al.* (unpublished) who found 25% of the termites at 1.0–2.0 m depth and 7% at 1.75–2.0 m.

The population figures for subterranean termites shown in Table 9.5 are partly underestimated because of the great depth to which some termites penetrate. Vertical distribution varies with species, soil type and season and no general correction factors can be applied.

There is considerable variation in the figures for abundance in natural ecosystems and it is not yet possible to make valid generalisations on the relative population density of termites in major biomes. Maximum populations (4450 per m^2) in rainforest were recorded by Strickland (1944)

Table 9.5. *Abundance and live weight biomass of termites in different ecosystems*

Ecosystem	Species or group of species	Type of nest	No. per m^2	g per m^2	Author
Warm temperate woodland					
Sclerophyll forest, South Australia	*Nasutitermes exitiosus*	Mound	600	3.0	Lee & Wood, 1971*b*
Semi-arid					
Grassland, North America	*Gnathamitermes tubiformans*	Subterranean (1)	0–9127	0–22.21	Bodine & Ueckert, 1975
		Mean over three years	2139	5.2	
Shrub–grassland, North America	*Heterotermes aureus*	Subterranean (1)	431		Haverty, Nutting & La Fage, 1975
Sahel savannah, West Africa	All species	Subterranean+Mounds (2)	229	1.0	Lepage, 1974*b*
Tropical savannah					
Savannah woodland, North Australia	All species	Subterranean (3)	2000		Lee & Wood, 1971*b*
Grass savannah, Central Africa	*Cubitermes exiguus*	Mound	612–701	1.3–1.9	Hébrant, in Bouillon, 1970
Savannah	*Apicotermes 'gurguliflex'*	Subterranean	70		Bouillon, 1962, 1964
N. Guinea savannah, West Africa	*Trinervitermes geminatus*	Mound	110–2860		Sands, 1965*a*, *b*
S. Guinea savannah, West Africa	All species	Subterranean+Mound (4)	4402	11.1	Wood *et al.*, 1977 and unpublished
Secondary S.G. savannah, West Africa	All species	Subterranean (5)	2966	3.6	Wood *et al.*, 1977
'Derived' savannah, West Africa	All species	Subterranean+Mound (6)	861	1.7	Josens, 1972[a]
Tropical forest					
Semi-deciduous forest, West Africa	All species	Subterranean (5)	3163	8.0	Wood & Johnson, unpublished
Riverine forest, Central Africa	All species	Subterranean+Mound	1000	11.0	Maldague, 1964
Rainforest, West Indies	All species	Subterranean (7)	4450		Strickland, 1944
Rainforest, South America	*Nasutitermes costalis*	Mound	87–104	0.1	Wiegert, 1970
Rainforest, Malaysia	Four species	Mound	1330	3.4	Masumoto, 1976
Agro-Ecosystems					
Grazed pasture, West Africa	All species	Subterranean (5)	2010	2.8	Wood *et al.*, 1977
Maize (first year), West Africa	All species	Subterranean (5)	1553	1.7	Wood *et al.*, 1977
Maize (8–24 years), West Africa	All species	Subterranean (5)	6825	18.9	Wood *et al.*, 1977

Depth of soil samples: (1) 30 cm, (2) 75 cm, (3) 8 cm, (4) 200 cm, (5) 100 cm, (6) 50–60 cm, (7) 7.5 cm.
Sampling methods for subterranean species: hand-excavated pits (Josens, 1972; Lepage, 1974*b*); core samples used in all other studies.
[a] Mean of facies D1 (open savannah woodland) and D2 (moderately open savannah woodland).

from soil cores only 7.5 cm deep which would probably sample less than half of the subterranean population and exclude mound-building and arboreal-nesting species. Mound and arboreal populations of a single species in Puerto Rico, *Nasutitermes costalis* (Holmgren), were estimated as 87–104 per m^2 by Wiegert (1970). In Congo Forest approximately one-third of the mound/subterranean population consisted of mound-building *Cubitermes* (Maldague, 1964) and in Malaysia, Matsumoto (1976) estimated four mound-building species at a density of 1330 per m^2, neither estimate including any arboreal species.

Maximum population densities in savannahs are the same order of magnitude. However, Malaisse, Freson, Goffinet & Malaisse-Mousset (1975), working in Miombo woodland in Zaire, recorded a population density of 15000 per m^2 and biomass of 108 g per m^2 for a single species of *Odontotermes*. These exceptionally high numbers could have arisen from biassed sampling of foraging concentrations and the biomass figure is questionable in view of the data for comparable *Odontotermes* species given in Table 9.6, where the maximum recorded mean individual weight was 3.0 mg. This would give a calculated biomass of 45 g per m^2, a much more realistic, if still high, figure which could be accepted in view of similarly high estimates of *Gnathamitermes tubiformans* (Buckley) from a semi-arid, short-grass, Texas rangeland ecosystem by Bodine & Ueckert (1975).

The latter authors used randomised soil cores 81.1 cm^2×30.2 cm and found that monthly populations and biomass fluctuated from zero to a maximum of 9127 per m^2 (22.2 g per m^2). The mean figures over a three-year period were 2139 per m^2 and 5.2 g per m^2, with a fluctuating decline during the study. Clearly, the termites cannot have disappeared completely even though 30 cm cores failed to sample them because an unknown population below this depth remained uncounted. It is implied in this paper though not actually stated, that *G. tubiformans* was the only species encountered since no other species were recorded or even mentioned in passing, but whether or not other species were present, the population density seems to be remarkably high for semi-arid conditions. These studies and the data listed in Tables 9.5 and 9.6, suggest upper limits for termite population and biomass in any ecosystem of around 15000 per m^2 and 50 g per m^2, respectively.

Within the African continent, there are indications (Tables 9.1 and 9.6) that in the succession from dry savannahs to humid rainforest, the number of species increases along with overall abundance and the increasing importance of detritus-feeders at all levels down to soil-feeders. Soil-feeders are absent (or very rare) in Sahel savannah, comprise 3.7% of the numbers (5.9% of biomass) in Guinea savannah, and 11.7% of numbers (9.5% of biomass) in 'derived' savannah. There are no comparable figures

Table 9.6. *Population density, live-weight biomass and respiration of termites in Sahel, southern Guinea and derived savannah in West Africa*

	Sahel savannah (Lepage, 1974*a*, *b*)			Southern Guinea savannah (Wood *et al.*, 1977, and unpublished)			Derived savannah (Josens, 1972)[a]		
Species	No. per m^2	g per m^2	l O_2 per m^2 year	No. per m^2	g per m^2	l O_2 per m^2 year	No. per m^2	g per m^2	l O_2 per m^2 year
RHINOTERMITIDAE									
Psammotermitinae									
Psammotermes hybostoma	105	0.14	0.26	—	—	—	—	—	—
Coptotermitinae									
Coptotermes spp.	6	0.01	0.02	—	—	—	—	—	—
TERMITINAE									
Macrotermitinae									
Ancistrotermes spp.	—	—	—	881	1.59	3.09	268	0.28	0.52
Macrotermes bellicosus	—	—	—	165	0.47[c]	1.90[c]	—	—	—
Macrotermes subhyalinus	39	0.60	1.33	37	0.74	1.72	—	—	—
Microtermes spp.	1	—	—	850	1.02	1.94	149	0.08	0.16
Odontotermes spp.	53	0.12	0.23	260	0.78	1.59	47	0.06	0.12
Pseudacanthotermes militaris	—	—	—	—	—	—	90	0.22	0.44
Apicotermitinae									
Adaiphrotermes sp.[b]	—	—	—	40	0.06	0.12	54	0.04	0.08
Aderitotermes sp.[b]	—	—	—	—	—	—	4	0.01	0.02
Anenteotermes sp.[b]	—	—	—	11	0.02	0.04	—	—	—
Astratotermes sp.[b]	—	—	—	—	—	—	1	—	—
Termitinae									
Amitermes evuncifer	—	—	—	21	0.12	0.25	6	0.04	0.08
Basidentitermes potens[b]	—	—	—	—	—	—	36	0.08	0.16
Cubitermes sp.[b]	—	—	—	104	0.56	1.18	—	—	—
Microcerotermes spp.	3	—	—	1148	1.15	2.21	25	0.02	0.04
Pericapritermes sp.[b]	—	—	—	7	0.02	0.04	—	—	—
Promirotermes holmgreni[b]	—	—	—	—	—	—	6	0.03	0.06
Nasutitermitinae									
Eutermellus sp.[b]	—	—	—	1	—	—	—	—	—
Trinervitermes geminatus	5	0.02	0.04	}	—	—	158	0.76	1.62
Trinervitermes occidentalis	—	—	—	}					
Trinervitermes oeconomus	1	—	—	} 483	2.27	4.65	7	0.03	0.06
Trinervitermes togoensis	—	—	—	}	—	—	13	0.06	0.12
Trinervitermes trinervius	18	0.07	0.15	}	—	—	3	0.02	0.04
TOTAL	231	0.96	2.03	4008	10.59	22.03	867	1.73	3.58

Respiration calculated from mean weight per individual and weight/respiration regression (see text).
Population density: for sampling details see text.
[a] Mean of facies D1 (open savannah woodland) and D2 (moderately open savannah woodland).
[b] Soil-feeding species.
[c] Estimates by N. M. Collins.

for rainforest, where the abundance of soil-feeders is often self-evident. Although Lepage (1972, 1974*a*, *b*) took account of catenary sequences in Sahel, the observations of Wood *et al.* (1977) and Josens (1971*a*, *b*, 1972) deliberately avoid these. The segregation of soil-feeders into dense concentrations in poorly drained sites in Guinea and Sudan savannahs is well known and the above remarks apply only to the much larger areas of 'typical' savannah.

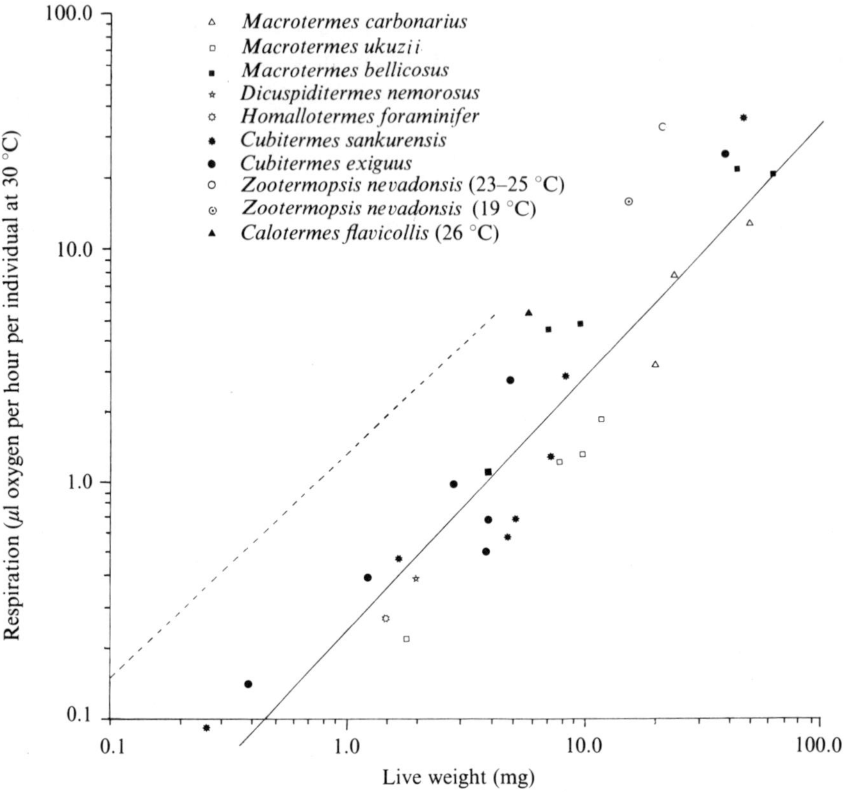

Fig. 9.3. Log–log plot of respiration rate at 30 °C as a function of live weight for different species of termites. Calculated regression line does not include *Calotermes* and *Zootermopsis*. Data for *Cubitermes exiguus* and *C. sankurensis* from Hébrant (1970*a*); *Macrotermes carbonarius*, *Dicuspiditermes nemorosus* and *Homallotermes foraminifer* from Matsumoto (1976); *Zootermopsis nevadensis* at 19 °C and *Kalotermes flavicollis* from Lüscher (1955*b*); *Z. nevadensis* at 23–25 °C from Hungate (1938); *Macrotermes bellicosus* from N. M. Collins (personal communication). Dotted line is log–log relationship obtained by Weigert & Coleman (1970) for respiration of *Nasutitermes costalis* at 21–25 °C as a function of body dry weight.

The influence of cultivation on termite populations in agro-ecosystems has been quantified only by Wood *et al.* (1977, and unpublished), in the southern Guinean savannah of Nigeria. Of the 23 species found in natural woodland, 15 disappeared within one year of clearing and cultivation. After eight or more years mechanical cultivation only four species survived. The reduction in number of species was associated with spectacular increases in *Amitermes evuncifer* Silvestri and *Microtermes* spp., from 20% of a primary woodland population of 4008 per m² (Table 9.6) up to 99.9%, in cultivated land, of a population of 6825 per m² (Table 9.5).

Respiration

Termites respire in two distinct environments: first, their foraging galleries and runways, where oxygen and carbon dioxide are at similar concentrations to the atmosphere or adjacent soil; second, the colony hive, where both temperature and carbon dioxide concentration are raised (Ruelle, 1964; Lee & Wood, 1971*b*). The methods and general problems of measuring respiration rate in ants and termites are reviewed by Peakin & Josens in Chapter 6 of this volume.

By assuming a respiratory quotient (RQ) of 1.0 (Lüscher, 1955 obtained values of 0.9 for *Zootermopsis nevadensis* (Hagen) and *Calotermes flavicollis* (Fabr.)) we have calculated a regression of oxygen consumption at 30 °C against body weight for various species (Fig. 9.3). The regression obtained (R = respiration in μl oxygen per individual per hour, W = individual weight in mg):

$$\log R = 1.0886 \log W - 0.6578 \quad (P \leqslant 0.01) \qquad (9.3)$$

does not include the estimates for *Calotermes* and *Zootermopsis* shown in Fig. 9.3. These species, which are primitive termites having symbiotic intestinal protozoa, appear to have a slightly higher weight-specific respiration rate than the other species, which are all Termitidae. In fact, the Termitidae should probably be divided into taxonomic or behavioural groups, such as fungus-growers, soil-feeders and grass-harvesters, with their own weight-specific respiration relationships (cf. Acari, as shown by Wood & Lawton, 1973).

The slope of 1.09 is at the upper limit of the range (0.6–1.0) for various invertebrates given by Keister & Buck (1964) and is higher than the value of 0.73 obtained by Wiegert (1970) for different castes of *Nasutitermes costalis*. The measurements on *N. costalis* were made at 21–25 °C and the regression

$$\log R = 0.73 \log W + 0.029 \qquad (9.4)$$

was obtained on the basis of body dry weight. The respiration is considerably higher than in our calculated regression but would be approximately coincident if calculated on the basis of body live weight (live weight/dry weight = 3.0–4.0; Josens, 1972; Matsumoto, 1976).

Assuming that respiration of 1 l of oxygen results in the release of 20.1 kJ the calculated regression between weight and respiration enables oxygen consumption and respiratory energy loss to be calculated for individual species and for entire termite communities (Tables 9.6, 9.7 and 9.8).

Table 9.7. *Standing crop, production and respiration of different species of termites*

	Population standing crop per m²				Production per m² per year[e]						
					Neuters		Alates				
Species	No.	g (live weight)	g (dry weight)	kJ[d]	g (dry weight)	kJ[d]	g (dry weight)	kJ[d]	Respiration kJ per m² per year	Production biomass ratios	Production efficiency (%)
RHINOTERMITIDAE											
Psammotermes hybostoma[a]	105	0.14	0.033	0.720	0.110	2.394	0.001	0.029	4.81	3.4	33.5
TERMITIDAE											
Macrotermitinae											
Ancistrotermes cavithorax[b]	188	0.20	0.048	0.975	—	9.207	0.024	0.745	7.95	10.2	55.6
Macrotermes subhyalinus[a]	39	0.60	0.096	2.088	0.445	9.684	0.124	3.842	30.89	6.5	30.4
Odontotermes smeathmani[a]	50	0.12	0.033	0.720	0.111	2.415	0.047	1.456	5.27	5.4	42.3
Nasutitermitinae											
Nasutitermes costalis[c]	118	—	0.075	2.331	—	—	—	1.398	22.56	—	—
Trinervitermes geminatus[a]	5	0.02	0.003	0.067	0.006	0.130	0.003	0.088	0.92	3.3	19.1
T. geminatus[b]	158	0.77	0.150	3.022	—	3.139	0.045	1.318	36.16	1.5	—
Trinervitermes trinervius[a]	17	0.07	0.011	0.239	0.018	0.393	0.019	0.557	2.68	3.9	26.2

[a] Lepage (1974*b*).
[b] Josens (1972, 1973).
[c] Wiegert (1970), not including data for eggs.
[d] Calculated from average calorific values (see text): neuters 21.8 kJ per g (dry weight); alates 31.0 kJ per g (dry weight) for Macrotermitinae and 29.3 kJ per g (dry weight) for all other species (Josens, 1972; Matsumoto, 1976).
[e] Calculated from mean weight per individual and relationship between live weight and respiration rate (see text) except for *N. costalis* which was measured directly; combustion of 1 l oxygen is equivalent to 20.1 kJ.

Table 9.8. *Biomass, consumption, production and respiration of termites in three savannah ecosystems in West Africa (data in Table 9.6) and rainforest in Malaysia (after Matsumoto, 1976)*

Ecosystem	Live weight (g per m²)	Annual consumption[b] g per m²	Annual consumption[b] kJ per m²	Annual production[c] (kJ per m²)	Annual respiration (kJ per m²)
Sahel savannah					
Soil feeders	0	0	0	0	0
Macrotermitinae	0.72	15.8	298	24	31
Others	0.24	2.6	49	4	10
Total	0.96	18.4	347	28	41
S. Guinea savannah					
Soil-feeders	0.66	—	136	11	28
Macrotermitinae	6.39	139.9	2635	209	272
Others	3.54	38.7	729	58	143
Total	10.59	178.6	3500	278	443
Derived savannah					
Soil-feeders	0.16	—	33	3	6
Macrotermitinae	0.64	14.0	264	21	25
Others	0.93	10.2	192	15	39
Total	1.73	24.2	489	39	70
Rainforest					
Soil-feeders[a]	1.62	—	—	—	—
Macrotermitinae[a]	1.62	35.5	669	53	72
Others[a]	1.83	20.0	377	30	72
Total	3.55	55.5	1046	83	144

[a] Soil-feeders not studied but many species present; one species of Macrotermitinae and three species of non-Macrotermitinae studied but many other species present (total species = 56; Abe & Matsumoto, personal communication).

[b] Consumption: calculated on basis of calorific value of plant tissue = 18.8 kJ per g; consumption by Macrotermitinae = 60 mg per g per day and by non-Macrotermitinae (i.e. 'others') = 30 mg per g per day; for 'soil-feeders' consumption based on same calorific requirements as other non-Macrotermitinae = 206 kJ per g per year.

[c] Production: based on live weight/dry weight = 4.0 and production: biomass ratio for Macrotermitinae = 6:1 and for soil-feeders and other non-Macrotermitinae = 3:1.

Production

There are two major forms of production in termite colonies, namely alates and neuters. Alates, which are usually produced annually over short periods of time usually during the rainy season, consist of sexually mature males and females with the potential for founding new colonies and represent a loss in biomass and energy to the parent colony. Nutting (1966*a*, *b*, 1969) recorded alates as making up 30% of the total individuals in *Pterotermes*, 14–16% in *Paraneotermes* and 39% in *Zootermopsis*. Lower values have been recorded for *Trinervitermes geminatus* Wasmann (1–5%: Sands, 1965*b*; Josens, 1972) and *Cubitermes exiguus* Mathot (10%: Bouillon, 1970). Roonwal (1970) recorded 28–43% for *Odontotermes obesus* (Rambur). Because alates are heavier than other castes and have a higher calorific value (Josens, 1972; Matsumoto, 1976; also footnote, Table 9.7) these percentages represent proportionally higher biomass and energy values. Sands (1965*b*) calculated an annual production of 240–470 alates per m^2 for *T. geminatus* in northern Nigeria, which is equivalent to 0.9–1.7 g per m^2. Rowan (1971, quoting a report by Boulton) recorded that a single colony of *Macrotermes falciger* (Gerstächer) produced 57.2 and 39.0 kg of alates in successive years which even at a postulated low mound density of one per hectare represents a biomass of 3.9–5.7 g per m^2. More detailed quantitative measurements of abundance, live weight, dry weight and production of neuters and alates, taken directly from the works of Josens (1972) and Lepage (1974*b*), are shown in Table 9.7. As far as alate production is concerned there are obvious anomalies in the case of Lepage's (1974*b*) figures for *Macrotermes subhyalinus*, *Odontotermes smeathmani* and *Trinervitermes* spp. which indicate alate production to be greater than or equal to standing crop biomass. This was explained by the author to be due to the fact that measurements of biomass and alate production were obtained independently and that the former was underestimated.

There have been very few measurements of the production of neuters apart from the figures summarised in Table 9.7. The problems associated with making these measurements are fully discussed by Nielsen & Josens in Chapter 3 who give data for *Ancistrotermes cavithorax* (Sjöstedt) and *Trinervitermes geminatus*. With the exception of *Psammotermes*, where alate production was admittedly underestimated, production of neuters is equivalent to that of alates (*Trinervitermes trinervius* (Rambur)) or greater by a factor of two (*T. geminatus*) to four (*M. subhyalinus*). Wiegert's (1970) assumption that the production of neuters approximates to one-third of the biomass is clearly erroneous. The calorific equivalents of production for these species have been calculated using the values given in the footnote to Table 9.7.

The production/biomass ratios in species other than Macrotermitinae are similar to those of other invertebrates (Phillipson, 1973) being approximately 3:1. Macrotermitinae appear to be a special case, having production/biomass ratios ranging from 5.4 to 10.2. These high values appear to be due to the high production of neuters, which may be a response to predation pressure.

The relationship between production and respiration is shown in Fig. 9.4 for the seven species listed in Table 9.7. Although the data are limited

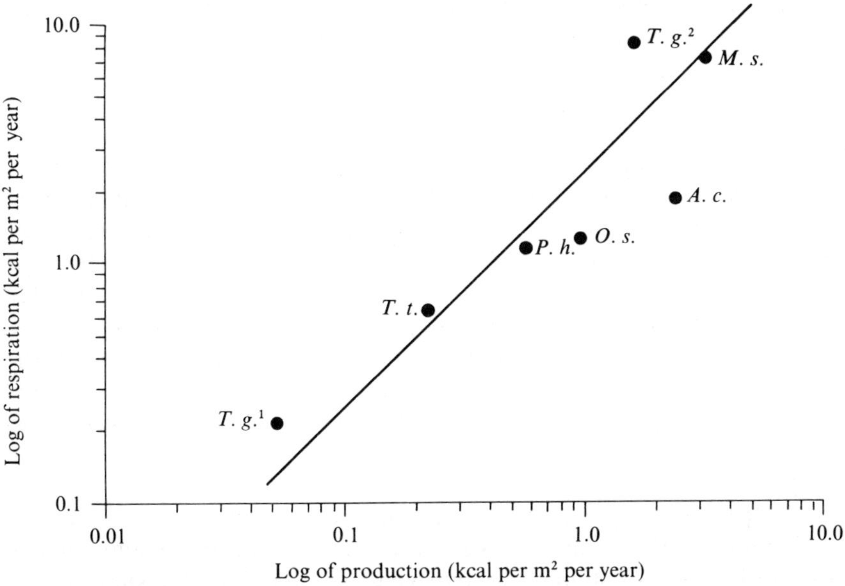

Fig. 9.4. Log–log plot of annual respiration and production for seven species of termites. Data from Table 9.7. The regression line is for a wide range of poikilotherms (McNeill & Lawton, 1970). *T. g.*[1] = *Trinervitermes geminatus* (Lepage, 1974*b*); *T. g.*[2] = *Trinervitermes geminatus* (Josens, 1972, 1973); *T. t* = *Trinervitermes trinervius*; *P. h.* = *Psammotermes hybostoma*; *O. s.* = *Odontotermes smeathmani*; *A. c.* = *Ancistrotermes cavithorax*; *M. s.* = *Macrotermes subhyalinus*.

there is a good agreement with the regression obtained by McNeill & Lawton (1970) for a wide range of poikilothermic animals.

Net population production efficiency is calculated as $100P/(P+R)$. McNeill & Lawton (1970) divided animals into two major groups: homoiotherms, with production efficiencies of 1.4–1.8% and poikilotherms, with production efficiencies ranging from 20–60%. Termites are within the latter range (Table 9.7) with the exception of *T. geminatus* in the Ivory Coast (Josens, 1972, 1973). Engelmann (1966) suggested that this apparent disadvantage of homoiotherms was possibly offset by the fact that their efficiency of digestion (approximately 70%) was greater than that of invertebrate poikilotherms (30% or less). As indicated in Chapters 4 and 6 of this volume, termites have the advantage of a very efficient digestive system (assimilation efficiency exceeding 60%) and as they also have high population production efficiencies they are ideally poised to convert plant material of one form or another into termite tissue.

We have not yet considered production of saliva. Salivary secretions used for feeding dependent castes are accounted for in alate and neuter production but the saliva used to cement soil particles for building various structures represents a loss which has not been taken into account. At the present time there are no estimates of the amount of saliva used for this purpose.

Consumption of food and production of faeces

Consumption was discussed in detail in Chapter 4 but there are considerable difficulties in interpreting laboratory and field measurements. Laboratory measurements have the inherent disadvantage of being unrealistic, particularly for social insects. On the other hand, field measurements, if based on removal of either naturally occurring or artificially provided substrates, have the disadvantage that some termites feed on a wide range of substrates in different situations and it is difficult to measure all the variables. For example, *Odontotermes smeathmani* in savannah woodland in Nigeria feeds predominately on woody litter but also on tree bark, grass litter, dung, to a small extent on leaf litter and probably on subterranean woody roots.

With these reservations and using data provided in Tables 4.1 and 4.2 of Chapter 4, this volume, we have adopted mean weight-specific consumption rates of 30 mg per g live weight per day for other sub-families and 60 mg per g live weight per day for Macrotermitinae. The latter figure may be an underestimate, as fungal metabolism appears to contribute more to the decomposition of ingested food than does termite metabolism.

Faeces production has been measured gravimetrically for wood-feeding species in laboratory cultures and has been used in conjunction with measurements of consumption to estimate assimilation (Chapter 4, this volume). An average value of 68% for assimilation efficiency will be used in this chapter.

Energetics of termite communities in different ecosystems

Termite communities

The specific parameters of respiration, production and consumption for individual species (Table 9.7) have been combined with population data from three savannah areas in West Africa (Table 9.6) and rainforest in Malaysia (Matsumoto, 1976) to give a preliminary picture of energy flow through termite populations in these ecosystems. The calculated data are shown in Table 9.8 although direct quantitative comparisons between ecosystems are precluded because of deficiencies in the population data. Soil-feeders, Macrotermitinae and species of other sub-families are considered separately. Energy consumption by soil-feeders has been assumed to be similar to that of species other than Macrotermitinae (i.e. 206 kJ per g).

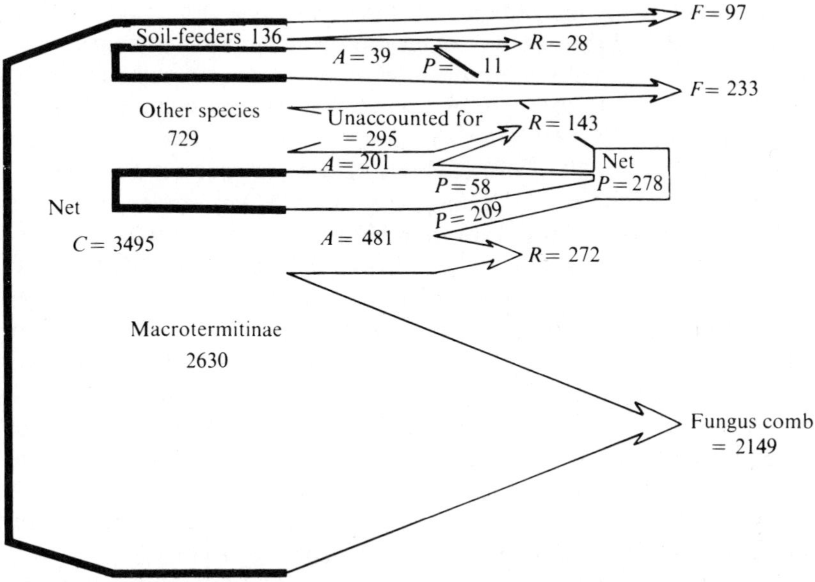

Fig. 9.5. Summary of energy consumption and conversion by termites in southern Guinea savannah; data from Table 9.8. *F* for 'other species' is based on an average assimilation efficiency of 68% (see text). *R* = respiration; *P* = production; *A* = assimilation; *F* = faeces production; *C* = consumption.

The relative importance of the Macrotermitinae in terms of their share of energy consumption appears to be inversely correlated with climatic humidity in West African savannahs (85.9%, 75.3% and 53.9% of the total energy consumed in Sahel, southern Guinea and derived savannah, respectively), whereas the soil-feeders show an opposite trend (0.0%, 3.9% and 6.8%, respectively). There are no data for West African rainforest but in Congo rainforest, Maldague (1964) found a total population of 1000 per m^2 (11.0 g per m^2) of which 337 per m^2 belonged to the soil-feeding species *Cubitermes fungifaber* (Sjöstedt). This species, having a live weight biomass of 10 mg and respiration rate of 2.5 μl oxygen per individual per hour (Fig. 9.3) would have an annual respiratory energy loss of 148 kJ; a value greatly exceeding that of soil-feeders in savannah and approximately equivalent to that of the four dominant mound-builders in Malayan rainforest.

We have attempted to summarise the consumption and conversion of energy for the termite community in southern Guinea savannah (Fig. 9.5), choosing this ecosystem because we believe that it has the most reliable quantitative population data. The most obvious feature is the considerable amount of energy (2.15 MJ per m^2 per year) being metabolised by the fungi in the fungus comb or returning to the ecosystem through decomposition of abandoned fungus combs; decomposition is discussed below and is thought to be relatively insignificant. There are no measurements of mean annual fungus comb biomass and respiration which enable us to estimate independently the annual respiratory energy loss from fungus combs. G. Rohrman (personal communication) found that oxygen consumption by the fungus comb of *Macrotermes ukuzii* was approximately 120 μl per g wet weight per hour at 30 °C although N. M. Collins (personal communication) showed that the newly deposited comb of *Macrotermes bellicosus* has a much higher respiration rate than older comb. R. A. Johnson (personal communication) has demonstrated seasonal changes in fungus comb biomass of *Microtermes* with a recorded maximum of 27 g dry weight per m^2. Using Rohrman's figure for respiration rate and a value of 50% water content (Roonwal, 1960) of the comb, this latter biomass of *Microtermes* fungus comb would consume 1.14 MJ per m^2 per year of respiratory energy. A second interesting feature is that 295 kJ per m^2 per year of the energy consumed by 'other' species has not been accounted for. This is because faeces production was calculated from consumption on the basis of an assimilation efficiency of 68% (p. 79, Chapter 4) and not by summing production (57.8 kJ per m^2) and respiration (142.7 kJ per m^2) which would have given an obviously low figure of 27.5% for assimilation efficiency ($A = 280.5$ kJ per m^2). Errors in calculations of production and respiration may have contributed to the unaccounted 295.5 kJ per m^2, but it is likely that the anaerobic metabolism of intestinal symbionts, resulting in the release of hydrogen (Hungate, 1939), is a major contributing factor.

The use of average weight-specific rates of consumption, production and respiration to compute energy flow through entire populations is open to criticism unless verified by field data and a comparison of both types of data is shown in Table 9.9. There is good agreement between calculated and measured data for Sahel savannah (Lepage, 1974*b*) but this may be more apparent than realistic as the calculated figures were obtained from population data that are probably underestimated. There is good agreement between calculated and measured data for southern Guinea savannah (Wood *et al.*, 1976, unpublished), where field measurements on wood, leaf and grass litter and their rates of consumption were made independently. However, no field estimates of consumption of bark, arboreal deadwood and subterranean woody roots were made, so that the difference between measured and calculated data may be greater than indicated. In derived savannah (Josens, 1972) the measured rate of consumption is 5.6 times greater than the calculated rate, but the difference should be considerably less as the measured rate of consumption was obtained from short-term baiting experiments which overestimate consumption rate (see Chapter 4, this volume), and there is reason to believe that the population density (and therefore the calculated consumption rate) was underestimated. In rainforest (Matsumoto, 1976) leaf consumption in the field was measured

Table 9.9. *Comparison of estimates of consumption by termite populations (excluding soil-feeders) based on field measurements and on calculation from mean weight-specific rates of consumption*

		Consumption by termites (g per m² and as a percentage of available litter)			
		Field Measurements		Calculated data	
Ecosystem[a]	g per m²	g per m²	% of litter	g per m²	% of litter
Sahel savannah					
Litter production: Total	125	12.5	10.0	18.4	14.7
S. Guinea savannah					
Litter production					
Woody litter	140[b]	86[b]	61.4	—	—
Leaves	230[b]	8[b]	3.5	—	—
Grass	155[c]	98[c]	63.2	—	—
Total	535	192	35.9	179	33.5
Derived savannah					
Litter production: Total	480	135	28.1	24.2	5.0
Rainforest					
Litter production					
Leaves	630	189	30.0	55.5	8.8
Other	430	—	—	—	—
Total	1060	—	—	—	—

[a] See Table 9.8 for authors and calculated data.
[b] N. M. Collins (personal communication)
[c] T. G. Wood & C. E. Ohiagu (unpublished).

by photometrically recording removal, a method which could have included removal by other leaf-eating insects and man, therefore, be an overestimate of consumption by termites.

Oxygen consumption of entire colonies in the field (Hébrant, 1970*a*, *b*; Wiegert, 1970) enables a comparison to be made between estimates of respiration obtained by this means and by the calculated method of multiplying weight-specific respiration rates by population density. Using the first method, Hébrant (1970*b*) obtained a respiration rate of 10.2 ml oxygen per hour for a colony of *Cubitermes exiguus*. On the basis of 650 mounds per hectare (Bouillon, 1970) there is an annual respiratory energy loss of 26.8 kJ per m², compared with calculated estimates of 37.7–72.8 kJ per m². By contrast, Wiegert (1970), found that field measurements of the annual respiration of an average nest of *Nasutitermes costalis* (170 kJ per day) were considerably greater than calculated estimates (74.1 kJ per day). He attributed the difference to the contribution of carbon dioxide liberated by microbial activity in the nest.

No valid comparisons can be made between Lepage's (1974*a*, *b*) field estimate of production (21.3 kJ per m² per year) and the calculated data (27.6 kJ per m² per year, Table 9.8) as some of Lepage's measurements (Table 9.7) were used to obtain mean production/biomass ratios which were subsequently used to calculate production.

Comparison with other animals

The significance of termites in ecosystems is well illustrated by comparing their biomass and respiration with that of grazing mammals, as Lee & Wood (1971*b*) did. The maximum biomass of ungulates recorded in a Tanzanian Game reserve by Lamprey (1964) was 12.3–17.5 g per m² and their respiration 649 kJ per m² per year. The corresponding figures for termites in southern Guinean savannah in Nigeria (Table 9.8) are 10.6 g per m² and 442 kJ per m² per year, which greatly exceeds the estimate by Child (1974) of 0.7 g per m² for above-ground mammalian biomass in a nearby game reserve, although this was in the less productive northern Guinean savannah.

In Senegal at the IBP Sahel savannah site of Fété Olé, termite biomass of 0.96 g per m² (Table 9.6) exceeded that of all other dominant above-ground arthropods (0.3 g per m²) and birds (0.3–0.6 g per m²) but was less than the combined biomass of domestic stock and herbivorous mammals, 2.0–3.0 g per m² (Lepage, 1974*b*; Bourlière, personal communication).

Mammals may be expected to have a greater weight-specific consumption than most termites, due to their higher metabolic requirements, except when the fauna is dominated by Macrotermitinae which channel much of their consumption into maintaining fungus gardens. There is little information by which consumption by herbivorous termites and mammals can be compared in the same ecosystem. In the Serengeti grasslands, annual grass production of 598 g per m² Sinclair (1975) was partitioned among the following: grass-eating mammals (13 species of ungulates, various smaller species) 158 g per m² (26.4%); grasshoppers, 46 g per m² (7.7%); 'decomposers', 115 g per m² (19.2%); fire, 319 g per m² (53.3%). In fire-protected grassland grazed by domestic cattle in Nigeria, the total grass production was 316 g per m² (Ohiagu & Wood, unpublished). Of this total, cattle consumed 140 g per m² (44.3%); grass-harvesting *Trinervitermes* took 8 g per m² (2.5%) and 'decomposers' 148 g per m² (46.8%). In both of these studies, 'decomposers' included termites, which in the Nigerian grassland were estimated to consume 98 g per m² of grass litter (31.0% of grass production).

In the semi-arid Texan rangeland studied by Bodine & Ueckert (1975) the cattle biomass was 5.6 g per m², compared with a mean termite biomass over three years of 5.2 g per m². The annual mean termite biomass declined during this period from 8.93 (11.51) g per m² to 2.80 (3.61) g per m², figures for growing seasons only being given in parentheses. Laboratory estimates of termite consumption during growing seasons were 41 g per m^2 equivalent to 686 kJ per m^2. Using litter bags, termite consumption was estimated to be 55% surface litter, and 38% litter incorporated in soil. Ungrazed plots treated with chlordane to eliminate termites showed higher standing crops of both grass and litter after 15 months (2227 kg per hectare and 737 kg per hectare, respectively). Ungrazed plots with termites had identical standing crops of grass to grazed plots without termites (1826 kg per hectare and 1829 kg per hectare) but litter in the termite plots was reduced by one-third (to 492 kg per hectare). Litter removal by termites was only slightly lower in grazed plots, but while termites appear to have consumed an equal amount of standing grass (approximately 20% of the standing crop) to cattle when not competing with them, the combined reduction in both grass and litter in grazed plots with termites was less than the sum of reductions in grazed, termite-free and ungrazed, termite-active plots. This suggests that foraging activity by termites might have been reduced by cattle trampling. Termite consumption of grass and litter was approximately 25% of the standing crop in ungrazed conditions, but only about 17% when cattle were present. These figures take no account of grass production stimulated by cropping.

Return of energy and nutrients to the ecosystem

Nutrients and unrespired energy can be returned to the ecosystem through constructional saliva, faeces and termite tissue (Wood, 1976) and the rates and relative importance of these avenues are largely dependent on the feeding habits and social behaviour of the termites. These returns will be discussed largely in relation to energy, carbon and nitrogen and their impact on other organisms.

Return via salivary secretions

We have already indicated that the amounts of saliva used for constructional purposes are unknown but must be considerable, particularly in those species of *Macrotermes* and *Odontotermes* whose mounds comprise several hundred kg per hectare (Lee & Wood, 1971*b*). Saliva incorporated into mounds and other structures is probably utilised by micro-organisms. Boyer (1955, 1959) and Meikeljohn (1965) found that *Macrotermes* mounds (where saliva is the sole cementing medium) contained a considerably greater number of micro-organisms than surrounding soils but nothing is known of their contribution to decomposition of saliva.

Returns via faeces

The chemical composition of faeces has already been discussed briefly above and in more detail by Lee & Wood (1971*a*, *b*), largely in relation to species feeding on grass and wood. Little is known of the properties of the faeces of soil-feeding species. The chemical and physical properties of faeces and their distribution in the form of gallery linings, structures in which they are incorporated with soil and discrete structures in which they are the dominant component determine their rate of decomposition. The entire nest system, the hive and associated galleries and runways, is maintained and protected against invasion of micro-organisms by intensive nest sanitation and against other invertebrates (with the exception of inquilines) by defensive behaviour. Sands (1969) noted that certain specific saprophytic fungi, particularly species of *Podaxon* and *Xylosphaera*, had been recorded from a wide range of termite nests. They probably have some effect on decomposition of faeces, particularly those species which habitually colonise abandoned fungus combs. However, it appears that the faecal material is not available to other organisms until the nest system, or parts of it are abandoned. Peripheral galleries and runways are abandoned and new ones constructed as food supplies diminish and new ones are exploited, but the hive may be maintained for many years. Estimates of the longevity of colonies range from less than 10 years for *Cubitermes fungifaber* (Noirot, 1970), *Tumulitermes hastilis* and *T. pastinator* (Wiliams, 1968), 30 years for *Amitermes hastatus* (Haviland) (Skaife, 1955), 50 years for *Nasutitermes exitiosus* (Ratcliffe *et al.*, 1952) and more than 80 years for certain *Macrotermes* (Grassé, 1949).

The only available information on decomposition of termite faeces is the work of Lee & Wood (1968, 1971*a*, *b*) on the wood-eating *Nasutitermes exitiosus*. This species builds mounds which are constructed largely of faeces (the internal nursery) or a mixture of soil and faeces (the outer wall surrounding the nursery) which is capped with a thin external layer of soil. When the colony dies the soil capping is rapidly eroded by rain leaving the rest of the mound exposed. A comparison of some of the chemical properties of faecal material in occupied mounds with that in mounds which had been abandoned for 11 years showed that little decomposition of faeces had occurred. This material, which consisted of 42–48% organic carbon, had several properties which undoubtedly contributed to its resistance to decomposition: low pH (3.7–4.1) which would inhibit bacterial activity, high C/N ratio, high lignin (30%) and low carbohydrate (8–10%) content and possibly concentrations of resins, phenols and other resistant wood components resulting from preferential digestion of the more readily degradable components. The C/N ratio in the abandoned mound was high (>60), although slighly lower than in occupied mounds and the slightly greater methoxyl content in the abandoned mound indicated that some decomposition of less complex organic materials had occurred with consequent increase in the percentage of lignin. The structure of the outer wall and nursery of the abandoned mound was intact and there were some fungal hyphae, particularly in the basal portion in contact with the soil. The probable fate of the organic matter in these and similar abandoned carton nests is combustion in bush fires.

Further investigations (Lee & Wood, 1971*b*) showed that the microbial population of the outer wall and nursery of occupied and abandoned mounds was very low. Total plate counts of bacteria gave populations ($\times 10^5$ per g dry weight) of 0.06 and 0.12 in the outer wall of occupied and abandoned mounds, respectively and 0.005 and 4.07 for the nursery. There was some colonisation of abandoned mounds by streptomyces ($\times 10^5$ per g dry weight: 0.001 in outer wall and nursery of occupied mounds, 0.35 and 7.0 in abandoned mounds) and to some extent by spore-forming fungi ($\times 10^5$ per g dry weight: 0.02 and 0.006 in the outer wall and nursery of occupied mounds and 0.08 and 1.16 in abandoned mounds). However, these organisms were largely inactive as respiration of samples from the outer wall and nursery of both occupied and abandoned mounds was so low that rates could not be reliably measured.

In addition, Lee & Wood (1971*b*) showed that inhibition of enzyme activity by compounds in the faeces, having properties similar to soil humic acids, may also contribute to inhibition of microbial activity.

Many other termites (Table 9.3) produce faeces with similar properties to those of *Nasutitermes exitiosus* and also construct parts of their nest system largely from faeces. Some of these structures can be expected to have similarly slow rates of decomposition. Others, such as those constructed by *Coptotermes lacteus* (Froggatt) and *C. acinaciformis*, have lower lignin/carbohydrate ratios and appear to decompose more rapidly (Lee & Wood, 1971*b*). There is no information on the decomposition of faecal material used to line walls of subterranean galleries.

The special case of the fungus-growing Macrotermitinae has been discussed in Chapter 4 in relation to utilisation of food and by Wood (1976) in relation to decomposition processes. The entire faecal production of these termites is used to construct fungus combs on which symbiotic *Termitomyces* grow. These apparently convert faeces into assimilable material available to termites which eat the matured parts of the fungus comb. What little is known of the processes involved and the relationships between termites and fungi was summarised by Sands (1969). The net result appears to be complete decomposition of faecal material with no direct return of faeces to the ecosystem, except on the death of the colony. Various organisms, including fungi and other insects, then rapidly break down the comb; these include termites belonging to the genus *Angulitermes* which specialise in decomposing antelope dung and the abandoned combs of *Macrotermes* (Sands, unpublished). The other exception to the fungus comb cycle occurs in large mounds of *Macrotermes bellicosus* containing fungus combs weighing up to 20 kg, where the fungus comb in a central part of the nest, appears to be inactive and remains uneaten by the termites (N. M. Collins, personal communication). This material would be available for use by other organisms when the *Macrotermes* colony died or became partly colonised by decomposer 'guest' species.

Returns via termite tissue

Direct quantitative comparisons between the three West African savannahs are precluded, because estimates of population density are incomplete or unreliable. However, it is valid to compare, within each ecosystem, the relative returns via faeces and termite tissue. Assuming that there are no faecal returns from Macrotermitinae, and excluding soil-feeders, calculations using the mean assimilation efficiency of 68% and the consumption figures given in Table 9.8 result in returns via faeces of 0.8 g (15.5 kJ) per m^2 in Sahel savannah, 12.4 g (233 kJ) per m^2 in Southern Guinea savannah and 3.3 g (61.5 kJ) per m^2 in derived savannah. Returns via termite tissue are 27.6 kJ per m^2, 277 kJ per m^2 and 38.9 kJ per m^2, respectively, indicating that this avenue of return is more important than faeces in Sahel and southern Guinea savannah but that returns via faeces may be relatively more important in derived savannah. However, the significance of faecal returns is diminished by the loss of an unknown proportion in bushfires and the fact that their nitrogen content (0.5%) is 11–25 times less than that of termite tissue (5.6–12.6%; Matsumoto, 1976). It is probably only in certain rainforests where soil-feeders and species building carton-type nests from faeces that returns via faeces are greater than via termite tissue. The latter can be accomplished via decomposition of corpses or predation.

Decomposition of corpses

Oophagy, cannibalism and necrophagy occurs within the colony and has been studied by Dhanarajan (unpublished) in laboratory colonies. In general it is at a low level, apart from oophagy which was estimated by Sands (1965*b*) at 30% in young colonies. When deaths from disease or injury rise in numbers, the corpses are walled off and left to decompose. A few foragers may be lost outside the nest and likewise parts of the annual alate swarms and these decompose if they escape predators. Compared with predation, the returns to the ecosystem through death and decomposition are probably insignificant, apart perhaps from the shed wings of swarming alates, and possibly the exoskeletal elements rejected by small predators, such as ants. Piles of major soldier heads of *Macrotermes* numbering hundreds are found around exit holes from *Megaponera* colonies.

The predation hazards confronting both neuter castes and alate reproductives that leave their parent colony are enormous. Even those termites that remain in or successfully return to a subterranean refuge are not immune, and the various forms of predation, both opportunist and specialised, are discussed below. Finally we consider a small extra pathway of energy recycling through termitophily and related phenomena.

Predation on alate reproductive colony-founding swarms

This type of predation is entirely opportunistic since the flights are seasonal and cannot provide a regular food supply. Nutting's (1969) review noted that arthropod predators included scorpions, solpugids, spiders, centipedes, dragonflies, cockroaches, mantids, crickets, beetles, flies and wasps. Curiously, ants were not included although Hegh (1922) and Wheeler (1936) discussed predation by ants on termites in some detail. It is likely that ants account for a significant proportion of the probable near-100% mortality of swarming alates. Basalingappa (1970*a*) estimated emergence of *Odontotermes assmuthi* Holmgren at $2.8\text{–}3.0 \times 10^4$ from a single hole and recorded *Camponotus* along with predaceous Heteroptera, Diptera and several vetebrates. This author considered colonisation success to be less than 0.5% and even when pairs have successfully dug into the soil they are immediately open to predation by several specialist predators (see below), such as ants of the genera *Paltothyreus* and *Megaponera*.

Vertebrate predators of termite swarms include fish, reptiles, amphibia, birds and mammals including man (Hegh, 1922; Snyder, 1956, 1961, 1968; Nutting, 1969). Of these, the birds are the best documented with varying estimates of the importance of termites in their diets. De Bout (1964) believed that survival of insectivorous palearctic migrants may depend on flights of alate termites. Williamson (1975) produced evidence that the Sahelian zone of West Africa was an important overwintering area for several migratory passerines from Europe and that the breeding populations of these species in Britain had declined during the Sahelian drought which ended in 1973. However, in the Sahel, flights of alates do not start until May which is long after the migrants have departed for their breeding areas in Europe and termite swarms can have little or no effect on their survival. They may, however, be more important to various inter-tropical migrants such as various raptors, rollers and bee-eaters, as Thiollay (1970) estimated that these birds take 10–30% of most swarms. This author distinguished between swarms in savannah that attract hundreds of birds and those in forest which attract small assemblies numbered in tens, and listed 150 species of birds preying on alate swarms, from the large storks and raptors, to small passerines, many normally frugivorous groups and nocturnal birds such as owls and nightjars. Stomach samples of a roller (*Eurystomus afer* (Latham)) contained up to 800 insects, mostly termites, and large raptors, up to 700 of the large alates of Macrotermitinae.

Thiollay (1970) noted that alate swarms occurred at the start of the breeding season for many tropical-breeding birds and suggested that their high fat and protein content would aid egg formation. Ward (1965) considered that the flight of alates, shortly after the first rains, provided a significant contribution to the diet of certain weaver birds (*Quelea*) in the Sahelian zone of West Africa, as the swarms came at a time when the normal diet of seeds was not available. Ward thought that the high protein content of alate termites would contribute to maturation of gonads and a change to breeding plumage. Net calorific values of alates have been estimated by several authors (Josens, 1972; Matsumoto, 1976; Wiegert, 1970) and average values of 31.3 kJ per g (dry weight) for Macrotermitinae and 29.3 kJ per g for other species have been adopted. Tihon (1946) reported that lightly roasted alates had 36.0% protein and 44.4% fat (dry weight), Cmelik (1969*b*) recorded 47% fat in alates of *Macrotermes falciger*, Basalingappa (1970*b*) recorded 52% and 42% in females and males, respectively, of *Odontotermes assmuthi*. Fox (1966) gave the protein content of alate *Macrotermes* as 18.8% and stated that alate termites had four to five times more fat and minerals than caterpillars and mature locusts. Rowan (1971) reported that the daily calorific requirements of *Passer domesticus* (L.) (mean weight 25 g) was 92 kJ per day, which could be provided by approximately 70 individuals of the average size of *Odontotermes* spp. Ward (1965) found that *Quelea* (mean weight 19 g) consumed an average of 46 alate *Odontotermes smeathmani* per day, a calorific requirement of approximately 59 kJ per bird per day. Lepage (1974*b*) estimated annual alate production by *O. smeathmani* in the Sahel at 1.46 kJ per m^2 (Table 9.7) and indicated that these flights occurred over a period of 31 days. During this time they would be capable of supporting a *Quelea* population of eight per hectare.

Opportunistic predation on foraging sterile castes

Opportunistic predation on neuters is widespread and the predators include all those groups described as attacking alates although the relative importance differs, for example, birds being less important and restricted to species which feed on the ground or arboreally. References and reviews are included in the zoogeographical chapters of Krishna & Weesner (1970), notably Araujo (1970), and in the synoptic works of Hegh (1922) and Snyder (1956, 1961, 1968).

Ants appear to be the dominant opportunistic predators. Sheppe (1970) found that the main opportunist ant predators of 11 species of termites (predominantly surface foragers including *Hodotermes*, *Trinervitermes* and several Macrotermitinae) in Zambia were *Myrmicaria eumenoides* Gerst. and *Pheidole megacephala* Fabr. Beetles of the genus *Zyas* (Staphylindae) preyed upon *Odontotermes* in the ventilation chimneys of their mound, eating the head and thorax in preference to the abdomen, suggesting that muscle protein was the main item of their diet. Various other arthropods were recorded as predators and in addition termites were observed entrapped in the slime of geoplanid Turbellaria which digested them externally and sucked out the liquified contents through a hole in the abdomen.

Kakaliev & Saparliev (1973) sampled 30 hectares in the 'termite quadrangle' of central Asia, finding 893 'nests' (29.8 per hectare containing 63 individuals per m^2) of *Anacanthotermes ahngerianus* (Jacobson) and recording over 70 species of predaceous ants. The most important ants were the fast-running *Cataglyphis setipes* Em. (3.3 nests per hectare; 2–3 individuals per m^2) and *Cataglyphis altisquamis foreli* Ruzs. (19.2 nests per hectare; 1.9 individuals per m^2). Bouillon (1970) noted that in South Africa the grass-harvesting *Hodotermes* and *Microhodotermes* are attacked by *Anoplolepis custodiens* (Smith) and *Iridomyrmex humilis* (Mayr), respectively to such effect that foraging is limited to an hour or two in any one location. Ohiagu & Wood (1976) observed that although the soldiers of grass-harvesting *Trinervitermes* were able to protect foraging parties against opportunistic ant predators (such as *Iridomyrmex*) when foraging holes were being closed, individual termites were often left 'stranded' on the soil surface and were rapidly consumed by *Iridomyrmex*.

Among vertebrate opportunists attacking the sterile castes, birds and mammals are best documented. Apart from the reviews already mentioned, some recent work has attempted to quantify certain aspects of the habit. Milstein (1964) recorded black and blue korhaann (*Afrotis afra* (L.), *Afrotis caerulescens* (Vieillot)) with 1900 and 1500 workers of *Hodotermes mossambicus* in their respective stomachs. A double-banded courser (*Rhinoptilus africanus* (Temminck)) had 368, and a tawny pipit (*Anthus novaeseelandiae* Vieillot), 34. Saayman (1966) examined stomach contents of *Bubulcus ibis* (L.), *Numida meleagris* (L.), and *Gallus gallus* (L.), finding up to 380 *H. mossambicus* each, equal to 130 kJ, about one-third of their daily energy requirements. Steyn (1967) found as many as 5100 workers of *Hodotermes* in one Guinea fowl (*Numida meleagris*). He estimated their protein content as 10%, fat 12%, equal to 1590 kJ, more than the probable daily requirement. Rowan (1971) stated that termites were used as food by 42.9% of all birds, 38.9% of passerines and 47.3% non-passerines. Aquatic and littoral birds tended to leave them alone, as did a few frugivores and nectar-feeders. Borret & Wilson (1971) found that

18–24% of two species of pipit (*Anthus* spp.) consumed foraging termites. Not all termites are acceptable as food by birds; Roonwal (1970) noted that foraging columns of *Hospitalitermes* (Nasutitermitinae) appeared unpalatable to domestic hens. In northern Ghana the Farafara tribe supply their domestic poultry with *Odontotermes* spp. (mainly *O. smeathmani*) trapped in earthenware pots filled with fresh cattle dung. The pots are inverted on the soil surface, shaded with a leafy branch and after a few days the contents consist more of foraging termites than dung. This and similar practices (Hegh, 1922) have never been quantified.

Many insectivorous mammals take foraging termites. Smithers (1971) examined the gut contents of Botswana mammals and found termites in elephant shrews (*Elephantulus* sp.), the hedgehog (*Erinaceus*, 98% of stomach content of one individual), the vervet monkey (*Cercopithecus*), wild cat (*Felis*), bat-eared fox (*Otocyon*, 26 out of 50 stomachs, mainly *Hodotermes mossambicus*), the cape fox (*Vulpes*, 3 out of 23 stomachs, *H. mossambicus*), the black-banded jackal (*Canis*, 9 out of 59 stomachs, *Macrotermes subhyalinus* and *H. mossambicus*), the side-striped jackal (2 out of 12), the small spotted genet (*Genetta*, 12 out of 78, *M. subhyalinus* 3, *Odontotermes* 1, *H. mossambicus* 7, unidentified 2), the rusty spotted genet (7 out of 30), the suricate (*Suricata*, 2 out of 17), the selous mongoose (*Paracynictis*, 9 out of 34, 6 *H. mossambicus*), the yellow mongoose (16 out of 50, 4 *H. mossambicus*), the white-tailed mongoose (*Ichneumia*), and the dwarf mongoose (*Helogale*, 10 out of 21, 1 *H. mossambicus*, 2 *M. subhyalinus*, 4 *Odontotermes* sp., 3 unidentified). Shcherbina & Sukhinin (1968) gave similar data for a range of vertebrates feeding mainly on *Anacanthotermes* in central Asia.

Clearly, surface-foraging termites, such as the harvester termites which feed in the open, or certain Macrotermitinae, such as *Macrotermes subhyalinus* or *Odontotermes* spp. which feed under thin surface sheets of soil or occasionally in the open, suffer most from opportunistic predation whether this is by ants, birds or mammals. There are no quantitative estimates of consumption by these opportunistic predators.

Specialised predation on foraging sterile castes

The more constantly available food supply represented by the non-reproductive castes has led to the evaluation of behavioural and morphological adaptations to this type of nutrition and in some cases to dependency. Such specialisations have diverged along two pathways to exploit the two main concentrations of termite activity, namely the foraging parties and the brood centre of the colony. The most conspicuous morphological adaptations have arisen in the latter connection, which involves the need for excavation of protective nest walls and resistance to defence mechanisms of the termites themselves. 'Hive' predators, which are discussed below, also tend to be nomadic since their feeding is commonly destructive, whereas predators on foraging parties require less in the way of digging equipment, but need the ability to harvest rapidly a sufficient amount of food before the rest flees underground. Since they compete with opportunist predators in non-destructive predation their activities tend to be territorial rather than nomadic. Both of these categories include invertebrate and vertebrate predators, and because of their specialisations, there is little overlap between them. Wheeler's (1936) classification of relationships between ants and termites did not take account of this distinction:

(i) *Cleptobiotic.* Ants that steal prey from other ants.

(ii) *Termitalestic.* Ants that live in termite mounds and prey on the brood – e.g. *Centromyrmex* (Ponerinae).

(iii) *Inquilines.* Ants that live in inhabited or abandoned termite mounds and may prey opportunistically on the termites.

(iv) *Termitharpactic.* Ants that habitually raid termites.

His fourth category includes representations of both types of exploitation, and the other groups overlap to a considerable extent.

Specialised, and apparently obligate, invertebrate predators of foraging termites seem to be confined to a few species of Ponerinae (ants). Emerson (1945) and earlier authors recorded concerted raids by *Termitopone commutata* (Roger) on foraging columns of *Syntermes* spp. in South America in which parties of over 100 ants carried off one or two termites each. The African termitophagous ponerine raider, *Megaponera foetens* Fabr., has been known for many years (Hegh, 1922). Fletcher (1973) described recruitment in this species which specialises in attacking foraging groups of Macrotermitinae. Work in progress by C. Longhurst (personal communication) in Nigeria in connection with the COPR/ABU Termite Research Project has shown that the most common target of *M. foetens* is the genus *Odontotermes*, with *Macrotermes* and *Ancistrotermes* used when the preferred genus is not foraging actively. Preliminary estimates of consumption of termites have been made for November 1974, April–May 1975 and October 1975. Expressed as termites per m^2 per day for the two main prey species, they are respectively: *Odontotermes pauperans* (Silvestri) 0.404, 0.855, 0.292; *Macrotermes bellicosus* 0.551, 0.414, 0.246. If this predation continued throughout the year, even at the lowest rates recorded, it would account for well over half of the standing crop biomass of both species in this area of southern Guinea savannah (Table 9.6). Characteristic of these ants is the speed with which a hunting column concentrates its attack on a foraging group of termites; the emergence of 200–300 ants from the nest to their return carrying up to eight termites each may take less than 30 min (C. Longhurst, personal communication; Sheppe, 1970).

The soldiers of Macrotermitinae seem to be ineffective in defending foraging parties against raiding *Megaponera* and therefore these termites are subject to considerable predation pressure in addition to that brought about by opportunistic predators. This may be one reason why Macrotermitinae appear to have a higher production/biomass ratio than other termites (Table 9.8). In contrast, we have noticed that when *M. foetens* encountered a foraging party of *Trinervitermes* the former were repelled and became disorganised.

A parallel type of feeding on foraging termites by a vertebrate was described by Kruuk & Sands (1972). The aardwolf (*Proteles cristatus* Sparrman) specialises almost entirely on those species of *Trinervitermes* that graze in the open at night in short-grass savannahs of East and Central Africa. It appears to locate its prey by scent and hearing and feeds by quickly lapping with a broad adhesive tongue. It runs from one prey location to another, apparently ceasing to feed when the distasteful soldiers predominate in the remainder of the foraging party. In East Africa its principal prey is *Trinervitermes bettonianus* (Sjöstedt) and according to Smithers (1971) *Trinervitermes rhodesiensis* (Sjöstedt) in Central Africa. The aardwolf does not occur in West Africa where the most abundant species of *Trinervitermes* form fast-moving foraging columns rather than grazing swarms covering a large area. Its intake has not been quantified but its 'home range' appears to be 1.5–4.0 km^2. Using a production/biomass ratio of 3:1 for *Trinervitermes*, the annual production by a modest population of 100 *Trinervitermes* per m^2 in an area of this size would supply approximately 21×10^3 MJ per km^2 which is approximately 20–50 times the annual metabolic requirements of a single aardwolf.

Specialised predation on colony brood centres and nest systems

Predation on the brood centre of the colony ('hive', 'mound' or 'nest') has only been perfected on a large scale by a group of invertebrates, the burrowing doryline ants of the genus *Dorylus*. These have entirely subterranean raiding columns and appear to bivouac in the nest systems of the termites on which they prey, but very little seems to be known of their biology. The reaction of termite colonies invaded by *Dorylus* spp. (various 'sub-genera' have been named, but the currently accepted position is to assign subterranean species to this genus, while the surface-raiding species belong to *Anomma* and other genera) varies in different groups. In general, *Macrotermes* spp. seem to stay in their nest systems and fight to a finish, as do *Cubitermes* spp., while *Trinervitermes*, *Pseudacanthotermes* and some soldierless genera such as *Alyscotermes* have been observed to leave their nests and migrate to the surface where the ants do not often follow. Sands (1965*a*) pointed out that *Sociotomie* (Grassé & Noirot, 1951*a*) usually results from *Dorylus* attacks, and Sands (1972*a*) described a suicidal defensive behaviour in soldierless termites presumably evolved in response to subterranean ant predation. Bodot (1961, 1967*a*) noted that *Macrotermes bellicosus* nests attacked by *Dorylus dentifrons* Wasm. seldom survived, and that this termite appeared to be the preferred prey. In the area of her study in the Ivory Coast, a large proportion of mounds of this species were unoccupied and this was attributed to doryline predation. It also seemed that the dorylines were more active or more successful in environments disturbed by clearing for cultivation, which could be a reflection of the lowered vitality of *M. bellicosus* colonies from shortage of food, or the greater ease of movement for the ants through the loosened soil. *Dorylus* spp. are adapted to penetrate soil spaces, as opposed to digging their own tunnels, and they probably approach a *Macrotermes* colony along the termite's own access galleries more often than by direct entry. The elaborate maze-like structures found in the 'basement' levels of some *Macrotermes* mounds are possibly adaptations against such attacks, although we have found in Nigeria that many dead young mounds bear unmistakable evidence of doryline attack in the ants' mandible marks around the enlarged openings to the queen cell (also recorded by Bodot, 1961, 1967*a*). However, Williams (1959*b*) indicated that *Anomma kohli* Wasm., which is a surface-raiding doryline, was less successful in preying on mounds of *Cubitermes ugandensis* Fuller and *Cubitermes testaceus* Williams, since the ruptured bodies of the workers appeared distasteful and impeded progress towards the brood. The nomadic behaviour of Doryline ants is well known, and is clearly necessitated by their destructive effects on their prey. Several years must pass for a flourishing *Macrotermes* colony to be replaced, and even with the maximum populations of such termites, the distances travelled by foraging *Dorylus* colonies must be considerable.

Vertebrate predators that specialise in digging out termite nests are all mammalian, with typically 'ant-eater' adaptations expressed in varying degrees. They have narrow elongate muzzles, long, adhesive, protrusible tongues, reduced dentition, thick skins, fur, scales, or spines, and powerful fossorial forelimbs and claws. In Africa and India the Pangolins (scaly anteaters, *Manis* spp.) feed on both ants and termites with a preference for the former. The aardvark (*Orycteropus afer* Pallas) of tropical Africa is reputed to be a major predator of termites. However, Smithers (1971) recorded nearly 100% ants in 7 out of 11 stomachs and in only two stomachs did termites constitute nearly 100% of the contents. Other vertebrates specialising in predation on brood centres include the armadillo in Texas (Kalmbach, 1943), ant-eaters (*Myrmecophaga* spp.) in South America (Araujo, 1970), the sloth-bear (*Melursus ursinus* (Shaw)) in India (Sen-Sarma, 1974), the numbat (*Myrmecobius f. fasciatus* Waterhouse) and echidna (*Tachyglossus aculeatus* (Shaw)) in Australia. There are no quantitative studies on these specialised vertebrate and invertebrate predators to enable us to gain even very approximate indications of their annual consumption of termites.

Lestobiosis, termitophily and related phenomena

Animals that live in the brood centre or galleries and prey on the inhabitants are a last form of specialised predation which has already been mentioned in connection with ants. Wheeler (1936) defined Termitolestic ants as adapted to this way of life, as exemplified by such genera as *Centromyrmex* and *Carebara*. Lestobiosis is by definition a phenomenon of eu-social organisms, but non-social animals with similar habits (synechthrans) occur in widely separated groups. Here, the borderline with termitophily is not sharp; possibly one evolutionary route to the latter is through the former. Some beetles, such as *Tetralobus* spp. (Elateridae) and *Orthogonius* (Carabidae), are found as larvae or adults in Macrotermitinae mounds and in the absence of other food are assumed to be predatory on the termites. A reduviid bug *Acanthaspis sulcipes* Signoret was recorded by Anwar (1970) as living throughout the year, ovipositing and developing in termite nest galleries (species unspecified), where it was observed to eat two to ten termite workers per hour as a larva, and 15–20 termites per hour as an adult bug.

Lestobiosis and related non-social phenomena have been studied very little and never quantified. The last stage of adaptation, termitophily, was reviewed by Kistner (1969), mainly from the point of view of the staphylinid Coleoptera. As he said, 'Our knowledge of termitophiles is still in the descriptive stage.' This is not the place to continue the review of the extensive literature on the subject especially since there has again been no quantification of the recirculation of termite tissue through this pathway.

We are most grateful to the following research workers who gave us permission to use their unpublished data: N. M. Collins, R. A. Johnson, G. Josens, M. Lepage, C. E. Ohiagu, D. E. Pomeroy and G. Rohrman.

10. The role of ants in ecosystems

J. PĘTAL

Ants as insects constructing nests

Ants live in societies occupying nests more or less isolated from the environmental conditions. About 80–90% of the members of a colony normally stay within the nest. Nests can regulate to some extent the micro-climatic conditions, which approximate the optimum; they also provide shelter from enemies and are the places where food is stored. Due to diversity in nest construction and flexibility in the adaptation of their construction to various situations, ants can occupy many different habitats and different niches within these (Francoeur & Maldague, 1966; Stawarski, 1966; Gaspar, 1970; Letendre & Pilon, 1973).

Types of nests

A nest is generally composed of a central part, where the queen lives, eggs are deposited and the brood is reared, and a periferal set of galleries and chambers occupied by workers; in some species nests are used for food storage and contain places where fungi and aphids are raised. Underground galleries may connect the nest to distant food resources. Ants can build their nests in a number of ways, from small ones with a dispersed system of chambers and galleries to large ones with a complex architecture.

Nests are constructed from the most readily available materials. So, in rocky environments nests are generally made of mineral materials, in meadows they are made of mineral material and parts of plants, in forests they consist mainly of woody materials. Salivary secretions are used as a cementing substance. Some tropical ants fasten living plant materials with bands of fibres secreted by the larvae (e.g. *Oecophylla*). Others, such as *Lasius fuliginosus* Latr., transform wood pulp into carton which is used in the construction of very complex nests. The chemical composition of this carton is different from that of the original material (Stumper, 1950): it contains 10 times more protein, three times more resin, twice as much lignin, while the cellulose content is considerably lower.

In rare cases, for instance in the camps of wandering ants (Dorylinae), the bodies of living workers form a cover for the queen, eggs, larvae and pupae.

Table 10.1 *Temperature in ant nests and in the same depth of soil in different habitats*

Species	Temperature (°C) In the nest	Temperature (°C) In the soil	Habitat	Type of nest and temperature regulation	Author
Myrmica rubra	23–24	22	Forest	Nests with mounds	Dlussky, 1967
Myrmica scabrinodis	25–25	22	Kursk, USSR		after Grinfeld,
Lasius niger	27–28	22			1939
Formica fusca	27–28	22			
Formica rufa	25–29	22			
Formica cinerea	29–30	22			
Formica exsecta	29–30	22			
Formica truncorum	29–32	22			
Formica cunicularia glauca	30–32	35	Grassland, Kursk, USSR	Nests in the soil	
Formica rufa polyctena	23–29	13.4	Forest, Godina Belgium	Nests with mounds	Raignier, 1948
F. rufa	25.3	10.1	Forest, Oberlausitz, *Pinus, Picea, Abies*	Nests with mounds	Hempel, 1963–64
	24.0	10.2	*Picea, Abies*		
	19	13.2	*Betula*		
	25–30				Kneitz, 1969
Camponotus acvapimensis	26±1	26±1	Savannah, Ivory Coast	Nests in the soil	Lévieux, 1973
Trachymyrmex septentrionalis	29.8–32.5	25–29	Rainforest, Florida	Nests in the soil	Weber, 1972*a*, *b*
Aeromyrmex ambiguus	28.5	12.2	Grassland, Brasil	Nests with mounds	Weber, 1972*a*, *b*
Atta colombica tonsipes	26£0.3	29–32	Forest, Panama	Nests in the soil with craters, ventilation-canal system	Weber, 1972*a*, *b*

Temperature

The thermal optimum in ant nests ranges between 19 and 32 °C for all geographical zones (Table 10.1). Species occurring in the same habitat may have different optima, but these are always higher than the soil temperature at the same depth. Temperature fluctuations related to the ambient temperature occur but the range of these fluctuations is always smaller (Raignier, 1948; Hampel, 1963–64; Dlussky, 1967; Kneitz, 1969; Ettershank, 1971; Gallé, 1972*a*).

The optimum temperature of the same species may be established at various levels depending on the nest aspect, insolation and kind of vegetation (Hampel, 1963–64). In a pine forest the nest temperature of *Formica rufa* L. is established during 19 days in spring and is 25.3 °C, while in a more shaded birch–fir forest it is established during 30 days and then at the lower level of 19.2 °C.

Temperature regulation also depends on the size of the colony; larger colonies maintain a higher temperature (Hampel, 1963–64; Kneitz, 1969).

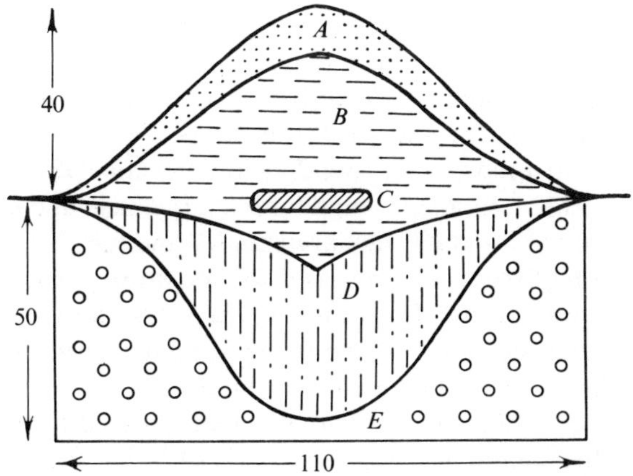

Fig. 10.1. Schematic cross-section of a *Formica rufa* nest. The dimensions are presented in centimetres. *A*, Fine plant material of the surface. *B*, Thick homogeneous material: small branches. *C*, Stump. *D*, Old heterogeneous material being decomposed. *E*, Galleries drained in the soil (redrawn from Raigner, 1948).

This is probably due to the physiological heat of the inhabitants. Raignier (1948) found that the temperature in a container with 12 ants per cm^2 at 34.8 °C increases to 38.6 °C during five hours.

Optimum temperature can be reached by modifications of nest construction. For instance, *Formica cunicularia glauca* Ruzsky builds its nests without mounds in an open, forestless slope where the soil temperature at a depth of 10 cm is 35 °C, and its nests with mounds in the forests, where the soil temperature at the same depth is 22 °C (Dlussky, 1967, after Grinfeld, 1939). Thermal regulation is also possible due to the stratified structure of the mound. Raignier (1948) has distinguished four zones in the nests of *F. rufa polyctena* Foerst. built of different materials (Fig. 10.1). The external zone is compact and composed of fine plant material, sand, and the needles of trees. It is impermeable and thicker at the top than along the sides. Below this layer there is the main part of the nest which is built of larger plant material, often including a stump or a piece of wood in the centre. Then, there is a transitional zone containing twigs, humus and little stones. And still lower there is the normal soil with galleries to a depth of 1 m. Such a structure makes it possible to eliminate the slightest changes in the temperature of the central part. The summer temperature is almost stationary there and remains between 23–30 °C, whilst the temperature of the air or the soil surface may drop below 20 °C, or the temperature of the nest surface may reach 70 °C.

According to Raignier (1948) one regulatory mechanism is the daily rhythm of evaporation and condensation of vapour. During the day, heat is absorbed due to evaporation, which protects the nest against overheating. At night, condensation of vapour and the formation of a blanket of mist over the nest protects it from heat loss from the lower parts.

In the heat economy of the nest the shape of the mound is also important. Ants can modify it depending on the insolation. In poorly-insolated habitats the mounds are high with a large base and steep slopes, while in well-insolated habitats they are smaller and flat. This enables the optimum utilisation of solar radiation and soil moisture; small, flat mounds absorb relatively more water from the soil than large and high ones. The relationship between the shape of the mound and insolation was noted by Dlussky (1967) for several species of *Formica.*

The temperature inside the nest can also be regulated through changes in the depth of the chambers. Lévieux (1972*a*) has found that the nest temperature of *Camponotus acvapimensis* Mayr in the African savannah averages 26 °C, with a tolerance of 1 °C in both the dry and the rainy seasons. During the period of low temperatures a large number of chambers with broods is located at a depth of 20 cm, where the soil temperature is 26 °C. During the warmer periods the central chambers are moved to a greater depth between 20–40 cm, so that the soil temperature of 26 °C can be retained. The depth of the chambers depends on the density of the overlying vegetation. The sparser the vegetation, the deeper the chambers, so that the brood may develop under approximately constant temperature conditions.

Moisture

It seems that the average moisture conditions in the nest are not as strictly regulated as the temperature. The nest moisture of *Camponotus aevapimensis* in savannah is 100% in both the dry and rainy seasons, but such is also the case with the soil moisture at the depth of the nest location (Lévieux, 1973). Also, in the nests of *Iridomyrmex purpureus* living in Australia the relative moisture differs only slightly from that of the soil and amounts of 68.3% (Ettershank, 1971). In the fungus chambers of *Atta sexdens* (L.) occurring in Brazil it is 36% during the three months of activity (Weber, 1972*b*). In the mounds of *Formica rufa* L. it ranges from 30 to 100%, the fluctuations being related to the temperature regulation.

Scherba (1959), however, has found that the total moisture in the mounds of *Formica ulkei* Emery was 27.23±0.66% at a depth of 5 cm where pupae develop, and 29.4±0.54% at a depth of 30 cm, where larvae develop. The range of fluctuations over the season was only 5.4±1.13%. At the same time the soil moisture fluctuated between 15.3 and 40.5% at a depth of 5 cm and between 21.6 and 40.7% at a depth of 30 cm. Thus, in the nests the minimum of the soil moisture was increased and the maximum was lowered.

Semi-desert ant species often remove vegetation from the nest site; this certainly lowers transpiration but increases the direct moisture loss from the soil surface. Rogers (1972) has reported that the diameter of the vegetation-free disc on the nest sites of *Pogonomyrmex occidentalis* (Cresson) range from 0.71 to 1.23 and is larger in the habitats with abundant vegetation. The proportion of water in the soil increases towards the centre of the disc and is 3.6% at a distance of 15 cm from the border and 4.2% in the centre, whilst beyond the disc it is 3.0% at a distance of 15 cm and 3.2% at a distance of 60 cm. These differences are statistically significant ($p < 0.05$).

Soil modification by ants

The construction of nests with above-ground mounds, and systems of underground chambers and galleries modifies the physical and chemical properties of both the soil used for construction and the adjacent soil. The soil is drained, its particles are moved and the walls of the chambers and galleries are fastened with the organic matter transformed by ants. Within the nest, food is stored, fungi and aphids are cultivated, faeces and the remains of the food brought from the outside are buried. The nest can be occupied for a long period, even for several years (*Formica rufa*, *Lasius flavus* (F.), *Atta* spp.) but ants can also move several times a year (*Myrmica* spp.). There are few publications on the effects of ants on the soil, and they are not always concerned with the same points; also, different methods are often used.

The range of the area under ant influence

The area influenced by ants is often rather large. In England on silty gravel nests of *Lasius flavus* occupy 10–11% of the area (Waloff & Blackith, 1962). In the lowland meadows and pastures of Poland several species, of *Myrmica*, with *Lasius niger* (L.) and *L. flavus*, occupy from 0.1 to 1.0% of the area of these habitats (Czerwiński, Jakubczyk & Pętal, 1971; Pętal, Jakubczyk, Nowak & Czerwiński, 1973) at a nest density of 0.1–0.3 per m^2. In Colorado pastures *Pogonomyrmex occidentalis* occupies 0.03–0.28% of the area (Rogers, 1972) at a nest density of 0.0003–0.0031 per m^2. The amount of soil moved to the surface there fluctuated from 0.28 to 0.8 g per m^2. Talbot (1953) has found that the amount of soil moved to the surface by *L. niger* in old fields in Michigan is 85.5 g per m^2. Lyford (1963) reports a much higher value of 60 g per m^2 of soil replaced by all ants on

Table 10.2. *Regulation of pH in the nest*

Species	pH (in H_2O) In the nest	pH (in H_2O) In the soil	Habitat	Author
Formica rufibarbis	6.85	6.60	Forest podzolic soil	Dlussky, 1967, after Grinfeld, 1941
	6.80	6.60		
Lasius niger	6.95	6.60		
	6.80	6.60		
	7.20	6.80		
Lasius flavus	7.00	6.80		
	6.80	6.70		
Lasius umbratus	7.55	6.90		
	7.60	6.50		
Myrmica sp.	7.80	7.90	Grassland soil	Dlussky, 1967, after Krupienikov, 1951
	7.90	8.00		
Lasius sp.	8.00	8.00		
	7.90	7.90		
L. flavus	7.30	7.00	Grasslands, Famenne	Gaspar, 1972
	7.40	7.45		
	6.80	6.70		
Myrmica sp.	5.70	5.50	Grasslands near Warsaw; black, muck-like soil	Czerwiński *et al.*, 1971
Lasius flavus	6.40	6.60		
Lasius niger	7.40	7.00		
Myrmica sp.	5.8–6.2	6.30	Silty, alluvial soil	
Myrmica sp.	6.60	6.70	Silty loam; alluvial soil	
Formica cunicularia	5.10	4.90	Sand dune	
Lasius flavus	7.00	5.80	Mountain pastures; brown pseudogley soil	Czerwiński, Jakubczyk, & Pętal, unpublished data
L. flavus	5.9–7.0	5.80		
Lasius niger	4.30	3.90	Sandy soil	
Myrmica sp.	4.00	2.80	Area polluted by sulphur mines; sandy loam	

podzolic soil in New England over the season. Sudd (1969) reports that ants of the genus *Myrmica*, *L. niger*, *Formica lemani* and *F. lugubris* can move to the surface from about 40 to more than 400 mg of soil per nest per day.

Physical changes

Physical changes are concerned with the soil profile. Smaller particles are raised from the deeper soil layers to the surface, while the organic matter is moved much lower down. Bulk density is considerably lowered; in the nests of *Pogonomyrmex occidentalis* located in sandy loam it decreases from 1.54 to 1.47 g per m^2 (Rogers, 1972). The system of chambers and galleries increases the porosity of the soil, which has as a result better aeration and infiltration of water. The high amount of organic matter increases the water-holding capacity. All these changes are favourable for the penetration of plant roots.

Chemical changes

Chemical changes are partly related to the physical changes. They result from the accumulation of organic matter in the nest and from decomposition processes. Many workers have observed the changes in pH within the nest as compared with the surrounding soil (Table 10.2). In alkaline soils the pH of ant nests decreases (Dlussky, 1967, after Krupienikov, 1951), while in acid soils it increases (Czerwiński *et al.*, 1971, unpublished data). In almost neutral soils only a slight decrease or increase in the pH of ant nests occurs (Dlussky, 1967, after Grinfel'd, 1941; Gaspar, 1972). The pH of ant nests depends on the soil type. Almost all types of a particular environment are characterised by similar trends in pH changes. There is, however, a specific difference in the intensity of these changes and it seems possible that ants can regulate to some extent the pH value in the nest, but the mechanism of this regulation is not known. Presumably, it would largely depend on the amount of exchangeable cations and on the organic matter: an increase in pH is related to the accumulation of exchangeable cations and a decrease is associated with the increase in organic matter content (Fig. 10.2).

Potassium compounds (K_2O) and phosphorus compounds (P_2O_5), which are readily assimilated by plants, and also, exchangeable cations (Ca^{2+}, Mg^{2+}, K^+, Na^+) come from the decomposition of organic matter accumulated in the nests. The content of potassium and phosphorus in the ant nests located in the meadows of Belgium (Gaspar, 1972) and Poland (Czerwiński *et al.*, 1971) is considerably higher than in the surrounding soil. Moreover, the lower their concentration is in the soil the higher it is in the nest. The concentration of exchangeable cations is also slightly higher or similar to that in the surrounding soil (Table 10.3). Higher concentrations of these substances are maintained in the nests for a longer time; even if the nest is abandoned by ants the level in the surrounding soil may still be exceeded two years later (Czerwiński, Jakubczyk & Pętal, 1969).

Decomposition of organic matter

The first stages of decomposition of organic matter, which are carried out by bacteria and fungi, are very intensive in ant nests. The abundance of

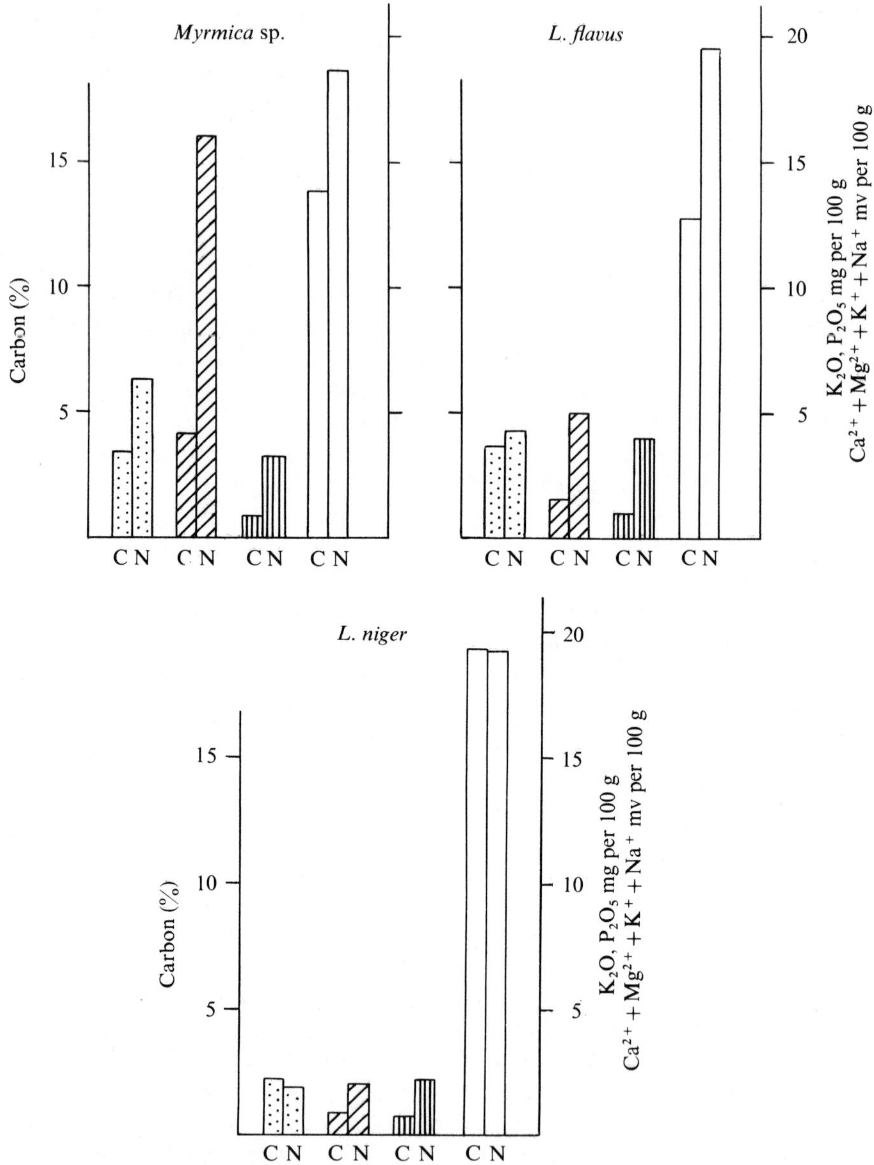

Fig. 10.2. Chemical properties of the soil in the nests of different species of ant; C – control soil, N – soil of the nest mound (redrawn from Jakubczyk, Czerwiński & Pętal, 1972). ⊡ = carbon; ▨ = K_2O; ▥ = P_2O_5; □ = Ca^{2+}, Mg^{2+}, K^{+}, Na^{+}.

Table 10.3. *Chemical properties of ant nests compared with those of adjacent soil samples (sample depth, 0–10 cm)*

	Content in mg per 100 g of soil determined by Enger's method				Exchangeable cations in mequiv. per 100 g of soil											
	K_2O		P_2O_5		Ca^{2+}		Mg^{2+}		K^+		Na^+		Carbon (%)			
Ant species	Nest	Soil	Nest	Soil	Nest	Soil	Nest	Soil	Nest	Soil	Nest	Soil	Nest	Soil	Habitat, type of soil	Author
Myrmica sp.	16.4	4.3	3.5	0.9	16.67	13.27	1.27	0.23	0.28	0.13	0.27	0.23	5.99	3.44	Uncultivated meadow; black, muck-like soil	Czerwiński *et al.*, 1971
Lasius flavus	5.3	1.3	3.5	1.0	17.81	12.17	1.39	0.38	0.18	0.07	0.40	0.20	4.98	3.78		
Lasius niger	1.6	0.4	2.3	0.9	17.81	17.96	1.55	1.17	0.11	0.07	0.30	0.22	2.31	2.79		
Myrmica sp.	10.6	9.1	5.1	3.6	24.81	28.04	5.74	4.93	3.63	0.33	0.31	0.33	5.05	3.79	Cultivated meadow, heavy, silty, alluvial soil	Czerwiński *et al.*, 1971
Myrmica sp.	8.2	6.7	8.4	5.0	—	—	—	—	—	—	—	—	7.16	6.96	Cultivated meadow; silty loam	Czerwiński *et al.*, 1971
L. flavus	—	—	—	—	16.52	11.60	3.68	2.76	0.87	0.23	0.39	0.36	—	—	Mountain pastures; brown pseudogley soil	Czerwiński *et al.*, unpublished data

Table 10.4. *The abundance of microflora in ant nests and in adjacent soil samples (in millions per cm³ of soil, determined by plating method)*

Ant species	Bacteria		Fungi		Actinomycetales		Habitat, type of soil	References
	Nest	Soil	Nest	Soil	Nest	Soil		
Myrmica spp.	686.5	55.8	1.46	0.40	0.181	1.19	Uncultivated meadow; black, muck-like soil	Czerwiński *et al.*, 1971
Lasius flavus	141.6	55.8	0.30	0.40	1.46	1.19		
Lasius niger	1547.1	55.8	4.70	0.40	0.21	1.19		
Myrmica spp.	628.7	140.8	2.12	0.65	0.93	2.10	Cultivated meadow; heavy, silt-like, alluvial soil	
Myrmica spp.	24.7	58.7	0.53	0.19	—	—	Cultivated meadow; silty loam	
L. flavus	26.0	16.0	0.22	0.12	2.10	17.30	Mountain pastures; brown, pseudogley soil	Czerwiński *et al.*, unpublished data

bacteria and fungi in the nest is higher than in the surrounding soil (Table 10.4). But the humification processes, carried out by Actinomycetales, are delayed. The intensification of rate of decomposition in the nest is higher on poorer sites and is accompanied by the accumulation of organic matter and nutrients. In nests abandoned by ants the reverse processes are observed; lack of organic matter input accounts for a decrease in the numbers of bacteria and fungi, while the number of Actinomycetales slowly increases (Czerwiński *et al.*, 1971).

There are considerable differences in the nutrient content and microflora abundance, between the nests of different ant species (Jakubczyk, Czerwiński & Pętal, 1972). This is probably due to the differences in their way of life, feeding and in nest construction (Fig. 10.3). The mainly predatory *Myrmica* spp. construct temporary nests and may move two to three times during a growing season. They accumulate large amounts of organic matter in their nests, which is followed by a large increase in microflora. Large amounts of released nutrients remain in the mound for about two years after the nest has been abandoned by ants. *Lasius niger* occupies one nest for a number of years and feeds on animal and plant food. High accumulation rates of organic matter collected from the surrounding area, in particular during the first half of summer, are accompanied by rapid mineralisation. The construction of the nest, however, allows for the leaching of the released nutrients. *L. flavus* also occupies the same nest over many years, feeding mainly on the honeydew of aphids raised on plant roots within the nest. The changes in the soil in the nest consist of: low microflora activity, a rate of mineralisation almost the same as in the surrounding soil, but very high humification. Large amounts of nutrients are accumulated in the nest.

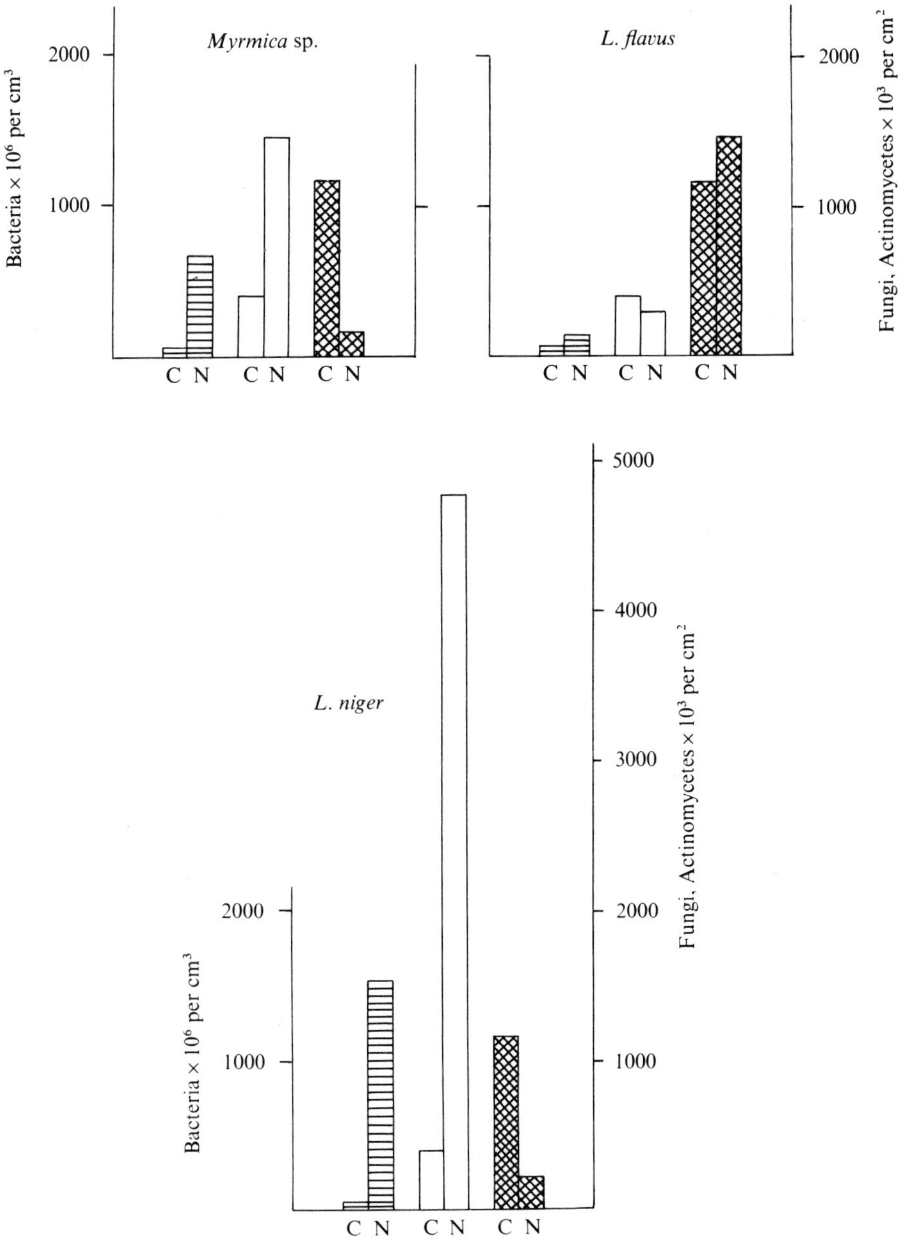

Fig. 10.3. Abundance of micro-organisms in different species of ants; C – control soil, N – soil of the nest mound redrawn from Jakubczyk, Czerwiński & Pętal, 1972). ⊟ = bacteria; □ = fungi; ⊠ = Actinomycetes.

Effect on the vegetation

The changes in the soil are followed by a characteristic plant succession. First, rhizomic plants grow on the mound; in a meadow with a *Deschampsietum* association these are *Carex panicae* L. and *Poa pratensis* L. (Pętal, Jakubczyk & Wójcik, 1970). Then come seedlings typical of more fertile sites, requiring drier and warmer soils. The temporary nests of *Myrmica* spp. do not cause a long succession: when the nest is abandoned by ants, the vegetation returns to its original state within two years.

Greater differences in vegetation are observed on the long-term nests of *Lasius flavus*. Gaspar (1972) has reported that these are inhabited by fewer species but that these species are favoured. Thus *Festuca rubra* L. and *Thymus* sp. occurred on the nest, while in the surrounding area they were not found, *Alopecurus pratensis* L., *Poa pratensis* L., *Dactylis glomerata* L. and others being recorded instead. The flowering period was delayed in the plants growing on the mounds.

Ant nests also influence the standing crop of plants and their yields (Rogers, 1972). An American species of the genus *Pogonomyrmex* clears the area around the nest of vegetation on a disc of radius 60 cm. At the disc edge, however, the vegetation is more abundant; its standing crop is much higher and amounts to 160 g dry weight per m^2, while beyond the influence of the nest it is 93.5 g dry weight per m^2. Thus, the losses resulting from removing the vegetation are compensated by a 63% increase.

Comparison with other soil animals

There is comparatively little information on the effect of animals on decomposition of organic matter and soil processes. Soil animals are divided into three groups: (1) those that break down organic matter without using much of it as a source of food and only slightly change its chemical properties (principally invertebrates feeding on dead and live plant material, such as Isopoda, Diptera and Coleoptera); (2) animals that intensify mineralisation processes and cause the accumulation of mineral nutrients in the transformed material; and (3) animals that intensify mineralisation and humification and mix organic matter with the mineral particles of soil (e.g. Enchytraeidae, Lumbricidae). They all facilitate the activity of microflora through breaking down matter, distributing it by means of their bodies and regulating its concentration by grazing or stimulating development (Macfadyen, 1967; Wallwork, 1970). In a number of species of Oligochaeta, Isopoda, Diplopoda, Chilopoda, insects and Gastropoda, Nielsen (1962) found the enzyme carbonase; these species hydrolyse starch and glycogen. Cellulose is decomposed by Gastropoda,

Lumbricidae and probably by Enchytraeidae. Though the initial stages of this hydrolysis are due to the microflora, the stimulation of its chemical activity is due to the soil and litter animals.

It is difficult to classify the activity of ants. They encourage the breakdown of organic matter and stimulate mineralisation by modifying the number of groups of microflora responsible for these processes. Probably specific substances such as the antibiotics recorded in some ant nests are also involved in this way (Gaspar, 1966). At the same time large amounts of organic matter are brought to the nest as food, from all trophic levels. Food debris can be distributed in the soil profile to a depth of 1 m or more.

There are a few data on the direct comparisons of physical and chemical changes caused by ants and other animals in the same soil habitat. They were collected in lowland meadows and mountain pastures in Poland, for ants and earthworms (Pętal *et al.*, 1973). Both these groups drain the soil of moisture, which increases the aeration and translocation of organic matter. The chemical and microbiological changes were compared in the ant nests and wormcasts in relation to the surrounding soil, the state of which was regarded as standard. The quality of the decomposition processes in the nests and wormcasts depends among other things on the food of these groups. Ants, mainly a predatory species, carry into the nests large amounts of energy-rich, protein food, and the earthworms feed on detritus. The groups use about 25 and 100 times more food, respectively, than their biomass. They intensify different stages of decomposition; ants mainly the first stages of mineralisation, and earthworms humification.

The contribution of these two groups of animals to processes within the ecosystem and the range of their influence are functionally equivalent in environments with different fertility, plant production, rate of decomposition of organic matter and trophic complexity. Along the increasing gradient of these indices the biomass of ants and the proportion of the area occupied by them both decrease, whilst the biomass of earthworms and the amount of soil transformed by them both increase (Table 10.5*a*). Also, the activity of micro-organisms in ant nests decreases considerably along the gradient of soil fertility, whilst it increases in the wormcasts (Table 10.5*b*). The changes in the chemical properties of soil, expressed as the nutrient content, are similar to those in the microflora (Table 10.5*c*). For different types of environments the effect of both the groups is complementary in nutrient cycling. In infertile soils with low organic matter decomposition rates and a more complicated trophic structure epigeic predators, such as ants, speed the return to the soil of nutrients accumulated in the bodies of other animals. By contrast, the role of earthworms is greater in fertile soils, with a more simplified trophic structure; they speed the release of nutrients from the organic matter as it is being decomposed. The changes caused by these two groups of

Table 10.5. *Influence of ants and earthworms on soil environment*

Complexity of the ecosystem structure ←; primary production →; decomposition rate →

		Meadows			Pastures	
		1	2	3	1	2
(a) Biomass of animals and percentage of transformed soil						
Formicidae	Biomass (mg dry weight per m^2 per year)	217	190	102	52.9	22.7
	% of the area	0.75	0.24	0.16	0.97	0.42
Lumbricidae	Biomass (mg dry weight per m^2 per year)	5300	7200	13400	4900	29100
	% of the soil	1.9	3.2	3.1	1.6	10.5

(b) Microflora activity

Changes of density (% per m^2)		Meadows 1	Meadows 2	Meadows 3	Pastures 1	Pastures 2
Formicidae	Bacteria	1064.3	81.1	18.9	121.3	3.8
	Actinomycetes	39.0	14.6	—	17.5	45.0
Lumbricidae	Bacteria	1164.7	1177.6	632.4	760.0	1619.0
	Actinomycetes	248.9	713.6	3397.6	700.8	1894.0

(c) Chemical properties of the soil

Changes in concentration (% per m^2)		Meadows 1	Meadows 2	Meadows 3	Pastures 1	Pastures 2
Formicidae	$K_2O + P_2O_5$	379.5	29.8	—	—	—
	$\Sigma = Ca^{2+} + Mg^{2+} + K^{+} + Na^{+}$	120.0	22.3	119.3	21.2	—
Lumbricidae	$K_2O + P_2O_5$	—	—	1565.5	1140.8	1638.0
	$\Sigma = Ca^{2+} + Mg^{2+} + K^{+} + Na^{+}$	—	—	375.0	273.0	851.0

animals occur in a large part of the soil and, as they occur successively, may be an important factor controlling the biological balance in the soil.

Ants as consumers in the ecosystem

The place of food consumption by ants in the general ecosystem flux

The role of ants as consumers is related to their considerable numbers in almost all ecosystems and to their high consumption which in proportion to their numbers is higher than in the populations of other invertebrates.

Ants take food from different trophic levels. The largest group consists of polyphagous species with a predominance of either plant or animal material in their food which may be dead or alive. There are also a number of highly specialised species, for instance, ants collecting seeds, cultivating fungi in the nest, feeding on the honeydew of aphids living on the green parts of plants or within the nest on the roots of plants growing on it. Ants are also social parasites in the nests of other ants and termites.

There are variations in the composition of the food carried into the nest (Table 10.6). It depends on the food supply and on the requirements of the colony (Lévieux, 1967*b*; Kajak *et al.*, 1972). During the period of intensive development of the larvae, animal food is collected. During the period when young workers are maturing, more honeydew is foraged (Ayre, 1959). Although only 10–20% of the colony take part in foraging (Golley & Gentry, 1964; Pętal, unpublished data) organic matter is carried into the nest in large quantities. An average-sized colony of *Formica polyctena* Foerst., which uses a territory of 0.27 hectares, forages about 6.1 million animals over the season, which corresponds to 259 MJ, and 155 l of honeydew, which corresponds to 686 MJ (Horstman, 1974). Attinae carry leaves, flowers, fruits seeds and faeces of insects as a substrate for the fungus gardens (Weber, 1972*a*; Werner & Murray, 1972; Werner, 1973). Weber (1972*a*) has found that in agricultural fields in Trinidad, Attinae populations with a standing crop of 10 kg per hectare have fungus gardens of 130 kg per hectare. He states (after Amante, 1967*b*) that colonies of *Atta sexdens* with a density of 10–18 combs per hectare consume 52.5 kg of plants daily, which is the normal consumption of three calves. Food is accumulated in the nest in greater quantities than the colony requires. Golley & Gentry (1964) have found 73–303 g of seeds in the nests of *Pogonomyrmex badius* Latreille, which is equivalent to 92 kJ per m^2.

Ants living in arid habitats have a caste of 'repletes'. These are the workers with enlarged abdomens, unable to move, which gather in their crops the honeydew carried into the nest by the foragers. This caste exists in *Myrmecocystus* from California and Mexico, in *Melophorus* from Australia, *Leptomyrmex* from Africa and also in the European species of *Proformica*.

The storage of food or its culture make ants independent of the actual food supply in the habitat and enable them to stay within the nest during unfavourable daily and seasonal changes in temperature and humidity. Due to this, ants have greater possibilities for the use of natural habitat resources than other invertebrates.

Table 10.6. *Material foraged by ants (%)*

Ant species	Plant material			Animal material					
	Seeds	Plant parts	Juices	Termites	Other insects	Faeces	Honey-dew	Un-identified	Author
Pogonomyrmexoccidentalis	39.0	34.0	—	—	20.0	3.0	—	4.0	Rogers, 1972
Pogonomyrmex rugosus	6.3	72.1	—	6.3	10.8	4.5	—	—	Whitford & Ettershank, 1972
Novomessor cocherelli	40.5	18.8	—	25.0	14.0	—	—	1.7	Whitford, 1973
Formica rufa	0.2	0.3	—	—	33.0	—	62.0	4.5	Wellenstein, 1952
Formica subnitens	—	—	—	—	39.0	—	50.0	11.0	Ayre, 1957
Myrmica spp.	—	—	—	—	50.0	—	—	50.0	Pętal, 1967
Formica exsecta	—	—	—	—	67.0	—	—	33.0	Pisarski, 1973
Formica polyctena	—	—	0.2	—	28.4	—	71.4	—	Horstman, 1974

Comparison of ants with other consumers

The energy flow through ant populations is higher than through populations of homoiothermic vertebrates occurring in the same habitat. In the population of *Pogonomyrmex badius* living in old fields of South Carolina it is 58–75 kJ per m^2, so it is higher than for a sparrow population (17 kJ per m^2) and a mouse population (31 kJ per m^2) in the same locality (Golley & Gentry, 1964). The total ingestion of primary consumers (savannah sparrow, old-field mouse, orthoptera, harvester ants and other herbivorous insects) is 546 kJ per m^2 which represents 12% of the above-ground primary production, 3.2% of which is due to the ant consumption (Wiegert & Evans, 1967).

The production of *P. occidentalis* occurring in the short-grass plains of Colorado costs 0.59−6.07 kJ per m^2 i.e. 0.3–0.2% of the primary production. These values are considerably higher for the species which are preferred by ants. Whitford (1973) reports that populations of harvester ants (*Pogonomyrmex rugosus* Emery and *Novomessor cocherelli* (André)) in Jordana Playa take more than 90% of the seed produced by *Erigonum trichopes*, 71% of the seed produced by *Boutsola barbata* and only 8% of the fruit produced by *Erigenum abortianum*.

The populations of dominant ant species living in the meadows of Poland (*Myrmica laevinodis* Nyl., *Myrmica ruginodis* Nyl., *Myrmica rugulosa*, *Lasius flavus* and *L. niger*) consume 24–280 kJ per m^2 i.e. 0.2–2.8% of primary production (Kajak, Breymeyer & Pętal, 1971), while those living in pastures use from 3.14–25.10 kJ per m^2 or 0.01–0.11% primary production (Pętal, 1974). The energy consumption of only one population of *Lasius alienus* (Foerst.) living in the sandy areas of Denmark reaches 272 kJ per m^2 (Nielsen, 1972*a*), and the consumption of *Formica polyctena* in the forests of Germany is 339 kJ per m^2 (Horstman, 1974).

Only a small part of the energy taken from the habitat is used for production (Table 10.7). For the population of *Pogonomyrmex badius* this is about 0.3% and *Myrmica* sp. 2–4%. The efficiency of larval biomass production is higher in laboratory cultures. The value of production in relation to food consumption varies from 2–50% in different species. Considerable seasonal fluctuations are observed. The coefficient of utilisation of protein food for biomass production is higher in spring than in summer, when the larvae tend to enter diapause (Brian, 1973*a*). Brian (1973*a*) has suggested that the larvae grow better and use proteins more efficiently if sugar is available in their diet. The presence of the queen in the colony accounts for an increase in food and water consumption, along with an increase in the growth rate of the larvae, but does not influence the ecological growth efficiency (NP/I, NP = net production, I = ingestion).

Table 10.7. *Ecological efficiencies*

	Assimilation efficiency (*A/I*)	Ecological growth efficiency (*NP/I*)	Tissue growth efficiency (*NP/A*)	Respiration coefficient (*R/NP*)	Author
Soil invertebrates					
saprovore	0.10–0.20	0.05–0.08	0.17–0.18	3.70–4.60	Reichle, 1971
herbivore	0.36–0.78	0.08–0.27	0.25–0.37	2.16–3.06	
predator	0.47–0.92	0.34	0.37	1.70–4.18	
Formicidae					
Myrmica laevinodis	—	0.026	—	—	Pętal, 1967
Pogonomyrmex badius	0.90	0.0026	0.003	34.3	Golley & Gentry, 1964
Pogonomyrmex occidentalis	—	—	0.10	8.91	Rogers, Lavigne & Miller, 1972
Lasius alienus	—	—	0.16	5.35	Nielsen, 1972*b*
Formica pratensis	0.707	0.018[a]	0.025	38.20	Horn-Mrozowska, 1974
F. polyctena	—	0.12–0.32[a]	—	—	Ayre, 1960
F. fusca	—	0.11–0.30[a]	—	—	Ayre, 1966
F. exsectoides	—	0.015–0.077[a]	—	—	Ayre, 1966
M. rubra	—	0.21–0.40[a]	—	—	Dlussky & Kupianskaja, 1972
M. lobicornis ussuriensis	—	0.10–0.20[a]	—	—	Dlussky & Kupianskaja, 1972
M. scabrinodis	—	0.16–0.18[a]	—	—	Dlussky & Kupianskaja, 1972
M. rubra (= *laevinodis*)	0.28–0.51[a]	0.30–0.50[a]	—	—	Brian, 1972*a*
Vertebrates					
Peromyscus polionotus	—	—	—	54.75	McNeil & Lawton, 1970
Passerculus sandwichensis	—	—	—	88.75	McNeil & Lawton, 1970

[a] I = protein food only. I = ingestion. A = assimilation. NP = net production. R = respiration.

Tissue growth efficiency (NP/A, A = assimilation) for ant populations is considerably lower than for soil invertebrates of all trophic levels and it ranges from 0.003 to 0.16; but the respiratory coefficient (R/NP, R = respiration) is very high. The methods of estimating production used to calculate this ratio are not always comparable with each other. So, its values fluctuate within rather large limits for different ant species and it is 5.35 for *Lasius alienus*, 8.91 for *Pogonomyrmex occidentalis* and 34.2 for *P. badius*. The results are higher still in laboratory cultures (Horn-Mrozowska, 1974). They are considerably higher than those for other invertebrates and similar to those of vertebrates.

This indicates that there are substantial losses of energy from ant populations. The organic matter carried into the nest as food is rapidly mineralised. As a result, nutrients are released, and their concentration in the nest soil increases. So, their recycling within the ecosystem is speeded up several times, and in the case of some elements, many times.

Effect of ants on the populations of other consumers

Ants influence the population of other consumers mostly within their foraging area which can be located in the soil, on the ground surface and also in different layers of the vegetation. There are differences in the size of the foraging area among different ant species. The African species *Anomma* and southern American *Eciton*, which do not build permanent nests, forage by moving from site to site in columns which reach up to several hundreds of metres in length and are composed of millions of individuals. The foraging areas of others surround their nests. Their size depends on the numbers in the colony, on the food abundance in the environment and also on the local intraspecific and interspecific relations among the ant societies. The larger ant species have delimited foraging areas which they defend against intruders. Other species occupy the free spaces between these areas which they may even enter, so that it may happen that the whole environment is occupied by ants.

Distribution pattern of foraging area

Interrelationships among the ant societies largely account for optimum utilisation of food resources in the environment. Both societies of the same species and different species compete for food.

Generally, a diversified habitat with a more abundant and diversified food supply is inhabited by a number of species with different alimentary regimes. They have a clumped distribution even at high nest density. Species domination, however, usually develops, expressed as the size of foraging areas. Such a structure was described by Yasuno (1964*a*) for ants in the meadows on Mount Hakkôda, using Bray's association index which expresses the frequency of species' encounters in samples of different sizes. He found that antagonistic behaviour was displayed during encounters between individuals of the dominant species (Fig. 10.4). The area studied was dominated by *Formica truncorum yessensis* Forel. The minimum area on which it encountered *Camponotus herculeanus japonicus* was 15 m^2, and *Formica fusca japonica*, 7.8 m^2. The latter two met on a surface of 3 m^2. Smaller species (*Tetramorium caespitum* Latr., *Pheidole forvida* and *Paratrechina flavipes*) inhabit the same areas as the larger dominants, but can be found together on a surface as small as 1 m^2.

When the foraging areas of certain species overlap, spatial avoidance by foragers is observed and subordination to the dominants consists of releasing prey objects without a struggle (Pisarski, 1973). Also, a shift in the period of subdominant activity often takes place (Baroni-Urbani, 1969; Lévieux, 1971*a*, 1972*b*). For instance, *Formica subpilosa*, which occurs on the same area as the dominant *Formica pratensis* Retz., shifts its maximum daily activity from the morning hours till afternoon (Stebaev

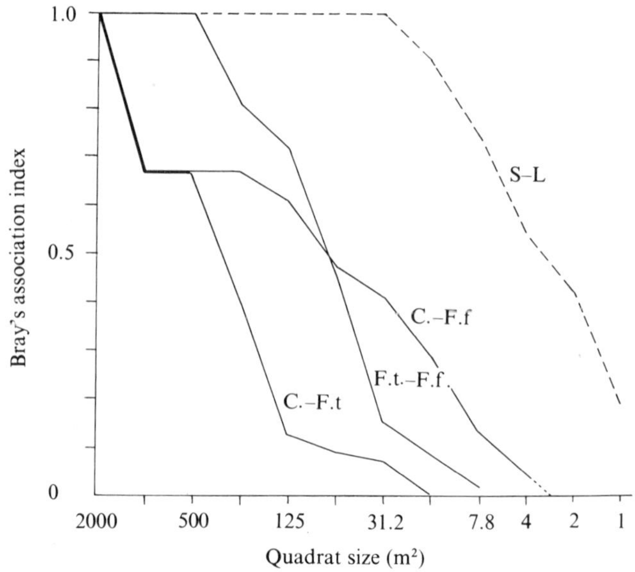

Fig. 10.4. Bray's association index between every two species, in relation to quadrat size (log scale). C – *Camponotus herculeanus japonicus*; F. t. – *Formica truncorum yessensis*; F. f. – *F. fusca japonica*; S – smaller species; L – larger species. (Redrawn from Yasuno, 1964*a*, *b*).

& Reznikova, 1972). Such relations enable the penetration of the environments by ants to be permanent.

Uniform and food-impoverished habitats are occupied by lower numbers of species with similar alimetary regimes. The distribution of nests depends in such cases on the nest density; it is clumped at lower densities and more uniform at higher densities. Using the nearest-neighbour method, Brian (1956*b*), Baroni-Urbani (1969), Gallé (1972*a*) and Yasuno (1964*a*) have all found that the lowest distances within the aggregations of the nests of many species occurred less frequently than would be expected in a random distribution, while the average distances occurred more frequently. According to these authors a tendency to overdispersion of the nests results from intraspecific competition.

The pattern of the nest distribution can be changed as a result of changes in food supply. The strongly clumped distribution of *Myrmica laevinodis* nests in a natural meadow abundant in food (index of dispersion, $S^2/\bar{x} = 1.66$) changed to near-random dispersion ($S^2/\bar{x} = 1.22$) in the next year when the food supply was reduced by 30%. Simultaneously, the number of nests slightly increased (from 60 to 62) in an area of 250 m^2 (Pętal, 1968). Adaptations in the pattern of the nest distribution enable ants to use the food available in the habitat more efficiently and to reduce the unfavourable results of competition among societies, which limit their reproduction and numbers.

Effect of ants on the Homoptera populations cultured

Between ants on the one hand, and Aphidae, Coccidae and Psyllidae on the other, reciprocal relations are formed; ants protect the cultivated populations and stimulate their development and the latter provide ants with food in the form of honeydew rich in various sugars, with amino acids and mineral salts. It originates from the sap sucked by these populations from the parenchyma or phloem of plants. Such cultures represent a protein reserve for ants (Pontin, 1958). It appears that ants protect the cultures against enemies by building a special earth construction (Lévieux, 1967*b*); they also move cultures to younger parts of the plants, with better sap composition (El-Ziadi & Kennedy, 1956; El-Ziadi, 1960) and keep the winter eggs of the aphids in their nests (Kloft, 1959).

Using radioactive phosphorus, ^{32}P, it has been found that under the influence of ants, aphids use twice as much plant sap and excrete twice as much honeydew (Banks & Nixon, 1958). Some sugars are converted into melezitose which can be harmful (Wat, 1963). The amount of honeydew taken by ants is very large. An average colony of *Formica polyctena* stores 290–320 kg of honeydew per season and *F. rufa*, 450–500 kg (Zoebelein, 1956*a*).

The development of Hemioptera populations cultivated by ants is more rapid and their numbers can be 10 times higher than without ants (Müller, 1958). However, the culture of aphids in the ant nests can be harmful for plants. Müller (1958) reports that the population of *Lachnus exsicator* cultivated by ants in a beech–pine forest causes the death of young beeches. As a result of the long-term influence of aphids the branches of old beeches also die and some deformations appear as they are infested by canker fungus (*Nectria ditissima*).

Aphid cultures also have an effect on the nutrient cycling of the ecosystem. This mainly concerns such elements as nitrogen, phosphorus, potassium, copper and zinc which occur in large quantities in the young plant tissues preferred by aphids (van Emden, Eastop, Hughes & Way, 1969). The increase in the number of aphids and in the amount of sap consumed by them, which is stimulated by ants, probably speeds the cycling of these elements.

Predation by ants and their role in controlling other invertebrates

The intensification of ant predation coincides with the period of growth of potential prey in the environment. The prey species are generally

Table 10.8. *Influence of ants on prey populations* (*after Kajak* et al., *1972*)

	Arthropod group		
	Auchenor-rhyncha	Diptera	Araneae (wandering)
Number produced (No. per m^2 per year)	1943	707	2500
Number killed (No. per m^2 per year)	840	226	1241
Elimination by ants %	43	32	49

invertebrates, especially forms new to the environment, such as larvae hatching from eggs, larvae immediately after metamorphosis and adults emerging from pupae. At that period a number of ant species change their diet considerably. Lévieux (1972*b*) has reported that in the rainy season 95% of the animal food in the diet of *Pheidole sculpturata*, living in the African savannah, is composed of small invertebrates, whilst in the dry season 50% of the food is composed of seeds. Also, in *Tetramorium sericiventre* the proportion of animal food increases from 81% to 89% in the rainy season. Similarly, the predation rate of *Myrmica laevinodis*, in the meadows of Poland, increases several times during the prey maximum in June and July (Pętal, Andrzejewska, Breymeyer & Olechowicz, 1971; Kajak *et al.*, 1972).

The contribution of different prey groups to the ant diet depends on their density in the environment. In savannah, permanently abundant ants and termites predominate in the diet of their predators. Moreover their proportion in both the dry and rainy seasons are similar. Thus in the diet of *Pheidole sculpturata* the contribution of other ants is 28 and 30%, whilst that of termites is 4 and 5%; and in the food of *Tetramorium sericiventre* ants represent 54 and 60% of the total prey (Lévieux, 1972*b*).

Myrmica laevinodis generally reduces the abundance of three groups of invertebrates predominating in meadowland: Araneae, Auchenorrhyncha and Diptera, which together constitute 78–88% of its prey (Pętal *et al.*, 1971; Kajak *et al.*, 1972). In different years the dietary contribution of these groups varies and is related to their production; the groups with higher production are more intensively reduced by ants (Table 10.8). Also, forest pests during the period of mass appearance may contribute to more than 90% of the prey of *Formica rufa*. Ants kill them in greater quantities than they need and store them in the vicinity of the nest. Pavan (1960*b*) reports that a population of ants of the group *Formica rufa* (*F. lugubris*, *F. rufa*, *Formica aquilonia*, *F. polyctena*), living in the alpine forests and composed of about 1000000 nests including some 300000 workers, each

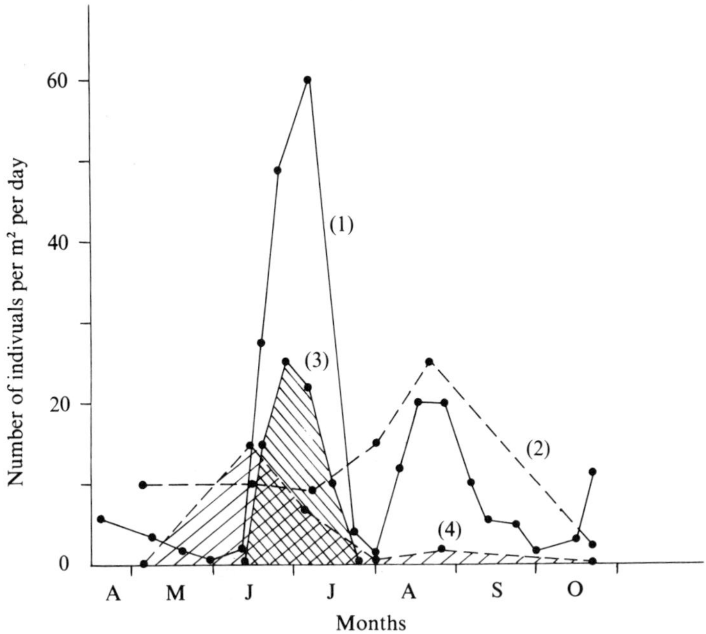

Fig. 10.5. Production dynamics of Auchenorrhyncha and Diptera and their predation by ants: (1) Auchenorrhyncha, (2) Diptera, (3) number of Auchenorrhyncha killed, (4) number of Diptera killed (redrawn from Kajak *et al.*, 1972).

kills 14×10^6 kg of insects in 200 days of activity. *Formica rufa rufopratensis*, with a density of six nests per hectare, ate 12300 larvae of *Tortrix viridiana* per day; as a result the number of pest pupae decreased by one-third. In consequence, in areas with high nest density 75–90% of trees were saved (Otto, 1958*c*).

Another forest pest, *Bupalus piniarius*, is greatly reduced by ants. *Formica rufa* consumes 28% of the population and *F. pratensis* 24% (Dlussky, 1967). *Formica polyctena*, at a density of eight nests per hectare, reduces the number of cocoons of another pest, *Lygaenomatus abietum* by 22–35% (Borchers *et al.*, 1960).

The effect of ants on insect species without a tendency to mass occurrence is less. Nevertheless, Otto (1965) has found that the numbers of predatory beetles and Hymenoptera decrease in the foraging areas of *F. polyctena* though the density of predatory bugs and spiders is not reduced.

The increase in ant predation also depends on intrapopulation factors such as an increase in the protein food required for the growth of larvae. In the population of *Myrmica laevinodis* (Kajak *et al.*, 1972) living in meadows, an intensive development of larvae occurs during June–July.

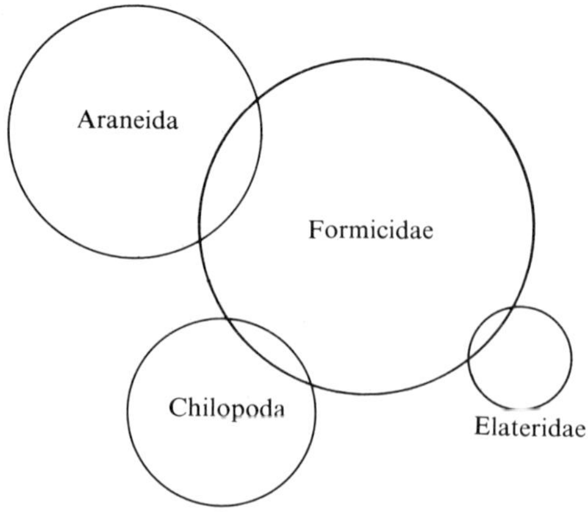

Fig. 10.6. Quantitative relations between dominant groups of predators (redrawn from Kaczmarek, 1963).

At this period the first hatching maximum of Auchenorrhyncha larvae is observed and the mergence of adult Diptera takes place. This first maximum is intensively exploited by ants, while the second maximum occurring during August and September, is neglected by them (Fig. 10.5).

Reduction of prey numbers in the environment during ant maximum requirements is followed by changes in the foraging organisation (Pętal, 1968). In *M. laevinodis* the number of foragers and the foraging area are first increased. Also, the prey species used are extended as more groups become available in the area and are included. In addition, the proportion of liquid food, from plants as well as animals, increases. The density of nests then decreases, presumably through mutual interference, and increases the efficiency of utilisation of food resources in the habitat. These adaptations often set a lower limit to ant numbers and production than to other predators.

Comparison of ants with other predatory invertebrates

Ants occupy an important position among other predators; this is due to their intense ecological activity. This results partly from their social life (which enables 80–90% of them to stay in the nest under ideal conditions) and partly from their complete lack of food specialisation.

In a number of habitats other groups of predators differently specialised (euzoophages, parazoophages) are suppressed by ants. Kaczmarek (1963) noted this in his study of the influence of physical environmental factors (shading, humidity, soil structure), and food relations and composition, on the soil predators of field and forest habitats. In these environments, although the number of ants depends on the habitat factors, the number

Table 10.9. *Production turnover and energy of ant populations available for other consumers*

Species	Turn-over $(P/B)^a$	Energy (kJ per m^2 per year)	Habitat	Author
Myrmica sabuleti	0.93	—	Southern England	Brian, 1972
Myrmica laevinodis	0.6–0.7	5.1–8.4	Meadows near Warsaw, Poland	Pętal, 1967
Lasius niger	0.3	0.42–0.88		Pętal, unpublished data
Lasius alienus	1.4	30.1	Sandy heath, Denmark	Recalculated from Nielsen, 1972*a*, *b*, and in press
Pogonomyrmex badius	0.4	1.51	Old field, South Carolina	Golley & Gentry 1964
Pogonomyrmex occidentalis	1.01	6.3–63.2	Short-grass plain, Colorado	Rogers, 1972

[a] P = production of population, B = standing crop of population.

of other predators, such as Araneida, Chilopoda and Elateridae, does not. The abundance of all predators depends on the abundance of potential prey in the environment, but the polyphagous ants predominate. They have an influence on the numbers of the remaining groups of predators, whose food niches overlap with their own, so that in these habitats ants effectively force other predators into one competitive system (Fig. 10.6).

Ants as food for other animals

The energy available for other consumers from the ant population is determined by its turnover. Presented as a ratio of production (P) to standing crop (B) (P/B), turnover indicates how large a part of the population is replaced by new individuals during a definite period of time. Turnover depends on the average ecological lifetime of the population and it does not fluctuate much as long as environmental conditions are equable. For ants of the temperate zone, which live from one to three years, the turnover value is probably lower than it is in the tropical zone. Delage-Darchen (1971) reports that the population of *Acrocoelia heliophila* in a savannah of the Ivory Coast, with a density of 13.3 individuals per m^2, is completely replaced every 1.5 months; several times per season.

The turnover values given in Table 10.9 are different for different species, which is partly a result of not always using comparable methods of production and biomass estimation. Hence, the amount of energy contained in the ant tissues available for other consumers can either be underestimated or slightly overestimated. Nevertheless, these are rather large values compared with, for instance, the consumption of birds and mice, which is 16.7 kJ per m² and 31.0 kJ per m², respectively, in the old fields of South Carolina (Wiegert & Evans, 1967).

The predators of ants are such invertebrates as Araneae, Thomisidae, Reduvidae, and also certain vertebrates: amphibians, reptiles, but primarily birds. Lévieux (1972*b*) has found that in the Ivory Coast savannah small soil ants, such as *Paltothyreus*, *Pheidole*, *Tetramorium* and *Crematogaster*, represent 63–90% of the food of small amphibians such as *Arthroleptis pecoilonotus*. Larger and more active species (*Camponotus acvapimensis* and *Megaponera ambigua*) are preyed upon by larger amphibians such as *Bufo regularis*. The impact of birds, however, is greater than that of amphibians; they prey upon ants foraging on plants and they greatly reduce the number of sexuals during the latter's mating flight. Lévieux (1972*b*) reports after Thiollay (1970) that birds eat 5–30% of sexuals (ants and termites) that have left the nest, and in this way they reduce the breeding potential of these populations substantially.

In colder climates Picidae eat large amounts of forest ants. De Bruyn, Goosen-De Roo, Hubregtse-Van den Berg & Feijin (1972) found that the percentage of small ants in the food of *Picus viridis* is highly correlated with their density in the habitat ($r = +0.94$). In winter this bird takes 5% of the population.

According to de Bruyn & Mabelis (1972) a considerable reduction of a *Formica* population may be caused by fighting between adjacent colonies and subsequent cannibalism; these occur at a period of food deficiency in the habitat. In English heathland after destruction of the vegetation by fire and in the spring, *Lasius alienus* and *Tetramorium caespitum* fight and eat intraspecifically (Brian *et al.*, 1965, 1966, 1967).

Regulation of ant populations

We shall emphasise here the differences in the method of number regulation found in social insects and in other insects.

Population dynamics in ants is modified by their social behaviour. It diminishes the influence of the habitat conditions. Due to their nesting behaviour and flexibility in nest construction, there are no great differences between the colony size of the same species in different climatic zones. The density of nests, however, is limited by the vegetation structure and the number of suitable nest sites, which also has an effect on the size of the colony and individuals (Brian, 1956*b*). Polyphagism and the bug cultures maintained by ants reduce their dependence on changes in the food supply. Intraspecific competition for food has a lower limiting effect than interspecific competition (Brian, 1965; Kaczmarek, 1963).

According to Brian (1965), predation also, which is particularly intensive during the mating flight, does not seem to have a significant effect on the population size of a stabilised colony. It seems probable that the size of ant populations mostly depends on the intrapopulation factors and not on environmental ones. This is supported by studies carried out on the species of the genus *Myrmica* (*M. laevinodis*, *Myrmica ruginodis* Nyl., *Myrmica scabrinodis* and *Myrmica sulcinodis*) which occurs on various kinds of lowland meadows and mountain pastures in Poland (Pętal, unpublished data). The size of these colonies ranged from several hundred to several thousand individuals in different habitats within nine years. The density of their nests was not as variable. There were significant differences in the food supply among the different habitats and also in the mean temperature over the growing season. The colony size was closely related to the age structure, i.e. the number of old workers (with a correlation coefficient of 0.96); then to the temperature ($r = 0.58$) and least to the food supply ($r = 0.48$).

These relationships are similar for production, i.e. the increase in the colony biomass. The density of nests does not seem to limit the colony size ($r = 0.62$), it limits, however, the density of foragers ($r = -0.58$), which suggests the existence of intraspecific competition for food.

Regulation of colony density

The longevity of ant colonies, and the relatively long period needed for maturation (i.e. reaching an appropriate size and developing the ability to produce sexuals), both ensure a certain stability for the population. A kind of aging process is observed in the community. It consists of an increase in the proportion of large colonies, accompanied by a decrease in the proportion of young colonies, without any significant variation in the density of nests. Scherba (1963) has found that a population of *Formica opaciventris* maintained its density at a level of 0.04 nests per m² during three years; the increase in the number of nests was 5–13%, while 8–9% of nests died and these were mostly new ones. It seems that young, poorly organised colonies die not only because of lower possibilities of adaptation to the habitat, but primarily because of intraspecific competition. The density of nests varies less significantly than the number of individuals.

The density of nests of populations of *Myrmica ruginodis*, *Formica fusca*, *Lasius flavus* in English bracken heath increased approximately twice during five years (Fig. 10.7). The fluctuations in the numbers of individuals in the colonies were several times higher (Fig. 10.8) and were not coordinated with the changes in the nest density (Pickles, 1940). Also, the population of *M. laevinodis* in uncultivated meadow in Poland maintained a density of 0.15–0.18 nests per m², while the colony size fluctuated between 500 and 2700 mg dry weight (Pętal, 1977*a*).

According to Wilson (1971) such an organisation of the population can

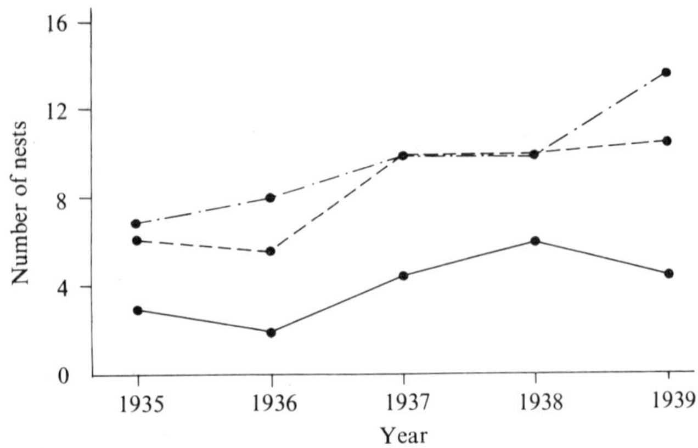

Fig. 10.7. The number of nests over five years of three species of ants on bracken heath in Yorkshire (redrawn from Pickles, 1940). –·–·–, *Myrmica ruginodis*; – – –, *Formica fusca*; ——, *Lasius flavus*.

serve as a homeostatic device protecting it against excessive fluctuations in numbers. Due to this device even a drastic reduction in the number of individuals is possible without changes in the population size and in the distribution in the habitat. When environmental conditions improve, the colonies become the nuclei from which a rapid restoration of individuals takes place. As a result, the fluctuations in ant populations are much smaller than in non-social insects.

Regulation of the number of individuals in a society

The number of individuals in the colony is mostly regulated by intra-population factors. In young colonies a period of exponential growth is observed, when the increase in the number of individuals is proportional to the number of workers. It is limited by worker mortality and natality. As it approaches maturity the numbers in a colony approximate to an asymptotic limit or even decrease. Density-dependent factors intensify; one is the change in the proportion of workers and sexuals produced. Brian (1957*a*, 1965) suggests that this is the main factor limiting the size of the society. His model of colony growth is based on the assumption that the production of both workers and sexuals is proportional to the numbers of workers in preceding generations. Consequently, it means that the change in one of the factors favouring the development of sexuals is sufficient to lower the production of workers. This, in turn, is followed by a decrease in the colony size, and thus its reproductive rate in the next year.

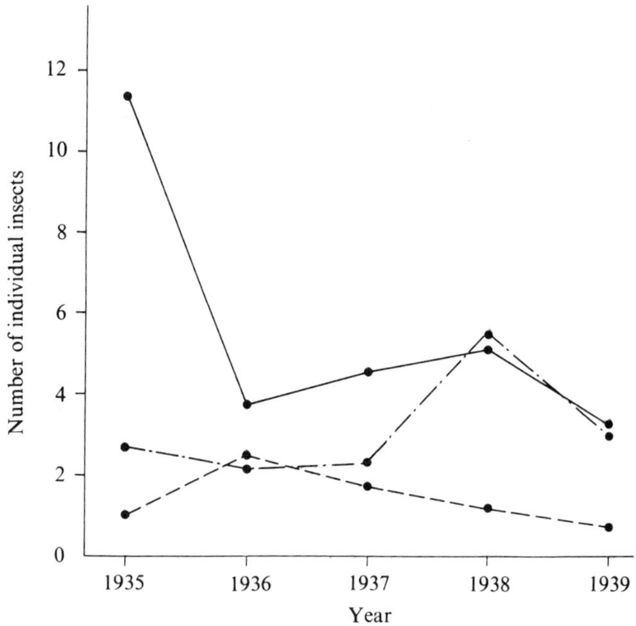

Fig. 10.8. Fluctuation of the numbers of individuals in the same three species of ants shown in Fig. 10.7 (redrawn from Pickles, 1940). –·–·–, *Myrmica ruginodis*; – – – –, *Formica fusca*; ——, *Lasius flavus*.

Brian's model (1965) is described by the equation:

$$dN_i/dt = aN_w^s - dNj/dt \quad (10.1)$$

or (because $dNj/dt = bN_w$),

$$dN_i/dt = aN_w^s - bN_w \quad (10.2)$$

N_w = number of old workers,
N_s = number of immature individuals giving rise to workers,
N_j = number of immature individuals giving rise to sexuals,
s = constant with value < 1,
a, b = fitted constants,
t = time.

Small variations in the coefficient b, which determine the production of sexuals, result in significant changes of the society size (Fig. 10.9). This kind of regulation of the size of societies can also be met in nature.

During eight years the number fluctuations of an average colony in a natural population of *Myrmica laevinodis* were due not only to the

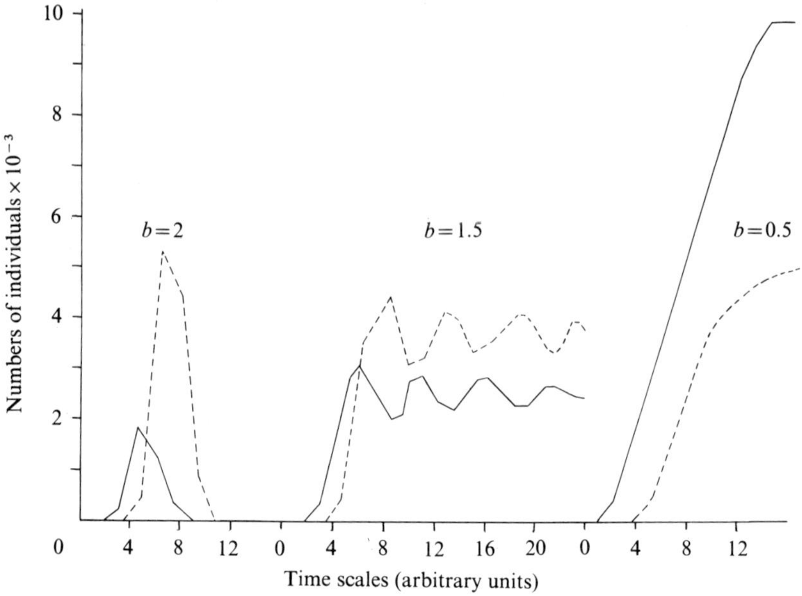

Fig. 10.9. Model population growth (number of workers) and sexual emission rates. All constants are fixed except *b*, the reproductive coefficient. As this is decreased the population programme changes from an explosive to a steady state and finally to one with a gradual approach to a steady state (redrawn from Brian, 1965). ——, workers; – – – –, sexuals.

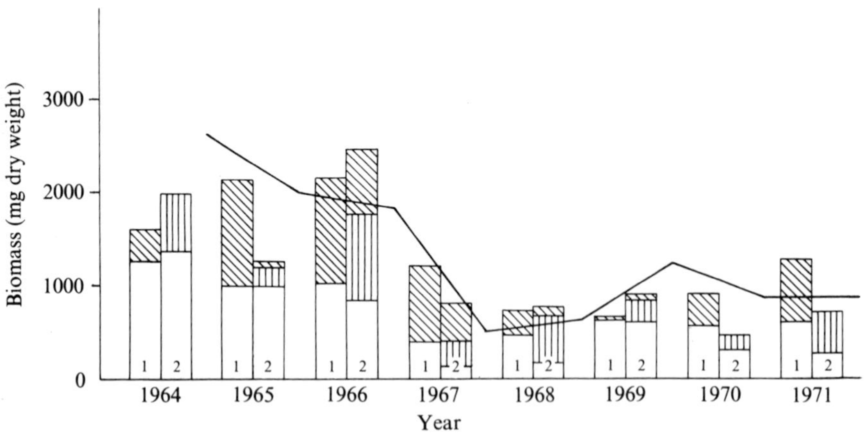

Fig. 10.10. Population dynamics of *Myrmica leavinodis* Nylander (= *M. rubra* (L.)) as dry weight in mg. Column 1 represents the biomass of old workers at the beginning of the growing season. The hatched area indicates the elimination of workers over the season, and the clear area indicates the number of workers at the end of the season. Column 2 represents the production of new generation biomass. The clear area indicates young workers, the vertically hatched area, the sexuals and the cross-hatched are, the elimination of the young generation.

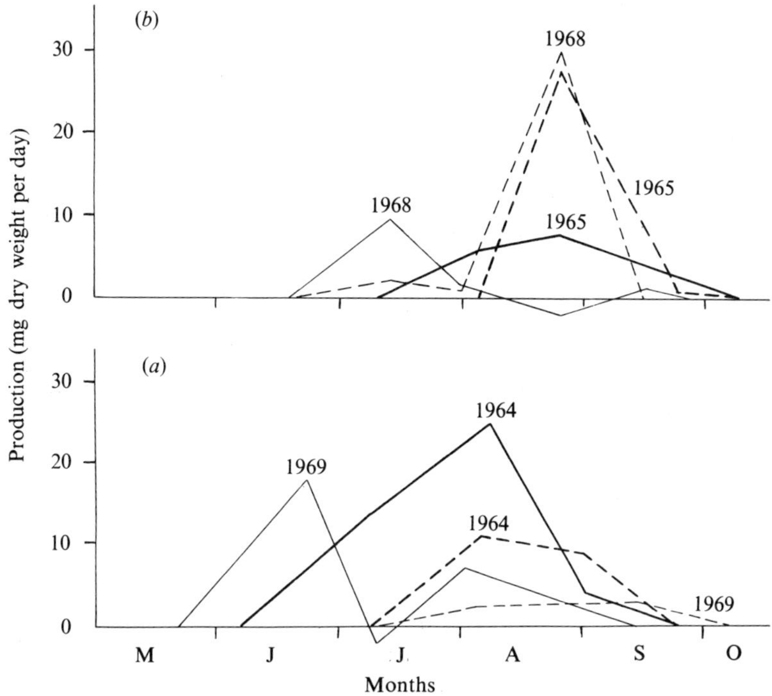

Fig. 10.11. Daily production of adult workers and sexuals (*a*) when the production process is not disturbed: high production of workers and then, sexuals (*b*) when the production process is delayed: low production of workers and higher of sexuals. ——, ——, workers; – – –, – – –, sexuals.

mortality of larvae and workers but primarily to modifications in the production of sexuals (Fig. 10.10). The contribution of the sexuals to the total production ranged from 16 to 62%. A year with high production of sexuals was followed by a decrease in the number in the colony in the next year and for even longer. The increased production of sexuals was caused by a delay of the maximum summer production, from early June until July. This was accompanied by elimination of small larvae, probably as a result of cannibalism, and the growth of larger larvae from which sexuals developed (Fig. 10.11).

This delayed production (affecting Brian's coefficient *b*) was induced by low temperature, precipitation, reduced food supply and also depended on the age structure of the colony. It is interesting that those colonies with two- and three-year-old workers produced fewer sexuals.

Another major factor of intrapopulation regulation in ants is cannibalism. It is mainly observed during the periods of food shortage in the environment. Eggs and larvae are eaten by the workers of the same colony and also workers from other colonies are killed in the struggles for territory. According to de Bruyn & Mabellis (1972) cannibalism may be regarded as a mechanism for adjusting the ratio of predator to prey in the environment.

Adaptability of ants to environmental pollution

Ants from a significant part of the fauna of environments even severely modified by man. They occur in towns, building nests even in the streets among the flagstones, and they also occur in buildings.

Their resistance to pollution is higher than that of other invertebrates. They exhibit a very high resistance to radioactivity. Torossian & Causse (1968) have found that short term but intensive radiation causes death more rapidly in *Dolichoderus quadripunctatus* than a less intensive but more prolonged radiation. Societies in which the queen is present are more resistant to radiation. Thus from a colony without a queen 50% of individuals disappeared within four days after being exposed to radiation of 150000 rad for 48 min. and within 16 days after a dose of 50000 rad for two hours, 24 min. Under the same radiation the lethal doses for colonies with queens were 28 and 30 days, respectively.

Le Masne & Bonavita-Cougourdan (1972) report that irradiation of the soil of a forest ecosystem in Cadarache with 84000–8400 rad during 11 months did not cause any disturbances in the numbers of ant colonies or in their activity. On the other hand radiation had an effect on the survival of some plant species (*Rosmarinus*), and caused reduction of the leaf surface (*Cornus mas*) and deformation (*Potentilla verna*).

Ants are also highly resistant to industrial pollution. A population of *Myrmica ruginodis* and of *Lasius niger* occurring in the areas polluted by a nitrogen plant can serve as an example (Pętal, Jakubczyk, Chmielewski & Tatur, 1975). In this area ants are a dominant group among other invertebrates which are very sparse. They modify the soil habitat, reducing the content of the main pollutant, nitrogen in nitrate form, in their nests. Each of these two species do this to a different degree. It seems just possible that they stimulate the micro-organisms that bind the mineral nitrogen, as their numbers are considerably higher in nests than in the adjacent soil (Fig. 10.12). The mechanism of such a stimulation is not known. Ants can, however, produce antibiotics or even insecticides (Büchel & Korte, 1960; Cavill & Hinterberger, 1960; Pavan, 1960*a*).

The high resistance of ants to environmental pollution results from their social life, their ability to modify the microhabitat of the nest, and it is also due to the division of labour, as a result of which only 10% of the colony is outside the nest and thus exposed to the direct effect of the environmental pollutants. The role of social hierarchy is also important,

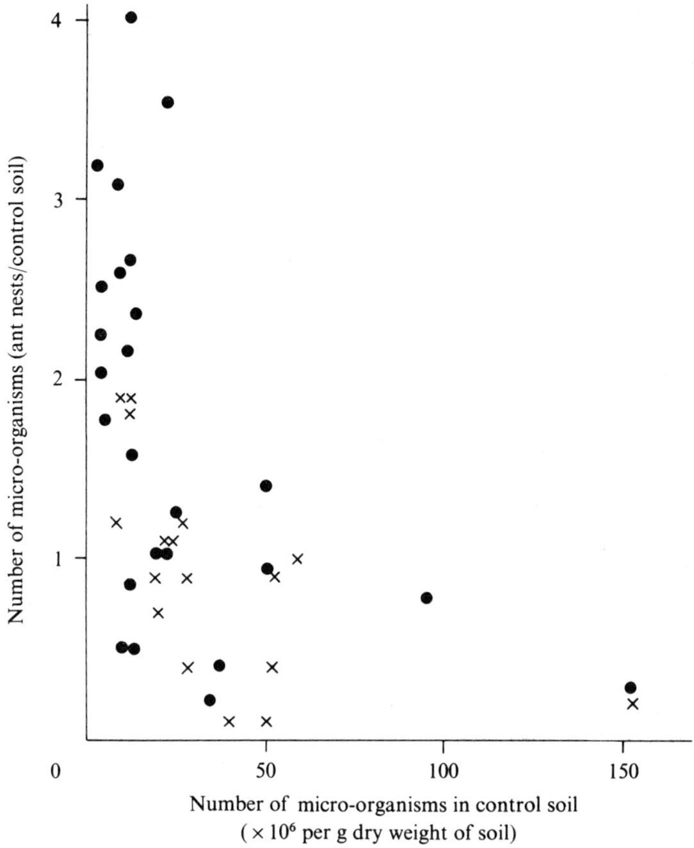

Fig. 10.12. Relationship between the number of micro-organisms utilising mineral nitrogen in the environment and the ratio in ant nests to control soil (redrawn from Pętal *et al.*, 1973). '●', *Lasius niger*; ' × ', *Myrmica* sp. ($P < 0.05$).

as the queen plays a significant role in the organisation of the colony processes. The ability to produce a number of substances modifying the development of other organisms, not met in other invertebrates, is probably also related to the social life. Hence, ants seem to be an interesting group for the study of adaptations to the greatly modified environments.

Comparison of various biomes

B. PISARSKI

Proportions of ant species in different biomes

Social insects are represented by a relatively small number of species (less than 10000 taxa), which is only about 1% of all insect species; due to their common occurrence and great numbers they are, nevertheless, one of the predominating groups in the majority of habitats. Taking into account the role of ants in so many biocoenoses, astonishingly few data exist on their communities in particular biotopes and biomes, particularly on their numbers and role in energy transformations. The available information can only be used for the illustration of some structures or phenomena but not for the comprehensive analysis of the problems discussed.

Ants, as a group of tropical origin, are represented by many species in the biomes of the equatorial zone. In the biomes located north and south of the equatorial zone the number of ant species declines with decreasing temperature of successive biomes. This decrease is particularly pronounced where mountain chains occur in an area, such as the Alps and the Carpathians in Europe or the Himalayas in Asia. The differences among the habitats and ecological niches of ants also have an effect on the number of ant species. For instance, many more species occur in forests (provided they are open to sunshine) than in steppes or deserts of the same latitude.

While the number of species decreases towards the north, the density of ant populations is maintained at a very high level in the majority of natural biocoenoses, including even those at the edge of tundra. The percentage contribution of ants to the fauna of many habitats of the temperate zone is slightly lower than in the tropics.

Ants in tropical rainforests

The Myrmecofauna of tropical rainforests is characterised by a huge abundance of species. For instance, about 700 species were found in the Congo river basin, but unfortunately there are no data on their density and biomass (Brown, 1973). Army ants and a large number of soil-living and arboreal forms are characteristic of biomes of this type. Among them there are many exclusive predators, including comparatively large numbers of species with pronounced food preferences; also, trophobiotic species are numerous. Some species form very large colonies. The biome of the rainforests of southern America has a distinct and very characteristic myrmecofauna. The most typical representatives living in this biome are army ants of the sub-family Ecitoninae, and fungus-growing ants of the tribe Attini, as well as arboreal ants of the genus *Azteca*.

The biomes of the rainforests of Africa and Asia are very similar to each other. Of course, there are significant differences in the myrmecofauna of these biomes but the majority of higher taxonomic units is common to both. The characteristic species of the rainforests of Africa and Asia are army ants of the sub-family Dorylinae and arboreal ants of the genera *Oecophylla* F. Sm. and *Polyrhachis* F. Sm. Characteristic genera of the rainforests of Africa are *Anomma* Fabr. and *Dorylus* Shuckh, both of the sub-family Dorylinae. Those characteristic of Asia are arboreal ants of the genus *Echinopla* F. Sm.

Ants of the steppe and desert

The myrmecofauna of the biomes of steppes and deserts is of a similar character. It is considerably less abundant that in the rainforests and the number of ant species in particular biomes does not exceed 100. In steppes and deserts, epigeic species predominate, which feed on the seeds of grasses, prey upon other animals and store 'honey' for the periods of summer drought. All species construct nests of earth, reaching the considerable depth of 3–5 m. Although the number of ant species occurring in these biomes is comparatively low, they have a dominant position due to their abundance. For instance, in the western Sahara, ants contribute 75% of the whole fauna, and in irrigated areas this may exceed 80% (Bernard, 1972).

Characteristic of the desert and steppe biomes of America are ants of the genera *Pogonomyrmex* Mayr and *Veromessor* For. feeding on seeds; also, ants of the genus *Myrmecocystus* Wasm. which store 'honey' for the periods of drought are characteristic.

Typical of the deserts and steppes of the Old World are granivorous ants of the genus *Messor* For. and predatory ants of the genera *Cataglyphis* Foerst. and *Proformica* Ruzs. The last genus also frequently stores 'honey'. The number of species in biomes of this type markedly decreases towards the north. For instance, in the western Sahara there are about 100 ant species, in the Turanian deserts about 50 and in the Gobi less than 20 species.

A rather considerable proportion of the myrmecofauna of the deserts and steppes of Australia is composed of primitive ants of the sub-families Myrmeciinae and Ponerinae. Characteristic of these biomes are large, predatory ants of the genus *Myrmecia* Fabr., granivorous ants of the genus *Meranoplus* F. Sm. and 'honey'-storing ants of the genus *Melophorus* Lub.

Ants of evergreen shrub

The biome of evergreen bushes forms a very diversified habitat, inhabited by ants in abundance. For instance, more than 100 ant species live in the European part of this biome. It has many features in common with the adjacent biomes of deserts or steppes and is inhabited by the representatives of the genera *Messor, Cataglyphis* and *Proformica*, but also the arboreal species of the genera *Crematogaster* Lund. and *Dolichoderus* Lund. are abundant. In addition, endemic genera such as *Gonioma* E. or *Oxyopomyrmex* And. are characteristic of this biome.

Ants of deciduous forest and taiga

Deciduous forest and taiga biomes occur almost exclusively in the northern hemisphere. Taiga is one of the most uniform habitats, with the result that its myrmecofauna is also uniform. There are considerable differences in the species composition of fauna and flora between the biomes of deciduous forests of different geographical regions. These differences also concern their myrmecofauna, which is not rich in species but it is frequently characterised by very high densities so that ants represent one of the predominant groups. These biomes are dominated by epigeic and arboreal species. Exclusively predatory species are rare, omnivores being dominant. The food of the latter is largely composed of the honeydew of aphids, and some species feed almost exclusively on honeydew. Many species construct nests with mounds of earth or organic remains. In the nests of this type ants can maintain an optimum temperature, usually higher than the ambient temperature, due to the appropriate slope of the walls of the mound and a system of aeration. Characteristic of both these types of biomes are ants of the genera *Myrmica* Latr., *Lasius* Fabr., *Formica* L. and of the sub-genus *Camponotus* s.s. The differences in the myrmecofauna between the biomes of deciduous forests and taiga are readily visible; but consist largely in the considerably lower number of genera and species in the taiga (about 30) than in the deciduous forests (in Europe, for instance, there are 80 species).

The myrmecofauna of the tundra biome is very sparse. Only two species have been recorded and they occur rarely (*Formica gagatoides* and less characteristically *Leptothorax acervorum*).

The importance of ants in different ecosystems

As has already been noted, social insects play an important role in almost all ecosystems. Studies of their numbers are, however, not numerous; moreover, most of them are conducted by different and non-comparable methods. The numbers of particular species in the ecosystem are frequently determined only on the basis of the number of nests, without considering the number of individuals in them. This often results in a false picture because the species with a large number of small colonies are given undue prominence. There are, for instance, some species of the genus *Myrmica* whose role in the ecosystem may not be as important as the role of the species living in large colonies, e.g. ants from the sub-genus *Formica.* The second, less frequently used, method is a quick trap method or pitfall traps. Both these methods are also selective because only foraging workers are trapped, and the proportion of these varies widely depending on the season as well as the species. The individuals which stay permanently inside the nest are not recorded. Data on the total number of ants and other animals living in the ecosystem are very scarce and they have only been collected in grassland ecosystems, similar to those concerned with the productivity of ants. With such limited data, which are frequently not comparable, it is possible to analyse the role of ants only in some selected ecosystems.

Ants occur in the tundra ecosystems and it is probable that their role is important but data are too scarce to be analysed.

In the coniferous forests of Siberia, eastern Europe, central Europe and mountain regions of Europe the species composition of the myrmecofauna is rather poor but it plays an important role. Generally, the density of ants, which are polyphagous, is high in these relatively poor ecosystems. They mainly feed on the honeydew of aphids but during the periods of the mass occurrence of other species they become zoophagous and significantly reduce these populations. In these ecosystems eurytopic species, such as *Lasius niger*, *Tetramorium caespitum*, *Myrmica laevinodis* and other *Myrmica* species generally predominate. But the greatest biotic impact on these ecosystems is due to the ants of the sub-genus *Formica* s.s. because of the large numbers of individuals in their colonies (see Appendix 1: Table 2).

Climatic and habitat conditions in the majority of deciduous forests of the temperate zone are similar. These forests and the fauna associated with them have evolved independently in different geographical regions. As a result, their fauna is composed of similar biotic elements but there are large differences in the species composition among particular geographical regions. The myrmecofauna of broadleaved forests is in general rather abundant and diversified, like the density, but in some types e.g. in humid and shaded mountain forests of *Fagetum carpaticum*, ants are very rare and their role is insignificant (Parapura & Pisarski, 1971).

Ecosystems rich in myrmecofauna but little studied are the forest–steppes and steppes of Asia (Nefedov, 1930; Yasuno, 1964*a*, *b*; Žigulskaja, 1968).

Ants in grassland

The best-known ecosystems are the grasslands of North America and Europe. Both the number of ant species and their density in these ecosystems are very different. They depend partly on the habitat conditions and partly on the intensity of management. The highest number of species has been found in dry and uncultivated grassland (Wengris, 1948; Kannowski, 1956; Parapura & Pisarski, 1971; Gallé, 1972*a*). An increase in moisture is followed by a decrease in the number of ant species. Agricultural treatments, such as heavy grazing, fertilisation and mowing, ploughing and reseeding, have similar results.

The number and biomass of ants in grassland ecosystems has also been estimated in relation to the numbers of other invertebrates. Although the whole fauna was not always studied, these investigations provide a picture of the relationships at least. The proportion of ants to other invertebrates in particular grasslands is characterised by a significant variability. This variability is primarily related to the origin of a particular ecosystem, and next to the habitat conditions and management. In the ecosystems of North America or Africa, which are located in the zone of natural grasslands, the predominant species are very well adapted to such environments. They are generally herbivorous or rather granivorous; predatory ants are significantly less abundant. Ants are frequently the predominant group in these environments because they are both primary and secondary consumers and in addition, during the periods of abundant food supply they store food for the remaining parts of the year.

The European grasslands are usually located in forest habitats and for this reason their myrmecofauna is mainly composed of the eurytopic species of forest origin. These are mainly omnivorous species feeding on other small invertebrates and honeydew. As they are secondary consumers, their contribution to the total number of invertebrates is proportionally lower.

In the grasslands of North America the biomass of foraging ants contribute from 1 to 15% of the total biomass of all epigeic invertebrates (Appendix 1: Table 3; Levis, 1971), while in the soil fauna ants represent 17% of the number of individuals of all invertebrates. Simultaneously, both their numbers and percentage contribution in the irrigated areas are higher than in the natural habitats, while in the cultivated areas they do not occur at all (Willard, 1973).

In European grasslands the whole invertebrate fauna has been studied (Breymeyer, 1971). These studies indicate that in the uncultivated meadows, ant biomass represents 1.02–1.33% of the biomass of all invertebrates, while in the cultivated areas their contribution is considerably lower and ranges from 0.11–0.81% (Appendix 1: Table 3). The consump-

Table 10.10. *Density, biomass and consumption of ants and spiders* (*Kajak* et al., *1971*)

Animal groups	Density (individuals per m²)	Biomass (mg per m²)	Consumption (mg per m² per year)
Ants	142.0	71.0	13400
Epigeic spiders	45.3	175.0	2630
Field layer spiders	52.0	78.3	2482

tion of ants in this area is very high, it ranges from 1.160 to 13.400 mg per m^2 per year. The consumption by ants in relation to their body weight is very high. This is readily seen when compared to the biomass and consumption of other predatory groups, e.g. spiders (Table 10.10).

In the grasslands of Africa the proportion of ants in the invertebrate fauna is of the same order or even higher than in the grasslands of North America. The percentage contribution of ants to the invertebrate fauna increases with an increase in the dryness of the habitat. This is particularly the case with the harvester ants of the genera *Messor* and *Monomorium*.

In the savannah of West Africa, where all the invertebrate fauna was investigated, ants represent 27% of individuals and 4% of biomass (Lamotte, 1947). In desert ecosystems of the Great Eastern Erg, where all the epigeic fauna was studied, foragers, despite their low density (0.1 workers per m^2), were 2.6 times more abundant than other epigeic animals (Bernard, 1972). But in the richer desert steppes of the Great Western Erg, covered with alfalfa, ants were more abundant. The number of foragers reached 135 individuals per m^2 and they were 10–420 times more numerous than other invertebrates (Bernard, 1972).

There is not much information on the myrmecofauna of crop fields, which are unfavourable environments for ants because of the treatments applied and, in particular, because of the frequent breaking up of the soil, which destroys the ant nests. In the crop fields of central Europe only eight ant species occur. These are all eurytopic species, *Lasius niger* being most frequently recorded (Appendix 1: Table 2). In the crops under study (rye and potato) ants contributed 3.9% of the number of soil insects. In perennial crops the myrmecofauna is more diversified and abundant, particularly in the zone of the steppe biomes, where the majority of ant species is well adapted.

Appendix 1

Table 1. *Adult populations in ant colonies*
by C. Baroni-Urbani

Table 2. *Density of ant colonies and individuals in various regions and biotopes*
by C. Baroni-Urbani and B. Pisarski

Table 3. *The contribution of ants to the invertebrate fauna of different ecosystems*
by B. Pisarski

Appendix 1

Appendix Table 1. *Adult populations in ant colonies*

Species	Locality	Number of adults per colony	Author
Ponerinae			
Proceratium silaceum	North America	9–60, mean = 28	Kennedy & Talbot, 1939
Apomyrma stygia	Central Africa	13–92	Brown *et al.*, 1970
Amblyopone pallipes	Canada	4–35, mean = 12	Francoeur, 1965
Rhytidoponera convexa	Australia	63–274	Whelden, 1957, 1958
Bothroponera silvestrii	Central Africa	< 61, mean = 35	Lévieux, 1967*a*
Mesoponera ambigua	Central Africa	< 101, mean = 70	Lévieux, 1967*a*
Mesoponera caffraria	Central Africa	< 121, mean = 70	Lévieux, 1967*a*
Megaponera foetens	Central Africa	< 701, mean = 500	Lévieux, 1967*a*
Hypoponera sp. near *coeca*	Central Africa	< 41, mean = 20	Lévieux, 1967*a*
Odontomachus assiniensis	Central Africa	250–300	Ledoux, 1952
Odontomachus troglodytes	Central Africa	< 501, mean = 350	Lévieux, 1967*a*
Paltothyreus tarsatus	Central Africa	< 451, mean = 350	Lévieux, 1967*a*
Heteroponera monticola	South America	30–45	Kempf & Brown, 1970
Dorylinae			
Anomma nigricans	Central Africa	1036800	Vosseler, 1905
Anomma wilverthi	Central Africa	$1.5\times10^7-2.2\times10^7$	Raigner & van Boven, 1955
Aenictus laeviceps	Philippines	60000–110000	Schneirla & Reves, 1966
Ecitoninae			
Neivamyrmex nigrescens	USA	80000–140000	Schneirla, 1958
Eciton hamatum	South America	$1.5\times10^5-2.5\times10^5$	Schneirla, 1957*a*
E. hamatum	South America	$1\times10^5-5\times10^5$	Rettenmeyer, 1963
Eciton burchelli	Central America	$3\times10^5-1.5\times10^6$	Schneirla, 1957*a*
E. burchelli	Central America	$1.5\times10^5-7\times10^5$	Rettenmeyer, 1963
Myrmicinae			
Myrmica brevinodis	USA	282–838	Kannowski, 1970
Myrmica fracticornis	USA	72–631	Kannowski, 1970
Myrmica punctiventris	USA	17–69	Kannowski, 1970
Myrmica emeryana	USA	35–561, mean = 255	Talbot, 1945
M. emeryana	USA	54–558	Kannowski, 1970
Myrmica sabuleti	Europe	106–6389	Brian, 1972
Myrmica ruginodis	Europe	305–2855, mean = 1216	Brian, 1950
M. ruginodis	Europe	351–3991	Brian, 1972
M. ruginodis	Europe	579±77	Stradling, 1970
Myrmica laevinodis (rubra)	Europe	954±144	Stradling, 1970
M. laevinodis (rubra)	Europe	1140–3216	Pętal, 1972
Myrmica rubra	Europe	1020 workers + 159 queens	Elmes, 1973
Pogonomyrmex marcusi	South America	162–389	Marcus & Marcus, 1951
Pogonomyrmex californicus	USA	3960–5263	Erickson, 1972
Pogonomyrmex occidentalis	USA	254–7506	Lavigne, 1969
P. occidentalis	USA	1548–4443, mean = 1155	Rogers *et al.*, 1972
Aphaenogaster treatae	USA	66–1662, mean = 682	Talbot, 1954
Aphaenogaster rudis	USA	11–950, mean = 280	Headley, 1949
A. rudis	USA	27–2083, mean = 326	Talbot, 1951
Stenamma impar	USA	5–109	Smith, 1957
Stenamma meridionale	USA	11–19	Talbot, 1957*b*
Stenamma diecki	USA	13–30	Cole, 1950
S. diecki	Canada	5–103, mean = 41	Francoeur, 1965
S. diecki	Canada	26–139, mean = 53	Letendre & Pilon, 1972
Cremastogaster dohrni	India	56947	Ayyar, 1937
C. dohrni	India	5690	Roonwal, 1954
Solenopsis invicta	USA	7620–48200	Wilson, N. L. *et al.*, 1971
Monomorium pharaonis	England	35–1400	Peacock, Waterhouse & Baxter, 1955

Appendix Table 1 (*cont.*)

Species	Locality	Number of adults per colony	Author
Tetramorium caespitum	England 1963	14068±1247	Brian *et al.*, 1967
T. caespitum	England 1964	7881±1150	Brian *et al.*, 1967
Leptothorax nylanderi	France	100–467	Plateaux, 1970
Leptothorax ambiguus	USA	6–106, mean = 37	Talbot, 1965
Leptothorax curvispinosus	USA	8–368, mean = 84	Headley, 1943
L. curvispinosus	USA	Mean = 62	Talbot, 1957*a*
Leptothorax longispinosus	USA	2–142, mean = 47	Headley, 1943
L. longispinosus	Canada	34–176, mean = 85	Letendre & Pilon, 1972
Leptothorax acervorum	Europe	21–804	Büschinger, 1966
L. acervorum	Europe	55–298	Büschinger, 1965
L. acervorum	Europe	28–223	Büschinger, 1971
Leptothorax muscorum	Europe	20–370	Büschinger, 1966
L. muscorum	Canada	66–109	Béique & Francoeur, 1968
Leptothorax kutteri	Europe	1–6	Büschinger, 1971
Leptothorax duloticus	USA	4–20, mean = 12	Talbot, 1957*a*
Epimyrma stumperi	Europe	3–42	Büschinger, 1971
Harpagoxenus sublaevis	Europe	4–96	Büschinger, 1966
Harpagoxenus americanus	USA	3–9	Talbot, 1957*b*
Smithistruma rostrata	USA	45–98	Talbot, 1957*b*
Sericomyrmex urichi	South America	200–1691	Weber, 1967*a*
Atta colombica	Central America	$1\times10^6-2.5\times10^6$	Martin *et al.*, 1967
Myrmeciinae			
Myrmecia gulosa	Australia	188–1586	Haskins & Haskins, 1950*a*
Myrmecia pilosula	Australia	553–862	Haskins & Haskins, 1950*a*
Myrmecia vindex	Australia	109–272	Haskins & Haskins, 1950*a*
Dolichoderinae			
Hypoclinea pustulata	USA (summer)	8–796, mean = 176	Kannowski, 1967
H. pustulata	USA (winter)	54–757, mean = 334	Kannowski, 1967
Hypoclinea plagiata	USA	28–378, mean = 115	Kannowski, 1967
Dorymyrmex emmaericaellus	South America	143–1000	Marcus & Marcus, 1951
Formicinae			
Aporomyrmex ampeloni	Europe	15–513	Faber, 1969
Camponotus maculatus	Central Africa	< 1801, mean = 1400	Lévieux, 1967*a*
Camponotus congolensis	Central Africa	< 1201, mean = 900	Lévieux, 1967*a*
Camponotus acvapimensis	Central Africa	< 1501, mean = 1200	Lévieux, 1967*a*
Camponotus compressiscapus	Central Africa	< 1501, mean = 1200	Lévieux, 1967*a*
Camponotus pennsylvanicus	USA	1943–2500	Pricer, 1908
Camponotus inaequalis	Central America	48–76	Smith, 1942
Camponotus noveboracensis	USA	677–818	Smith, 1942
Camponotus santosi	Central America	103–364	Smith, 1942
Polyrhachis debilis	New Guinea	300–350, mean = 325	Wilson, 1959
Polyrhachis viscosa	Central Africa	< 901, mean = 600	Lévieux, 1967*a*
Prenolepis imparis	USA	48–2208, mean = 1582	Talbot, 1943
Lasius niger	Europe	5462±812	Stradling, 1970
Lasius alienus	Europe	9700–18000	Nielsen, 1972*b*
Lasius flavus	England	2000–10000	Odum & Pontin, 1961
Formica incerta	USA	222–2350, mean = 855	Talbot, 1948
Formica pallidefulva	USA	106–1667, mean = 713	Talbot, 1948
Formica exsectoides	USA	41366–238510	Cory & Haviland, 1938
Formica subnitens	Canada	20118	Ayre, 1957
Formica rufa	Europe	$2\times10^4-8.94\times10^6$	Gösswald, 1951*b*

Appendix 1 Table 2. *Density of colonies and individuals in various regions and biotopes*

Locality	Biotope	Colonies per 100 m²	Workers per m²	No. of species	Author
Finland	< 50% lichen-covered rocks, south	3	—	7	Oinonen, 1956
	Idem, north slopes	5	—	7	
	> 50% lichen-covered rocks, south	7.5	—	17	
	Idem, north slopes	10.6	—	17	
	< 50% moss-covered rocks, south	12.3	—	16	
	Idem, north slopes	5.6	—	12	
	Heathy woodland	1–84	—	8	
	Wet pine peat-moors	24–96	—	15	
Scotland	Acid grassland	5–11	—	4	Brian, 1956*b*
England	Lowland heath	1.75±0.35	336±34	*Tetramorium caespitum*	Brian *et al.*, 1965
	Lowland heath	—	430±106	*Lasius alienus*	
	Grassland	60	—	*Lasius flavus*	Waloff & Blackith, 1962
	Grassland	—	420–1130	*L. flavus*	Odum & Pontin, 1961
Russia	Forest steppe, chernozem soil	41.5	—	24	Nefedov, 1930
	Forest steppe, alkaline soil	34	—	24	
	Forest steppe, saline soil	25.7	—	24	
Holland	Sandy soil	0–24	—	10	Quispel, 1941
	Oak stands	3–44	—	11	
	Planted pine stands	0–16	—	3	
	Dry oak–birch wood, sandy soil	17–82	—	16	Westhoff & Westhoff-DeJoncheere, 1942
	Idem, on loamy soil	2–4	—	2	
	Moist oak–birch wood	24–48	—	10	
	Inland birch wood	35–39	—	13	
	Dune birch wood	21–26	—	14	
	Dune oak wood	3–17	—	7	
	Inland sand dunes with pines	5–21	—	9	

	Dry pine afforestation little bottom growth	3–7	—	3	
	Idem, rich vegetation bottom growth	7–17	—	4	
	Moist pine afforestation	8–38	—	8	
	Pine–oak forest	16–43	—	11	
	Fir-trees afforestation without bottom growth	0–2	—	2	
	Idem, with vegetation bottom growth	3–10	—	2	
	Larch afforestation	1–19	—	5	
	American oak afforestation	3–11	—	7	
	Beech without bottom growth	0	—	0	
	Idem, with bottom growth	0–2	—	3	
	Natural beech wood	0	—	0	
	Moist oak–ash wood	0–8	—	8	
	Alder woods	0–10	—	2	
Belgium	*Pinus* forest	27	—	16	Gaspar, 1971
	Oak forest	15	—	17	
	Grassland	113	—	9–10	
Germany	Main region	4.7	—	46	Gösswald, 1932
	General	40	—	*Formica rufa*	Gösswald, 1957*b*
	Cloppenburg	7	—	*F. rufa*	Bruns, 1960
	Oak wood	6–8	—	*F. rufa*	Otto, 1958*c*
Alps	Primary forest	1–2	—	*F. rufa*	Adeli, 1962
	Plantations	6–7	—	*F. rufa*	
Siberia	Forest	1.8	4237	*F. rufa*	Marikovsky, 1962
Eastern Siberia	Steppe	2–143	1–370	30	Žigulskaja, 1968
Central Italy	Macchia glades	64.9	—	10	Baroni-Urbani, 1968
	Oak macchia	114.9	—	12	
	Old field	71.4	—	6	
	Alkaline meadow	47.6	—	7	
	Coastal dunes	113.7	—	6	
Apennines	*Brachypodium* grassland	45.4	—	5	Baroni-Urbani, 1969
	Festuca grassland	65.2	—	4	

Appendix 1 Table 2 (*cont.*)

Locality	Biotope	Colonies per 100 m²	Workers per m²	No. of species	Author
Germany	Black Forest 1966	0.159	—	Five of the *Formica rufa* group	Klimetzek, 1970
	Black Forest 1969	0.126	—		Klimetzek, 1972
	Different woods and years	0.08–0.165	—		Wellenstein, 1967
	Mixed wood	0.381	—	*F. rufa*	Eckstein, 1937
Denmark	Sandy heath	—	4590±212	*Lasius alienus*	Nielsen, 1974*b*
		—	7470±940	*Tetramorium caespitum*	
		—	1590±240	*Lasius niger*	
Poland	River banks	32	—	6	Parapura & Pisarski, 1971
	Pastures	36	—	10	
	Mixed beech forest	0	—	0	
	Mountain meadow	40	—	7	
	Lower meadow	40	—	26	
	Meadow	10	—	3	Pętal *et al.*, 1970
Eastern Europe	*Picea* forest	440	—	3	Wengris, 1948
	Mixed forest	1140	—	13	
	Pinus forest	1160	—	19	
	Wet meadow	5	—	5	
	Dry meadow	26	—	3	
Hungary	Flooded grassland, NW exposure 1969	210	—	6	Gallé, 1972*b*
	1970	220	—	6	
	SE exposure 1969	220	—	4	
	1970	320–660	—	6–9	
USSR, central	Kara-kum, sand without plants	0.26±0.15	—	*Cataglyphis pallida*	Dlussky, 1974
	sand with plants	0.955	—	5	
	Takyr (= clay with or without plants)	1.74	—	9	
	Well-shaded forest	30.95	—	11	Malozemova, 1972

USSR, Kazakhstan	Shaded forest	20.05	—	26	
	Cut forest	23.87	—	29	
	Moist forest	23.65	—	14	
	Wet forest	18.27	—	6	
Africa, Ivory Coast	Iron sands, savannah	55.1	—	13	Lévieux, 1967*a*
	Black soils, savannah	68.8	—	13	
	Hydromorphic soils, savannah	62.6	—	> 14	Lévieux, 1972
Africa, Sahara	Desert	0.06	—	12	Délye, 1968
		0.10	—	11	
Canada, Québec	Sugar maple stands	18–80	—	Four *Stenamma* spp.	Francoeur, 1966*a*
		56–228	—	11	Francoeur, 1966*b*
	Picea wood	132	—	6	Béique & Francoeur, 1968
	Mixed wood	250	—	21	Francoeur, 1965
USA, Michigan	Bogs	36	72.7	*Hypoclinea pustulata*	Kannowski, 1967
	Bogs	12	13.4	*H. plagiata*	
	Abandoned fields	5.7	75	*Aphaenogaster treatae*	Talbot, 1954
	Old field	113.2	—	16	Talbot, 1953
	Low field	1295	—	20	Talbot, 1965
USA, Missouri	Mixed wood	185	615.6	16	Talbot, 1957*b*
USA, Ohio	*Robinia* wood	1110	—	10	Headley, 1952
	Mixed wood	125	—	5	Headley, 1943
USA, Wyoming	Insular meadow	0.7–4.0	—	*Formica opaciventris*	Scherba, 1963
USA, Utah	General	1158	115852	*Pogonomyrmex occidentalis*	Weber, 1959
USA, California	Desert	15	—	*Veromessor pergandei*	Tevis, 1958
USA, Texas		21	—	*Pogonomyrmex barbatus*	Box, 1960
USA, S. Carolina	Sandy grassland	27	1215	*P. badius*	Golley & Gentry, 1964
USA, Colorado	Meadow, 1965	34	—	8	Conklin, 1972
	Moderately grazed area	0.31±0.18	—	*Pogonomyrmex occidentalis*	Rogers *et al.*, 1972
	Lightly grazed area	0.28±0.32	—		
	Ungrazed area	0.23±0.32	—		
	Heavily grazed area	0.03±0.08	—		
USA, Louisiana	Pasture	0.96	—	*Solenopsis invicta*	Baroni-Urbani & Kannowski, 1974

Appendix 1 Table 3. *The contribution of ants to the invertebrate fauna of different ecosystems*

Ecosystem type	Soil type	Primary productivity g per m² per year	Invertebrate numbers No. per m²	Ants: Numbers	Ants: % of total	Invertebrate biomass g per m² year	Ants: Biomass	Ants: % of total	Method used	Author and locality
Natural pasture										
mountain	Brown	570	—	90.7	—	5.41	0.072	1.3	—	Kajak, 1974
sheep fold	Pseudogley	900	—	31.5	—	22.46	0.025	0.1	—	Poland, Jaworki
Cultivated meadows										
five years old	Heavy, silty	564	—	132.9	—	8.32	0.067	0.8	—	Breymeyer, 1971
eight years old	alluvial soil	585	—	46.0	—	13.38	0.023	0.2	—	Poland, Kazun
Natural meadows	Black soil	196	—	142.0	—	6.91	0.071	1.0	—	Breymeyer, 1971 Poland, Strzeleckie Meadows
Grassland										
natural		—	—	120.0	—	—	—	—	Soil cores	Willard, 1973
irrigated		—	1060	188.0	17.2	—	—	—		Canada, Matador
True prairie										
ungrazed	Silty	—	—	—	—	0.126	0.0177	14.0	Trap	Levis, 1971
grazed		—	—	—	—	0.183	0.0073	—		USA, Osage
Mixed prairies										
ungrazed	Silty clay	—	—	—	—	0.363	0.0036	1.0	Trap	Levis, 1971
grazed		—	—	—	—	0.407	0.0081	2.0		Cottonwood, USA
Plains										
ungrazed	Silty clay	—	—	—	—	0.420	0.0013	3.0	Trap	Levis, 1971
grazed	loam	—	—	—	—	0.581	0.0023	4.0		USA, Pantex
Short-grass										
ungrazed	Fine sandy	—	—	—	—	0.401	0.0036	9.0	Trap	Levis, 1971
grazed	loam	—	—	19.9	—	0.323	0.0023	7.0		USA, Pawnee Rogers *et al.*, 1972
Desert grassland										
ungrazed	Sandy loam	—	—	347.8	—	0.044	0.0007	15.0	Trap	Levis, 1971
grazed		—	—	—	—	0.050	0.0006	11.0		USA, Jornada, Whitford, 1973
Savannah	Sand	—	900	250.0	27.0	25	1	4.0	—	Lamotte, 1947, Mont Guinée
Desert steppe	Sand	—	0.3–1.0	30–135	96.4–99.7	—	—	—	—	Bernard, 1972 Algeria, Great Western Erg
Crop fields, grain, potatoes	Loam	—	0.087[a]	0.0034	3.9	—	—	—	—	Honczarenko, 1964 Poland

[a] Insects.

Appendix 2. List of IBP/PT projects involving social insects (from IBP News No. 13)

Country	Title	Site	Leader
Australia (PT/55)	Soil fauna studies; role of termites, especially *Nasutitermes exitiosus*, in soil organic cycle in dry sclerophyll forest	—	K. E. Lee CSIRO Division of Soils, Adelaide, Australia
France (PT/5)	Productivity of a dry bush savannah; primary production of grassland areas; secondary production of small mammals, birds, ants and termites	Senegal (40 km south-east of Richard Toll)	F. Bourlière Faculté de Médecine, 45 rue des Saints-Pères, 75 Paris 6, France
(PT/6)	Productivity of a savannah area in Côte d'Ivoire, near to the border of the tropical forest; primary production of grass areas; secondary production of small mammals, birds, amphibia, reptiles, arthropods, earthworms and other soil organisms	Between Banda and Lamto, Côte d'Ivoire	M. Lamotte, Faculté des Sciences, Laboratoire de Zoologie, 24 rue Lhomond, 75 Paris 5, France
Poland (PT/15)	Energy flow through populations of *Myrmica laevinodis* (Hymenoptera)	Strzeleckie meadows in the Kampinos Forest	J. Pętal, Institute of Ecology, Nowy Swiat 72, Warsaw, Poland
(PT/16)	The effect of ants on the rate of soil mineralisation in various meadow habitats	Strzeleckie meadows	J. Pętal, as above
Romania (PT/3)	Studies on the zoocenoses of forest ecosystems. The project will include work on the following: Microfera, Thermytha, Insect pests, Formicidae, Othoptera	—	Mihai Ionescu, Min. Invatamint, Faculty of Biology, Cluj, Romania
South Africa (PT/2)	The energy cycle of a savannah ecosystem	Transvaal Bushveld	J. J. Van Wyk, Dept of Botany, Potchefstroom University for Christian Higher Education, Potchefstroom, Transvaal, South Africa
(PT/4)	Studies on the energy flow in natural sub-tropical, woodland-savannah communities in the eastern Transvaal	Timbavati Private Nature Reserve	The Director of Nature Conservation, Transvaal Provincial Administration, Private Bag 209, Pretoria, South Africa
UK (PT/9)	Productivity of southern heath ecosystems, including studies on nutrient cycles and major arthropod populations	Hartland Dorset	M. V. Brian, Institute of Terrestrial Ecology, Furzebrook Research Station, Wareham, Dorset, England
USA Desert Biome	Foraging activity of the leaf-cutter ant, *Acromyrmex versicolor* in relation to season, weather and colony condition	Santa Pinta	F. W. Werner, University of Arizona, Tucson, Arizona, USA

Appendix 2

Country	Title	Site	Leader
	Systems analysis of the presaharan ecosystem of southern Tunisia: ant population, biomass, activity, food, foraging and respiration	Southern Tunisia	R. J. Muir, H. Heatwole and E. Davison, Utah State University and Tunisian Ministry of Agriculture
USSR (PT/9)	Optimal density and optimal structure of animal populations	All Geographical zones of USSR	S. S. Schwarz, Institute of Plant and Animal Ecology, Ural Branch, USSR, Ac. Sci., Sverdlovsk

Bibliography

Abushama, F. T. (1967). The role of chemical stimuli in the feeding behaviour of termites. *Proceedings of the Royal Entomological Society, London, (A)*, **42**, 77–82.

Adamson, A. M. (1943). Termites and the fertility of soils. *Tropical Agriculture*, **20**(6), 107–12.

Adeli, E. (1962). Zur Öekologie der Ameisen im Gebiet des Urwaldes Rotwald (Niederösterreich). *Zeitschrift für angewandte Entomologie*, **49**, 290–6.

Adler, E. (1961). Über den Stand der Ligninforschung. *Papier*, **15**, 604–9.

Ahmad, M. (1950). The phenology of termite generalbased on imago-worker mandibles. *Bulletin of the American Museum of Natural History*, **95**, 43–86.

Alibert, J. (1963). Echanges trophallactiques chez un termite supérieur. Contamination par le phosphore radio-actif de la population d'un nid de *Cubitermes fungifaber. Insectes sociaux*, **10**, 1–12.

Alibert, J. (1964). L'évolution dans le temps des meules à champignons construites par les termites. *Comptes rendus hebdomadaires des séances de l'Académie des sciences, Paris*, **258**, 5260–3.

Alibert, J. (1968). Influence de la société et de l'individu sur la trophallaxie chez *Calotermes flavicollis* (Fabr.) et *Cubitermes fungifaber* (Isoptera). *Colloques Internationaux du Centre National de la Recherche Scientifique, Paris, 21–24 Novembre, 1967*, No. **173** (*L'effet de groupe chez les animaux*), 237–88.

Alibert-Bertho, J. (1969). La trophallaxie chez le termite à cou jaune, *Calotermes flavicollis* (Fabr.), étudiée à l'aide de radio-éléments. I. Participation des différentes castes à la trophallaxie 'globale'. Les catégories d'aliments transmis. *Annales des sciences naturelles*, (*Zoologie*), Sér. 12, **11**, 235–325.

Allen, T. C., Smythe, R. G. & Coppel, H. C. (1964). Response of twenty-one termite species to aqueous extracts of wood invaded by the fungus *Lenzites trabea* Pers. ex Fr. *Journal of Economic Entomology*, **57**, 1009–11.

Allison, M. J. (1969). Biosynthesis of amino acids by ruminal microorganisms. *Journal of Animal Science*, **29**, 797–807.

Amante, E. (1966). Saúva desfolha 745 milhoes de toneladas em un só ano. *Informação Agricola, Rio de Janeiro, Ministry of Agriculture*, Ano **11** (No. 3), 4.

Amante, E. (1967*a*). Prejuizos causados pela formiga saúva em plantaçoes de *Eucalyptus* e *Pinus* no estado de São Paulo. Silvicultura em São Paulo, *Revista Tecnica Servicio Flora*, **6** (26 September), 355–63.

Amante, E. (1967*b*). A formiga saúva *Atta capiguara*, praga das pastagens. *Biológico*, **33**, 113–20.

Ausat, A., Cheema, P. S., Koshi, T., Perti, S. L. & Ranganathan, S. K. (1962). Laboratory culturing of termites. In *Proceedings of the New Delhi symposium, 1960. Termites in the humid tropics*, pp. 121–5. Paris: UNESCO.

Autuori, M. (1940). Algumas obsevacoes sobre formigas cultivadoras de fungo (Hymenoptera: Formicidae). *Revista Entomologica, Rio de Janeiro*, **11**, 215–26.

Autuori, M. (1942). Contribução para o conhecimento da Saúva (*Atta* spp.). (II). O Saúveiro inicial (*Atta sexdens rubropilosa* Forel, 1908). *Archivos do Instituto biológico, São Paulo*, **13**, 67–86.

Autuori, M. (1950). Contribução para o conhecimento da Saúva (*Atta* spp.). (V). Numero de formas aladas e redução dos saúveiros iniciais. *Archivos do Instituto biológico, São Paulo*, **19**, 325–31.
Autuori, M. (1956). Contribução opara o conhecimento da Saúva (*Atta* spp.). (VI). Infestação residual da saúva. *Archivos do Instituto biológico, São Paulo*, **23**, 109–16.
Ayre, G. L. (1957). Ecological notes on *Formica subnitens* Creighton (Hymenoptera, Formicidae). *Insectes sociaux*, **4**, 173–6.
Ayre, G. L. (1958). Some meteorological factors affecting the foraging of *Formica subnitens* Creighton (Hymenoptera: Formicidae). *Insectes sociaux*, **5**, 147–57.
Ayre, G. L. (1959). Food habits of *Formica subnitens* Creighton (Hymenoptera: Formicidae) at Westbank, British Colombia, *Insectes sociaux*, **6**, 105–14.
Ayre, G. L. (1960). Der Einfluss von Insektennahrung auf das Wachstum von Waldameisenvolkern. *Die Naturwissenschaften*, **47** (21), 502–3.
Amante, E. (1972). Influência de Alguns Fatores Microclimáticos sobre a Formiga Saúva *Atta laevigata* (F. Smith, 1958), *Atta sexdens rubropilosa* Forel, 1908, *Atta bisphaerica* Forel, 1908 e *Atta capiguara* Gonçalves, 1944 (Hymenoptera, Formicidae). In *Formigueiros Localizados no Estado de São Paulo*. São Paulo. ii+175 pp., 27 figures, 55 tables, 5 maps, 10 charts.
Anderson, J. M. (1973). The breakdown and decomposition of sweet chestnut (*Castanea sativa* Mill) and beech (*Fagus silvatica* L.) leaf litter in two deciduous woodland soils: 1. Breakdown, leaching, and decomposition. *Oecologia*, **12**, 251–74.
Anderson, J. M. & Healey, I. N. (1972). Seasonal and inter-specific variation in major components of the gut contents of some woodland Collembola. *Journal of Animal Ecology*, **41**(2), 359–68.
Andrew, B. J. (1930). Method and rate of protozoan refaunation in the termite *Termopsis angusticollis* Hagen. *University of California Publications in Zoology*, **33**, 449–70.
Andrew, B. J. & Light, S. F. (1929). Natural and artificial production of so-called 'mitotic flares' in the intestinal flagellates of *Termopsis angusticollis*. *University of California Publications in Zoology*, **31**, 433–40.
Anwar, M. (1970). *Acanthaspis sulcipes* Signoret as predator of termites: a new record. *Entomologist*, **103**, 281–2.
Araujo, R. L. (1970). Termites of the neotropical region. In *Biology of termites*, ed. K. Krishna & F. M. Weesner, vol. 2, pp. 527–76. New York & London: Academic Press.
Arnold, G. (1926). A monograph of the Formicidae of South Africa. Appendix. *Annals of the South African Museum*, **23**, 191–295.
Arora, G. L. & Gilotra, S. K. (1960). The biology of *Odontotermes obesus* Rambur. *Panjab University Research Bulletin*, **10**, 247–55.
Auclair, J. L. (1963). Aphid feeding and nutrition. *Annual Review of Entomology*, **8**, 439–90.
Ayre, G. L. (1962). Problems in using the Lincoln index for estimating the size of ant colonies. *Journal of the New York Entomological Society*, **70**, 159–66.
Ayre, G. L. (1963). Feeding behaviour and digestion in *Camponotus herculeanus* (L.) (Hym. formicidae). *Entomologia experimentalis et applicata*, **6**, 165–70.
Ayre, G. L. (1966). Colony size and food consumption of three species of *Formica*. *Entomologia experimentalis et applicata*, **9**, 461–7.
Ayre, G. L. (1967). The relationships between food and digestive enzymes in five species of ants (Hymenoptera: Formicidae). *Canadian Entomologist*, **99**, 408–11.

Ayyar, P. N. K. (1937). A new carton-building species of ant in South India, *Crematogaster dohrni artifex* Mayr. *Journal of the Bombay Natural History Society*, **39**, 291–308.

Bailey, N. T. J. (1951). On estimating the size of mobile populations from recapture data. *Biometrika*, **38**, 293–306.

Balogh, J. & Loksa, I. (1956). Untersuchungen über die Zoozönose des Luzernenfeldes. *Acta zoologica Academiae scientiarum hungaricae*, **2**, 17–114.

Balzam, N. (1933). Recherches sur le métabolisme chimique et énergétique au cours du développement des insectes. II. Relation entre la chaleur dégagée et les échanges respiratoires au cours du développement post-embryonnaire des insectes. *Archives internationales de physiologie*, **37**, 317–28.

Banks, C. J. & Nixon, H. L. (1958). Effects of the ant *Lasius niger* (L.), on the feeding and excretion of the bean aphid, *Aphis fabae* Scop. *Journal of Experimental Biology*, **35**, 703–11.

Barbier, M. & Delage, B. (1967). Le contenu des glandes pharyngiennes de la fourmi *Messor capitatus* Latr. (Hymenoptera: Formicidae). *Comptes rendus hebdomadaires des séances de l'Académie des sciences, Paris*, **264**, 1520–2.

Baroni-Urbani, C. (1965). Sull'attività di foraggiamento notturna del *Camponotus nylanderi. Insectes sociaux*, **12**, 253–64.

Baroni-Urbani, C. (1968). Studi sulla mirmecofauna d'Italia. V. Aspettilecologici della riviera del M. Conero. *Bollettino di zoologia, pubblicato dall'Unione zoologica italiana*, **35**, 39–76.

Baroni-Urbani, C. (1969). Ant communities of the high-altitude Appenine grasslands. *Ecology*, **50**, 488–92.

Baroni-Urbani, C. (1972). An example of the usefulness of simulation techniques in ecological researches: the study of food consumption of a large ant community. *Ekologia polska*, **20**, 33–42.

Baroni-Urbani, C. (1973). Simultaneous mass recruitment in exotic ponerine ants. *Proceedings, VIIth International Congress of the International Union for the Study of Social Insects, London, 10–15 September, 1973*, 12–15.

Baroni-Urbani, C. & Kannowski, P. B. (1974). Patterns of the red imported fire ant settlement of a Louisiana pasture: some demographic parameters, interspecific competition and food sharing. *Environmental Entomology*, **3**, 755–60.

Barrer, P. M. & Cherrett, J. M. (1972). Some factors affecting the site and pattern of leaf-cutting activity in the ant *Atta cephalotes* (L.). *Journal of Entomology*, (*A*), **47**, 15–27.

Basalingappa, S. (1970*a*). The differential occurrence of fat in the various castes and undifferentiated instars of the termite *Odontotermes assmuthi*: the result of discriminate feeding. *Journal of Animal Morphology and Physiology*, **17**, 106–10.

Basalingappa, S. (1970*b*). Environmental hazards to reproductives of *Odontotermes assmuthi* Holmgren. *Indian Zoologist*, **1**, 45–50.

Batra, L. R. & Batra, S. W. T. (1966). Fungus-growing termites of tropical India and associated fungi. *Journal of the Kansas Entomological Society*, **39**, 725–38.

Bausenwein, F. (1960). Untersuchungen über sekretorische Drüsen der Kopf und Brust – abschnittes in der *Formica rufa* Gruppe. *Acta Societatis entomologicae Čechoslovenicae*, **57**, 31–57.

Beard, R. L. (1974). Termite biology and bait-block method of control. *The Connecticut Agricultural Experiment Station, New Haven, Bulletin*, **748**, 8.

Becker, G. (1961). Beobachtungen und Versuche über den Beginn der Kolonie-

Entwicklung von *Nasutitermes ephratae* Holmgren (*Isoptera*). *Zeitschrift für angewandte Entomologie*, **49**, 78–93.
Becker, G. (1962). Laboratoriumsprüfung von Holz und Holzschutzmitteln mit der sudasiatischen Termite *Heterotermes indicola* Wasmann. *Holz als Roh- und Werkstoff*, **20**, 476–86.
Becker, G. (1964). Termiten-anlockende Wirkung einiger bei Basidiomyceten-Angriff in Holzentstehender Verbindungen. *Holzforschung*, **18**, 168–72.
Becker, G. (1965*a*). Versuche über den Einfluss von Braunfäulepilzen auf Wahl und Ausnutzung der Holznahrung durch Termiten. *Material und Organismen*, **1**, 95–156.
Becker, G. (1965*b*). Feuchtigkeitseinfluss auf Nahrungswahl und -verbrauch einiger Termiten-Arten. *Insectes sociaux*, **12**, 151–84.
Becker, G. (1967). Die Temperature-Abhängigkeit der Frasstatigkeit einiger Termitenarten. *Zeitschrift für angewandte Entomologie*, **60**, 97–125.
Becker, G. (1969). Rearing of termites and testing methods used in the laboratory. In *Biology of termites*, ed. K. Krishna & F. M.Weesner, vol. 1, pp. 351–85. New York & London: Academic Press.
Becker, G. (1971). Physiological influences on wood-destroying insects of wood compounds and substances produced by microorganisms. *Wood Science and Technology*, **5**, 236–46.
Becker, G. & Kerner-Gang, W. (1964). Schädigung und Förderung von Termiten durch Schimmelpilze. *Zeitschrift für angewandte Entomologie*, **53**, 429–48.
Becker, G. & Seifert, K. (1962). Über die chemische Zusammensetzung des Nest- und Galeriematerials von Termiten. *Insectes sociaux*, **9**, 273–87.
Beckwith, T. D. & Rose, E. J. (1929). Cellulose digestion by organisms from the termite gut. *Proceedings of the Society for Experimental Biology and Medicine*, **27**, 4–6.
Behr, E. A., Behr, C. T. & Wilson, L. F. (1972). Influence of wood hardness on feeding by the eastern subterranean termite, *Reticulitermes flavipes* (Isoptera: Rhinotermitidae). *Annals of the Entomological Society of America*, **65**, 457–60.
Béique, R. & Francoeur, A. (1968). Les fourmis de la pessière à Cladonia II. Etude quantitative d'une pessière naturelle. *Revue d'écologie et de biologie du sol*, **5**, 523–31.
Benemann, J. R. (1973). Nitrogen fixation in termites. *Science*, **181**, 164–5.
Bennett, C. F. (1968). Human influences on the ecology of Panama. *Bulletin of the Ecological Society of America*, **49**, 58.
Bequaert, J. (1912). Notes biologiques sur quelques fourmis et termites du Congo Belge. *Revue zoologique africaine*, **2**, 396–431.
Bequaert, J. (1926). Medical and economic entomology. In *Medical report of the Hamilton Rice seventh expedition to Amazon, in conjunction with the department of tropical medicine of Harvard University 1924–25*, pp. 155–257. Cambridge, Massachusetts: Harvard University Press.
Berg, K. (1900). Termitariophile. *Communicaciones del Museo nacional de Buenos Aires*, **1**, 212–15.
Bernard, F. (1972). Premiers résultats de dénombrement de la faune par carrés en Afrique du Nord. *Buelletin de la Société d'histoire naturelle de l'Afrique du Nord, Alger*, **63**(1–2), 1–13.
Berthet, P. (1964). L'activité des Oribatides (Acari: Oribatei) d'une chengie. *Mémoires de l'Institute royale des sciences naturelles de Belgique*, **152**, 1–152.
Bialaszewicz, K. (1933). Recherches sur le métabolisme chimique et énergétique au cours du développement des insectes. I. Thermogénèse pendant la période

de croissance larvaire et pendant la métamorphose de '*Lymantria dispar*' L. *Archives internationales de physiologie*, **37**, 1–15.

Bigger, M. (1966). The biology and control of termites damaging field crops in Tanganyika. *Bulletin of Entomological Research*, **56**, 417–44.

Bitancourt, A. A. (1941). Expressão matematica do crescimento de formigueiros de *Atta sexdens rubropilosa* representado pelo aumento do numero de olheiros. *Archivos do Instituto biológica, São Paulo*, **12**(16), 229–36.

Bletchley, J. D. (1969). Seasonal differences in nitrogen content of Scots Pine (*Pinus sylvestris*) sapwood and their effects on the development of the larvae of the common furniture beetle (*Anobium punctatum* De G.). *Journal, Institute of Wood Science*, **4**, 43–7.

Blum, M. S. & Brand, J. M. (1972). Social insect pheromones: their chemistry and function. *American Zoologist*, **12**, 553–76.

Bodenheimer, F. S. (1937). Population problems of social insects. *Biological Review*, **12**, 393–430.

Bodine, M. C. & Ueckert, D. N. (1975). Effect of desert termites on herbage and litter in a shortgrass ecosystem in west Texas. *Journal of Range Management*, **28**, 353–8.

Bodot, P. (1961). La destruction des termitières de *Bellicositermes natalensis* par une fourmi: *Dorylus* (*Typlopone*) *dentifrons* Wasmann. *Comptes rendus hebdomadaires des séances de l'Académie des sciences, Paris*, **253**, 3053.

Bodot, P. (1966). Etudes écologiques et biologiques des termites des savanes de Basse Côte d'Ivoire. Doctoral thesis, University of Aix-Marseille, Aix-en-Provence, 197 pp.

Bodot, P. (1967*a*). Etudes écologiques des termites des savanes de Basse Côte d'Ivoire. *Insectes sociaux*, **14**, 229–58.

Bodot, P. (1967*b*). Cycles saisonniers d'activité collective des termites des savanes de Basse Côte d'Ivoire. *Insectes sociaux*, **14**, 359–88.

Bodot, P. (1969). Composition des colonies de termites: ses fluctuations au cours du temp. *Insectes sociaux*, **16**, 39–54.

Bohart, G. E. & Knowlton, G. F. (1953). Notes on food habits of the western harvester ant (Hymenoptera: Formicidae). *Proceedings of the Entomological Society of Washington*, **55**, 151–3.

Bonetto, A. A. (1959). *Las hormigas 'cortadoras' de la Provincia de Santa Fe, Argentina*, pp. 1–79.

Borchers, K., Bruns, H., Gösswald, K., Ohnesorge, P., Reiche, G. A., Schrader, A. & Schwerdtfeger, F. (1960). Erfahrungen mit der Kleinen Roten Waldameise (*Formica polyctena* bzw. *F. rufa*) bei der Bekämpfung von Forstchädlingen. *Aus dem Walde*, **4**, 1–99.

Borrett, P. P. & Wilson, K. J. (1971). Comparative feeding ecology of *Anthus novaeseelandiae* and *Anthus vaalensis* in Rhodesia. *Ostrich*, Supplement **8**, 333–41.

Bouillon, A. (1964). Etude de la composition des sociétés dans trois espèces d'*Apicotermes* Holmgren (Isoptera, Termitinae). In *Etudes sur les termites africains*, ed. A. Bouillon, pp. 181–96. Paris: Masson et Cie.

Bouillon, A. (1970). Termites of the Ethiopian region. In *Biology of termites*, ed. K. Krishna & F. M. Weesner, vol. 2, pp. 153–280. New York & London: Academic Press.

Bouillon, A. (1974). Durée de développement et saisons dans la sexualisation des Termitidae (Isoptera). *Revue zoologique africaine*, **88**, 65–80.

Bouillon, A. & Kidieri, S. (1964). Répartition des termitières de *Bellicositermes bellicosus rex* Grassé et Noirot dans l'Ubangi d'après les photos aériennes. Corrélations écologiques qu'elle révèle. In *Etudes sur les termites africaines*, ed. A. Bouillon, pp. 373–7. Paris: Masson et Cie.

Bouillon, A. & Lekie, R. (1964). Populations, rythme d'activité diurne et cycle de croissance du nid de *Cubitermes sankurensis* Wasmann (Isoptera, Termitinae). In *Etudes sur les termites africains*, ed. A. Bouillon, pp. 197–213. Paris: Masson et Cie.

Bouillon, A., Lekie, R. & Mathot, G. (1962). Etudes sur les termites africains. 1. Distribution spatiale et essai sur l'origine et la dispersion des espèces du genre *Apicotermes* (Termitinae). *Studia Universitatis Lovanium Faculté des Sciences*, **15**, 1–35.

Bouillon, A. & Mathot, G. (1964). Observations sur l'écologie et le nid de *Cubitermes exiguus* Mathot. Description de nymphes-soldats et d'un pseudimago. In *Etudes sur les termites africains*, ed. A. Bouillon, pp. 215–30. Paris: Masson et Cie.

van Boven, J. K. A. (1970). Le polymorphisme des ouvrières de *Megaponera foetens* Fabr. *Publicatiës van het Natuurhistorisch genootschap in Limburg*, **20**, 5–9.

Box, T. S. (1960). Notes on the harvester ant, *Pogonomyrmex barbatus* var. *molefacieus*, in south Texas. *Ecology*, **41**, 381–2.

Boyd, N. D. & Martin, M. M. (1975). Faecal proteinases of the fungus growing ant, *Atta texana:* their fungal origin and ecological significance. *Journal of Insect Physiology*, **21**(11), 1815–20.

Boyer, P. (1948). Sur les matériaux composant la termitière géante de *Bellicositermes rex. Comptes rendus hebdomadaires des séances de l'Académie des sciences, Paris*, **227**, 488–90.

Boyer, P. (1955). Première études pédologiques et bactériologiques des termitières. *Comptes rendus hebdomadaires des séances de l'Académie des sciences, Paris*, **240**, 569–71.

Boyer, P. (1959). De l'influence des termites de la zone intertropicale sur la configuration de certains sols. *Revue de géomorphologie dynamique*, **10**, 41–4.

Breymeyer, A. (ed.). (1971). Productivity investigation of two types of meadows in the Vistula valley. *Ekologia polska*, **19**, 93–201.

Breznak, J. A., Brill, W. J., Mertins, J. W. & Coppel, H. C. (1973). Nitrogen fixation in termites. *Nature, London*, **244**, 577–80.

Brian, M. V. (1950). The stable winter population structure in species of *Myrmica. Journal of Animal Ecology*, **19**, 119–23.

Brian, M. V. (1951). Summer population changes of the ant *Myrmica. Physiologia comparata et oecologia*, **2**, 248–62.

Brian, M. V. (1952). The structure of a dense natural ant population. *Journal of Animal Ecology*, **21**, 12–24.

Brian, M. V. (1953*a*). Oviposition by workers of the ant *Myrmica. Physiologia comparata et oecologia*, **3**, 25–36.

Brian, M. V. (1953*b*). Brood-rearing in relation to worker number in the ant *Myrmica. Physiological Zoology*, **26**, 355–66.

Brian, M. V. (1955). Food collection by a Scottish ant community. *Journal of Animal Ecology*, **24**, 336–51.

Brian, M. V. (1956*a*). Studies of caste differentiation in *Myrmica rubra* (L.). 4. Controlled larval nutrition. *Insectes sociaux*, **3**, 369–94.

Brian, M. V. (1956*b*). The natural density of *Myrmica rubra* and associated ants in west Scotland. *Insectes sociaux*, **3**, 473–87.

Brian, M. V. (1957*a*). The growth and development of colonies of the ant *Myrmica. Insectes sociaux*, **4**, 177–90.

Brian, M. V. (1957*b*). Serial organization of brood in *Myrmica. Insectes sociaux*, **4**, 191–210.

Brian, M. V. (1964). Ant distribution in a southern English heath. *Journal of Animal Ecology*, **33**(3), 451–61.

Brian, M. V. (1965). *Social insect populations*. New York & London: Academic Press. 135 pp.

Brian, M. V. (1969). Male production in the ant *Myrmica rubra* (L.). *Insectes sociaux*, **16**, 249–68.

Brian, M. V. (1970). Measuring population and energy flow in ants and termites. In *Methods of study in soil ecology. Proceedings of the Paris symposium*, ed. J. Phillipson, pp. 231–4. Paris: UNESCO.

Brian, M. V. (1971). Ants and termites. In *Methods of study in quantitative soil ecology: population, production and energy flow*, IBP Handbook No. 18, ed. J. Phillipson, pp. 247–61. Oxford: Blackwell Scientific Publications.

Brian, M. V. (1972). Population turnover in wild colonies of the ant *Myrmica. Ekologia polska*, **20**, 43–53.

Brian, M. V. (1973*a*). Feeding and growth in the ant *Myrmica. Journal of Animal Ecology*, **42**, 37–53.

Brian, M. V. (1973*b*). Temperature choice and its relevance to brood survival and caste determination in the ant *Myrmica rubra* (L.). *Physiological Zoology*, **46**, 245–52.

Brian, M. V. (1973*c*). Caste control through worker attack in the ant *Myrmica. Insectes sociaux*, **20**(2), 87–102.

Brian, M. V. & Brian, A. D. (1951). Insolation and ant population in the west of Scotland. *Transactions of the Royal Entomological Society of London*, **102**, 303–30.

Brian, M. V. & Elmes, G. (1974). Production by the ant *Tetramorium caespitum* in a southern English heath. *Journal of Animal Ecology*, **43**, 889–903.

Brian, M. V., Elmes, G. & Kelly, A. F. (1967). Populations of the ant *Tetramorium caespitum* Latreille. *Journal of Animal Ecology*, **36**, 337–42.

Brian, M. V. & Hibble, J. (1964). Studies of caste differentiation in *Myrmica rubra* (L.). 7. Caste bias, queen age and influence. *Insectes sociaux*, **11**, 223–38.

Brian, M. V., Hibble, J. & Stradling, D. J. (1965). Ant pattern and density in a southern English heath. *Journal of Animal Ecology*, **34**, 545–55.

Brian, M. V., Hibble, J. & Kelly, A. F. (1966). The dispersion of ant species in a southern English heath. *Journal of Animal Ecology*, **35**, 281–90.

Brian, M. V. & Pętal, J. (eds.). (1972). *Productivity investigations on social insects and their role in the ecosystems. Ekologia polska*, **20**, 184 pp.

Brody, S. (1945). *Bioenergetics and growth, with special reference to the efficiency complex in domestic animals*. New York: Wiley.

Brown, B. & Smith, R. (1954). The relationship of the concentration of acetic acid in normal and defaunated termites. *Mendelian*, **25**, 19–20.

Brown, H. D. (ed.). (1969). *Biochemical calorimetry*. New York: Academic Press.

Brown, W. L., Jr (1958). Contributions toward a reclassification of the Formicidae. II. Tribe Ectatommini. *Bulletin of the Museum of Comparative Zoology*, **118**, 175–362.

Brown, W. L., Jr (1959). A revision of the Dacetine ant genus *Neostruma. Breviora, Museum of Comparative Zoology*, **107**, 1–13.

Brown, W. L., Jr (1965). Contribution to a reclassification of the Formicidae. IV. Tribe Typhlomyrmecini. *Psyche*, **72**, 65–78.

Brown, W. L., Jr (1973). A comparison of the hylaean and Congo–West African rain forest ant Faunas. In *Tropical forest ecosystems in Africa and South America: a comparative review*, ed. B. J. Meggers, E. S. Ayensu & W. D. Duckworth, pp. 161–85. Washington D.C.: Smithsonian Institute Press.

Brown, W. L., Jr & Kempf, W. W. (1969). A revision of the neotropical Dacetine ant genus *Acanthognathus*. *Psyche*, **76**, 87–109.

Brown, W. L., Jr, Gotwald, W. H., Jr & Lévieux, J. (1970). A new genus of ponerine ants from West Africa (Hymoptera: Formicidae) with ecological notes. *Psyche*, **77**(3), 259–75.

Browning, B. L. (1952). Introduction and miscellaneous determinations. In *Wood chemistry*, ed. L. E. Wise & E. C. Jahn, vol. **2**, 1119–37. New York: Reinhold Publishing Corporation.

Bruniquel, S. (1972). La ponte de la fourmi *Aphoenogaster subterranea* (Latreille): oeufs reproducteurs – oeufs alimentaires. *Comptes rendus hebdomadaires des séances de l'Académie des sciences, Paris*, **275**(3), 397–9.

Bruns, H. (1960). Die künstliche Ansiedling und Entwicklung von Kolonien der Roten Waldameise (*Formica polyctena* bzw. *rufa*) in dem Cloppenburger Schadgebiet der Kl. Fichtenblattwespe (*Pristiphora abietina*) 1952–59. *Aus dem Walde*, **4**, 26–72.

de Bruyn, G. J., Goosen-De Roo, L., Hubregtse-Van den Berg, A. I. M. & Feijen, H. R., (1972). Predation of ants by woodpeckers. *Ekologia polska*, **20**, 83–91.

de Bruyn, G. J. & Kruk-de Bruin, M. (1972). The diurnal rhythm in a population of *Formica polyctena* Först. In *Productivity investigations on social insects and their role in the ecosystems*, eds. M. V. Brian & J. Pętal, pp. 117–27. Warsaw: Polish Academy of Sciences.

de Bruyn, G. J. & Mabelis, A. A., (1972). Predation and aggression as possible regulatory mechanisms in *Formica*. *Ekologia polska*, **20**, 93–101.

Büchel, K. H. & Korte, F. (1960). Zur chemie der Iridolactone. *IX Internationaler Kongress für Entomologie, Wien*, **3**(3), 60–5.

Bucher, E. H. & Zuccardi, R. B. (1967). Significacion de los hormigueros de *Atta vollenweideri* Forel como alteradores del suelo en la provincia de Tucuman. *Acta zoologica lilloana*, **23**, 83–96.

Buchli, H. (1950). Recherche sur la fondation et le développement des nouvelles colonies chez le termite lucifuge (*Reticulitermes lucifugus* (Rossi)). *Physiologia comparata et oecologia*, **2**, 145–60.

Buchli, H. (1956). Die Neotenie bei *Reticulitermes*. *Insectes sociaux*, **3**, 131–43.

Buchli, H. (1958). L'origine des castes et les potentialités ontogéniques des termites européens du genre *Reticulitermes* Holmgren. *Annales des sciences naturelles* (*Zoologie*), Sér. 11, **20**, 263–429.

Buchli, H. (1960). Le premier accouplement at la fécondité de la jeune reine imaginale chez *Reticulitermes lucifugus santonensis* Feyt. *Vie et milieu*, **11**, 494–9.

Buchli, H. (1961). Les relations entre la colonie maternelle et les jeunes imagos ailés de *Reticulitermes lucifugus*. *Vie et milieu*, **12**, 627–32.

Buchner, P. (1965). *Endosymbiosis of animals with plant microorganisms*, revised English version. New York, London & Sydney: Interscience Publishers. 909 pp.

Bugnion, E. (1930). Les pièces buccales, le sac infrabuccal et le pharynx des fourmis. *Bulletin, Société entomologique d'Egypte*, **40**, 85–210.

Burgett, D. M. & Young, R. G. (1974). Lipid storage by honey ant repletes. *Annals of the Entomological Society of America*, **67**(5), 743–4.

Burtt, B. D. (1942). Some East Africa vegetation communities. *Journal of Ecology*, **30**, 65–146.

Büschinger, A. (1965). *Leptothorax* (*Mychothorax*) *kutteri* n.sp., eine sozialparasitische Ameise. *Insectes sociaux*, **12**, 327–34.

Büschinger, A. (1966). Untersuchungen an *Harpagoxenus sublaevis* Nyl. (Hymenoptera, Formicidae). I. Freilandbeobachtungen zu Verbreitung und Lebensweise. *Insectes sociaux*, **13**, 5–16.

Büschlinger, A. (1970). Zur Frage der Monogynie oder Polygynie bei *Myrmecina graminicola* (Latreille). *Insectes sociaux*, **18**, 177–81.

Büschinger, A. (1971). Zur Verbreitung und Lebensweise sozialparasitischer Ameisen des schweiser Wallis. *Zoologischer Anzeiger*, **186**, 47–59.

Büschinger, A. (1973). Transport und Ansetzen von Larven an Beutestücke bei der Ameise *Aphaenogaster subterranea* (Latreille) (Hymenoptera, Formicidae). *Zoologischer Anzeiger*, **190**(1–2), 63–6.

Butler, C. G. (1965). Die Wirkung von Königinnen. Extrakten verschiedener sozialer Insekten auf die Aufzucht von Königinnen und die Entwicklung der Ovarien von Arbeiterinnen der Honigbiene (*Apis mellifera*). *Zeitschrift für Bienenforschung*, **8**, 143–7.

Butterworth, D. & MacNulty, B. J. (1966). Testing materials for resistance to termite attack. Part 1. A quantitative field test. *Material und Organismen*, **1**, 173–84.

Calvet, E. & Prat, H. (1956). *Microcalorimetrie: applications physicochimiques et biologique*. Paris: Masson et Cie.

Calvet, E. & Prat, H. (1963). *Progress in microcalorimetry*. London: Pergamon Press.

Campbell, W. G. (1952). The biological decomposition of wood. In *Wood chemistry*, eds. L. E. Wise & E. C. Jahn, vol. 2, pp. 1061–116. New York: Reinhold Publishing Corporation.

Carter, F. L., Dinus, L. A. & Smythe, R. V. (1972*a*). Fatty acids of the eastern subterranean termite *Reticulitermes flavipes* (Isoptera: Rhinotermitidae). *Annals of the Entomological Society of America*, **65**, 655–8.

Carter, F. L., Dinus, L. A. & Smythe, R. V. (1972*b*). Effect of wood decayed by *Lenzites trabea* on the fatty acid composition of the eastern subterranean termite, *Reticulitermes flavipes*. *Journal of Insect Physiology*, **18**, 1387–93.

Carter, F. L. & Smythe, R. V. (1973). Effect of sound and *Lenzites*-decayed wood on the amino acid composition of *Reticulitermes flavipes*. *Journal of Insect Physiology*, **19**, 1623–9.

Cartwright, K. St. G. & Findlay, W. P. K. (1958). *Decay of timber and its prevention*, 2nd edn. London: Her Majesty's Stationery Office. 332 pp.

Castle, G. B. (1934). The damp-wood termites of western United States, genus *Zootermopsis* (formerly *Termopsis*). In *Termites and termite control*, ed. C. A. Kofoid, 2nd edn, pp. 273–310. Berkeley: University of California Press.

Cavill, G. & Hinterberger, H. (1960). Dolichoderine ant extractives. *IX Internationaler Kongress für Entomologie, Wien*, **3**(3), 53–9.

Chang, F. & Friedman, S. (1971). A developmental analysis of the uptake and release of lipids by the fat-body of the tobacco hornworm, *Manduca sexta*. *Insect Biochemistry*, **1**, 63–80.

Chauvin, R. (1966). Un procédé pour récolter automatiquement les proies que les *Formica polyctena* rapportent au nid. *Insectes sociaux*, **13**, 59–68.

Chauvin, R. (1967). Un procédé pour recolter tout ce que les *Formica polyctena* rejettent du nid. *Insectes sociaux*, **14**, 41–6.

Chauvin, R., Courtois, G. & Lecomte, J. (1961). Sur la transmission d'isotopes radio-actifs entre deux fourmilières d'espèces differentes (*Formica rufa* et *Formica polyctena*). *Insectes sociaux*, **8**, 99–107.

Chen, S. C. (1937*a*). Social modification of the activity of ants in nest-building. *Physiological Zoology*, **10**, 420–36.

Chen, S. C. (1937*b*). The leaders and followers among ants in nest-building. *Physiological Zoology*, **10**, 437–55.

Cherrett, J. M. (1968). The foraging behaviour of *Atta cephalotes* (L.) (Hymenoptera, Formicidae). I. Foraging pattern and plant species attacked in tropical rain forest. *Journal of Animal Ecology*, **37**, 387–403.

Cherrett, J. M. (1972). Some factors involved in the selection of vegetable substrate by *Atta cephalotes* (L.) (Hymenoptera: Formicidae) in tropical rain forest. *Journal of Animal Ecology*, **41**, 647–60.

Cherrett, J. M., Pollard, G. V. & Turner, J. A. (1974). Preliminary observations on *Acromyrmex landolti* (For.) and *Atta laevigata* (Fr. Smith) as pasture pests in Guyana. *Tropical Agriculture*, **51**, 69–74.

Cherrett, J. M. & Seaforth, C. E. (1970). Phytochemical arrestants for the leaf-cutting ants, *Atta cephalotes* (L.) and *Acromyrmex octospinosus* (Reich), with some notes on the ants' response. *Bulletin of Entomological Research*, **59**, 615–25.

Chew, R. M. (1959). Estimation of ant colony size by the Lincoln index method. *Journal of the New York Entomological Society*, **67**, 157–61.

Child, H. J. (1934). The internal anatomy of termites and the histology of the digestive tract. In *Termites and termite control*, ed. C. A. Kofoid, 2nd edn, pp. 58–88. Berkeley: University of California Press.

Child, G. S. (1974). *Kainji Lake Research Project Nigeria. An Ecological Survey of the Borgu Game Reserve. F1: SF/NIR 24 Technical Report*, **4**. Rome: Food and Agriculture Organisation/United Nations Development Project.

Clark, P. J. & Evans, F. C. (1954). Distance to nearest neighbour as a measure of spatial relationships in populations. *Ecology*, **35**, 445–53.

Clark, W. H. & Comanor, P. L. (1973). A quantitative examination of spring foraging of *Veromessor pergandei* (Mayr) in northern Death Valley, California (Hymenoptera: Formicidae). *American Midland Naturalist*, **90**, 467–74.

Cleveland, L. R. (1923). Symbiosis between termites and their intestinal protozoa. *Proceedings of the National Academy of Sciences of the United States of America*, **9**, 424–8.

Cleveland, L. R. (1924). The physiological and symbiotic relationships between the intestinal protozoa of termites and their host, with special reference to *Reticulitermes flavipes* (Kollar). *Biological Bulletin, Marine Biological Laboratory, Woods Hole*, **46**, 178–227.

Cleveland, L. R. (1925*a*). The ability of termites to live perhaps indefinitely on a diet of pure cellulose. *Biological Bulletin, Marine Biological Laboratory, Woods Hole*, **48**, 289–93.

Cleveland, L. R. (1925*b*). The feeding habit of termite castes and its relation to their intestinal flagellates. *Biological Bulletin, Marine Biological Laboratory, Woods Hole*, **48**, 295–308.

Cleveland, L. R. (1926). Symbiosis among animals with special reference to termites and their intestinal flagellates. *Quarterly Review of Biology*, **1**, 51–60.

Cleveland, L. R. (1928). Further observations and experiments on the symbiosis between termites and their intestinal protozoa. *Biological Bulletin, Marine Biological Laboratory, Woods Hole*, **54**, 231–7.

Cleveland, L. R., Hall, S. R., Sanders, E. P. & Collier, J. (1934). The wood-feeding roach *Cryptocercus*, its protozoa, and the symbiosis between protozoa and roach. *Memoirs of the American Academy of Arts and Sciences*, **17**(2), 185–342.

Cmelik, S. H. W. (1969*a*). The neutral lipids from various organs of the termite *Macrotermes goliath. Journal of Insect Physiology*, **15**, 839–49.

Cmelik, S. H. W. (1969*b*). Composition of the neutral lipids from termite queens. *Journal of Insect Physiology*, **15**, 1481–7.

Cmelik, S. H. W. (1971). Composition of the lipids from the guts of termite queens. *Journal of Insect Physiology*, **17**, 1349–58.

Cmelik, S. H. W. (1972). Some properties of the neutral lipids from the workers of the termite *Macrotermes falciger. Insect Biochemistry*, **2**, 361–6.

Cmelik, S. H. W. & Douglas, C. C. (1970). Chemical composition of 'fungus gardens' from two species of termites. *Comparative Biochemistry and Physiology*, **36**, 493–502.

Coaton, W. G. H. (1947). The Pienaars River complex of wood-eating termites. *Journal of the Entomological Society of Southern Africa*, **9**, 136–77.

Coaton, W. G. H. (1948*a*). The snouted harvester termite (*Trinervitermes*). *Farming in South Africa*, **23**, 97–108.

Coaton, W. G. H. (1948*b*). The harvester termite. *Farming in South Africa*, **23**, 259–67.

Coaton, W. G. H. (1958). *The Hodotermitid harvester termites South Africa. Union of South Africa Department of Agriculture Bulletin*, **375**.

Cole, A. C. (1934). A brief account of aestivation and over-wintering of the occident ant *Pogonomyrmex occidentalis* (Cresson), in Idaho. *Canadian Entomologist*, **66**, 193–8.

Cole, A. C. (1939). The life history of a fungus-growing ant of the Mississippi gulf coast. *Lloydia*, **2**, 153–60.

Cole, A. C. (1950). Some observations on *Stenamma* Westwood. *Journal of the Tennessee Academy of Science*, **25**, 297.

Collins, M. S. (1951). Variations in the fat body of *Reticulitermes flavipes* (Kollar). *Anatomical Record*, **3**, 477.

Collins, M. S. (1969). Water relations in termites. In *Biology of termites*, ed. K. Krishna & F. M. Weesner, vol. 1, pp. 433–58. New York & London: Academic Press.

Conklin, A. (1972). A study of ant populations at the plains–foothill border, Colorado. *Southwestern Naturalist*, **17**, 43–54.

Cook, B. J. & Eddington, L. C. (1967). The release of triglycerides and free fatty acids from the fat body of the cockroach, *Periplaneta americana. Journal of Insect Physiology*, **13**, 1361–72.

Cook, S. F. (1932). The respiratory gas exchange in *Termopsis nevadensis. Biological Bulletin, Marine Biological Laboratory, Woods Hole*, **63**, 246–57.

Cook, S. F. (1943). Nonsymbiotic utilization of carbohydrates by the termite, *Zootermopsis angusticollis. Physiological Zoology*, **16**, 123–8.

Cook, S. F. & Scott, K. G. (1933). The nutritional requirements of *Zootermopsis* (*Termopsis*) *angusticollis. Journal of Cellular and Comparative Physiology*, **4**, 95–110.

Cook, S. F. & Smith, R. E. (1942). Metabolic relations in the termite–protozoa symbiosis: temperature effects. *Journal of Cellular and Comparative Physiology*, **4**, 95–110.

Cory, E. N. & Haviland, E. E. (1938). Population studies of *Formica exsectoides* Ford. *Annals of the Entomological Society of America*, **31**, 50–6.

Costello, D. F. (1944). Natural vegetation of abandoned plowed land in mixed prairie association of northeastern Colorado. *Ecology*, **25**, 312–26.

Cowling, E. B. (1961). *Comparative biochemistry of the decay of sweetgum sapwood by white-rot and brown-rot fungi. United States Department of Agriculture Technical Bulletin*, No. **1258**, 79 pp.

Cowling, E. B. & Merrill, W. (1966). Nitrogen in wood and its role in wood deterioration. *Canadian Journal of Botany*, **44**, 1539–54.

Côté, W. A., Jr (1968*a*). Chemical composition of wood. In *Principles of wood science and technology, I, Solid wood*, ed. F. F. P. Kollmann & W. A. Côté, Jr, pp. 55–78. New York: Springer-Verlag.

Côté, W. A., Jr (1968*b*). Biological deterioration of wood. In *Principles of wood science and technology, I, Solid wood*, ed. F. F. P. Kollmann & W. A. Côté, Jr, pp. 97–135. New York: Springer-Verlag.

Crawley, W. C. (1914). A revision of the genus *Leptothorax* Mayr, in the British Isles. *Entomologist's Record and Journal of Variation*, **36**, 89–96.

Crescitelli, F. (1935). The respiratory metabolism of *Galleria mellonella* (bee moth) during pupal development at different temperatures. *Journal of Cellular and Comparative Physiology*, **6**, 351–68.

Cumber, R. A. (1968). Notes on a colony of *Orectognathus antennatus*, Fr. Smith (Hymenoptera: Myrmicidae) from the Coromandel peninsula. *New Zealand Entomologist*, **4**, 43–4.

Czerwiński, Z., Jakubczyk, H. & Pętal, J. (1969). The influence of ants of the genus *Myrmica* on the physico-chemical and microbiological properties of soil within the compass of ant hills in Strzeleckie Meadows. *Polish Journal of Soil Science*, **3**, 51–8.

Czerwiński, Z., Jakubczyk, H. & Pętal, J. (1971). Influence of ant hills on the meadow soils. *Pedobiologia*, **11**, 277–85.

Dadd, R. H. (1973). Insect nutrition: current developments and metabolic implications. *Annual Review of Entomology*, **18**, 381–420.

Damaschke, K. & Becker, G. (1964). Korrelation der Atmungsintensität von Termiten zu Änderungen der Impulsfolgefrequenz der atmospherics. *Zeitschrift für Naturforschung*, **19*b***, 157–60.

Day, M. F. (1938). Preliminary observations on the environment of *Eutermes exitiosus* Hill (Isoptera). *Australian Journal of the Council for Scientific and Industrial Research*, **11**, 317–27.

De Bout, A.-F. (1964). Termites et densité d'oiseaux. In *Etudes sur les termites africains*, ed. A. Bouillon, pp. 273–83. Paris: Masson et Cie.

Dejean, A. & Passera, L. (1974). Ponte des ouvrières et inhibition royale chez la fourmi *Temnothorax recedens* (Nyl.) (Formicidae, Myrmicinae). *Insectes sociaux*, **21**(4), 343–56.

Delage, B. (1966). Sur une fonction particulière des glandes pharyngiennes des fourmis. *Comptes rendus hebdomadaires des séances de l'Académie des sciences, Paris*, **263**, 1743–4.

Delage, B. (1968). Recherches sur les fourmis moissoneuses du Bassin Aquitain: éthologie, physiologie de l'alimentation. *Annales des sciences naturalles (Zoologie)*, Sér. 12, **10**(2), 197–265.

Delage-Darchen, B. (1971). Contribution a l'étude écologique d'une savane de Côte d'Ivoire (Lamto): les fourmis des strates herbacées et arborées. *Biologia Gabonica*, **7**(4), 461–96.

Delage-Darchen, B. (1974). Polymorphismus in der Ameisengattung *Messor* und ein Vergleich mit *Pheidole*. In *Soziapolymorphismus bei Insekten*, ed. G. H. Schmidt, pp. 590–603. Stuttgart: Wissenschaftliche Verlagsgesellschaft.

Deligne, J. (1966). Caractères adaptis au régime alimentaire dans la mandibule des termites (Insectes, Isopteres). *Comptes rendus hebdomadaires des séances de l'Académie des sciences, Paris*, **263**, 1323–5.

Délye, G. (1968). Recherches sur l'écologie, la physiologie et l'éthologie des fourmis du Sahara. Thèses Faculté de Science, Université d'Aix-Marseille, Aix-en-Provence, 155 pp.

Dibley, G. C. & Lewis, T. (1972). An ant counter and its use in the field. *Entomologia experimentalis et applicata*, **15**, 499–508.

Dickman, A. (1931). Studies on the intestinal flora of termites with reference to their ability to digest cellulose. *Biological Bulletin, Marine Biological Laboratory, Woods Hole*, **61**, 85–92.

Dlussky, G. M. (1965). Metody kolicestvennovo uceta pocvobitajuscich murav'ev. *Zoologicheskiĭ zhurnal*, **44**, 716–27.

Dlussky, G. M. (1967). *Murav'i roda formika.* Moskva: Izdatel'stvo 'Nauka'. Moskva, pp. 236.

Dlussky, G. M. (1974). Ants of the barkhan–takyr complex of central Kara-kum. *Biologicheskiĭ Nauki*, **9**, 19–23. (In Russian.)

Dlussky, G. M. & Kupianskaya, A. N. (1972). Consumption of protein food and growth of *Myrmica* colonies. *Ekologia polska*, **20**, 73–82.

Dobrzański, J. (1966). Contribution to the ethology of *Leptothorax acervorum. Acta biologiae experimentalis*, **26**, 71–8.

Dole, V. P. (1964). Fat as an energy source. In *Fat as a tissue*, ed. K. Rodahl, pp. 250–9. New York: McGraw-Hill.

Donisthorpe, H. St. J. K. (1913). Myrmecophilous notes for 1912. *Entomologist's Record and Journal of Variation*, **25**, 61–8.

Donisthorpe, H. St. J. K. (1915). *British ants, their life-history and classification.* Plymouth: W. Brendon & Son. xv+379 pp., 18 plates.

Donisthorpe, H. St. J. K. (1936). The oldest insect on record. *Entomologist's Record and Journal of Variation*, **48**, 1–2.

Dreyer, W. A. (1932). The effect of hibernation and seasonal variation of temperature on the respiratory exchange of *Formica ulkei* Emery. *Physiological Zoology*, **5**, 301–31.

Dropkin, V. H. (1946). The use of mixed colonies of termites in the study of host–symbiont relations. *Journal of Parasitology*, **32**, 247–51.

Drummond, H. (1888). *Tropical Africa.* London.

Echols, H. W. (1966). Assimilation and transfer of Mirex in colonies of Texas leaf-cutting ants. *Journal of Economic Entomology*, **59**, 1336–8.

Eckstein, K. (1937). Die Nester der Waldameisen *Formica rufa* L., *Formica truncicola* Nyl., und *Formica exsecta* (Nyl.). *Mitteilungen aus Forstwirtschaft und Forstwissenschaft*, **8**, 635–85.

Eddy, T. A. (1970). Foraging behavior of the western harvester ant *Pogonomyrmex occidentalis*, (Hymenoptera: Formicidae) in Kansas. Ph.D. Thesis, Kansas State University, Manhattan, 151 pp.

Eidmann, H. (1944). Die Ameisenfauna von Fernando Po. *Zoologische Jahrbücher*, **76**, 413–90.

Eisner, T. (1957). A comparative morphological study of the proventriculus of ants (Hymenoptera: Formicidae). *Bulletin of the Museum of Comparative Zoology*, **116**(8), 437–41.

Eisner, T. & Brown, W. L. (1958). The evolution and social significance of the ant proventriculus. *Proceedings of the 10th International Congress of Entomology, Montreal*, **2**, 503–8.

Eisner, T. & Happ, G. M. (1962). The infrabuccal pocket of a formicine ant: a social filtration device. *Psyche*, **69**, 107–16.

Eisner, T. & Wilson, E. O. (1958). Radioactive tracer studies of food transmission in ants. *Proceedings of the 10th International Congress of Entomology, Montreal*, **2**, 509–13.

Elliot, G. (1904). Anglo–French Niger–Tchad Boundary Commission. *Geographical Journal*, **24**(5), 505–23.

Elmes, G. W. (1973). Observations of the density of queens in natural colonies of *Myrmica rubra* (L.) (Hymenoptera: Formicidae). *Journal of Animal Ecology*, **42**, 761–72.

Elton, C. (1932). Territory among wood ants (*Formica rufa* L.) at Picket Hill. *Journal of Animal Ecology*, **1**, 69–76.

Emden, H. F. van, Eastop, V. F., Hughes, R. D. & Way, M. J. (1969). The ecology of *Myzus persicae*. *Annual Review of Entomology*, **14**, 197–270.

Emden, H. F. van & Williams, G. F. (1974). Insect stability and diversity in agro-ecosystems. *Annual Review of Entomology*, **19**, 455–75.

Emerson, A. E. (1935). Termitophile distribution and quantitative characters as indicators of physiological speciation in British Guiana termites. *Annals of the Entomological Society of America*, **28**, 369–95.

Emerson, A. E. (1938). Termite nests. A study of the phylogeny of behavior. *Ecological Monographs*, **8**, 247–84.

Emerson, A. E. (1939). Populations of social insects. *Ecological Monographs*, **9**, 287–300.

Emerson, A. E. (1945). The neotropical genus Syntermes (Isoptera: Termitidae). *Bulletin of the American Museum of Natural History*, **83**, 427–71.

Emerson, A. E. (1956). Regenerative behaviour and social homeostasis of termites. *Ecology*, **37**, 248–58.

Emmel, T. C. (1967). Ecology and activity of leaf-cutting ants (*Atta* sp.). In *Advanced zoology: insect ecology in the tropics*, pp. 125–31. San José, Costa Rica: Organisation for Tropical Studies.

Emmert, W. (1968). Die Postembryonalentwicklung sekretorischer Kopfdrüsen von *Formica pratensis* Retz. and *Apis mellifera* L. (Ins., Hym.). *Zeitschrift für Morphologie und Ökologie der Tiere*, **63**, 1–62.

Emmert, W. (1969). Entwicklungsleistungen fragmentierten Labialdrüsen Imaginalanlagen von *Formica pratensis* Retz. (Hym.). *Wilhelm Roux Archiv für Entwicklungsmechanik der Organismen*. **162**, 97–113.

Engelmann, M. D. (1966). Energetics, terrestrial field studies, and animal productivity. *Advances in Ecological Research*, **3**, 73–115.

Erickson, J. M. (1972). Mark–recapture techniques for population estimates of *Pogonomyrmex* ant colonies: an evaluation of the P^{32} technique. *Annals of the Entomological Society of America*, **65**, 57–61.

Escherich, K. (1917). *Die Ameise*. Brainsweig: F. Vieweg & Sohn. 348 pp.

Esenther, G. R. (1969). Termites in Wisconsin. *Annals of the Entomological Society of America*, **62**, 1274–84.

Ettershank, G. (1971). Some aspects of the ecology and nest microclimatology of the meat ant *Iridomyrmex purpureus* (Sm.). *Royal Society of Victoria Proceedings*, **84**(1), 137–52.

Ettershank, G. & Whitford, W. G. (1973). Oxygen consumption of two species of *Pogonomyrmex* harvesting ants (Hymenoptera: Formicidae). *Comparative Biochemistry and Physiology*, **46A**, 605–11.

Faber, W. (1969). Beiträge zur Kenntnis sozialparasitischer Ameisen. 2. *Aporomyrmex ampeloni* nov. gen., nov. spec. (Hym. Formicidae), ein neuer permanenter Sozialparasit bei *Plagiolepis vindobonensis* Lomnicki aus Österreich. *Pflanzenschutz, Berlin*, **39**, 39–100.

Ferrar, P. & Watson, J. A. L. (1970). Termites (Isoptera) associated with dung in Australia. *Journal of the Australian Entomological Society*, **9**, 100–2.

Fielde, A. M. (1904). Observations on ants in their relation to temperature and submergence. *Biological Bulletin, Marine Biological Laboratory, Woods Hole*, **7**, 170–4.

Fielde, A. M. (1905). Three odd incidents in ant-life. *Proceedings of the Academy of Natural Sciences of Philadelphia*, **56**, 639–41.

Fisser, M. G. (1970). Reliability of harvester ant mound counts from aerial photographs. *The Journal of the Colorado–Wyoming Academy of Science*, **7**, 42–3.

Fletcher, D. J. C. (1973). 'Army Ant' behaviour in the Ponerinae: a reassessment. *Proceedings, VIIth International Congress of the International Union for the study of Social Insects, London, 10–15 September 1973*, 116–21.

Fölster, H. (1964). Die Pedi-Sedimente der Seedsudanesischen Pediplane. Herkunft und Bodenbildung. *Pédologie*, **14**, 68–84.

Forbes, J. & McFarlane, A. M. (1961). The comparative anatomy of digestive glands in the female castes and the males of *Camponotus pennsylvanicus* (De Geer) (formicidae; Hymenoptera). *Journal of the New York Entomological Society*, **69**, 92–103.

Forel, A. (1920). *Les fourmis de la Suisse. Notices anatomiques et Physiologiques. Architecture. Distribution géographique. Nouvelles expériences et observations de moeurs*, 2nd ed. La Chaux-de-Fonds. xvi+333 pp.

Forel, A. (1928). *The social world of the ants*. New York & London: G. P. Putnam's Sons Ltd. pp. 445, 551.

Fox, F. W. (1966). Studies on the chemical composition of foods commonly used in South Africa. *Publications of the South African Institute of Medical Research, Johannesburg*.

Francoeur, A. (1965). Ecologie des populations de fourmis dans un bois de chênes rouges et d'érables rouges. *Naturaliste canadien*, **92**, 263–76.

Francoeur, A. (1966*a*). Le genre '*Stenamma*' Westwood au Québec. *Annales de la Société entomologique de Québec*, **11**, 115–19.

Francoeur, A. (1966*b*). La faune myrmécologique de l'érablière à sucre (*Aceretum saccharophori* Dansereau) de la région de Québec. *Naturaliste canadien*, **93**, 443–72.

Francoeur, A. & Maldaque, M. (1966). Classification des micro-milieux de nidification des fourmis. *Naturaliste canadien*, **93**, 473–8.

Free, J. B. (1956). A study of the stimuli which release the food begging and offering responses of worker honeybees. *British Journal of Animal Behaviour*, **4**, 94–101.

Free, J. B. (1957). The transmission of food between worker honeybees. *British Journal of Animal Behaviour*, **5**, 41–7.

Fujii, N. (1964). Studies on the free amino acids in Formosan termites (*Coptotermes formosanus* Shiraki). *Bulletin of the Faculty of Agriculture, University of Miyazaki*, **9**, 213–16.

Fuller, C. (1915). Observations on some South African termites. *Annals of the Natal Museum*, **3**(2), 329–505.

Fyfe, R. V. & Gray, F. J. (1938). The humidity of the atmosphere and the moisture conditions within mounds of *Eutermes exitiosus* Hill. *Council for Scientific and Industrial Research, Melbourne*, Pamphlet No. **82**, 1–82.

Gabe, M. & Noirot, C. (1960). Particularités histochimiques du tissu adipeux royal des termites. *Bulletin Société zoologique de France*, **85**, 376–82.

Gallardo, A. (1916). Notas acerca de la hormiga *Trachymyrmex pruinosus* Emery. *Anales del Museo nacional de historia natural de Buenos Aires*, **28**, 241–52.

Gallardo, J. M. (1951). Sobre un Teiidae (Reptilia, Sauria) poco conocido para la fauna Argentina. *Communicaciones del Instituto nacional de investigacion de las ciencias naturales v museo de ciencias naturales 'Bernardino Rivadavia'*. **2**, 8.

Gallé, L. (1972*a*). Study of ant-population in various grassland ecosystems. *Acta biologica, Szeged*, **18**(1–4), 159–64.

Gallé, L. (1972*b*). Formicidae populations of the ecosystems in the environs of Tiszafüred. *Tiscia, Szeged*, **7**, 59–68.

Gallé, L. (1973). Thermoregulation in the nest of *Formica pratensis* Retz. (Hymenoptera: Formicidae). *Acta biologica, Szeged*, **19**, 139–42.

Gara, R. I. (1970). Report of forest entomology consultant (United Nations Development Project 80) IICA, Turrialba, Costa Rica. Mimeo.

Gaspar, Ch. (1966). Les fourmis et l'agriculture (Hymenoptera: Formicidae). *Annales de Gembloux*, **72**, 235–43.

Gaspar, Ch. (1970). *Les formicides de la Famenne une monographie faunique régionale. Faculté des Sciences Agronomiques de l'État, Gembloux*, 219 pp.

Gaspar, Ch. (1971). Les formicides de la Famenne. I. Une étude zoo-sociologique. *Revue d'écologie et de biologie du sol*, **8**(4), 553–607.

Gaspar, Ch. (1972). Action des fourmis du genre *Lasius* dans l'écosystème prairie. *Ekologia polska*, **20**, 145–52.

Gay, F. J. & Calaby, J. H. (1970). Termites from the Australian region. In *Biology of termites*, ed. K. Krishna & F. M. Weesner, vol. 2, pp. 393–448. New York & London: Academic Press.

Gay, F. J., Greaves, T., Holdaway, F. G. & Wetherley, A. H. (1955). *Standard laboratory colonies of termites for evaluating the resistance of timber, timber preservatives and other materials to termite attack. Bulletin, Commonwealth Scientific and Industrial Research Organization*, **277**, 60 pp.

Gay, F. J., Greaves, T., Holdaway, F. G. & Wetherley, A. H. (1975). *The development and use of field testing techniques with termites in Australia. Bulletin, Commonwealth Scientific and Industrial Research Organisation*, **280**, 31 pp.

Geyer, J. W. C. (1951). The reproductive organs of certain termites, with notes on hermaphrodites of *Neotermes. Entomology Memoirs. Department of Agriculture, Union of South Africa*, **2**, 232–325.

Ghidini, G. M. (1939). Studi sulle termiti. 6. Ricerche sul quoziente respiratorio nelle diverse caste di *Reticulitermes lucifugus. Rivista di biologia coloniale*, **2**, 385–99.

Gilbert, L. I. (1967). Lipid metabolism and function in insects. In *Advances in insect physiology*, ed. J. W. L. Beament *et al.*, vol. 4, 69–211. New York & London: Academic Press.

Gilmour, D. (1940). The anaerobic gaseous metabolism of the termite *Zootermopsis nevadensis* (Hagen). *Journal of Cellular and Comparative Physiology*, **15**, 331–42.

Glancey, B. M., Stringer, C. E. & Bishop, P. M. (1973). Trophic egg production in the imported fire ant *Solenopsis invicta* (Hymenoptera: Formicidae). *Journal of the Georgia Entomological Society*, **8**(3), 217–20.

Glancey, B. M., Stringer, C. E., Craig, C. H., Bishop, P. M. & Martin, B. B. (1973). Evidence of a reptile caste in the fire ant *Solenopsis invicta*. *Annals of the Entomological Society of America*, **66**(1), 233–4.

Glover, P. E., Trump, E. C. & Wateridge, L. E. D. (1964). Termitaria and vegetation patterns on the Loita plains of Kenya. *Journal of Ecology*, **52**, 367–77.

Goetsch, W. (1947). Ein neuentdeckter Wirkstoff (Vitamin-T-Komplex). *Experientia*, **3**, 326–7.

Golley, F. B. & Gentry, J. B. (1964). Bioenergetics of the southern harvester ant *Pogonomyrmex badius*. *Ecology*, **45**, 217–25.

Goodland, R. J. A. (1965). On termitaria in a savanna ecosystem. *Journal of Zoology*, **43**, 641–50.

Goodman, D. (1975). The theory of diversity–stability relationships in ecology. *Quarterly Review of Biology*, **50**(3), 237–66.

Gösswald, K. (1932). Ökologische Studien über die Ameisenfauna des mittleren Maingebietes. *Zeitschrift für wissenschaftliche Zoologie, A*, **142**, 1–156.

Gösswald, K. (1951*a*). Über den Lebensablauf von Kolonien der Roten Waldameise. *Zoologische Jahrbücher*, **80**, 27–73.

Gösswald, K. (1951*b*). *Die rote Waldameise im Dienste der Waldhygiene*. Lüneburg: Metta Kinau Verlag.

Gösswald, K. (1957*a*). Über die biologischen Grundlagen der Zucht und Answeiselung junger Königinnen der kleinen roten Waldameise nebst praktischen Erfahrungen. *Waldhygiene*, **2**, 33–53.

Gösswald, K. (1957*b*). Bildung von Ablegern der kleinen roten Waldameise auf der Grundlage einer Massenzucht von Königinnen (Verfahren II). *Waldhygiene*, **2**, 54–72.

Gösswald, K. (1958). Neue Erfahrungen über Einwirkung der roten Waldameise auf den Massenwechsel von Schadinsekten sowie einige methodische Verbesserungen bei ihrem praktischen Einsatz. *Proceedings of the 10th International Congress of Entomology, Montreal*, **4**, 567–71.

Gösswald, K. & Bier, K. (1954). Untersuchungen zur Kastendetermination in der Gattung *Formica*. 3. Die Kastendetermination von *Formica rufarufo-pratensis minor* Gössw. *Insectes sociaux*, **1**, 229–46.

Gösswald, K. & Bier, K. (1957). Untersuchungen zur Kastendetermination in der Gattung *Formica*. 5. Der Einfluss der Temperatur auf die Eiablage und Geschlechtsbestimmung. *Insectes sociaux*, **4**, 335–48.

Gösswald, K. & Kloft, W. (1956). Untersuchungen über die Verteilung von radioaktiv markiertem Futter im Volk der Kleinen roten Waldameise (*Formica rufopratensis minor*). *Waldhygiene*, **1**, 200–2.

Gösswald, K. & Kloft, W. (1963). Tracer experiments on food exchange in ants and termites. In *Radiation and radioisotopes applied to insects of agricultural importance*, pp. 25–42. Vienna: International Atomic Energy Agency.

Gösswald, K. & Schmidt, G. H. (1960). Untersuchungen zum Flugelabwirf und Begattungs Verhalten einiger *Formica*-Arten (Ins. Hym.) im Hinblick auf ihre systematische Differenzierungen. *Insectes sociaux*, **7**, 297–321.

Graf, I. (1964). Untersuchungen zur Verdauungsphysiologie von *Formica polyctena* Foerst. *Experientia*, **20**, 330–1.

Graf, I. & Hölldobler, B. (1964). Untersuchungen zur Frageder Holzverwertung als Nahrung bei holzzerstörenden Rossameisen (*Camponotus ligniperda* Latr. und *Camponotus herculeanus* (L.)) unter Berücksichtigung der Cellulase-Aktivität. *Zeitschrift für angewandte Entomologie*, **55**, 77–80.

Grassé, P. P. (1936). Les termites en Afrique occidentale francaise. Leur importance économique. Les moyens de lutte. *Revue de pathologie végétale et d'entomologie agricole de France*, **23**, 265–306.

Grassé, P. P. (1944). Recherches sur la biologie des termites champignonnistes (Macrotermitinae). *Annales des sciences naturelles* (*Zoologie*), **6**, 97–171.

Grassé, P. P. (1945). Recherches sur la biologie des termites champignonnistes (Macrotermitinae). *Annales des sciences naturelles* (*Zoologie*), Sér. 11, **7**, 115–46.

Grassé, P. P. (1949). Ordre des Isoptères ou termites. In *Traité de zoologie*, ed. P. P. Grassé, vol. IX, pp. 408–544. Paris: Masson et Cie.

Grassé, P. P. (1950). Termites et sols tropicaux. *Revue internationale de botanique appliquée et d'agriculture tropicale*, 1950, 549–54.

Grassé, P. P. (1959). Un nouveau type de symbiose: la meule alimentaire des termites champignonnistes. *Nature, Paris*, **3293**, 385–9.

Grassé, P. P. & Chauvin, R. (1944). L'effet de groupe et la survie des neutres dans les sociétés d'insectes. *Revue scientifique*, **82**, 461–4.

Grassé, P. P. & Gharagozlou, I. (1963). L'ergastoplasme et la genèse des protéines dans le tissu adipeux royal du termite à cou jaune. *Comptes rendus hebdomadaires des séances de l'Académie des sciences, Paris*, **257**, 3346–548.

Grassé, P. P. & Gharagozlou, I. (1964). Sur une nouvelle sorte de cellules du tissu adipeux royal de *Calotermes flavicollis* (Insecte isoptère): l'endolophocyte. *Comptes rendus hebdomadaires des séances de l'Académie des sciences, Paris*, **258**, 1045–7.

Grassé, P. P. & Noirot, C. (1945). La transmission des flagellés symbiotiques et les aliments des termites. *Bulletin biologique de la France et de la Belgique*, **79**, 273–92.

Grassé, P. P. & Noirot, C. (1946). La production des sexués néoténiques chez la termite à cou jaune (*Calotermes flavicollis* (F.)): inhibition germinale et inhibition somatique. *Comptes rendus hebdomadaires des séances de l'Académie des sciences, Paris*, **223**, 869–71.

Grassé, P. P. & Noirot, C. (1948). La 'climatisation' de la termitière par ses habitants et le transport de l'eau. *Comptes rendus hebdomadaires des séances de l'Académie des sciences, Paris*, **227**, 869–71.

Grassé, P. P. & Noirot, C. (1951*a*). La Sociotomie: migration et fragmentation de la termitière chez les *Anoplotermes* et les *Trinervitermes*. *Behaviour*, **3**, 146–66.

Grassé, P. P. & Noirot, C. (1951*b*). Orientation et routes chez les termites. Les 'balisage' des pistes. *Année psychologique*, 50e année, 273–80.

Grassé, P. P. & Noirot, C. (1958*a*). La meule des termites champignonnistes et sa signification symbiotique. *Annales des sciences naturelles* (*Zoologie*), Sér. 11, **20**(2), 113–28.

Grassé, P. P. & Noirot, C. (1958*b*). Le comportement des termites à l'égard de l'air libre. L'atmosphère des termitières et son renouvellement. *Annales des sciences naturelles* (*Zoologie*), Sér. 11, **20**(2), 1–28.

Grassé, P. P. & Noirot, C. (1959*a*). Rapports des termites avec les sols tropicaux. *Revue de géomorphologie dynamique*, **10**, 35–40.

Grassé, P. P. & Noirot, C. (1959*b*). L'évolution de la symbiose chez les Isoptères. *Experientia*, **15**, 365–72.

Grassé, P. P. & Noirot, C. (1960). Rôle respectif des mâles et des femelles dans la formation des sexués néoténiques chez *Calotermes flavicollis. Insectes sociaux*, **7**, 109–23.

Grassé, P. P. & Noirot, C. (1961). Nouvelles recherches sur la systématique et l'éthologie des termites champignonnistes du genre *Bellicositermes* Emerson. *Insectes sociaux*, **8**, 311–59.

Greaves, T. (1962). Studies of the foraging galleries and the invasion of living trees by *Coptotermes acinaciformis* and *C. brunneus* (Isoptera). *Australian Journal of Zoology*, **10**(4), 630–51.

Greaves, T. (1964). Temperature studies of termite colonies in living trees. *Australian Journal of Zoology*, **12**(2), 250–62.

Greaves, T. (1967). Experiments to determine the population of tree-dwelling colonies of termites (*Coptotermes acinaciformis* (Froggatt) and *C. frenchi* (Hill)). *Technical Papers. Division of Entomology, CSIRO Australia*, **7**, 19–33.

Green, J. W. (1963). Wood cellulose. In *Methods in carbohydrate chemistry*, ed. R. L. Whistler, vol. 3, pp. 9–21. New York & London: Academic Press.

Gregg, J. H. (1947). A microrespirometer. *Review of Scientific Instruments*, **18**, 514–16.

Gregg, J. H. (1950). Oxygen utilization in relation to growth and morphogenesis of the slime mould *Dictyostelium discoideum. Journal of Experimental Zoology*, **114**, 173–96.

Gregg, J. H. (1966). A microrespirometer capable of quantitative substrate mixing. *Experimental Cell Research*, **42**, 260–4.

Gregg, J. H. & Lints, F. A. (1967). A constant-volume respirometer for *Drosophila* imagoes. *Comptes rendus des travaux du Laboratoire Carlsberg*, **36**, 25–34.

Grinfel'd, É. K. (1939). Ékologiya murav'ev zapovednika 'Les na Vorskle' i ego okrestnostei. *Uchennỳe zapiski LGU*, **28**, 207–57.

Grinfel'd, É. K. (1941). Vozdeistvie murav'ev na reaktsiyu pochvỳ. *Zoologicheskiĭ zhurnal*, **20**(1), 100.

Grodzinski, W., Kłekowski, R. Z. & Duncan, A. (1975). *Methods for ecological bioenergetics*, IBP Handbook No. 24. Oxford: Blackwell Scientific Publications.

Hampel, H. (1963–64). Der Einfluss des Biotops auf dem Wärmhaushalt bei *Formica rufa* L. *Entomologische Abhandlungen*, **29**, 519–32.

Hangartner, W. (1969). Carbon dioxide, releaser for digging behaviour in *Solenopsis geminata* (Hymenoptera: Formicidae). *Psyche*, **76**, 58–67.

Harris, G. C. (1952). Wood resins. In *Wood chemistry*, ed. L. E. Wise & E. C. Jahn, vol. 1, pp. 590–617. New York: Reinhold Publishing Corporation.

Harris, L. D. (1969). A consideration of the nutrient 'sumping' activities of leaf-cutter ants (*Atta* sp.) in the new world tropics. Research report in *Advanced population biology*. San José, Costa Rica: Organization for Tropical Studies.

Harris, L. E. (1966). *Biological energy interrelationships and glossary of energy terms. National Academy of Sciences, National Research Council*, Publication **1411**. Washington: Printing & Publishing Office, National Academy of Sciences. 35 pp.

Harris, W. V. (1961). *Termites: their recognition and control.* London: Longmans Green & Co. xii+187 pp.

Harris, W. V. (1962). Termites in Europe. *New Scientist*, **13**, 614–17.

Harris, W. V. (1963). *Exploration du Parc National de la Garamba.* Part 42, *Isoptera.* Hayez: Brussels. 43 pp.

Harris, W. V. (1969). *Termites as pests of crops and trees.* London: Commonwealth Institute of Entomology.
Harris, W. V. (1970). Termites of the Palaearctic Region. In *Biology of termites*, ed. K. Krishna & F. M. Weesner, vol. 2, pp. 295–313. New York & London: Academic Press.
Harris, W. V. & Sands, W. A. (1965). The social organization of termite colonies. *Symposia of the Zoological Society of London*, **14**, 113–31.
Hartwig, E. K. (1956). The determination of the population distribution of *Trinervitermes* nests as a basis for control measures. *Bollettino del Laboratorio di zoologia generale e agraria della R. Scuola superiore d'agricoltura Portici*, **33**, 629–39.
Hartwig, E. K. (1966). The nest and control of *Odontotermes latericius* (Haviland) (Termitidae: Isoptera). *South African Journal of Agricultural Science*, **9**, 407–18.
Haskins, C. P. & Haskins, E. F. (1950*a*). Notes on the biology and social behaviour of the archaic ponerine ants of the genera *Myrmecia* and *Promyrmecia. Annals of the Entomological Society of America*, **43**, 461–91.
Haskins, C. P. & Haskins, E. F. (1950*b*). Note on the method of colony foundation of the Ponerine ant *Brachyponera* (*Euponera*) *lutea* Mayr. *Psyche*, **57**, 1–9.
Haskins, C. P. & Whelden, R. M. (1954). Note on the exchange of ingluvial food in the genus *Myrmecia. Insectes sociaux*, **1**, (1), 33–7.
Haverty, M. I. (1974). The significance of the subterranean termites, *Heterotermes aureus* (Snyder), as a detritivore in a desert grassland ecosystem. Ph.D. dissertation, University of Arizona, Tucson, Arizona, 89 pp.
Haverty, M. L., La Fage, J. P. & Nutting, W. L. (1974). Seasonal activity and environmental control of foraging of the subterranean termite, *Heterotermes aureus* (Snyder), in a desert grassland. *Life Sciences*, **15**, 1091–1101.
Haverty, M. L. & Nutting, W. L. (1974). Natural wood-consumption rates and survival of a dry-wood and a subterranean termite at constant temperature. *Annals of the Entomological Society of America*, **67**, 153–7.
Haverty, M. L. & Nutting, W. L. (1975). A simulation of wood consumption by the subterranean termite *Heterotermes aureus* (Snyder), in an Arizona desert grassland. *Insectes sociaux*, **22**(1), 93–102.
Haverty, M. L. & Nutting, W. L. & La Fage, J. P. (1975). Density of colonies and spatial distribution of foraging territories of the desert subterranean termite, *Heterotermes aureus* (Snyder). *Environmental Entomology*, **4**, 105–9.
Hayashida, K. (1960). Studies on the ecological distribution of ants in Sapporo and its vicinity. *Insectes sociaux*, **7**, 125–62.
Headley, A. E. (1943). Population studies of two species of ants, *Leptothorax longispinosus* Roger and *Leptothorax curvispinosus* Mayr. *Annals of the Entomological Society of America*, **36**, 743–53.
Headley, A. E. (1949). A population study of the ant *Aphaenogaster fulva ssp. aquia* (Buckley). *Annals of the Entomological Society of America*, **42**, 265–72.
Headley, A. E. (1952). Colonies of ants in locust wood. *Annals of the Entomological Society of America*, **45**, 435–42.
Healey, I. N. & Russell-Smith, A. (1971). Abundance and feeding preferences of fly larvae in two woodland soils. In *IV Colloquium pedobiologiae, Paris*, pp. 177–90.
Healey, M. J. R. (1962). Some basic statistical techniques in soil zoology. In *Progress in soil zoology*, ed. P. W. Murphy, pp. 3–9. London: Butterworths.
Heath, G. W., Edwards, C. A. & Arnold, M. K. (1964). Some methods for asses-

sing the activity of soil animals in the breakdown of leaves. *Pedobiologia*, **4**, 80–7.

Heath, H. (1931). Experiments in termite caste development. *Science*, **73**, 431.

Hébrant, F. (1964). Mesures de la consommation d'oxygène chez *Cubitermes exiguus* Mathot (Isoptera, Termitinae). In *Etudes sur les termites africains*, ed. A. Bouillon, pp. 153–71. Paris: Masson et Cie.

Hébrant, F. (1965). Etude de l'influence du poids des individus et de l'humidité du milieu sur la consommation d'oxygène d'ouvriers de *Cubitermes exiguus* Mathot (Isoptera, Termitinae). *Compte rendu du 5e Congrès UIEIS Toulouse, 1965*, 107–15.

Hébrant, F. (1970*a*). Etude du flux énergétique chez deux espèces du genre *Cubitermes* Wasmann (Isoptera, Termitinae), termites humivores des savanes tropicales de la région Éthiopienne. Thèse de Docteur en Sciences, Université Catholique de Louvain, Louvain.

Hébrant, F. (1970*b*). Circadian rhythm of respiratory metabolism in whole colonies of the termite, *Cubitermes exiguus*. *Journal of Insect Physiology*, **16**, 1229–35.

Hegh, E. (1922). *Les termites*. Brussels: Imprimerie Industrielle et Financière. 756 pp.

Heim, R. (1957). A propos du *Rozites gongylophora* A. Möller. *Review of Applied Mycology*, **22**, 293–9.

Hendee, E. C. (1933). The association of the termites, *Kalotermes minor*, *Reticulitermes hesperus*, and *Zootermopsis angusticollis* with fungi. *University of California Publications in Zoology*, **39**, 111–34.

Hendee, E. C. (1934). The association of termites and fungi. In *Termites and termite control*, ed. C. A. Kofoid, 2nd edn, pp. 105–16. Berkeley: University of California Press.

Hendee, E. C. (1935). The role of fungi in the diet of the common damp-wood termite, *Zootermopsis angusticollis*. *Hilgardia*, **9**, 499–525.

Herter, K. (1924). Untersuchungen über den Temperatursinn einiger Insekten. *Zeitschrift für vergleichende Physiologie*, **1**, 221–88.

Herter, K. (1925). Temperaturoptimum und relative Luftfeuchtigkeit bei *Formica rufa* L. *Zeitschrift für vergleichende Physiologie*, **2**, 226–32.

Herzig, J. (1938) Ameisen und Blattlaus. *Zeitschrift für angewandte Entomologie*, **24**, 367–435.

Hesse, P. R. (1955). A chemical and physical study of the soil of termite mounds in East Africa. *Journal of Ecology*, **43**, 449–61.

Hewitt, P. H. & Nel, J. J. C. (1969*a*). Toxicity and repellency of *Chrysocoma tenuifolia* (Berg) (Compositae) to the harvester termite *Hodotermes mossambicus* (Hagen) (Hodotermitidae). *Journal of the Entomological Society of Southern Africa*, **32**, 133–6.

Hewitt, P. H. & Nel, J. J. C. (1969*b*). The influence of group size on the sarcosomal activity and the behaviour of *Hodotermes mossambicus* alate termites. *Journal of Insect Physiology*, **15**, 2169–77.

Hewitt, P. H., Nel, J. J. C. & Conradie, S. (1969). Preliminary studies on the control of caste formation in the harvester termite *Hodotermes mossambicus* (Hagen). *Insectes sociaux*, **16**, 159–72.

Hewitt, P. H., Nel, J. J. C. & Schoeman, I. (1971). Influence of group size on water imbibition by *Hodotermes mossambicus* alate termites. *Journal of Insect Physiology*, **17**, 587–600.

Hewitt, P. H., Nel, J. J. C. & Schoeman, I. (1972). Some aspects of the biochemistry and behaviour of *Trinervitermes trinervoides* (Sjöstedt) reproductives. *Journal of the Entomological Society of Southern Africa*, **35**, 111–17.

Hewitt, P. H., Retief, L. W. & Nel, J. J. C. (1974). Aryl-β-glycosidases in the heads of workers of the termite, *Trinervitermes trinervoides. Insect Biochemistry*, **4**, 197–203.

Hewitt, P. H., Watson, J. A. L., Nel, J. J. C. & Schoeman, I. (1969). Control of the change from group to pair behaviour by *Hodotermes mossambicus* reproductives. *Journal of Insect Physiology*, **18**, 143–50.

Hill, A. V. (1911). A new form of differential microcalorimeter, for the estimation of heat produced in physiological, bacteriological or ferment actions. *Journal of Physiology*, **43**, 261–85.

Hill, G. F. (1921). The white ant pest in northern Australia. *Bulletin of the Advisory Council of Science and Industry, Melbourne*, **21**, 1–26.

Hocking, B. (1970). Insect associations with the swollen thorn acacias. *Transactions of the Royal Entomological Society, London*, **122**, 211–55.

Hodgson, E. S. (1955). An ecological study of the behaviour of the leaf-cutting ant *Atta cephalotes. Ecology*, **36**, 293–304.

Holdaway, F. G. & Gay, F. J. (1948). Temperature studies of the habitat of *Eutermes exitiosus* with special reference to the temperatures within the mound. *Australian Journal of Scientific Research, B*, **1**, 464–93.

Holmquist, A. M. (1928). Studies in arthropod hibernation. II. The hibernation of the ant *Formica ulkei* Emery. *Physiological Zoology*, **1**, 325–57.

Holt, S. J. (1955). On the foraging activity of the wood ant. *Journal of Animal Ecology*, **24**, 1–34.

Honczarenko, J. (1964). Badania nad entomofauną glebową w różnych typach płodozmianów. *Polskie pismo entomologiczne, Wrocław*, **5**(1–2), 57–69.

Honigberg, B. M. (1970). Protozoa associated with termites and their role in digestion. In *Biology of termites*, ed. K. Krishna & F. M. Weesner, vol. 2, pp. 1–36. New York & London: Academic Press.

Hopkins, B. (1966). Vegetation of the Olokemeji Forest Reserve, Nigeria. IV. The litter and soil with special reference to their seasonal changes. *Journal of Ecology*, **54**, 687–703.

Horn-Mrozowska, E. (1974). Elementy bilansu energetycznego doswiadczelnego gniazda *Formica pratensis* Retzius (Hymenoptera: Formicidae). Ph.D. thesis, Nencki's Institute of Experimental Biology, Polish Academy of Sciences, Warsaw, 102 pp.+33 tables+52 figures.

Horstman, K. (1974). Untersuchungen über den Nahrungserwerb der Waldameisen (*Formica polyctena* Foerst.). In Eichenwald. III. Jahresbilanz. *Oecologia*, **15**, 187–204.

Howse, P. E. (1970). *Termites: a study in social behaviour.* London: Hutchinson. 150 pp.

Hrdý, I., Novák, V. J. A. & Skrobal, D. (1960). Influence of the queen inhibitory substance of honeybee on the development of supplementary reproductives in the termite *Kalotermes flavicollis*. In *The ontogeny of insects* (Acta Symposii de Evolutione Insectorum, Prague, 1959), ed. I. Hrdý, pp. 172–4. London: Academic Press.

Hrdý, I. & Zelaný, J. (1967). Preference for wood of different degrees of dampness in some termites from Cuba (Isoptera). *Acta entomologica, Bohemoslov*, **64**, 352–63.

Hungate, R. E. (1936). Studies on the nutrition of *Zootermopsis*. I. The role of bacteria and molds in cellulose decomposition. *Zentralblatt für Bakteriologie, Parasitenkunde, Infektionskrankheiten und Hygiene*, Abt. 2, **94**, 240–9.

Hungate, R. E. (1938). Studies on the nutrition of *Zootermopsis*. II. The relative importance of the termite and the protozoa in wood digestion. *Ecology*, **19**, 1–25.

Hungate, R. E. (1939). Experiments on the nutrition of *Zootermopsis*. III. The anaerobic carbohydrate dissimilation by the intestinal protozoa. *Ecology*, **20**, 230–45.

Hungate, R. E. (1940). Nitrogen content of sound and decayed coniferous woods and its relation to loss in weight during decay. *Botanical Gazette*, **102**, 382–92.

Hungate, R. E. (1941). Experiments on the nitrogen economy of termites. *Annals of the Entomological Society of America*, **34**, 467–89.

Hungate, R. E. (1943). Quantitative analyses on the cellulose fermentation by termite protozoa. *Annals of the Entomological Society of America*, **36**, 730–9.

Hungate, R. E. (1944). Termite growth and nitrogen utilization in laboratory cultures. *Proceedings and Transactions of the Texas Academy of Science*, **27**, 91–8.

Hungate, R. E. (1946*a*). Studies on cellulose fermentation. II. An anaerobic cellulose decomposing actinomycete, *Micromonospora propionici* n. sp. *Journal of Bacteriology*, **51**, 51–6.

Hungate, R. E. (1946*b*). The symbiotic utilization of cellulose. *Journal of the Elisha Mitchell Scientific Society*, **62**, 9–24.

Hungate, R. E. (1955). Mutualistic intestinal protozoa. In *Biochemistry and physiology of protozoa*, ed. S. H. Hunter & A. Lwoff, vol. 2, pp. 159–99. New York: Academic Press.

Hungate, R. E. (1966). *The rumen and its microbes*. New York & London: Academic Press. 533 pp.

Ishay, J. & Ikan, R. (1969). Gluconeogenesis in the Oriental hornet *Vespa orientalis* F. *Ecology*, **49**, 169–71.

Ito, M. (1973). Seasonal population trends and nest structure in a polydomous ant, *Formica* (*Formica*) *yessensis* Forel. *Journal of the Faculty of Science, Hokkaido University*, Ser. VI, **19**, 270–93.

Ivlev, V. G. (1934). Eine mikromethode zur Bestimmung des Kaloriegehalts von Nahrostoffen. *Biochemische Zeitschrift*, **275**, 49–55.

Jacoby, M. (1953). Die Erforschung des Nestes der Blattschneiderameise *Atta sexdens rubropilosa* Forel. I. *Zeitschrift für angewandte Entomologie*, **34**, 145–69.

Jacoby, M. (1955). Die erforschung des nestes der Blattschneiderameise *Atta sexdens rubropilosa* Forel. II. *Zeitschrift für angewandte Entomologie*, **37**, 129–52.

Jakubczyk, H., Czerwiński, Z. & Pętal, J. (1972). Ants as agents of the soil habitat changes. *Ekologia polska*, **20**, 153–61.

Janet, C. (1899). *Essai sur la constitution morphologique de la tête de l'insecte*. Paris: Carré et Naud.

Janet, Ch. (1904). *Observations sur les fourmis*. Paris: Carré et Naud. 68 pp., 7 plates.

Janzen, D. H. (1967). Interaction of the bull's-horn acacia (*Acacia cornigera* L.) with an ant inhabitant (*Pseudomyrmex ferruginea* F. Smith) in eastern Mexico. *Kansas University Science Bulletin*, **47**, 315–558.

Janzen, D. H. (1969). Allelopathy by myrmecophytes: the ant *Azteca* as an allelopathic agent of *Cecropia*. *Ecology*, **50**, 147–53.

Janzen, D. H. (1973*a*). Dissolution of mutualism between *Cecropia* and its *Azteca* ants. *Biotropica*, **5**, 15–28.

Janzen, D. H. (1973*b*). Evolution of polygnous obligate acacia – ants in western Mexico. *Journal of Animal Ecology*, **42**, 727–50.

Jensen, T. F. & Nielsen, M. G. (1975). The influence of body size and temperature on worker ant respiration. *Natura Jutlandica*, **18**, 21–5.

Jermyn, M. A. (1955). Cellulose and hemicelluloses. In *Modern methods of plant analyses*, ed. K. Paech & M. V. Tracey, pp. 197–225. Berlin, Göttingen & Heidelberg: Springer-Verlag.

Joachim, A. W. R. & Kandiah, S. (1940). Studies on Ceylon soils. XIV. A comparison of soils from termite mounds and adjacent land. *The Tropical Agriculturist and Magazine of the Ceylon Agricultural Society*, **95**, 333–8.

Jonkman, J. C. M. (1971). *El Ysau*. Universidad Nacional de Asuncion, Instituto de Ciencias, Paraguay. 23 pp. (Mimeo.)

Josens, G. (1971*a*). Le renouvellement des meules à champignons construites par quatre Macrotermitinae (Isoptères) des savanes de Lamto-Pacobo (Côte d'Ivoire). *Comptes rendus hebdomadaires des séances de l'Académie des sciences, Paris*, Sér. D, **272**, 3329–32.

Josens, G. (1971*b*). Recherches écologiques dans la savane de Lamto (Côte d'Ivoire): données préliminaires sur le peuplement en termites. *Terre et la vie*, **25**, 255–72.

Josens, G. (1971*c*). Variations thermiques dans les nids de *Trinervitermes geminatus* Wasmann, en relation avec le milieu extérieur dans la savane de Lamto (Côte d'Ivoire). *Insectes sociaux*, **18**(1), 1–14.

Josens, G. (1972). Etudes biologiques et écologiques des termites (Isoptera) de la savane de Lamto-Pakobo (Côte d'Ivoire). Doctoral thesis, University of Brussels, Brussels.

Josens, G. (1973). Observations sur les bilans énergétiques dans deux populations de termites à Lamto (Côte d'Ivoire). *Annales de la Société royale zoologique de Belgique*, **103**, 169–76.

Kaczmarek, W. (1963). An analysis of interspecific competition in communities of the soil macrofauna of some habitats in the Kampinos National Park. *Ekologia polska*, **11**(17), 421–83.

Kaiser, P. (1953). *Anoplotermes pacificus*, eine mit Pflanzenwurzeln vergelleschaftet lebende Termite. *Mitteilungen aus dem Zoologischen Museum in Hamburg*, **52**, 77–92.

Kajak, A. (ed.). (1974). Analysis of a sheep pasture ecosystem in the Pieniny mountains (the Carpathians). XVII. Analysis of the transfer of carbon. *Ekologia polska*, **22**, 711–32.

Kajak, A., Breymeyer, A. & Pętal, J. (1971). Productivity investigation of two types of meadows in the Vistula Valley. XI. Predatory arthropods. *Ekologia polska*, **19**, 223–33.

Kajak, A., Breymeyer, A., Pętal, J. & Olechowicz, E. (1972). The influence of ants on the meadow invertebrates. *Ekologia polska*, **20**, 163–71.

Kakaliev, K. & Saparliev, K. (1973). Enquiry into some problems in the ecology of termite predators. In *Termite studies with a review of termite control measures. Proceedings of the All-Union Congress in Ashkabad, 1973 ' Ylym' Ashkabad*, ed. A. O. Tashliev, pp. 215.

Kalmbach, E. R. (1943). *The armadillo, its relation to agriculture and game*. Texas: Game, Fish and Oyster Commission.

Kalshoven, L. G. E. (1930). De biologie van de Djatermiet (*Kalotermes tectonae* Damm) in verband met Zijn bestrijding. *Mededelingen van het Instituut voor plantenziekten, Buitenzorg*, **76**, 1–154.

Kalshoven, L. G. E. (1956). Observations on *Macrotermes gilvus* Holmgr. in Java.

3. Accumulations of finely cut vegetable matter in the nests. *Insectes sociaux*, **3**, 455–61.

Kalshoven, L. G. E. (1958). Observations on the black termites, *Hospitalitermes* spp. of Java and Sumatra. *Insectes sociaux*, **5**, 9–30.

Kannowski, P. B. (1956). The ants of Ramsey County, North Dakota. *American Midland Naturalist*, **56**(1), 168–85.

Kannowski, P. B. (1959). The use of radioactive phosphorus in the study of colony distribution of the ant *Lasius minutus*. *Ecology*, **40**, 162–5.

Kannowski, P. B. (1963). The flight activities of formicine ants. *Symposia genetica et biologica italica*, **12**, 74–102.

Kannowski, P. B. (1967). Colony populations of two species of *Dolichoderus* (Hymenoptera, Formicidae). *Annals of the Entomological Society of America*, **60**, 1246–52.

Kannowski, P. B. (1970). Colony populations of five species of *Myrmica*. *Proceedings of the Entomological Society of America, North Central Branch*, **25**, 119–25.

Karpiński, J. J. (1956). Mrówki w biocenozie Bialowieskiego Parku Narodowego. *Rocznik nauk leśnych*, **14**, 201–21.

Kâto, M. (1939). The diurnal rhythm of temperature in the mound of an ant, *Formica truncorum truncorum var. yesseni* Forel, widely distributed at Mt. Hakkôda. *Science Reports of the Tôhoku Imperial University*, **14**, 53–64.

Katzin, L. K. & Kirby, H., Jr (1939). The relative weights of termites and their protozoa. *Journal of Parasitology*, **25**, 444–5.

Kay, C. A. & Whitford, W. G. (1975). Influences of temperature and humidity on oxygen consumption of five Chihuahuan Desert ants. *Comparative Biochemistry and Physiology*, **52A**, 281–6.

Keay, R. W. J. & Aubreville, A. (1959). *Vegetation map of Africa, south of the Tropic of Cancer.* Oxford: Oxford University Press.

Keister, M. & Buck, J. (1964). Respiration: some exogenous and endogenous effects on the rate of respiration. In *Physiology of insecta 3*, ed. W. T. Edmondson & G. G. Winberg, pp. 617–58. New York & London: Academic Press.

Kemp, P. B. (1955). The termites of north-eastern Tanganyika: their distribution and biology. *Bulletin of Entomological Research*, **46**, 113–35.

Kempf, W. W. (1962). Retoques à classifição das formigas neotropicais do gênero *Heteroponera* Mayr. *Papéis avulsos do Departamento de zoologia Secretaria de agricultura, São Paulo*, **15**, 29–47.

Kempf, W. W. & Brown, W. L., Jr (1970). Two new ants of tribe Ectatommini from Colombia. *Studia entomologica*, **13**, 311–20.

Kennedy, C. H. & Talbot, M. (1939). Notes on the hypogaeic ant, *Proceratium silaceum* Roger. *Proceedings of the Indiana Academy of Science*, **48**, 202–10.

Kennington, G. S. (1957). Influence of altitude and temperature upon rate of oxygen consumption of *Tribolium confusum* Duval and *Camponotus pennsylvanicus modoc* Wheeler. *Physiological Zoology*, **30**, 305–14.

Khamala, C. P. M. & Büschinger, A. (1971). Effect of temperature and season on food transmission: activity of 3 ant species as shown by radioactive tracers. *Zeitschrift für angewandte Entomologie*, **67**, 337–42.

Kirchner, W. (1964). Jahreszyklische untersuchungen zur Reservestoffspeicherung und Überlebensfähigkeit adulter Waldameisenarbeiterinnen. *Zoologische Jahrbücher (Physiologie)*, **71**, 1–72.

Kirk, T. K. (1971). Effects of microorganisms on lignin. *Annual Review of Phytopathology*, **9**, 185–210.

Kirk, T. K. & Harkin, J. M. (1972). *Lignin biodegradation and the bioconversion of wood. American Institute of Chemical Engineers, Annual Meeting*, Paper No. **30a**, 11 pp.

Kistner, D. H. (1968). A taxonomic revision of the termitophilous tribe Termitopaedini, with notes on behavior, systematics, and post-imaginal growth (Coleoptera: Staphylinidae). *Miscellaneous Publications of the Entomological Society of America*, **6**, 141–96.

Kistner, D. H. (1969). The biology of termitophiles. In *Biology of termites*, ed. K. Krishna & F. M. Weesner, vol. 1, pp. 525–57. New York & London: Academic Press.

Kistner, D. H. (1973). The termitophilous Staphylinidae associated with *Grallatotermes* in Africa; their taxonomy, behavior, and a survey of their glands of external secretion. *Annals of the Entomological Society of America*, **66**, 197–222.

Kleiber, M. (1961). *The fire of life. An introduction to animal energetics.* New York: Wiley.

Klimetzek, D. (1970). Zur Bedeutung des Kleinstandorts für die Verbreitung hügelbauender Waldameisen der *Formica rufa*-Gruppe. *Zeitschrift für angewandte Entomologie*, **66**, 84–95.

Klimetzek, D. (1972). Veränderungen in einem natürlichen Vorkommen hügelbauender Waldameisen der *Formica rufa*-Gruppe (Hymenoptera: Formicidae) im Verlauf von drei Jahren. *Insectes sociaux*, **19**, 1–5.

Kloft, W. (1959). Versuch einer Analyse der trophobiotischen Beziehungen von Ameisen zu Aphiden. *Biologisches Zentralblatt*, **78**, 863–70.

Kneitz, G. (1967). Untersuchungen zum Atmungsstoffwechsel der Arbeiterinnen von *Formica polyctena* Foerst. (Hym., Formicidae). *Proceedings, Vth Congress of the International Union for the Study of Social Insects, Toulouse, 1965*, 277–91.

Kneitz, G. (1969). Temperaturprofile in Waldameisennestern. *Proceedings, VIth Congress of the International Union for the Study of Social Insects, Berne, 15–20 September, 1969*, 95–100.

Kofoid, C. A. & Bowe, E. E. (1934). Biological tests of untreated and treated woods. In *Termites and termite control*, ed. C. A. Kofoid, pp. 517–44. Berkeley: University of California Press.

König, H. (1958). Die atmosphärische Impulsstrahlung. *Medizin-Meteorologischer*, **13**, 157–60.

Korovkina, N. M. (1971). pH of termite intestine. *Proceedings of the XIIIth International Congress of Entomology, Moskva, 2–9 August, 1968*, **1**, 402–3.

Kovoor, J. (1964*a*). Modifications chimiques d'une sciure de bois de peuplier sous l'action d'un termitidé: *Microcerotermes edentatus* Wasmann. *Comptes rendus hebdomadaires des séances de l'Académie des sciences, Paris*, **258**, 2887–9.

Kovoor, J. (1964*b*). Modifications chimiques provoquées par un termitidé (*Microcerotermes edentatus* Was.) dans du bois de peuplier sain ou partiellement dégradé par des champignons. *Bulletin biologique de la France et de la Belgique*, **98**, 491–510.

Kovoor, J. (1967*a*). Présence d'acides gras volatils dans la panse d'un termite supérieur (*Microcerotermes edentatus* Was., Amitermitinae). *Comptes rendus hebdomadaires des séances de l'Académie des sciences, Paris*, Sér. D, **264**, 486–8.

Kovoor, J. (1967*b*). Le pH intestinal d'un termite supérieur (*Microcerotermes edentatus* Was., Amitermitinae). *Insectes sociaux*, **14**, 157–60.

Kovoor, J. (1967*c*). Etude radiographique du transit intestinal chez un termite supérieur. *Experientia*, **23**, 820–3.

Kovoor, J. (1970). Présence d'enzymes cellulolytiques dans l'intestin d'un termite supérieur, *Microcerotermes edentatus* Was. *Annales des sciences naturelles (Zoologie)*, **12**, 65–71.

Krishna, S. S. & Singh, N. B. (1968). Sugar-dye movement through the alimentary canal of *Odontotermes obesus* (Isoptera: Termitidae). *Annals of the Entomological Society of America*, **61**, 230.

Krishna, K. & Weesner, F. M. (eds.) (1969). *Biology of termites*, vol. 1. New York & London: Academic Press.

Krishna, K. & Weesner, F. M. (eds.) (1970). *Biology of termites*, vol. 2. New York & London: Academic Press.

Krishnamoorthy, R. V. (1960). The digestive enzymes of the termite *Heterotermes indicola*. *Journal of Animal Morphology and Physiology*, **7**, 156–61.

Krogh, A. (1914). On the rate of development and CO_2 production of chrysalids of *Tenebrio molitor* at different temperatures. *Zeitschrift für allgemeine Physiologie*, **16**, 178–90.

Krupienikov, N. A. (1951). Nablyudeniya nad vliyaniem nasekomȳkh na pochvu. *Byulleten' Moskovskogo obschestva ispȳtateleĭ priørodȳ*, **56**(1), 45–8.

Kruuk, H. & Sands, W. A. (1972). The aardwolf (*Proteles cristatus* Sparrman, 1783) as predator of termites. *East African Wildlife Journal*, **10**, 211–27.

Kürschner, I. (1971). Zur Anatomie von *Formica pratensis* Retzius, 1783. *Beiträge zur Entomologie*, **21**(1/2), 191–210.

Kusnezov, N. (1949). Sobre la reproduccion de las formas sexuales en *Solenopsis patagonica* Emery. *Acta zoologica lilloana*, **8**, 281–90.

Kusnezov, N. (1951). El genero '*Pogonomyrmex*' Mayr. *Acta zoologica lilloana*, **9**, 227–333.

Kutter, H. (1952). Über *Plagiolepis xene* Stärcke. *Mitteilungen der Schweizerischen entomologischen Gesellschaft*, **25**, 57–72.

Kutter, H. (1956). Beiträge zur Biologie palearktischer *Coptoformica*. *Mitteilungen der Schweizerischen entomologischen Gesellschaft*, **29**, 1–18.

Kutter, H. & Stumper, R. (1969). Hermann Appel, ein leidgeadelter Entomologe (1892–1966). *Proceedings, VIth Congress of the International Union for the Study of Social Insects, Berne, 15–20 September, 1969*, 275–79.

Labrador, J. R., Martinez, I. J. Q. & Mora, A. (1972). *Acromyrmex landolti* Forel, plaga del Pasto Guinea (*Panicum maximum*) en el estado Zulia. *Jornal de Agronomia, Universidad de Zulia*, **8**, 12 pp.

La Fage, J. P., Nutting, W. L. & Haverty, M. I. (1973). Desert subterranean termites: a method for studying foraging behavior. *Environmental Entomology*, **2**, 954–56.

Lamotte, M. (1947). Recherches écologiques sur le cycle saisonnier d'une savane guinéene. *Bulletin de la Société zoologique de France, Paris*, **72**, 88–90.

Lamotte, M. (1975). The structure and function of a tropical savanna ecosystem. In *Tropical ecological systems: trends in terrestrial and aquatic research*, ed. F. Golley & E. Medina, pp. 179–222. New York & Berlin: Springer-Verlag.

Lamprey, H. P. (1964). Estimation of the large mammal densities, biomass and energy exchange in the Tarangire Game Reserve and the Masai Steppe in Tanganyika. *East African Wildlife Journal*, **2**, 1–47.

Länge, R. (1967). Die Nahrungsverteilung unter den Arbeiterinnen des Waldameisenstaates. *Zeitschrift für Tierpsychologie*, **24**(5), 513–45.
Lasker, R. (1959). Cellulose digestion in insects. In *Marine boring and fouling organisms*, ed. D. L. Ray, pp. 348–55. Seattle: University of Washington Press.
Lasker, R. & Giese, A. C. (1956). Cellulose digestion by the silverfish *Ctenolepisma lineata*. *Journal of Experimental Biology*, **33**, 542–53.
Lavette, A. (1964). La digestion du bois par les flagellés symbiotiques des termites: cellulose et lignine. *Comptes rendus hebdomadaires des séances de l'Académie des sciences, Paris*, **258**, 2211–13.
Lavette, A. (1970). A propos de la digestion de bois par voie fermentaire et sur la présence de la succino-déshydrogénase chez les flagellés termiticoles. *Comptes rendus hebdomadaires des séances de l'Académie des sciences, Paris*, Sér. D, **270**, 2002–4.
Lavigne, R. J. (1969). Bionomics and nest structure of *Pogonomyrmex occidentalis*. *Annals of the Entomological Society of America*, **62**, 1166–75.
Law, J. H. & Regnier, F. E. (1971). Pheromones. *Annual Review of Biochemistry*, **40**, 533–48.
Leach, J. G. & Granovsky, A. A. (1938). Nitrogen in the nutrition of termites. *Science*, **87**, 66–7.
Lebrun, D. (1963). Implantation de glandes de la mue de *Periplaneta americana* dans des soldats de *Calotermes flavicollis* (Fabr.). *Comptes rendus hebdomadaires des séances de l'Académie des sciences, Paris*, **257**, 3487–8.
Lebrun, D. (1972). Effets de l'implantation de glandes mandibulaires sur le differentiation imaginale de *Calotermes flavicollis* (Fabr.). *Comptes rendus hebdomadaires des séances des l'Académie des sciences, Paris*, Sér. D, **274**, 2077–9.
Lebrun, D. (1973). Pheromones et determinisme des castes de *Calotermes flavicollis* (Fabr.). *Proceedings, VIIth International Congress, of the International Union for the Study of Social Insects, London, 10–15 September, 1973*, 220–4.
Le Cren, E. D. (1965). A note on the history of mark–recapture population estimates. *Journal of Animal Ecology*, **34**, 453–4.
Ledoux, A. (1950). Recherche sur la biologie de la fourmi fileuse *Oecophylla longinoda* (Latr.). *Annales des sciences naturelles* (*Zoologie*), **12**, 313–461.
Ledoux, A. (1952). Recherches préliminaires sur quelques points de la biologie d'*Odontomachus assiniensis* Latr. *Annales des sciences naturelles* (*Zoologie*), **14**, 231–48.
Lee, K. E. & Wood, T. G. (1968). Preliminary studies of the role of *Nasutitermes exitiosus* (Hill) in the cycling of organic matter in a yellow podzolic soil under dry sclerophyll forest in South Australia. *Transactions of the 9th International Congress of Soil Science, Adelaide*, **2**, 11–18.
Lee, K. E. & Wood, T. G. (1971*a*). Physical and chemical effects on soils of some Australian termites and their pedological significance. *Pedobiologia*, **11**, 376–409.
Lee, K. E. & Wood, T. G. (1971*b*). *Termites and soils*. New York & London: Academic Press.
Lehmann, J. (1974). Ist der Nahrungspilz der pilzzuchtenden Blattschneiderameisen und Termiten ein Aspergillus? *Waldhygiene*, **10**(8), 252–5.
Lehninger, A. L. (1970). *Biochemistry*. New York: Worth Publishers Inc. 833 pp.
Le Masne, G. (1953). Observations sur les relations entre le couvain et les adultes chez les fourmis. *Annales des sciences naturelles* (Zoologie), **15**, 1–56.

Le Masne, G. & Bonavita-Cougourdan, A. (1972). Premiers résultats d'une irradiation prolongée au césium sur les populations de fourmis en Haute Provence. *Ekologia polska*, **20**, 129–44.

Le Masne, G. & Torossian, C. (1965). Observations sur le comportement du Coléoptère myrmécophile *Amorphocephalus coronatus* Germar (Brenthidae) hôte des *Camponotus. Insectes sociaux*, **12**, 185–94.

Leopold, B. (1952). Studies on lignin. XIV. The composition of Douglas fir wood digested by the West Indian dry-wood termite (*Cryptotermes brevis* (Walker)). *Svensk Papperstidning*, **55**, 784–6.

Lepage, M. (1972). Recherches écologiques sur une savane sahelienne du Ferlo septentrional, Sénégal: données préliminaires sur l'écologie des termites. *Terre et la vie*, **26**, 388–409.

Lepage, M. (1974*a*). Recherches écologiques sur une savane sahélienne du Ferlo septentrional, Sénégal: influence de la sécheresse sur la peuplement en termites. *Terre et la vie* **28**, 76–94.

Lepage, M. (1974*b*). Les termites d'une savane sahélienne (Ferlo Septentrional, Sénégal): peuplement, populations, consommation, rôle dans l'écosystème. Doctoral Thesis, University of Dijon, Dijon.

Leslie, P. H. (1952). The estimation of population parameters from data obtained by means of the capture–recapture method. II. The estimation of total numbers. *Biometrika*, **39**, 363–88.

Leston, D. (1971). The Ectatommini of Ghana. *Journal of Entomology*, **40**, 117–20.

Leston, D. (1973). The natural history of some West African insects. *Entomologist's Monthly Magazine*. **108**, 110–22.

Letendre, M. & Pilon, J.-G. (1972). Écologie des populations de *Leptothorax longispinosus* Roger et *Stenamma diecki* Emery dans les peuplements forestiers des basses laurentides, Québec. *Naturaliste canadien*, **99**, 73–82.

Letendre, M. & Pilon, J.-G. (1973). Nids et micromilieux de nidification utilisés par les fourmis dans les peuplements forestiers des basses laurentides, Québec (Hymenoptera: Formicidae). *Naturaliste canadien*, **100**, 237–46.

Leung, W.-T. W. (1968). *Food composition table for use in Africa*. Maryland & Rome: United States Department of Health, Education and Welfare & Food and Agriculture Organization, United Nations. ix+306 pp.

Levi, M. P., Merrill, W. & Cowling, E. B. (1968). Role of nitrogen in wood deterioration. VI. Mycelial fractions and model nitrogen compounds as substrates for growth of *Polyporus versicolor* and other wood-destroying and wood-inhabiting fungi. *Phytopathology*, **58**, 626–34.

Lévieux, J. (1966). Traits généraux du peuplement en fourmis terricoles d'une savane de Côte d'Ivoire. *Comptes rendus hebdomadaires des séances de l'Académie des sciences, Paris*, **262**, 1583–5.

Lévieux, J. (1967*a*). Recherches écologiques dans la savane de Lamto (Côte d'Ivoire): données préliminaires sur le peuplement en fourmis terricoles. *Terre et la vie*, **3**, 278–96.

Lévieux, J. (1967*b*). La place de *Camponotus aevapimensis* Mayr. (Hyménoptère, Formicidae) dans la chaine alimentaire d'une savane de Côte d'Ivoire. *Insectes sociaux*, **14**, 313–22.

Lévieux, J. (1969). L'échantillonage des peuplements de fourmis terricoles. In *Problèmes d'écologie: l'échantillonage des peuplements animaux des milieux terrestres*, ed. M. Lamotte & F. Bourlière, pp. 289–300. Comité Français du Programme Biologique International.

Lévieux, J. (1971*a*). Mise en évidence de la structure des nids et de l'implantation des zones de chasse de deux espèces de *Camponotus* (Hymenoptera: Formicidae) à l'aide de radio-isotopes. *Insectes sociaux*, **18**, 29–48.

Lévieux, J. (1971*b*). Données écologiques et biologiques sur le peuplement en fourmis terricoles d'une savane préforestière de Côte d'Ivoire. Doctoral thesis, University of Paris, Paris, 282 pp.

Lévieux, J. (1972*a*). Le microclimat des nids et des zones de chasse de *Camponotus aevapimensis* Mayr. *Insectes sociaux*, **19**, 63–79.

Lévieux, J. (1972*b*). Le rôle des fourmis dans les réseaux trophiques d'une savane préforestière de Côte d'Ivoire. *Annales de l'Université d'Abidjan*, Sér. *E*, **5**(1), 143–240.

Lévieux, J. (1972*c*). Quelques remarques au sujet des méthodes d'échantillonnage des peuplements de formis terricoles. *Ekologia polska*, **20**, 1–7.

Lévieux, J. (1973). Etude du peuplement en fourmis terricoles d'une savane préforestière de Côte d'Ivoire. *Revue d'écologie et de biologie du sol*, **10**(3), 379–428.

Levis, J. K. (1971). The grassland biome: a synthesis of structure and function, 1970. In *Preliminary analysis of structure and function in grasslands*, ed. Norman R. French, pp. 317–87. Fort Collins, Colorado: Colorado State University.

Lewis, T., Pollard, G. V. & Dibley, G. C. (1974). Micro-environmental factors affecting diel patterns of foraging in the leaf-cutting ant *Atta cephalotes* (L.) (Formicidae: Attini). *Journal of Animal Ecology*, **43**, 143–53.

Light, S. F. & Weesner, F. M. (1947). Methods for culturing termites. *Science*, **106**, 131.

Light, S. F. & Weesner, F. M. (1951). Further studies on the production of supplementary reproductives in *Zootermopsis* (Isoptera). *Journal of Experimental Zoology*, **117**, 397–414.

Light, S. F. & Weesner, F. M. (1955). The incipient colony of *Tenuirostritermes tenuirostris* (Desneaux). *Insectes sociaux*, **2**, 135–46.

Lincoln, F. B. & Mulay, A. S. (1929). The extraction of nitrogenous materials from pear tissues. *Plant Physiology*, **4**, 233–50.

Lints, C. V., Lints, F. A. & Zeuthen, E. (1967). Respiration in *Drosophila*. 1. Oxygen consumption during development of the egg in genotypes of *Drosophila melanogaster* with contributions to the gradient diver technique *Comptes rendus des travaux du Laboratoire Carlsberg*, **36**, 35–66.

Lockwood, L. L. (1973). Distribution density and dispersion of two species of *Atta* (Hymenoptera, Formicidae) in Guanacaste Province, Costa Rica. *Journal of Animal Ecology*, **42**, 803–18.

Lofgren, C. S., Banks, W. A. & Glancey, B. M. (1975). Biology and control of imported fire ants. *Annual Review of Entomology*, **20**, 1–30.

Loos, R. (1964). A sensitive anemometer and its use for the measurement of air currents in the nests of *Macrotermes natalensis* (Haviland). In *Etudes sur les termites africains*, ed. A. Bouillon, pp. 363–72. Paris: Masson et Cie.

Løvlie, A. & Zeuthen, E. (1962). The gradient diver, a recording instrument for gasometric micro-analysis. *Comptes rendus des travaux du Laboratoire Carlsberg*, **32**, 513–34.

Løvtrup, S. & Larsson, E. (1965). Electromagnetic recording diver balance. *Nature, London*, **208**, 1116–17.

Lubbock, J. (1883). *Fourmis, abeilles et guêpes. Etudes expérimentales sur l'organisation et les moeurs des sociétés d'insectes hyménoptères*. Paris: Germer Baillière et Cie. 1, xi+195 pp., 13 plates.

Lugo, A. E., Farnworth, E. G., Pool, D., Jerez, P. & Kaufman, G. (1973). The impact of the leaf cutter ant *Atta colombica* on the energy flow of a tropical wet forest. *Ecology*, **54**, 1292–301.

Lund, E. E. (1930). The effect of diet upon the intestinal fauna of *Termopsis. University of California Publications in Zoology*, **36**, 81–96.

Lüscher, M. (1952). Die produktion und Elimination von Ersatzgeschlechstieren bei der Termite *Kalotermes flavicollis* (Fabr.). *Zeitschrift für vergleichende Physiologie*, **34**, 123–41.

Lüscher, M. (1953). The termite and the cell. *Scientific American*, **188**(5), 75.

Lüscher, M. (1955*a*). Zur Frage der Übertragung sozialer Wirkstoffe bei Termiten. *Die Naturwissenschaften*, **42**, 186.

Lüscher, M. (1955*b*). Der Sauerstoffverbrauch bei Termiten und die Ventilation des Nestes bei *Macrotermes natalensis* (Haviland). *Acta tropica*, **12**, 289–307.

Lüscher, M. (1956*a*). Die Entstehung von Ersatzgeschlechtstieren bei der Termite *Kalotermes flavicollis* (Fabr.). *Insectes sociaux*, **3**, 119–28.

Lüscher, M. (1956*b*). Die Lufterneuerung im Nest der Termite *Macrotermes natalensis* (Haviland). *Insectes sociaux*, **3**, 273–76.

Lüscher, M. (1958). Experimentelle Erzeugung von Soldaten bei der Termite *Kalotermes flavicollis. Die Naturwissenschaften*, **45**, 69–70.

Lüscher, M. (1960). Hormonal control of caste differentiation in termites. *Annals of the New York Academy of Sciences*, **89**, 549–63.

Lüscher, M. (1961*a*). Social control of polymorphism in termites. In *Insect polymorphism*, ed. J. S. Kennedy, pp. 57–67. *Symposium of the Royal Entomological Society of London*, No. **1**.

Lüscher, M. (1961*b*). Air-conditioned termite nests. *Scientific American*, **205**, 138–45.

Lüscher, M. (1964). Die spezifische Wirkung männlicher und weiblicher Ersatzgeschlechtstiere auf die Entstehung von Ersatzgeschlechtstieren bei der Termite *Kalotermes flavicollis* (Fabr.). *Insectes sociaux*, **11**, 79–90.

Lüscher, M. (1972). Environmental control of juvenile hormone (JH) secretion and caste differentiation in termites. *General and Comparative Endocrinology*, Supplement **3**, 509–14.

Lüscher, M. (1973). The influence of the composition of experimental groups on caste development in *Zootermopsis* (Isoptera). *Proceedings, VIIth International Congress of the International Union for the Study of Social Insects, London, 10–15 September, 1973*, 253–6.

Lüscher, M. & Springhetti, A. (1960). Untersuchungen über die Bedeutung der Corpora Allata für die Differenzierung der Kasten bei der Termite *Kalotermes flavicollis* (F.). *Journal of Insect Physiology*, **5**, 190–212.

Lyford, W. H. (1963). *Importance of ants to brown podzolic soil genesis in New England. Harvard Forest Paper*, No. **7**. Cambridge, Massachusetts: Harvard University.

Macfadyen, A. (1961). A new system for continuous respirometry of small air-breathing invertebrates in near-natural conditions. *Journal of Experimental Biology*, **38**, 323–41.

Macfadyen, A. (1971). The soil and its total metabolism. In *Methods of study in quantitative soil ecology: population, production and energy flow*, IBP Handbook No. 18, ed. J. Phillipson, pp. 1–13. Oxford: Blackwell Scientific Publications.

Macfadyen, W. A. (1950). Vegetation patterns in the semi-desert plains of British Somaliland. *Geographical Journal*, **116**, 199–211.

Macfadyen, G. (1967). Methods of investigation of productivity of invertebrates in terrestrial ecosystems. In *Secondary productivity of terrestrial ecosystems*, ed. K. Petrusewicz, vol. II, 383–412. Warszawa –Kraków: Państwowe Wydawnictwo Naukowe.

Madge, D. S. (1969). Litter disappearance in forest and savanna. *Pedobiologia*, **9**, 288–99.

Maignier, R. (1966). *Review of Research on Laterites*. Paris: UNESCO.

Malisse, F., Freson, R., Goffiret, G. & Malaisse-Mousset, M. (1975). Litter fall and litter breakdown in Miombo. In *Tropical ecological systems. Trends in terrestrial and aquatic research*, ed. F. Golley & E. Medina, pp. 137–52. Berlin & New York: Springer-Verlag.

Maldague, M. E. (1959). Analyses de sols et matériaux de termitières du Congo Belge. *Insectes sociaux*, **6**, 343–59.

Maldague, M. E. (1964). Importance des populations de termites dans les sols équatoriaux. *Transactions of the 8th International Congress of Soil Science, Bucharest, 1964*, **3**, 743–51.

Maldague, M. E. (1967). Aspects faunistiques de la fertilité des sols forestiers et spécialement des sols forestiers équatoriaux. Doctoral thesis, University of Louvain, Louvain, 265 pp.

Malozemova, L. A. (1972). Ants of steppe forests, their distribution by habitats and perspectives of their utilization for protection of forests (North Kazakhstan). *Zoological Journal*, **51**, 57–68. (In Russian with English summary.)

Maltais, J. B. & Auclair, J. L. (1952). Occurrence of amino acids in the honeydew of the crescent-marked lily aphid *Myzus circumflexus* (Buck.). *Canadian Journal of Zoology*, **30**, 191–3.

Mannesmann, R. (1969). Vergleichende Untersuchungen über den Einfluss der temperatur auf die Darm-Symbionten von Termiten und über die regulatorischen Mechanismen bei der Symbiose, Teil I. *Zeitschrift für angewandte Zoologie*, **56**, 385–440.

Mannesmann, R. (1972). A comparison between cellulolytic bacteria of the termites *Coptotermes formosanus* Shiraki and *Reticulitermes virginicus* (Banks). *International Biodeterioration Bulletin*, **8**, 104–11.

Manzanilla, E. B. & Ynalvez, L. A. (1958). Essential amino acid content of some tropical termites. *Philippine Agriculturist*, **42**, 36.

Marcus, H. (1949). Como las hormigas evitan el incesto. *Folia Universitaria, Cochabamba*, **3**, 95–6.

Marcus, H. & Marcus, E. E. (1951). Los nidos y los organos de estridulacion y de equilibrio de *Pogonomyrmex marcusi* y de *Dorymyrmex emmaericaellus* (Kusn). *Folia Universitaria, Cochabamba*, **5**, 117–43.

Marikovsky, P. I. (1962). On intraspecific relations of *Formica rufa* L. *Entomological Review*, **1**, 47–51.

Markham, C. P. (1966). The surface activity, behaviour and energy influx rhythms of leaf-cutting ants, *Acromyrmex* sp. and *Atta cephalotes*. Research Report in *Tropical biology: an ecological approach*, February–March, 1966. San José, Costa Rica: Organization for Tropical Studies.

Markin, G. P. (1970). Food distribution within laboratory colonies of the Argentine ant, *Iridomyrmex humilis* (Mayr). *Insectes sociaux*, **17**(2), 127–57.

Markin, G. P., Dillier, J. H., Hill, S. O., Blum, M. S. & Hermann, H. R. (1971). Nuptial flight and flight ranges of the imported fire ant, *Solenopsis saevissima richteri* (Hymenoptera: Formicidae). *Journal of the Georgia Entomological Society*, **6**(3), 145–56.

Martin, J. S. (1969). Studies on assimilation, mobilization, and transport of lipids by the fat body and haemolymph of *Pyrrhocoris apterus*. *Journal of Insect Physiology*, **15**, 2319–44.

Martin, J. S. & Martin, M. M. (1970). The presence of protease activity in the rectal fluid of attine ants. *Journal of Insect Physiology*, **16**, 227–32.

Martin, M. M. (1970). The biochemical basis of the fungus–attine symbiosis. *Science*, **169**, 16–20.

Martin, M. M. (1974). Biochemical ecology of the attine ants (Hymenoptera: Formicidae). *Accounts Chemical Research*, **7**(1), 1–5.

Martin, M. M., Boyd, N. D., Gieselmann, M. J. & Silver, R. G. (1975). Activity of faecal fluid of a leaf-cutting ant toward plant cell wall polysaccharides. *Journal of Insect Physiology*, **21**, 1887–92.

Martin, M. M., Carls, C. A., Hutchins, R. F. N., MacConnell, J. G., Martin, J. S. & Steiner, O. D. (1967). Observations on *Atta colombica tonsipes* (Hymenoptera: Formicidae). *Annals of the Entomological Society of America*, **60**, 1329–30.

Martin, M. M., Carman, R. M. & MacConnell, J. G. (1969). Nutrients derived from the fungus cultured by the fungus-growing ant *Atta colombica tonsipes*. *Annals of the Entomological Society of America*, **62**, 11–13.

Martin, M. M., Gieselmann, M. J. & Martin, J. S. (1973). Rectal enzymes of attine ants. α-amylase and chitinase. *Journal of Insect Physiology*, **19**, 1409–16.

Martin, M. M., MacConnell, M. G. & Gale, G. R. (1969). The chemical basis for the attine ant–fungus symbiosis: absence of antibiotics. *Annals of the Entomological Society of America*, **62**, 386–8.

Martin, M. M. & Martin, J. S. (1970). The biochemical basis for the symbiosis between the ant, *Atta colombica tonsipes*, and its food fungus. *Journal of Insect Physiology*, **16**, 109–19.

Martin, M. M. & Martin, J. S. (1971). The presence of protease activity in the rectal fluid of primitive attine ants. *Journal of Insect Physiology*, **17**, 1897–906.

Martin, M. M. & Weber, N. A. (1969). The cellulose utilizing capability of the fungus cultured by the attine ant *Atta colombica tonsipes*. *Annals of the Entomological Society of America*, **62**, 1386–7.

Maschwitz, U. (1966). Das Speichelsekret der Wespenlarven und seine biologische Bedeutung. *Zeitschrift für vergleichende Physiologie*, **53**(3), 228–52.

Maschwitz, U., Hölldobler, B. & Möglich, M. (1974). Tandemlaufen als Rekrutierungsverhalten bei *Bothroponera tesserinoda* Forel. *Zeitschrift für Tierpsychologie*, **35**, 113–23.

Maschwitz, U., Koob, K. & Schlidknecht, H. (1970). Ein Beitrag zur Funktion der Metathoracaldrüse der Ameisen. *Journal of Insect Physiology*, **16**, 387–404.

Marton, J. (Symposium Chairman). (1966). Lignin structure and reactions. In *Advances in chemistry series 59*, ed. R. F. Gould. Washington D.C.: American Chemical Society Publications. 267 pp.

Mathot, G. (1964). Rythme de nutrition chez *Cubitermes exiguus* Mathot, (Isoptera, Termitinae). In *Etudes sur les termites africains*, ed. A. Bouillon, pp. 263–71. Paris: Masson et Cie.

Mathot, G. (1967). Premier essai de détermination de facteurs écologiques corrélatifs à la distribution et l'abondance de *Cubitermes sankurensis* (Isoptera, Termitidae, Termitinae). *Proceedings, Vth Congress of the International Union for the Study of Social Insects, Toulouse*, 1965, 117–29.

Matsumoto, T. (1976). The role of termites in an equatorial rain forest ecosystem of West Malaysia. 1. Population density, biomass, carbon, nitrogen and calorific content and respiration rate. *Oecologia*, **22**(2), 153–78.

Matsumura, F., Coppel, H. C., & Tai, A. (1968). Isolation and identification, of termite trail-following pheromone. *Nature, London*, **219**, 963–4.

Mauldin, J. K. & Rich, N. M. (1975). Rearing two subterranean termites, *Reticulitermes flavipes* and *Coptotermes formosanus* on artifical diets. *Annals of the Entomological Society of America*, **68**, 454–6.

Mauldin, J. K. & Smythe, R. G. (1973). Protein-bound amino acid content of normally and abnormally faunated Formosan termites. *Journal of Insect Physiology*, **19**, 1955–60.

Mauldin, J. K., Smythe, R. G. & Baxter, C. E. (1972). Cellulose catabolism and lipid synthesis by the subterranean termite. *Coptotermes formosanus. Insect Biochemistry*, **2**, 209–17.

May, R. M. (1973). *Stability and complexity in model ecosystems*. Princeton: Princeton University Press.

Maynard, L. A. (1937). *Animal nutrition*. New York: McGraw-Hill. 483 pp.

McBee, R. H. (1959). Termite cellulase. In *Marine boring and fouling organisms*, ed. D. L. Ray, pp. 342–7. Seattle: University of Washington Press.

McBrayer, J. F. (1973). Exploitation of deciduous leaf litter by *Apheloria montana* (Diplopoda: Eurydesmidae). *Pedobiologia*, **13**, 90–8.

McMahan, E. A. (1963). A study of termite feeding relationships using radio-isotopes. *Annals of the Entomological Society of America*, **56**, 74–82.

McMahan, E. A. (1966*a*). Studies of termite wood-feeding preferences. *Proceedings, Hawaiian Entomological Society*, **19**, 239–50.

McMahan, E. A. (1966*b*). Food transmission within the *Cryptotermes brevis* colony (Isoptera: Kalotermitidae). *Annals of the Entomological Society of America*, **59**, 1131–7.

McMahan, E. A. (1969). Feeding relationships and radio-isotope techniques. In *Biology of termites*, ed. K. Krishna & F. M. Weesner, vol. 1, pp. 387–406. New York & London: Academic Press.

McNeill, S. & Lawton, J. H. (1970). Annual production and respiration in animal populations. *Nature, London*, **225**, 472–4.

Meiklejohn, J. (1965). Microbiological studies on large termite mounds. *Rhodesia, Zambia and Malawi Journal of Agricultural Research*, **3**, 67–79.

Menozzi, C. (1922). Nota su un nuovo genere ed una nuova specie di formica parassita. *Atti della Societa italiana di scienze naturali*. **61**, 3–7.

Menozzi, C. (1924). Res Mutinenses (Hymenoptera, Formicidae). *Atti della Societa dei naturalisti e matematici, Modena*, **3**(55), 22–47.

Menozzi, C. (1936). Nuovi contributi alla conoscenza della fauna delle isole italiane dell'Egeo (Form.). *Bollettino di zoologia generale e agraria della Facolta agraria in Portici*, **29**, 262–311.

Menozzi, C. (1940). Contributo alla fauna della Tripolitania. *Bollettino del Laboratorio di entomologia agraria 'Filippo Silvestri', Portici*, **31**, 244–73.

Meyer, D. & Lüscher, M. (1973). Juvenile hormone activity in the haemolymph and the anal secretion of the queen of *Macrotermes subhyalinus* (Rambur) (Isoptera, Termitidae). *Proceedings, VIIth International Congress of the International Union for the Study of Social Insects, London, 10–15, September, 1973*, pp. 268–73.

Meyer, J. A. (1960). Résultats agronomiques d'un essai de nivellement des termitières réalisé dans la Cuvette centrale Congolaise. *Bulletin agricole du Congo belge*, **51**, 1047–59.

Miller, E. M. (1942). The problem of castes and caste differentiation in *Prorhinotermes simplex* (Hagen). *Bulletin of the University of Miami*, **15**, 1–27.

Miller, E. M. (1969). Caste differentiation in the lower termites. In *Biology of termites*, ed. K. Krishna & F. M. Weesner, vol. 1, pp. 283–310. New York & London: Academic Press.

Milne, G. (1947). A soil reconnaissance journey through parts of Tanganyika Territory. December 1935 to February 1936. *Journal of Ecology*, **35**, 192–264.

Milstein, P. Le G. (1964). Birds and agriculture. *Lantern* **14**, 40–4.

Misra, J. N. & Ranganathan, V. (1954). Digestion of cellulose by the mound building termite *Termites* (*Cyclotermes*) *obesus* (Rambur). *Proceedings of the Indian Academy of Sciences*, **39**, 100–13.

Möeller, A. (1893). *Die Pilzgärten einiger südamerikanischer Ameisen.* Heft VI, *Schimpers Botanische Mitteilungen aus den Tropen*, 127 pp.

Montagner, H. (1966). Le mécanisme et les conséquences des comportements trophallactiques chez les guêpes du genre *Vespa*. Thèses, Faculté des Sciences de l'Université de Nancy, Nancy, 143 pp.

Montalenti, G. (1927). Sull' allevomento di termiti senza i protozoi del l'ampolla cecale. *Accademia Nazionale di Lincei, Rome, Classe de Scienze Fische, Matematiche e Naturali, Rendiconti*, Ser. 6, **6**, 529–32.

Montalenti, G. (1932). Gli enzimi degerenti e l'assorbimento delle sostanze solubile nell'intestino delle termiti. *Archivio zoologico italiano*, **16**, 859–64.

Moore, B. P. (1965). Pheromones and insect control. *Australian Journal of Science*, **28**, 243–45.

Moore, B. P. (1969). Biochemical studies in termites. In *Biology of termites*, ed. K. Krishna & F. M. Weesner, vol. 1, pp. 407–32. New York & London: Academic Press.

Moore, B. P. & Brown, W. V. (1971–72). Identification of termite scent-trail pheromone. *Annual Report, Commonwealth Scientific and Industrial Research Organization* (*Canberra*), 39.

Moore, S. & Stein, W. H. (1951). Chromatography of amino acids on sulphonated polystyrene resins. *Journal of Biological Chemistry*, **192**, 663–81.

Moran, M. R. (1959). Changes in the fat content during metamorphosis of the meal worm, *Tenebrio molitor* Linnaeus. *Journal of the New York Entomological Society*, **67**, 213–22.

Mortreuil, M. & Brader, I. M. (1962). Marquage radioactif des fourmis dans les plantations d'ananas. *Compte rendu Coloque 'Radioisotopes and radiation in entomology'*, p. 39. Vienne: International Atomic Energy Agency.

Moser, J. C. (1967). Trails of the leaf-cutters. *Natural History*, **76**, 32–5.

Moser, J. C. & Blum, M. S. (1963). Trail marking substance of the Texas leaf-cutting ant: source and potency. *Science*, **140**, 1228.

Müller, H. (1958). Zur Kenntnis der Schäden, die Lachniden an ihren Wirtsbäumen hervorrufen Konnen. *Zeitschrift für angewandte Entomologie*, **42**, 284–91.

Naarman, H. (1963). Untersuchungen über Bildung und Weitergabe von Drüsensekreten bei *Formica* (Hymenoptera, Formicidae) mit Hilfe der Radioisotopenmethode. *Experientia*, **19**, 412–13.

Nagin, R. (1972). Caste determination in *Neotermes jouteli* (Banks). *Insectes sociaux*, **19**, 39–61.

Nefedov, N. (1930). A quantitative study of the ant population of Troitsk Forest-Steppe Reserve. *Bulletin de l'Institut de recherches biologiques et de la Station biologique a l'Université de Perm*, **7**, 259–91. (In Russian with English summary.)

Nel, J. J. C. (1968*a*). Aggressive behaviour of the harvester termites *Hodotermes mossambicus* (Hagen) and *Trinervitermes trivervoides* (Sjöstedt). *Insectes sociaux*, **15**, 145–56.

Nel, J. J. C. (1968*b*). Verspreiding van die oesgate en grondhopies van die grasdraertermiet, *Hodotermes mossambicus* (Hagen) (Isoptera: Hodotermitidae). *South African Journal of Agricultural Science*, **11**, 173–82.

Nel, J. J. C. (1970). Aspects of the behaviour of workers of the harvester termite, *Hodotermes mossambicus* (Hagen), in the field (Isopt., *Hodotermitidae*). *Journal of the Entomological Society of Southern Africa*, **33**, 23–34.

Nel, J. J. C. & Hewitt, P. H. (1969). A study of the food eaten by a field population of the harvester termite, *Hodotermes mossambicus* (Hagen) (Isoptera, Hodotermitidae), and its relation to population density. *Journal of the Entomological Society of Southern Africa*, **32**, 123–31.

Nel, J. J. C., Hewitt, P. H. & Joubert, L. (1970). The collection and utilization of redgrass, *Themeda triandra* (Forsk.) by laboratory colonies of the harvester termite, *Hodotermes mossambicus* (Hagen) and its relation to population density. *Journal of the Entomological Society of Southern Africa*, **33**, 331–40.

Nel, J. J. C., Hewitt, P. H., Smith, L. J. & Smith, W. T. (1969). The behaviour of the harvester termite, *Hodotermes mossambicus* (Hagen) in a laboratory colony. *Journal of the Entomological Society of Southern Africa*, **32**, 9–24.

Nielsen, C. O. (1962). Carbohydrases in soil and litter invertebrates. *Oikos*, **13**, 200–15.

Nielsen, M. G. (1972*a*). An attempt to estimate energy flow through a population of workers of *Lasius alienus* (Först.), (Hymenoptera, Formicidae). *Natura Jutlandica*, **16**, 97–107.

Nielsen, M. G. (1972*b*). Production of workers in an ant nest. *Ekologia polska*, **20**, 65–71.

Nielsen, M. G. (1972*c*). Aboveground activity of the ant *Lasius alienus* (Först.) (Hymenoptera: Formicidae). *Natura Jutlandica*, **16**, 83–94.

Nielsen, M. G. (1974*a*). The use of Lincoln Index for estimating the worker population of *Lasius alienus* (Först.), (Hymenoptera, Formicidae). *Natura Jutlandica*, **17**, 87–90.

Nielsen, M. G. (1974*b*). Number and biomass of worker ants in a sandy heath area in Denmark. *Natura Jutlandica*, **17**, 93–5.

Nixon, G. E. J. (1951). *The association of ants with aphids and coccids. Commonwealth Institute of Entomology, London.* 36 pp.

Noirot, C. (1952). Les soins et l'alimentation des jeunes chez les termites. *Annales des sciences naturelles* (*Zoologie*), Sér. 11, **14**, 405–14.

Noirot, C. (1955). Recherches sur le polymorphisme des termites supérieurs. *Annales des sciences naturelles* (*Zoologie*), Sér. 11, **17**, 400–595.

Noirot, C. (1969*a*). Glands and secretions. In *Biology of Termites*, ed. K. Krishna & F. M. Weesner, vol. 1, pp. 89–123. New York & London: Academic Press.

Noirot, C. (1969*b*). Formation of castes in the higher termites. In *Biology of termites*, ed. K. Krishna & F. M. Weesner, vol. 1, pp. 311–50. New York & London: Academic Press.

Noirot, C. (1970). The nests of termites. In *Biology of termites*, ed. K. Krishna & F. M. Weesner, vol. 2, pp. 73–125. New York & London: Academic Press.

Noirot, C. & Noirot-Timothée, C. (1967). L'épithélium absorbant de la panse d'un termite supérieur. Ultrastructures et rapport avec la symbiose bactérienne. *Annales de la Société entomologique de France*, **3**, 577–92.

Noirot, C. & Noirot-Timothée, C. (1969). The digestive system. In *Biology of termites*, ed. K. Krishna & F. M. Weesner, vol. 1, pp. 49–88. New York & London: Academic Press.

Nutting, W. L. (1956). Reciprocal protozoan transfaunations between the roach,

Cryptocercus, and the termite, *Zootermopsis*. *Biological Bulletin, Marine Biological Laboratory, Woods Hole*, **110**, 83–90.

Nutting, W. L. (1965). Observations on the nesting site and biology of the Arizona damp-wood termite, *Zootermopsis laticeps* (Banks) (Hodotermitidae). *Psyche*, **73**, 113–25.

Nutting, W. L. (1966*a*). Colonizing flights and associated activities of termites. I. The desert damp-wood termite *Paraneotermes simplicicornis* (Kalotermitidae). *Psyche*, **73**(2), 131–49.

Nutting, W. L. (1966*b*). Distribution and biology of the primitive dry-wood termite, *Pterotermes occidentis* (Walker) (Kalotermitidae). *Psyche*, **73**(3), 165–79.

Nutting, W. L. (1969). Flight and colony foundation. In *Biology of termites*, ed. K. Krishna & F. M. Weesner, vol. 1, pp. 233–82. New York & London: Academic Press.

Nutting, W. L. (1970). Composition and size of some termite colonies in Arizona and Mexico. *Annals of the Entomological Society of America*, **63**, 1105–10.

Nutting, W. L., Haverty, M. I. & Lafage, J. P. (1973). Foraging behaviour of two species of subterranean termites in the Sonoran Desert of Arizona. *Proceedings, VIIth International Congress of the International Union for the Study of Social Insects, London 10–15 September, 1973*, pp. 298–301.

Nye, P. H. (1955). Some soil forming processes in the humid tropics. IV. The action of the soil fauna. *Journal of Soil Science*. **6**, 73–83.

Odum, E. P. (1959). *Fundamentals of ecology*, 2nd edn. Philadelphia & London: W. B. Saunders Company. 546 pp.

Odum, E. P. & Pontin, A. J. (1961). Population density of the underground ant, *Lasius flavus*, as determined by tagging with P^{32}. *Ecology*, **42**, 186–8.

Ohiagu, C. E. & Wood, T. G. (1976). A method for measuring rate of grass-harvesting by *Trinervitermes geminatus* Wasmann (Isoptera, Nasutitermitinae) and observation on its foraging behaviour in southern Guinea Savanna, Nigeria. *Journal of Applied Ecology*, **13**, 705–13.

Oinonen, E. A. (1956). Kallioiden muurahaisita ja niiden osuudesta kallioiden metsittymiseen Etelä-Suomessa. *Acta entomologica fennica*. **12**, 1–215, 9 tables.

Ollier, C. D. (1959). A two-cycle theory of tropical pedology. *Journal of Soil Science*, **10**, 137–48.

O'Neal, J. & Markin, G. P. (1973). Broad nutrition and parental relationships of the imported fire ant *Solenopsis invicta* (Hymenoptera, Formicidae). *Journal of the Georgia Entomological Society*, **8**(4), 294–303.

Oshima, M. (1919). Formosan termites and methods of preventing their damage. *Philippine Journal of Science*, **15**, 319–83.

O'Toole, C. (1973). The size and composition of a migrant colony of *Monomorium pharaonis* (L.). *Entomologist*, **106**, 225–7,

Otto, D. (1958*a*). Die Ortstreue der Blattlausbesucher von *Formica rufa* L. *Waldhygiene*, **2**, 114–18.

Otto, D. (1958*b*). Über die Homologieverhältnisse der Pharynx- und Maxillardrüsen bei Formicidae und Apidae (Hymenoptera). *Zoologischer Anzeiger*, **161**(9/10), 216–26.

Otto, D. (1958*c*). Zur Schutzwirkung der Waldameisenkolonien gegen Eichenschädlinge. *Waldhygiene*, **2**, 137–42.

Otto, D. (1965). Der Einfluss der Roten Waldameise (*Formica polyctena* Först.) auf die Zusonnensetzung der Insektenfauna (Ausschliesslich gradierende Arten). *Collana Verde*, **16**, 250–63.

Paparo, K. & Forbes, J. (1971). The salivary glands of *Atta cephalotes*. Unpublished paper, Fordham University, New York.

Parapura, E. & Pisarski, B. (1971). Mrówki (Hymenoptera, Formicidae) Bieszczadów. *Fragmenta faunistica*, **17**, 319–56.

Pardee, A. B. (1949). Measurements of oxygen uptake under controlled pressures of carbon dioxide. *Journal of Biological Chemistry*, **179**, 1085–91.

Parker, G. H. (1925). The weight of vegetation transported by tropical fungus ants. *Psyche*, **32**, 227–8.

Parsons, L. R. (1968). Aspects of leaf-cutter ant behaviour: *Acromyrmex octospinosa* and *Atta cephalotes*. Research Report in *Tropical biology: an ecological approach*. July–August. San José, Costa Rica: Organization for Tropical Studies.

Passera, L. (1966). La ponte des ouvrières de la fourmi *Plagiolepis pygmaea* Latr. (Hymenoptera, Formicidae): oeufs reproducteurs et oeufs alimentaires. *Comptes rendus hebdomadaires des séances de l'Académie des sciences, Paris*, **263**, 1095–8.

Passera, L. (1969). Biologie de la reproduction chez *Plagiolepis pygmaea* Latr. et ses deux parasites sociaux *Plagiolepis grassei* Le Mas. Pas. et *Plagiolepis xene* St. (Hymenoptera: Formicidae). Thèse, Faculté des Sciences de l'Université de Toulouse, Toulouse.

Passera, L., Bitsch, J. & Bressac, C. (1968). Observations histologiques sur la formation des oeufs alimentaires et des oeufs reproducteurs chez les ouvrières de *Plagiolepis pygmaea* Latr. (Hymenoptera: Formicidae). *Comptes rendus hebdomadaires des séances de l'Académie des sciences, Paris*, **266**, 2270–2.

Pasteels, J. M. (1965). Polyéthisme chez les ouvriers de *Nasutitermes lujae* (Termitidae, Isoptères). *Biologia Gabonica*, **1**, 191–205.

Pasteels, J. M. (1969). Les glandes tégumentaires des staphylins termitophiles. III. Les Aleocharinae des genres *Termitopullus* (Corotocini, Corotocina), *Perinthodes, Catalina* (Termitonannini, Perinthina), *Termitusa* (Termitohospitini, Termitusina). *Insectes sociaux*, **16**, 1–26.

Pathak, A. N. & Lehri, L. K. (1959). Studies on termite nests. I. Chemical, physical and biological characteristics of a termitarium in relation to its surroundings. *Journal of the Indian Society of Soil Science*, **7**, 87–90.

Paulsen, R. (1966). Trehalase in den Labialdrüsen von *Formica polyctena* F. (Hymenoptera, Formicidae). *Naturwissenschaften*. **53**, 337–8.

Paulsen, R. (1969). Zur Funktion der Propharynx Postpharynx- und Labialdrüsen von *Formica polyctena* Först. (Hymenoptera, Formicidae). Dissertationen zur Erlangung des Doktorgrades bei der Naturwissenschaftlichen Fakultät, Wurzburg, 90 pp.

Paulsen, R. (1971). Characterisation of trehalase from labial glands of ants (*Formica polyctena*). *Archives of Biochemistry and Biophysics*, **142**, 170–6.

Pavan, M. (1959*a*). Attivita Italiana per le lotta biologica con formiche del gruppo *Formica rufa* centro gli insetti daunosi alle foreste. *Ministero dell' Agricultura & delle Foreste Collana Verde*, **4**, 1–80.

Pavan, M. (1959*b*). La dendrolasina. *Notiziario Forestale e Montano*, **64–65**, 1737–40.

Pavan, M. (1960*a*). Estrazione e purificazione di alcuni componenti delle secrezioni defensive di arthropodi. *IX Congrés International d'Entomologie, Vienne*, 276–83.

Pavan, M. (1960*b*). Les fourmis dans la défense biologique des forêts. Résultats. Programmes d'activité internationale. *Collana Verde*, **7**, 148–57.

Peacock, A. D. (1950). Studies in pharaoh's ant, *Monomorium pharaonis* (L.) 4. Egg production. *Entomologists' Monthly Magazine*, **86**, 294–8.

Peacock, A. D., Waterhouse, F. L. & Baxter, A. T. (1955). Studies in pharaoh's ant, *Monomorium pharaonis* (L.) 10. Viability in regard to temperature and humidity. *Entomologists' Monthly Magazine*, **91**, 37–42.

Peakin, G. J. (1960). The influence of humidity on the organization of ant colonies both in the field and in the laboratory. Ph.D. thesis, University of London, London.

Peakin, G. J. (1965). Food reserves in the reproductive castes of *Lasius flavus* (Fab.). (Hymenoptera). *Proceedings of the XIIth International Congress of Entomology, London, 1964*, 303.

Peakin, G. J. (1972). Aspects of productivity in *Tetramorium caespitum* L. *Ekologia polska*, **20**, 55–63.

Peakin, G. J. (1973). The measurement of the costs of maintenance in terrestrial poikilotherms: a comparison between respirometry and calorimetry. *Experientia*, **29**, 801–2.

Pence, R. J. (1956). The tolerance of the drywood termite, *Kalotermes minor* (Hagen) to desiccation. *Journal of Economic Entomology*, **49**, 553–4.

Pendleton, R. L. (1941). Some results of termite activity in Thailand soils. *Thai Science Bulletin*, **3**(2), 29–53.

Peregrine, D. J. & Mudd, A. (1974). The effect of diet on the composition of the postpharyngeal glands of *Acromyrmex octospinosus* (Reich). *Insectes sociaux*, **21**(4), 417–24.

Peregrine, D. J., Mudd, A. & Cherrett, J. M. (1973). Anatomy and preliminary chemical analysis of the post-pharyngeal glands of the leaf-cutting ant *Acromyrmex octospinosus* (Reich) (Hymenoptera, Formicidae). *Insectes sociaux*, **20**(4), 355–63.

Peregrine, D. J., Percy, H. C. & Cherrett, J. M. (1972). Intake and possible transfer of lipid by the post-pharyngeal glands of *Atta cephalotes* (L.). *Entomologia experimentalis et applicata*, **15**, 248–9.

Pętal, J. (1967). Productivity and the consumption of food in the *Myrmica laevinodis* Nyl. population. In *Secondary productivity of terrestrial ecosystems*, ed. K. Petrusewicz, vol. II, pp. 841–57. Warszawa–Kraków: Państwowe Wydawnictwo Naukowe.

Pętal, J. (1968). Wpływ zasobności pokarmowej środowiska na rozwój populacji *Myrmica laevinodis* Nyl. (Formicidae). [The influence of the food resources of the habitat on the development of a population of *Myrmica laevinodis* Nyl. (Formicidae).] *Ekologia polska*, Ser. *B*, **14**, 287–96.

Pętal, J. (1972). Methods of investigating the productivity of ants. *Ekologia polska*, **20**, 9–22.

Pętal, J. (1974). The effect of pasture management on ant population. *Ekologia polska*, **22**(3/4), 679–92.

Pętal, J., Andrzejewska, L., Breymeyer, A. & Olechowicz, E. (1971). Productivity investigation of two types of meadows in the Vistula Valley. X. Role of the ants as predators in a habitat. *Ekologia polska*, **19**, 213–22.

Pętal, J., Jakubczyk, H., Chmielewski, K. & Tatur, A. (1975). Response of ants to environmental pollution. In *Progress in soil zoology*, ed. J. Vaněk, pp. 363–73. Prague: Academia (Czechoslovakia Academy of Sciences).

Pętal, J., Jakubczyk, H., Novak, E. & Czerwiński, Z. (1977). Effects of ants and earthworms on soil habitat modification. In *Soil organisms as components of ecosystems*. Proceedings VI International Soil Zoology Colloquium, ed. U. Lohm & T. Persson. *Ecological Bulletin* (*Stockholm*), **25**, in press.

Pętal, J., Jakubczyk, H. & Wójcik, Z. (1970). L'influence des fourmis sur la modification des sols et des plantes dans le milieu des praires. In *Methods of study in soil ecology, Proceedings of the Paris symposium*, ed. J. Phillipson, pp. 235–40. Paris: UNESCO.

Pętal, J. & Pisarski, B. (1966). Metody ilósciowe stosowane w badaniach myrmekologicznych. *Ekologia polska*, Ser. *B*. **12**, 363–76.

Petrusewicz, K. (1967). Suggested list of more important concepts in productivity studies (definitions and symbols). In *Secondary productivity of terrestrial ecosystems*, ed. K. Petrusewicz, vol. II, pp. 51–8. Warszawa–Kraków: Państwowe Wydawnictwo Naukowe.

Petrusewicz, K. & Macfayden, A. (1970). *Productivity of terrestrial animals. Principles and methods*, Oxford: Blackwell Scientific Publications. IBP Handbook No. 13.

Petersen-Braun, M. & Buschinger, A. (1975). Entstehung und Funktion eines thorakalen Kropfes bei Formiciden Königinnen. *Insectes sociaux*, **22**(1), 51–66.

Phillipson, J. (1962). Respirometry and the study of energy turnover in natural systems with particular reference to harvest spiders (Phalangiida). *Oikos*, **13**, 311–18.

Phillipson, J. (1973). The biological efficiency of protein production by grazing and other land-based systems. In *The biological efficiency of protein production*, ed. J. G. W. Jones, pp. 217–35. London: Cambridge University Press.

Pickavance, J. R. (1970). A new approach to the immunological analysis of invertebrate diets. *Journal of Animal Ecology*, **39**, 715–24.

Pickens, A. L. (1932). Observations on the genus *Reticulitermes* Holmgren. *Pan-Pacific Entomologist*, **8**, 178–80.

Pickens, A. L. (1934). The biology and economic significance of the western subterranean termite, *Reticulitermes hesperus*. In *Termites and termite control*, ed. C. A. Kofoid, 2nd edn, pp. 157–83. Berkeley: University of California Press.

Pickles, W. (1935). Populations, territory and interrelations of the ants *Formica fusca*, *Acanthomyops niger* and *Myrmica scabrinodis* at Garforth (Yorkshire). *Journal of Animal Ecology*, **4**, 22–31.

Pickles, W. (1940). Fluctuations in the populations, weights and biomasses of ants at Thornhill, Yorkshire, from 1935 to 1937. *Transactions of the Royal Entomological Society of London*, **90**, 467–85.

Pisarski, B. (1962). Sur *Sifolinia pechi* Samš trouvée en Pologne. *Bulletin de l'Académie polonaise des sciences*, Classe II, **10**, 367–9.

Pisarski, B. (1973). *Struktura społeczna* Formica (C) exsecta *Nyl.* (*Hymenoptera, Formicidae*) *i jej wpływ na morfologię, ekologię i etologię gatunku. PAN, Instytut Zoologiczny*, 134 pp.

Plateaux, L. (1970). Sur le polymorphisme social de la fourmi *Leptothorax nylanderi* (Förster). I. Morphologie et biologie comparées des castes. *Annales des sciences naturelles* (*Zoologie*), **12**, 373–478.

Ploey, J. de (1964). Nappes de gravats et couvertures argilo-sableuses au Bas-Congo; leur genèse et l'action des termites. In *Etudes sur les termites africains* ed. A. Bouillon, pp. 399–414. Paris: Masson et Cie.

Pochon, J., Barjac, H. de & Roche, A. (1959). Recherches sur la digestion de la cellulose chez le termite *Sphaerotermes sphaerothorax. Annales de l'Institut Pasteur*, **96**, 352–5.

Pollard, G. V. (1973). Some factors affecting the foraging behaviour in leaf-cutting ants of the genera *Atta* and *Acromyrmex*. Ph.D. thesis, University of the West Indies, St. Augustine, Trinidad.

Pomeroy, D. E. (1976*a*). Studies on a population of large termite mounds in Uganda. *Ecological Entomology*, **1**(part 1), 49–61.

Pomeroy, D. E. (1976*b*). Some effects of mound-building termites on soils in Uganda. *Journal of Soil Science*, **27**, 377–94.

Pontin, A. J. (1958). A preliminary note on the eating of aphids by ants of the genus *Lasius* (Hymenoptera, Formicidae). *Entomologist's Monthly Magazine* **94**, 9–11.

Pontin, A. J. (1961). Population stabilization and competition between the ants *Lasius flavus* (F.) and *L. niger* (L.) *Journal of Animal Ecology*, **30**, 47–54.

Pontin, A. J. (1963). Further considerations of competition and the ecology of the ants *Lasius flavus* (F.) and *L. niger* (L.). *Journal of Animal Ecology*, **32**, 565–74.

Potts, R. C. & Hewitt, P. H. (1972). Some properties of an aryl-β-glucosidase from the harvester termite, *Trinervitermes trinervoides. Insect Biochemistry*, **2**, 400–8.

Potts, R. C. & Hewitt, P. H. (1973). The distribution of intestinal bacteria and cellulose activity in the harvester termite *Trinervitermes trinervoides* (Nasutitermitinae). *Insectes sociaux*, **20**, 215–20.

Potts, R. C. & Hewitt, P. H. (1974*a*). The partial purification and some properties of the cellulase from the termite *Trinervitermes trinervoides* (Nasutitermitinae). *Comparative Biochemistry and Physiology*, **47*B***, 317–26.

Potts, R. C. & Hewitt, P. H. (1974*b*). Some properties and reaction characteristics of the partially purified cellulase from the termite *Trinervitermes trinervoides* (Nasutitermitinae). *Comparative Biochemistry and Physiology*, **47*B***, 327–37.

Prat, H. (1954). Micro calorimetric analysis of the variations in thermogenesis in different insects. *Canadian Journal of Zoology*, **32**, 172–97.

Prat, H. (1956). Régimes de la thermogénèse chez la Blatte américaine: *Periplaneta americana* (L.): effets d'excitations olfactives; influence de la décapitation. *Revue canadienne de biologie*, **14**, 360–98.

Pricer, J. L. (1908). The life history of the carpenter ant. *Biological Bulletin*, **14**, 177–218.

Pullar, J. D. (1969). Methods of calorimetry (A) direct. In *Nutrition of animals of agricultural importance.* Part I. *The science of nutrition of farm livestock*, ed. Sir David Cuthbertson, pp. 471–90. London: Pergamon Press.

Quispel, A. (1941). De verspreiding van de mierenfauna in het Nationale Park De Hoge Veluwe. *Nederland boschbouwtijdschrift*, **14**.

Raigner, A. (1948). L'économie thermique d'une colonie polycalique de la fourmis du bois (*Formica rufa polyctena* Först.). *Cellule, Louvain*, **51**, 281–368.

Raigner, A. & van Boven, J. K. A. (1952). Quelques aspects nouveaux de la taxonomie et de la biologie des *Dorylus* africains. *Annales des sciences naturelles* (*Zoologie*), **14**, 397–403.

Raigner, A. & von Boven, J. K. A. (1955). Etude taxonomique, biologique et biomètrique des *Dorylus* du sous-genre *Anomma* (Hymenoptera, Formicidae). *Annales du Musée royale du Congo belge*, Sér. 4, *Sciences zoologiques*, **2**, 1–359.

Randall, M. & Doody, T. (1934). Hydrogen-ion concentration in the termite intestine. In *Termites and termite control*, ed. C. A. Kofoid, 2nd edn, pp. 99–104. Berkeley: University of California Press.

Rao, K. P. (1962). Occurrence of enzymes for protein digestion in the termite *Heterotermes indicola.* In *Proceedings of the New Delhi symposium 1960. Termites in the humid tropics*, pp. 71–2. Paris: UNESCO.

Ratcliffe, F. N., Gay, F. J. & Greaves, T. (1952). *Australian termites. The biology, recognition and economic importance of the common species.* Melbourne: Commonwealth Scientific and Industrial Research Organization. 124 pp.

Ratcliffe, F. N., & Graves, T. (1940). The subterranean foraging galleries of *Coptotermes lacteus* (Frogg.). *Journal of the Council for Scientific and Industrial Research, Australia*, **13**, 150–61.

Reichle, D. E. (1971). Energy and nutrient metabolism of soil and litter invertebrates. In *Productivity of forest ecosystems*, ed. P. Duvigneaud, pp. 465–77. Paris: UNESCO.

Rennie, P. J. (1965). The determination of nitrogen in woody tissue. *Journal of the Science of Food and Agriculture*, **16**, 629–38.

Retief, L. W. & Hewitt, P. H. (1973*a*). Digestive carbohydrases of the harvester termite *Hodotermes mosambicus*: α-glycosidases. *Journal of Insect Physiology*, **19**, 105–13.

Retief, L. & Hewitt, P. H. (1973*b*). Purification and properties of trehalase from the harvester termite, *Trinervitermes trinervoides. Insect Biochemistry*, **3**, 345–51.

Retief, L. W. & Hewitt, P. H. (1973*c*). Digestive β-glycosidases of the harvester termite, *Hodotermes mossambicus*: properties and distribution. *Journal of Insect Physiology*, **19**, 1837–47.

Rettenmeyer, C. W. (1963). Behavioral studies of army ants. *Kansas University Science Bulletin*, **44**, 281–465.

Ricks, B. L. & Vinson, S. B. (1972). Digestive enzymes of the imported fire ant, *Solenopsis richteri* (Hymenoptera, Formicidae). *Entomologia experimentalis et applicata*, **15**, 329–34.

Rickson, F. R. (1971). Glycogen plastids in Mullerian body cells of *Cecropia peltata*, a higher green plant. *Science*, **173**, 344–7.

Ritter, F. J. & Coenen-Saraber, C. M. A. (1969). Food attractants and a pheromone as trail-following substances for the Saintonge termite. *Entomologia experimentalis et applicata*, **12**, 611–22.

Ritter, H. T. M. (1955). A study of the biology of certain intestinal associates of *Reticulitermes flavipes*. Ph.D. dissertation, Lehigh University, Bethlehem, Pennsylvania, 166 pp.

Robinson, J. B. D. (1958). Some chemical characteristics of 'termite soils' in Kenya coffee fields. *Journal of Soil Science*, **9**, 58–65.

Robinson, T. (1963). *The organic constituents of higher plants, their chemistry and interrelationships*. Minneapolis: Burgess Publishing Company. 306 pp.

Rockwood, L. L. (1973). Distribution, density and dispersion of two species of *Atta* (Hymenoptera, Formicidae) in Guanacaste Province, Costa Rica. *Journal of Animal Ecology*, **42**, 803–18.

Roessler, E. S. (1932). A preliminary study of the nitrogen needs of growing *Termopsis. University of California Publications in Zoology*, **36**, 357–68.

Roger, J. (1859). Beiträge zur Kenntniss der Ameisenfauna der Mittelmeerländer. Erstes Stück. *Berliner entomologische Zeitschrift*, **3**, 225–59.

Rogers, L. E. (1972). *The ecological effects of the western harvester ant* (Pogonomyrmex occidentalis) *in the shortgrass plains ecosystem. USA/IBP Grassland Biome, Technical Report*, **206**.

Rogers, L. E. (1974). Foraging activity of the western harvester ant in the short grass plains ecosystem. *Environmental Entomology*, **3**, 420–4.

Rogers, L., Lavigne, R. & Miller, J. L. (1972). Bioenergetics of the western harvester ant in the shortgrass plains ecosystem. *Environmental Entomology*, **1**(6), 763–8.

Roonwal, M. L. (1954). On the structure and population of the nest of the common Indian tree ant, *Crematogaster dohrni rogenhoferi* (Mayr) (Hymenoptera, Formicidae). *Journal of the Bombay Natural History Society*, **52**, 354–64.

Roonwal, M. L. (1960). Biology and ecology of oriental termites No. 5. Mound structures, nest and moisture content of fungus combs in *Odontotermes obesus*, with a discussion on the association of fungi with termites. *Record of the Indian Museum*, **58**, 131–50.

Roonwal, M. L. (1970). Termites of the oriental region. In *Biology of termites*, ed. K. Krishna & F. M. Weesner, vol. 2, pp. 315–91. New York & London: Academic Press.

Roth, M. (1964). Adaptation de la thermogénèse à la température ambiante et effet d'économie thermique de groupe chez l'abeille (*Apis mellifera* L.) *Comptes rendus hebdomadaires des séances de l'Académie des sciences, Paris*, **258**, 5534–7.

Rowan, M. K. (1971). The foods of South African birds. *Ostrich*, Supplement **8**, 343–56.

Roy-Noel, J. (1974). Recherches sur l'écologie des Isoptères de la presqu'île du Cap-Vert (Sénégal). Deuxième partie: les espèces et leur écologie. *Bulletin de l'Institut français d'Afrique Noire*, **36**, 525–609.

Rudman, P. & Gay, F. J. (1967). The causes of natural durability in timber. XX. The cause of variation in the termite resistance of jarrah (*Eucalyptus marginata* Sm.). *Holzforschung*, **21**, 21–3.

Ruelle, J. E. (1962). Etude de quelques variables du microclimat du nid de *Macrotermes natalensis* (Hav.) (Isoptera, Termitidae) en rapport avec le déclenchement de l'essaimage. Doctoral thesis, University of Kinshasa, Kinshasa, 120 pp.

Ruelle, J. E. (1964). L'architecture du nid de *Macrotermes natalensis* et son sens functionnel. In *Etudes sur les termites africains*, ed. A. Bouillon, pp. 325–51. Paris: Masson et Cie.

Ruelle, J. E. (1970). A revision of the termites of the genus *Macrotermes* from the ethiopian region (Isoptera, Termitidae). *Bulletin of the British Museum (Natural History), Entomology*, **24**, 363–444.

Ruppli, E. (1969). Die Elimination überzähliger Ersatzgeschlechtstiere bei der Termite *Kalotermes flavicollis* (Fabr.). *Insectes sociaux*, **16**, 235–48.

Ruppli, E. & Lüscher, M. (1964). Die Elimination überzähliger Ersatzgeschlechtstiere bei der Termite *Kalotermes flavicollis* (Fabr.) (Vorläufige Mitteilunge). *Revue suisse de zoologie*, **71**, 626–32.

Russell, P. G. (1955). Determination of glycogen content during metamorphosis of the mealworm (*Tenebrio molitor* Linnaeus). *Journal of the New York Entomological Society*, **63**, 107–10.

Saayman, J. E. (1966). A study of the diet and parasites of *Ardeala* (*Bubulcus*) *ibis*, *Numida meleagris* and *Gallus domesticus* from Eastern Cape Province, South Africa. Ph.D. thesis, University of South Africa, Pretoria.

Saeki, I., Sumimoto, M. & Kondo, T. (1971). The role of essential oil in resistance of coniferous woods to termite attack (*Coptotermes formosanus*). *Holzforschung*, **25**, 57–60.

Saeki, I., Sumimoto, M. & Kondo, T. (1973). The termiticidal substances from wood of *Chamaecyparis pisifera* D. Don. *Holzforschung*, **27**, 93–6.

Salt, G. (1952). The arthropod population of the soil in some East African pastures. *Bulletin of Entomological Research*, **43**, 203–20.

Sanders, C. J. & Baldwin, W. F. (1969). Iridium-192 as a tag for carpenter ants of the genus *Camponotus* (Hymenoptera, Formicidae). *Canadian Entomologist*, **101**, 416–18.

Sands, W. A. (1956). Some factors affecting the survival of *Odontotermes badius*. *Insectes sociaux*, **3**, 531–36.

Sands, W. A. (1960). The initiation of fungus comb construction in laboratory colonies of *Ancistrotermes guineensis* (Silvestri). *Insectes sociaux*, **7**, 251–63.

Sands, W. A. (1961*a*). Foraging behaviour and feeding habits of five species of *Trinervitermes* in West Africa. *Entomologia experimentalis et applicata*, **4**, 277–88.

Sands, W. A. (1961*b*). Nest structure and size distribution in the genus *Trinervitermes* (Isoptera, Termitidae, Nasutitermitinae), in West Africa. *Insectes sociaux*, **8**, 177–86.

Sands, W. A. (1965*a*). Mound population movements and fluctuation in *Trinervitermes ebenerianus* Sjöstedt (Isoptera, Termitidae, Nasutitermitinae). *Insectes sociaux*, **12**, 49–58.

Sands, W. A. (1965*b*). Termite distribution man-modified habitats in West Africa, with special reference to species segregation in the genus *Trinervitermes* (Isoptera, Termitidae, Nasutitermitinae). *Journal of Animal Ecology*, **34**, 557–71.

Sands, W. A. (1965*c*). *A revision of the Nasutitermitinae (Isoptera, Termitidae) of the Ethiopian zoogeographical region. Bulletin of the British Museum (Natural History), Entomology*, Supplement **4**, 1–172.

Sands, W. A. (1966). Remarks under 'Discussion: origins of social behaviour pp. 107–8'. In Kennedy, J. S., Some outstanding questions in insect behaviour. In *Insect behaviour*, ed. P. T. Haskell, pp. 97–112. *Symposia of the Royal Entomological Society of London*, No. **3**.

Sands, W. A. (1969). The association of termites and fungi. In *Biology of termites*, ed. K. Krishna & F. M. Weesner, vol. 1, pp. 495–524. New York & London: Academic Press.

Sands, W. A. (1972*a*). The soldierless termites of Africa (Isoptera: Termitidae). *Bulletin of the British Museum (Natural History), Entomology*, Supplement **18**, 1–243, 9 plates, 661 text figures.

Sands, W. A. (1972*b*). Problems in attempting to sample tropical subterranean termite populations. *Ekologia polska*, **20**, 23–31.

Sannasi, A. (1968). Vitamin A in the vision of termites. *Indian Journal of Experimental Biology*, **6**, 260–1.

Sannasi, A. (1969). Studies of an insect mycosis. II. Biochemical changes in the blood of the queen of the mound-building termite *Odontotermes obesus* accompanying fungal infection. *Journal of Invertebrate Pathology*, **13**, 11–14.

Sannasi, A. & George, C. J. (1972). Termite queen substance: 9-Oxodec-*trans*-2-enoic acid. *Nature, London*, **237**, 457.

Sannasi, A., Sen-Sarma, P. K., George, C. J. & Basalingappa, S. (1972). Juvenile hormone activity from various sources of termite castes and their fungus gardens. *Insectes sociaux*, **19**, 81–5.

Saunders, C. J. (1972). Seasonal and daily activity patterns of carpenter ants (*Camponotus spp.*) in north-western Ontario (Hymenoptera: Formicidae). *Canadian Entomologist*, **104**, 1681–7.

Scherba, G. (1959), Moisture regulation in mound nests of the ant *Formica ulkei*. *American Midland Naturalist*, **61**, 499–509.

Scherba, G. (1963). Population characteristics among colonies of the ant *Formica opaciventris* Emery (Hymenoptera: Formicidae). *Journal of the New York Entomological Society*, **71**, 219–32.

Schildknecht, H. & Koob, K. (1970). Plant bioregulators in the metathoracic glands of myrmicine ants. *Angewandte Chemie, New York*, **9**, 173.

Schildknecht, H. & Koob, K. (1971). Myrmicacin, the first insect herbicide. *Angewandte Chemie, New York*, **10**, 124–5.

Schmidt, G. H. (1966). Einfluss der temperatur auf den atmungsstoffwechsel der puppen von *Formica polyctena* (Hymenoptera, Insecta). *Helgoländer wissenschaftliche Meeresuntersuchungen*, **14**, 369–80.

Schmidt, G. H. (1967). Veränderungen im Gehalten energieliefernden Reservestoffen während der Kasten differenzierung von *Formica polyctena* Foerst. Ins. (Hym.). *Biologisches Zentralblatt*, **86**, 5–66.

Schmidt, G. H. (1968). Einfluss von Temperatur und Luftfeuchtigkeit auf die Energiebilanz während der Metamorphose verschiedener Kasten von *Formica polyctena* Foerst. (Hym.). *Zeitschrift für angewandte Entomologie*, **61**, 61–109.

Schmidt, G. H. (1974). Steuerung der Kastenbildung und Geschlechtsregulation im Waldameisenstaat. In *Sozialpolymorphismus bei Inseckten*, ed. G. H. Schmidt, pp. 404–512. Stuttgart: Wissenschaftliche Verlagsgesellschaft.

Schneider, P. (1966). Versuche zur Frage der Futterverteilung und Wasseraufnahme bei *Formica polyctena* Först. *Insectes sociaux*, **13**(4), 297–304.

Schneider, P. (1972). Versuche zur Frage der individuellen Futterverteilung bei der kleinen Roten Waldameise (*Formica polyctena* Foerst.). *Insectes sociaux*, **19**(3), 279–99.

Schneirla, T. C. (1949). Army-ant life and behavior under dry-season conditions. 3. The course of reproduction and colony behavior. *Bulletin of the American Museum of Natural History*, **94**, 5–81.

Schneirla, T. C. (1956). A preliminary survey of colony division and related processes in two species of terrestrial army ants. *Insectes sociaux*, **3**, 43–69.

Schneirla, T. C. (1957*a*). A comparison of species and genera in the ant subfamily Dorylinae with respect to functional pattern. *Insectes sociaux*, **4**, 259–98.

Schneirla, T. C. (1957*b*). Theoretical consideration of cyclic processes in doryline ants. *Proceedings of the American Philosophical Society*, **101**, 106–33.

Schneirla, T. C. (1958). The behaviour and biology of certain nearctic army ants. Last part of the functional season, south eastern Arizona. *Insectes sociaux*, **5**, 215–55.

Schneirla, T. C. (1971). *Army ants: a study in social organization*, ed. H. Topoff. San Francisco: Freeman & Co.

Schneirla, T. C., Brown, R. Z. & Brown, F. C. (1954). The bivouac or temporary nest as an adaptive factor in certain terrestrial species of army ants. *Ecological Monographs*, **24**, 269–96.

Schneirla, T. C. & Reves, A. Y. (1966). Raiding and related behaviour in two surface adapted species of the Old World doryline ant, *Aenictus*. *Animal Behaviour*, **14**, 132–48.

Schubert, W. J. (1965). *Lignin biochemistry*. New York & London: Academic Press. 131 pp.

Schumacher, A. M. & Whitford, W. G. (1974). Spatial and temporal variation in Chihuahuan desert ant faunas. *Southwestern Naturalist*, **21**, 1–8.

Schuster, R. (1956). Der Anteil der Oribatiden an der Zergetsungsvorgängen im Boden. *Zeitschrift für Morphologie and Ökologie der Tiere*, **45**, 1–33.
Scurfield, G. & Nicholls, P. W. (1970). Amino-acid composition of wood proteins. *Journal of Experimental Botany*, **21**, 857–68.
Seifert, K. (1962). Die chemische Veränderung der Holzzellwand–Komponenten unter dem Einfluss tierischer und pflanzlicher Schädlinge. 4. Mitteilung: Die Verdauung von Kiefern- und Rotbuchenholz durch die Termite *Kalotermes flavicollis* (Fabr.). *Holzforschung*, **16**, 161–8.
Seifert, K. & Becker, G. (1965). Der chemische Abbau von Laub- und Nadelholzarten durch verschiedene Termiten. *Holzforschung*, **19**, 105–11.
Sen-Sarma, P. K. (1974). Ecology of biogeography of the termites of India. In *Ecology and biogeography in India*, ed. M. S. Mani, pp. 421–72. The Hague: D. W. Junk.
Sen-Sarma, P. K. & Kloft, W. (1965). Trophallaxis in pseudoworkers of *Kalotermes flavicollis* (Fabricius) (Insecta: Isoptera: Kalotermitidae) using radioactive I^{131}. *Proceedings of the Zoological Society, Calcutta*, **18**, 41–6.
Shcherbina, E. I. & Sukhinin, A. N. (1968). Znachenie termitov v pitanii nekotorykh pozvonochnykh zhivotnykh. In *Termity i merv borbý s nimi*, ed. A. N. Luppova, pp. 126–33. Izdatel'stvo Ylym, Ashkabad.
Sheppe, W. (1970). Invertebrate predation on termites of the African savanna. *Insectes sociaux*, **17**, 205–18.
Shrikhande, J. G. & Pathak, A. N. (1948). Earthworms and insects in relation to soil fertility. *Current Science*, **17**, 327–8.
Siddorn, J. W. (1962). Weather factors in the ecology of insects. *Biology and Human Affairs*, **27**, 1–9.
Sinclair, A. R. E. (1975). The resource limitation of trophic levels in tropical grassland ecosystem. *Journal of Animal Ecology*, **44**, 497–520.
Skaife, S. H. (1954). Caste differentiation among termites. *Transactions of the Royal Society of South Africa*, **34**(2), 345–53.
Skaife, S. H. (1955). *Dwellers in darkness: an introduction to the study of termites.* London: Longmans Green & Co. 134 pp.
Skellam, J. G. (1952). Studies in statistical ecology. Part I. Spatial pattern. *Biometrika*, **39**, 346–62.
Skinner, H. A. ed. (1962). *Experimental thermochemistry*. New York: Wiley.
Skwarra, E. (1929). *Formica fusca-picea* Nyl. als Moorameise. *Zoologischer Anzeiger*, **82**, 46–55.
Sleeman, J. R. & Brewer, R. (1972). Micro-structures of some Australian termite nests. *Pedobiologia*, **12**, 347–73.
Slowtzoff, B. (1909). Über den Gaswechsel der Insekten und diessen Beziehung zur Temperatur der Luft. *Biochemische Zeitschrift*, **19**, 497–504.
Smalley, A. E. (1960). Energy flow of a saltmarsh grasshopper population. *Ecology*, **41**, 672–7.
Smith, A. (1971). *Matto-Grosso, last virgin land.* London: Royal Society and Royal Geographical Society.
Smith, F. (1858). Notes and observations on the (British) Aculeate Hymenoptera. *Entomologist's Annual*, 34–46.
Smith, Falconer (1942). Polymorphism in *Camponotus. Journal of the Tennessee Academy of Science*, **17**, 367–73.
Smith, M. R. (1931). A revision of the genus *Strumigenys* of America, north of Mexico, based on a study of the workers. *Annals of the Entomological Society of America*, **24**, 686–710.

Smith, M. R. (1957). Revision of the genus *Stenamma* Westwood in America North of Mexico. *American Midland Naturalist*, **57**, 133–74.
Smithers, R. H. N. (1971). *The Mammals of Botswana. Museum Memoir* No. **4**. Salisbury: Trustees of the National Museums of Rhodesia. 340 pp.
Smythe, R. V. (1972). Feeding and survival at constant temperatures by normally and abnormally faunated *Reticulitermes virginicus* (Isoptera: Rhinotermitidae). *Annals of the Entomological Society of America*, **65**, 756–7.
Smythe, R. V., Allen, T. C. & Coppel, H. C. (1965). Response of the eastern subterranean termite to an attractive extract from *Lenzites trabea*-invaded wood. *Journal of Economic Entomology*, **58**, 420–3.
Smythe, R. V. & Mauldin, J. K. (1972). Soldier differentiation, survival, and wood consumption by normally and abnormally faunated workers of the Formosan termite, *Coptotermes formosanus. Annals of the Entomological Society of America*, **65**, 1001–4.
Snyder, T. E. (1948). *Our enemy the termite.* New York: Comstock.
Snyder, T. E. (1956). Annotated, subject-heading bibliography of termites 1350 B.C. to A.D. 1954. *Smithsonian Miscellaneous Collections, Washington*, **130**, 1–305.
Snyder, T. E. (1961). Supplement to the annotated, subject-heading bibliography of termites 1955 to 1960. *Smithsonian Miscellaneous Collections, Washington*, **143**, 1–137.
Snyder, T. E. (1968). Second supplement to the annotated, subject-heading bibliography of termites 1961 to 1965. *Smithsonian Miscellaneous Collections, Washington*, **152**, 1–188.
Snyder, T. E. & Popenoe, E. P. (1932). The founding of new colonies by *Reticulitermes flavipes* (Kollar). *Proceedings of the Biological Society of Washington*, **45**, 153–8.
Southwood, T. R. E. (1966). *Ecological methods, with particular reference to the study of insect populations.* London: Methuen. 391 pp.
Speck, U., Becker, G. & Lenz, M. (1971). Ernährungsphysiologische Untersuchungen an Termiten nach selecktiver medikamentöser Ausschaltung der Darmsymbionten. *Zeitschrift für angewandte Zoologie*, **58**, 475–91.
Spector, W. S., ed. (1956). *Handbook of biological data.* Philadelphia: Saunders. 584 pp.
Springhetti, A. (1969). Il controllo sociale della differenziazione degli alati in *Kalotermes flavicollis* (Fabr.) (Isoptera). *Annali dell'Universita de Ferrara*, Nuova Serie, Sezione III, *Biologia Animale*, **3**, 73–96.
Springhetti, A. (1970). Influence of the king and queen on the differentiation of soldiers in *Kalotermes flavicollis* (Fabr.) (Isoptera). *Monitore zoologio italiano*, Nuova Serie, **4**, 99–105.
Springhetti, A. (1972). I feromoni nella differenziazione delle caste di *Kalotermes flavicollis* (Fabr.) (Isoptera). *Bollettino di zoologia*, **39**, 83–7.
Springhetti, A. (1973). Group effects in the differentiation of the soldiers of *Kalotermes flavicollis* (Fabr.) (Isoptera). *Insectes sociaux*, **20**, 333–42.
Stahel, G. & Geijskes, D. C. (1939). Über den Bau der Nester von *Atta cephalotes* (L.) und *Atta sexdens* (L.). (Hymenoptera, Formicidae). *Revista de entomologia*, **10**, 27–8.
Stahel, G. & Geijskes, D. C. (1943). A biologia da saúva e o seu combate na Guiana Holandes. *Secr. Agric. Dir. Publ. Agron. São Paulo*, p. 33.
Stawarski, I. (1966). Typy gniazd mrówek i ich zwiazki z siedliskiem na terenach

południowej Polski. Zeszyty Przyrodnicze Opolskiego Towarzystwa Przyjaciół Nauk. *Opole*, **6**, 93–157.

Stebaev, I. V. & Reznikova, J. S. (1972). Two interaction types of ants living in steppe ecosystem in south Siberia, USSR. *Ekologia polska*, **20**, 103–9.

Stecher, P. G., ed. (1968). *The Merck Index, an encyclopedia of chemicals and drugs*, 8th edn. Rahway, New Jersey: Merck & Co., Inc. 1713 pp.

Steiner, A. (1924). Über den sozialen Warmehaushalt der Waldameise (*Formica rufa* var. *rufo-pratensis* For.). *Zeitschrift für vergleichende Physiologie*, **2**, 23–56.

Steiner, A. (1929). Temperaturuntersuchungen in Ameisennestern mit Erdkuppeln, in Nest von *Formica exsecta* Nylander und in Nestern unter Steinen. *Zeitschrift für vergleichende Physiologie*, **9**, 1–66.

Stevens, R., ed. (1967). *Dictionary of organic compounds*, 4th edn, third supplement. New York: Oxford University Press. 279 pp.

Steyin, J. J. (1954). The pugnacious ant (*Anoplolepis custodiens* Smith) and its relation to the control of citrus scales at Letaba. *Memoirs of the Entomological Society of Southern Africa*, **3**, 1–96.

Steyn, P. (1967). Crop content of crowned Guinea-fowl *Numida meleagris*. *Ostrich*, **38**, 286.

Stoops, G. (1964). Application of some pedological methods to the analysis of termite mounds. In *Etudes sur les termites africains*, ed. A. Bouillon, pp. 379–98. Paris: Masson et Cie.

Stoops, G. (1968). Micromorphology of some characteristic soils of the lower Congo (Kinshasa). *Pédologie*, **18**, 110–49.

Stradling, D. J. (1968). Some aspects of the ecology of ants at Newborough Warren National Nature Reserve. Ph.D. thesis, University of Wales, Bangor.

Stradling, D. J. (1970). The estimation of worker ant populations by the mark–release–recapture method: an improved marking technique. *Journal of Animal Ecology*, **39**, 575–91.

Strickland, A. H. (1944). The arthropod fauna of some tropical soils. *Tropical Agriculture*, **21**, 107–14.

Strickland, A. H. (1945). A survey of the arthropod soil and litter fauna of some forest reserves and cacao estates in Trinidad (British West Indies). *Journal of Animal Ecology*, **14**, 1–11.

Striganova, B. R. (1971). Vozrastnyye izmeneniya aktivnosti pitaniya u kivsyakov (Juloidea). *Zoolohichnyĭ zhurnal Ukrayinȳ*, **50**, 1472–6.

Strong, F. (1965). Detection of lipids in the honeydew of an aphid. *Nature, London*, **205**, 1242.

Stuart, A. M. (1969). Social behavior and communication. In *Biology of termites*, ed. Krishna, K. & Weesner, F. M., vol. 1, pp. 193–232. New York & London: Academic Press.

Stumper, R. (1923). Sur la composition chimique des nids de l'*Apicotermes occultus* Silv. *Comptes rendus hebdomadaires des séances de l'Académie des sciences, Paris*, **177**, 409–11.

Stumper, R. (1950). Études myrmecologiques. VIII. Examène chimique et microbiologique de quelques nids de *Lasius fuliginosus* Latr. *Archives. Institut Grand-Ducal de Luxembourg* (N.S.), **19**, 243–50.

Stumper, R. & Kutter, H. (1957). Sur l'éthologie de la fourmi à miel *Proformica nasuta* Nyl. *Bulletin de la Société des naturalistes luxembourgeois*, **60**, 87–97.

Sudd, J. H. (1960). The foraging method of pharaoh's ant, *Monomorium pharaonis* (L.). *Animal Behaviour*, **8**, 67–75.

Sudd, J. H. (1967). *An introduction to the behaviour of ants.* London: Arnold. 200 pp.
Sudd, J. H. (1969). The excavation of soil by ants. *Zeitzchrift für Tierpsychologie,* **26**(3), 257–76.
Swiętosławski, W. (1946). *Microcalorimetry.* New York: Reinhold Publishing Company.
Tai, A., Matsumura, F. & Coppel, H. C. (1971). Synthetic analogues of the termite trail-following pheromone, structure and biological activity. *Journal of Insect Physiology,* **17**, 181–8.
Talbot, M. (1943). Population studies of the ant *Prenolepis imparis* (Say). *Ecology,* **24**, 31–44.
Talbot, M. (1945). Population studies of the ant *Myrmica schencki* sp. *emeryana* (Forel). *Annals of the Entomological Society of America,* **38**, 365–72.
Talbot, M. (1948). A comparison of two ants of the genus *Formica. Ecology,* **29**, 316–25.
Talbot, M. (1951). Population and hibernating condition of the ant *Aphaenogaster* (*Attomyrma*) *rudis* (Emery). *Annals of the Entomological Society of America,* **44**, 302–7.
Talbot, M. (1953). Ants of an old field community on the Edwin S. George Reserve, Livingston County, Michigan. *Contributions from the Laboratory of Vertebrate Biology of the University of Michigan,* **63**, 1–13.
Talbot, M. (1954). Populations of the ant *Aphaenogaster* (*Attomyrma*) *treatae* (Forel) on abandoned fields on the Edwin S. George Reserve. *Contributions from the Laboratory of Vertebrate Biology of the University of Michigan,* **69**, 1–9.
Talbot, M. (1957*a*). Population studies of the slave making ant *Leptothorax duloticus* and its slave *Leptothorax curvispinosus. Ecology,* **38**, 449–56.
Talbot, M. (1957*b*). Populations of ants in a Missouri woodland. *Insectes sociaux,* **4**, 375–84.
Talbot, M. (1965). Populations of ants in a low field. *Insectes sociaux,* **12**, 19–48.
Taylor, I. R. & Crescitelli, F. (1937). Measurement of heat production of small organisms. *Journal of Cellular and Comparative physiology,* **10**, 93–112.
Taylor, I. R. & Steinbach, H. B. (1931). Respiratory metabolism during pupal development of *Galleria mellonella* (bee moth). *Physiological Zoology,* **4**, 604–19.
Taylor, R. W. & Lowery, B. B. (1972). The New Guinean species of the ant genus *Orectognathus* Fr. Smith. *Journal of the Australian Entomological Society,* **11**, 306–10.
Terron, G. (1968). Composition des colonies de *Tetraponera anthracina* Santschi. *Annales de la Faculté des sciences du Cameroun,* **1**, 89–100.
Terron, G. (1969). Déscription de *Tetraponera ledouxi*, espèce nouvelle du Cameroun, parasite temporaire de *Tetraponera anthracina* Santschi. *Bulletin de l'Institut français d'Afrique Noire,* **31**, 629–42.
Terron, G. (1972). Observations sur les mâles ergatoïdes et les mâles ailés chez une fourmi du genre *Technomyrmex* Mayr. *Annales de la Faculté des sciences du Cameroun,* **10**, 107–20.
Tetrault, P. A. & Weis, W. L. (1937). Cellulose decomposition by a bacterial culture from the intestinal tract of termites. *Journal of Bacteriology,* **33**, 95–6.
Tevis, L., Jr (1958). Interrelations between the harvester ant *Veromessor pergandei* (Mayr) and some desert ephemerals. *Ecology,* **39**, 695–704.
Thiollay, J. M. (1970). L'exploitation par les oiseaux des essaimages de fourmis

et termites dans une zone de contact savane-forêt en Côte-d'Ivoire. *Alauda*, **38**, 255–73.

Tiemann, H. D. (1951). *Wood technology, constitution, properties, and uses*, 3rd edn. New York, Toronto & London: Pitman Publishing Corporation. 396 pp.

Tihon, L. (1946). A propos des termites au point de vue alimentaire. *Bulletin agricole du Congo Belge*, **37**, 865–8.

Torossian, C. (1967). Recherches sur la biologie et l'éthologie de *Dolichoderus quadripunctatus* (L.). *Insectes sociaux*, **14**, 105–22.

Torossian, C. & Causse, R. (1968). Effets des radiations gamma sur la fertilité et la longévité des colonies de *Dolichoderus quadripunctatus*. Compte rendu du Colloque '*Isotopes and radiation in entomology*', pp. 155–64. Vienne: International Atomic Energy Agency.

Toth, L. (1953). Nitrogen active microorganisms living in symbiosis with animals and their role in the nitrogen metabolism of the host animal. *Archiv für Mikrobiologie*, **18**, 242–4.

Tracey, M. V. & Youatt, G. (1958). Cellulase and chitinase in two species of Australian termites. *Enzymologia*, **19**, 70–2.

Trager, W. (1932). A cellulase from the symbiotic intestinal flagellates of termites and of the roach, *Cryptocercus punctulatus*. *Biochemical Journal*, **26**, 1762–71.

Trager, W. (1934). The cultivation of a cellulose-digesting flagellate, *Trichomonas termopsidis*, and of certain other termite protozoa. *Biological Bulletin, Marine Biological Laboratory, Woods Hole*, **66**, 182–90.

Tricart, J. (1957). Observations sur le rôle ameublisseur des termites. *Revue de géomorphologie dynamique*, **8**, 170–2.

Troll, C. (1936). Termitensavannen. In *Landekundliche Vorschrift Festschrift für Norbert Krebs*, pp. 275–312. Stuttgart: Engelhorn.

Tsoumis, G. (1968). *Wood as raw material*. Oxford: Pergamon Press. 276 pp.

Urich, F. W. (1895). Notes on the fungus growing and eating habits of *Sericomyrmex opacus* Mayr. *Transactions of the Royal Entomological Society of London*, **43**, 77–8.

Uttangi, J. C. & Joseph, K. J. (1962). Flagellate symbionts (protozoa) of termites from India. In *Proceedings of the New Delhi symposium, 1960. Termites in the humid tropics*, pp. 155–161. Paris: UNESCO.

Vandel, A. (1927). Observations sur les moeurs d'une fourmi parasite: *Epimyrma vandeli* Santschi. *Bulletin de la Société entomologique de France*, **32**, 289–95.

Vanderplank, F. L. (1960). The bionomics and ecology of the red tree ant, *Oecophylla* sp. and its relationship to the coconut bug *Pseudotheraptus wayi* Brown (Coreidae). *Journal of Animal Ecology*, **29**, 15–33.

Vaz-Ferreira, R., Covelo De Zolessi L. & Achaval, F. (1970). Oviposicion y desarrollo de Ofidos y Lacertilios en Hormigucros de *Acromyrmex*. *Physis* (*Buenos Aires*) **29**, 431–59.

Verron, H. (1963). Rôle des stimulis chimiques dans l'attraction sociale chez *Calotermes flavicollis* (Fabr.). *Insectes sociaux*, **10**, 167–355.

Vinson, S. B. (1968). The distribution of an oil, carbohydrate and protein food source to members of the imported fire ant colony. *Journal of Economic Entomology*, **61**, 712–14.

Vosseler, J. (1905). Die Ostafrikanische Treiberameise. *Der Pflanzer*, **1**, 289–302.

Vrkoč, J., Ubik, K., Dolegš, L. & Hrdý, I. (1973). On the chemical composition of frontal gland secretion in termites of the genus *Nasutitermes* (*N. costalis* and *N. rippertii, Isoptera*). *Acta entomologica Bohemoslovaca*, **70**, 74–80.

Wadsö, I. (1974). A microcalorimeter for biological analysis. *Science Tools*, **21**, 18–21.

Wagtendonk, W. J. & Soldo, A. T. (1970). Nitrogen metabolism in protozoa. In *Comparative biochemistry of nitrogen metabolism*, ed. J. W. Campbell, vol. **1**, pp. 1–56. New York & London: Academic Press.

Walker, D. J. (1965). Energy metabolism and rumen micro-organisms. In *Physiology of digestion in the ruminant*, ed. R. W. Dougherty *et al.*, pp. 296–310. London: Butterworths.

Wallis, D. I. (1961). Food-sharing behaviour of the ants *Formica sanguinea* and *Formica fusca. Behaviour*, **17**, 17–47.

Wallis, D. I. (1962). The relationship between hunger, activity and worker function in an ant colony. *Proceedings of the Zoological Society of London*, **139**, 589–605.

Wallwork, J. A. (1970). *Ecology of soil animals*. London: McGraw-Hill. 283 pp.

Waloff, N. (1957). The effect of the number of queens of the ant *Lasius flavus* (Fab.) (Hym. Formicidae) on their survival and on the rate of development of the first brood. *Insectes sociaux*, **4**, 391–408.

Waloff, N. & Blackith, R. E. (1962). The growth and distribution of the mounds of *Lasius flavus* (F.) (Hymenoptera: Formicidae) in Silwood Park, Berkshire. *Journal of Animal Ecology*, **31**, 421–37.

Wang, Y. J. & Happ, G. M. (1974). Larval development during the nomadic phase of a Nearctic army ant *Neivamyrmex nigrescens* (Cresson) (Hymenoptera, Formicidae). *International Journal of Insect Morphology and Embryology*, **3**(1), 73–86.

Wanyoni, K. & Lüscher, M. (1973). The action of juvenile hormone analogues on caste development in Zootermopsis (Isoptera). *Proceedings, VIIth International Congress of the International Union for the Study of Social Insects, London, 10–15 September, 1973*, 392–5.

Ward, P. (1965). Feeding ecology of the black-faced dioch *Quelea quelea* in Nigeria. *Ibis*, **107**, 173–214.

Wassmann, E. (1909). Zur Kenntniss der Ameisen und Ameisengäste von Luxemburg. III. Verzeichnis der Ameisen von Luxemburg, mit biologischen Notizen. *Archives. Institut Grand-Ducal de Luxembourg*, **4**, 1–114.

Watanabe, T. & Casida, J. E. (1963). Response of *Reticulitermes flavipes* to fractions from fungus-infected wood and synthetic chemicals. *Journal of Economic Entomology*, **56**, 300–7.

Waterhouse, D. F., Hackman, R. H. & McKellar, J. W. (1961). An investigation of chitinase activity in cockroach and termite extracts. *Journal of Insect Physiology*, **6**, 96–112.

Watkins, J. F. & Rettenmeyer, C. W. (1967). Effects of army ant queens on longevity of their workers (Formicidae: Dorylinae). *Psyche*, **74**, 228–33.

Watson, J. A. L. (1969). *Schedorhinotermes derosus*, a harvester termite in northern Australia (Isopt., Rhinotermitidae). *Insectes sociaux*, **16**, 173–8.

Watson, J. A. L. & Gay, F. J. (1970). The role of grain-eating termites in the degradation of a mulga ecosystem. *Search*, **1**, 43.

Watson, J. A. L., Hewitt, P. H. & Nel, J. J. C. (1971). The water-sacs of *Hodotermes mossambicus. Journal of Insect Physiology*, **17**, 1705–9.

Watson, J. P. (1960). Some observations on soil horizons and insect activity in granite soils. *Proceedings of the 1st Federated Scientific Congress of Rhodesia and Nyasaland*, **1**, 271–6.

Watson, J. P. (1962). The soil below a termite mound. *Journal of Soil Science*, **13**, 46–51.

Watson, J. P. (1974). Calcium carbonate in termite mounds. *Nature, London*, **247**, 74.

Way, M. J. (1954). Studies of the life history and ecology of the ant *Oecophylla longinoda* (Latr.). *Bulletin of Entomological Research*, **45**, 93–112.

Way, M. J. (1963). Mutualism between ants and honeydew-producing Homoptera. *Annual Review of Entomology*, **8**, 307–44.

Weber, N. A. (1938*a*). The biology of the fungus-growing ants. Part IV. Additional new forms. Part V. The Attini of Bolivia. *Revista de entomologia*, **9**, 154–206.

Weber, N. A. (1938*b*). The food of the giant toad, *Bufo marinus* (L.) in Trinidad and British Guiana with special reference to the ants. *Annals of the Entomological Society of America*, **31**, 499–503.

Weber, N. A. (1956). Symbiosis between fungus-growing ants and their fungus. *1955 Yearbook, American Philosophical Society*, April 1956, 153–7.

Weber, N. A. (1957*a*). Fungus growing ants and their fungi: *Cyphomyrmex costatus. Ecology*, **38**, 480–94.

Weber, N. A. (1957*b*). The nest of an anomalous colony of the arboreal ant *Cephalotes atratus. Psyche*, **64**, 60–9.

Weber, N. A. (1958). Evolution in fungus-growing ants. *Proceedings of the Tenth International Congress of Entomology*, **2**, 459–73.

Weber, N. A. (1959). The stings of the harvesting ant, *Pogonomyrmex occidentalis* (Cresson), with a note on populations (Hymenoptera). *Entomological News*, **70**, 85–90.

Weber, N. A. (1964*a*). A five-year colony of a fungus-growing ant, *Trachymyrmex zeteki. Annals of the Entomological Society of America*, **57**, 85–9.

Weber, N. A. (1964*b*). Termite prey of some African ants. *Entomological News*, **75**, 197–204.

Weber, N. A. (1966*a*). The fungus-growing ants. *Science*, **153**, 587–604.

Weber, N. A. (1966*b*). Fungus-growing ants and soil nutrition. *Actas Primero Coloquium Latino Americano Biologico Suelo Monografia*, **1**, 221–56. Centro Cooperativo Ciente Americano Latina, UNESCO, Montevideo, Uruguay.

Weber, N. A. (1967*a*). Growth of a colony of the fungus-growing ant, *Sericomyrmex urichi. Annals of the Entomological Society of America*, **60**, 1328–9.

Weber, N. A. (1967*b*). The growth of young *Acromyrmex* colonies in their first year (Hymenoptera: Formicidae). *Annals of the Entomological Society of America*, **60**, 506–8.

Weber, N. A. (1968). Biomass of fungus-growing ants (Attini) and their gardens. *Proceedings of the 2nd Latin American Congress on Soil Biology.* UNESCO.

Weber, N. A. (1969). Ecological relations of three *Atta* species in Panama. *Ecology*, **50**, 141–7.

Weber, N. A. (1972*a*). *Gardening ants, the attines. Memoirs* **92**, Philadelphia: American Philosophical Society. xvii+146 pp.

Weber, N. A. (1972*b*). The attines: the fungus-culturing ants. *American Scientist*, **60**, 448–56.

Weber, N. A. (1972*c*). The fungus-culturing behavior of ants. *American Zoologist*, **12**, 577–87.

Weesner, F. M. (1955). The reproductive system of young primary reproductives of *Tenuirostritermes tenuirostris* (Desneux). *Insectes sociaux*, **2**, 323–45.

Weesner, F. M. (1956). The biology of colony foundation in *Reticulitermes hesperus* Banks. *University of California Publications in Zoology*, **61**, 253–314.

Weir, J. S. (1972). Spatial distribution of elephants in an African National Park in relation to environmental sodium. *Oikos*, **23**, 1–12.
Weir, J. S. (1973). Air flow, evaporation and mineral accumulation in mounds of *Macrotermes subhyalinus* (Rambur). *Journal of Animal Ecology*, **42**, 509–20.
Wellenstein, G. (1952). Zur Ernährungsbiologie der Roten Waldameise. *Zeitschrift für Pflanzendranken und Pflanzenschutz*, **59**, 430–51.
Wellenstein, G. (1954). Die Insektenjagd der Roten Waldameise. *Zeitschrift für angewandte Entomologie*, **36**, 185–217.
Wellenstein, G. (1957). Die Beeinflussung der förstlichen Arthropodenfauna durch Waldameisen (*Formica rufa* Gruppe) Teil I. *Zeitschrift für angewandte Entomologie*, **41**, 368–84.
Wellenstein, G. (1967). Zur Frage der Standortansprüche hügelbauender Waldameisen, (*Formica rufa* Gruppe). *Zeitschrift für angewandte Zoologie*, **59**, 139–66.
Wengris, J. (1948). Badania nad rozmieszczeniem mrowisk w zaleznośći od warunków ekologicznych. *Studia Societatis scientiarum torunensis*, Section *E*, **1**, 1–79.
Went, F. W., Wheeler, J. & Wheeler, G. C. (1972). Feeding and digestion in some ants (*Veromessor* and *Manica*). *Bioscience*, **22**, 82–8.
Werner, F. G. (1973). *Foraging activity of the leaf-cutter ant,* Acromyrmex versicolor, *in relation to season, weather and colony condition. US/IBP Desert Biome Research Memorandum*, RM **73–28**, Report. **3**.
Werner, F. G. & Murray, S. L. (1972). *Demography, foraging activity of leaf-cutter ants,* Acromyrmex versicolor, *in relation to colony size and location, season, vegetation, and temperature. US/IBP Desert Biome Research Memorandum*, RM **72–33**.
West Eberhard, M. J. (1975). The evolution of social behaviour by kin selection. *Quarterly Review of Biology*, **50**(1), 1–33.
Westhoff, V. & Westhoff-De Joncheere, J. N. (1942). Verspreiding en nestoecologie van de mieren in de nederlandsche bosschen. *Tijdschrift over plantenziekten*, **9**, 1–76.
Weyrauch, W. (1942). Las hormigas cortadoras de hojas del valle Chanchamayo. *Boletín de la Dirección de agricultura, ganadería y colonización, Lima*, **15**, 204–59.
Wheeler, W. M. (1907). The fungus-growing ants of North America. *Bulletin of the American Museum of Natural History*, **23**, 669–807.
Wheeler, W. M. (1908). Honey ants, with a revision of the American Myrmecocysti. *Bulletin of the American Museum of Natural History*, **24**, 345–97.
Wheeler, W. M. (1910). *Ants, their structure, development and behaviour.* New York: Columbia University Press.
Wheeler, W. M. (1916). *Prodiscothyrea*, a new genus of ponerine ants from Queensland. *Transactions of the Royal Society of South Australia*, **40**, 33–7.
Wheeler, W. M. (1918*a*). The Australian ants of the ponerine tribe Cerapachyini. *Proceedings of the American Academy of Arts and Science*, **53**, 215–65.
Wheeler, W. M. (1918*b*). The ants of the genus *Ophisthopsis* Emery. *Bulletin of the Museum of Comparative Zoology*, **62**, 341–62.
Wheeler, W. M. (1923*a*). The occurrence of winged females in the ant genus *Leptogenys* Roger, with descriptions of new species. *American Museum Novitates*, **90**, 1–16.
Wheeler, W. M. (1923*b*). Ants of the genera *Myopias* and *Acanthoponera*. *Psyche*, **30**, 175–92.

Wheeler, W. M. (1924). The Formicidae of the Harrison Williams Galapagos Expedition. *Zoologica*, **5**, 101–22.
Wheeler, W. M. (1925). A new guest-ant and other new Formicidae from Barro Colorado island, Panama. *Biological Bulletin*, **49**, 150–81.
Wheeler, W. M. (1930). A second note on *Gesomyrmex*. *Psyche*, **37**, 35–40.
Wheeler, W. M. (1936). Ecological relations of ponerine and other ants to termites. *Proceedings of the American Academy of Arts and Science*, **71**, 159–243.
Wheeler, W. M. (1937). *Mosaics and other anomalies among ants*. Cambridge, Massachusetts: Harvard University Press. 95 pp.
Wheeler, W. M. (1942). Studies of neotropical ant-plants and their ants. *Bulletin of the Museum of Comparative Zoology*, **90**, 1–262.
Whelden, R. M. (1957). Notes on the anatomy of *Rhytidoponera convexa* Mayr. ('*violacea* Forel'). *Annals of the Entomological Society of America*, **50**, 271–82.
Whelden, R. M. (1958). Additional notes on *Rhytidoponera convexa* Mayr. *Annals of the Entomological Society of America*, **51**, 80–4.
White, W. F. (1895). *Ants and their ways*, 2nd edn revised. London: The Religious Tract Society.
Whitford, W. G. (1973). *Demography and bioenergetics of herbivorous ants in a desert ecosystem as a function of vegetation, soil type and weather variables. US/IBP Desert Biome Research Memorandum*, RM **73–29**, 63 pp.
Whitford, W. G. (1975). Factors affecting foraging activity in Chihuahuan desert harvester ants. *Environmental Entomology*, **4**, 689–96.
Whitford, W. G. & Ettershank, G. (1972). *Demography and role of herbivorous ants in a desert ecosystem as functions of vegetation, soil and climatic variables. US/IBP Desert Biome Research Memorandum*, RM **72–32**.
Wiegert, R. G. (1970). Energetics of the nest-building termite *Nasutitermes costalis* (Holm.) in a Puerto Rican forest. In *A tropical rain forest. A study of irradiation and ecology at El Verde, Puerto Rico*, ed. H. T. Odum, vol. 1, pp. 57–64. Division of Technical Information, United States Atomic Energy Commission.
Wiegert, R. G. & Coleman, D. C. (1970). Ecological significance of low oxygen consumption and high fat accumulation by *Nasutitermes costalis* (Isoptera: Termitidae). *Bioscience*, **20**, 663–5.
Wiegert, R. G. & Evans, F. C. (1967). Investigations of secondary productivity in grasslands. In *Secondary productivity in terrestrial ecosystems*, ed. K. Petursewicz, vol. II, pp. 499–518. Warszawa–Kraków: Państwowe Wydawnictwo Naukowe.
Wildermuth, V. L. & Davis, E. G. (1931). The red harvester ant and how to subdue it. *Farmer's Bulletin, USDA*, **1668**, 1–12.
Wilkinson, W. (1962). Dispersal of alates and establishment of new colonies in *Cryptotermes havilandi* (Sjöstedt) (Isoptera, Kalotermitidae). *Bulletin of Entomological Research*, **53**, 265–86.
Willard, J. R. (1973). Soil invertebrates: III. Collembola and minor insects: populations and biomass. Matador Project. Technical Report.
Willard, J. R. & Crowell, H. H. (1965). Biological activities of the harvester ant, *Pogonomyrmex owyheei*, in central Oregon. *Journal of Economic Entomology*, **58**(3), 484–9.
Williams, E. C. (1941). An ecological study of the floor fauna of the Panama rain forest. *Bulletin of the Chicago Academy of Science*, **6**, 63–124.
Williams, M. A. J. (1968). Termites and soil development near Brock's Creek, Northern Territory. *Australian Journal of Science*, **31**, 153–4.

Williams, O. L. (1934). Some factors limiting the distribution of termites. In *Termites and termite control*, ed. C. A. Kofoid, pp. 42–9. Berkeley: University of California Press.

Williams, R. M. C. (1959*a*). Flight and colony foundation in two *Cubitermes* species (Isoptera, Termitidae). *Insectes sociaux*, **6**, 203–218.

Williams, R. M. C. (1959*b*). Colony development in *Cubitermes ugandensis* Fuller (Isoptera: Termitidae). *Insectes sociaux*, **6**, 291–304.

Williams, R. M. C. (1965). Infestation of *Pinus caribaea* by the termite *Coptotermes niger* Snyder. *Proceedings of the XIIth International Congress of Entomology, London, 1964*, 675–6.

Williams, R. M. C. (1966). The East African termites of the genus *Cubitermes* (Isoptera: Termitidae). *Transactions of the Royal Entomological Society of London*, **118**, 73–216.

Williams, R. M. C. (1973). *Evaluation of field and laboratory methods for testing termite resistance of timber and building materials in Ghana, with relevant biological studies. Centre for Overseas Pest Research, Tropical Pest Bulletin*, **3**. London: Her Majesty's Stationary Office.

Williamson, K. (1975). Recent population fluctuations of some British migrant birds which winter in the Sahel Zone. *Bulletin of the British Ecological Society*, VI(4), 28.

Wilson, E. O. (1952). Notes on *Leptothorax bradleyi* Wheeler and *L. wheeleri* M. R. Smith. *Entomological News*, **43**, 67–71.

Wilson, E. O. (1953*a*). The ecology of some North American dacetine ants. *Annals of the Entomological Society of America*, **46**, 479–95.

Wilson, E. O. (1953*b*). The origin and evolution of polymorphism in ants. *Quarterly Review of Biology*, **28**, 136–56.

Wilson, E. O. (1957*a*). The discovery of cerapachyine ants on New Caledonia, with the description of new species of *Phyracaces* and *Sphinctomyrmex*. *Breviora, Museum of Comparative Zoology*, **74**, 1–9.

Wilson, E. O. (1957*b*). The *tenuis* and *selenophora* groups of the ant genus *Ponera*. *Bulletin of the Museum of Comparative Zoology*, **116**, 355–86.

Wilson, E. O. (1958*a*). Studies on the ant fauna of Melanesia. III. *Rhytidoponera* in western Melanesia and the Moluccas. *Bulletin of the Museum of Comparative Zoology*, **119**, 303–20.

Wilson, E. O. (1958*b*). The beginnings of nomadic and group-predatory behaviour in the ponerine ants. *Evolution*, **12**, 24–31.

Wilson, E. O. (1958*c*). Observations on the behaviour of cerapachyine ants. *Insectes sociaux*, **5**, 129–40.

Wilson, E. O. (1959). Some ecological characteristics of ants in New Guinea rain forests. *Ecology*, **40**, 437–47.

Wilson, E. O. (1962*a*). The Trinidad cave ant *Erebomyrma* (= Spelaeomyrmex) *urichi* (Wheeler), with a comment on cavernicolous ants in general. *Psyche*, **69**, 62–72.

Wilson, E. O. (1962*b*). Behavior of *Daceton armigerum* (Latreille), with a classification of self-grooming movements in ants. *Bulletin of the Museum of Comparative Zoology*, **127**, 403–21.

Wilson, E. O. (1964). The ants of the Florida keys. *Breviora, Museum of Comparative Zoology*, **210**, 1–14.

Wilson, E. O. (1966). Behaviour of social insects. *Symposia of the Royal Entomological Society of London*, **3**, 81–96.

Wilson, E. O. (1971). *The insect societies*. Cambridge, Massachusetts: The Belknap Press of Harvard University Press. 548 pp.

Wilson, E. O. & Eisner, T. (1957). Quantitative studies of liquid food transmission in ants. *Insectes sociaux*, **4**, 157–66.

Wilson, E. O., Eisner, T., Wheeler, G. C. & Wheeler, J. (1956). *Aneuretus simoni* Emery, a major link in ant evolution. *Bulletin of the Museum of Comparative Zoology*, **115**, 81–99.

Wilson, N. L., Dillier, J. H. & Markin, G. P. (1971). Foraging territories of imported fire ants. *Annals of the Entomological Society of America*, **64**, 660–5.

Wise, L. E. (1952). Miscellaneous extraneous components of wood. In *Wood chemistry*, ed. L. E. Wise & E. C. Jahn, vol. 1, pp. 638–88. New York: Reinhold Publishing Corporation.

Wlodawer, P., Ląglwińska, E. & Barańska, J. (1966). Esterification of fatty acids in the wax moth haemolymph and its possible role in lipid transport. *Journal of Insect Physiology*, **12**, 547–60.

Wolcott, G. N. (1946). Factors in the natural resistance of woods to termite attack. *Caribbean Forester*, **7**, 121–34.

Wood, T. G. (1971). The effects of soil fauna on the decomposition of *Eucalyptus* leaf litter in the Snowy Mountains, Australia. In *Organisms du sol et production primaire*, 4th Colloqium pedobiologiae, Dijon 1970, ed. INRA, pp. 349–58. Paris: INRA.

Wood, T. G. (1974). Field investigations on the decomposition of leaves of *Eucalyptus delegatensis* in relation to environmental factors. *Pedobiologia*, **14**(6), 343–71.

Wood, T. G. (1975). The effects of clearing and grazing on the termite fauna (Isoptera) of tropical savannas and woodlands. In *Progress in soil zoology*, ed. J. Varěk, pp. 409–17. Prague: Academia (Czechoslovak Academy of Sciences).

Wood, T. G. (1976). The role of termites in decomposition processes. In *The role of terrestrial and aquatic organisms in decomposition processes*, J. M. Anderson & A. Macfadyen, pp. 145–68, Oxford: Blackwell Scientific publications.

Wood T. G. & Ohiagu, C. E. (1976). A preliminary assessment of the significance of grass-eating termites (Isoptera) in pastures in northern Nigeria. *Samaru Agricultural Newsletter*, **18**, 22–30.

Wood, T. G., Johnson, R. A. & Ohiagu, C. E. (1977). *Proceedings of the 3rd International Conference on Tropical Ecology, Lumumbashi, Zaire, 7–13 April 1975.* Populations of termites (Isoptera) in natural and agricultural ecosystems in southern Guinea savanna near Mokwa, Nigeria. *Geo. Eco. Trop.* (In press).

Wood, T. G. & Lawton, J. H. (1973). Experimental studies on the respiratory rates of mites (Acari) from beech-woodland leaf litter. *Oecologia*, **129**, 169–91.

Wood, T. G. & Lee, K. E. (1971). Abundance of mounds and competition among colonies of some Australian termite species. *Pedobiologia*, **11**, 341–66.

Woodwell, G. M. & Botkin, D. B. (1970). Metabolism of terrestrial eco-systems by gas exchange techniques: the Brookhaven approach. In *Analysis of temperate forest ecosystems*, ed. D. E. Reichle, pp. 73–85. Berlin: Springer-Verlag.

Wray, D. L. (1938). Notes on the southern harvester ant (*Pogonomyrmex badius* Latr.) in North Carolina. *Annals of the Entomological Society of America*, **31**, 196–201.

Wüst, M. (1973). Stomodeal und Proctodeale Sekrete von Ameisenlarven und ihre

Biologische Bedeutung. *Proceedings, VIIth International Congress of the Union for the Study of Social Insects, London, 10–15 September 1973*, 412–17.
Yasuno, M. (1963). The study of the ant population in the grassland at Mt. Hakkôda. I. The distribution and nest abundance of ants in the grassland. *Ecological Review*, **16**, 83–91.
Yasuno, M. (1964*a*). The study of the ant population in the grassland of Mt. Hakkôda. II. The distribution pattern of ant nests at the Kayano Grassland. *Science Reports of the Tôhoku University*, Ser. IV, **30**, 43–55.
Yasuno, M. (1964*b*). The study of the ant population in the grassland at Mt. Hakkôda. III. The effect of the slave making ant, *Polyergus samurai*, upon the nest distribution pattern of the slave ant, *Formica fusca japonica*. *Science Reports of the Research Institutes, Tôhoku University*. Ser. IV, (*Biol.*), *Sendai*, **30**, 167–70.
Yung, É. (1900). Combien y a-t-il de fourmis dans une fourmilière? (*Formica rufa*). *Archives des sciences physiques et naturelles*, **10**(4), 46–56.
Zeuthen, E. (1953). Oxygen uptake as related to body size in organisms. *Quarterly Review of Biology*, **28**, 1–12.
Zeuthen, E. (1964). Microgasometric methods: cartesian divers. In *Second international congress of histo- and cytochemistry*, ed. T. M. Schiebler, A. G. E. Pearse & H. H. Wolff, pp. 70–80. Berlin: Springer-Verlag.
El-Ziadi, S. (1960). Further effects of *Lasius niger* (L.) on *Aphis fabae* Scopoli. *Proceedings of the Royal Entomological Society of London*, (*A*), **35**, 30–8.
El-Ziadi, S. & Kennedy, J. S. (1956). Beneficial effects of the common garden ant *Lasius niger* (L.) on the black bean aphid, *Aphis fabae* Scopoli. *Proceedings of the Royal Entomological Society*, **34**, 61–5.
Žigulskaja, Z. A. (1968). Naselenie murav'ev (Formicidae) stepnykh landshaftov Tuvy. In *Zhivotnoe naselenie pochv v bezlesnykh biogeocenozakh Altae-Sajanskoï Gornoï sistemy*, ed. I. V. Stebaev, pp. 115–39. Novosibirsk: Novosibirskiï Gosudarstvennyï Universitet.
Zoebelein, G. (1956*a*). Der Honigtau als Nahrung der Insekten. Teil I. *Zeitschrift für angewandte Entomologie*, **38**, 369–416.
Zoebelein, G. (1956*b*). Der Honigtau als Nahrung der Insekten. Teil II. *Zeitschrift für angewandte Entomologie*, **39**, 129–67.
Zylberberg, L., Jeantet, A. Y. & Delage-Darchen, B. (1974). Particularités structurales de l'intima cuticulaire des glandes post-pharyngiennes des fourmis. *Journal de Microscopie*, **21**(3), 331–42.

Index

Acacia drepanolobium, 23
Acanthomyops claviger, 238, 241, 242
Acanthotermes spp., 50, 63
Acrocoelia heliophila, 317
Acromyrmex (Atta) discigera, 83
Acromyrmex landolti, 87, 98
Acromyrmex octospinosus, 87, 244
Acromyrmex spp., 87, 90
Acromyrmex versicolor, 87, 90
Actinomycetales, 302, 303
Adaiphrotermes spp., 21, 33, 63–4, 269
Aderitotermes spp., 33, 269
Aenictus laeviceps, 334
Aenictus spp., 86
aerial photography, for estimating mound density, 6–7
Aeromyrmex ambiguus, 294
Ahamitermes spp., 57, 69, 225
alates, of ants, 53
alates, of termites, 62, 123, 213, 221, 222, 232; chemistry of, 202, 205, 207, 209; as prey, 285–6; production of, 161, 273–6; respiratory data, 159; swarming, as a means of energy loss, 53; trophallaxis, 215–16
alimentary canal, of termites, 65
Allodotermes tenax, 19
Alyscotermes spp., 290
Amblyopone pallipes, 334
amino acid composition: of ant secretions, 237; of termites, 202, 210
Amitermes atlanticus, 126, 223
Amitermes evuncifer, 9, 10, 122, 269, 270
Amitermes hastatus, 224, 225, 282; cannibalism, 228–9, 230; egg production of, 46, 223; food and foraging of, 63, 68, 72
Amitermes laurensis, 8, 251, 252
Amitermes minimus, 197
Amitermes spp., 9, 50, 65, 110, 184, 264
Amitermes unidentatus, 250, 252
Amitermes ritiosus, 254, 261
Amitermitinae, 217, 251
amylase, presence in different ant spp., 238
Anacanthotermes ahngerianus, 187, 287
Anacanthotermes spp., 58, 288
Anacanthotermes turkestanicus, 187
Ancistrotermes cavithorax, 21, 33, 129, 274; energy budget of, 161–2; food consumption of, 76; fungus combs of, 20, 68; production of, 50–1, 53, 272, 275
Ancistrotermes guineensis, 224
Ancistrotermes latinotus, 19
Ancistrotermes spp., 50, 77, 269, 289; mound structure and soil composition of, 250, 252, 254, 256
Anenteotermes spp., 64, 269
Angulitermes spp., 283–4
Anomma kohli, 291
Anomma nigricans, 334
Anomma spp., 84, 290, 311, 327
Anomma wilverthi, 14, 334
Anoplolepis custodiens, 287
Anoplotermes pacificus, 263, 264
antibiotics, found in ant nests, 305
Aphaenogaster fulva, 124
Aphaenogaster rudis, 334
Aphaenogaster subterranea, 235, 236
Aphaenogaster treatae, 334, 339
Apicotermes desneuxi, 15, 47
Apicotermes gurgulifex, 15, 47, 267
Apicotermes spp., 15, 257
Apicotermitinae, 57, 65, 269
Apomyrma stygia, 334
Aporomyrmex ampeloni, 335
Apterostigma spp., 86–7
arboreal-nesting ants, 87, 326, 328
arboreal-nesting termites, 70–1, 268, 269
Archotermopsis spp., 65
assimilation, 45, 81; in ants, 81, 310; in termites, 78–80, 161, 162, 201–4, 210–11, 277; *see also* nutrient dynamics
Astratotermes spp., 33, 269
Atta capignara, 87, 98
Atta cephalotes, 37, 97, 123, 129, 243–4; food and foraging of, 87, 88, 89, 90, 103
Atta colombica tonsipes, 14, 37, 240, 294, 335; digestive enzymes of, 238, 241, 242; feeding habits of, 97, 103; fungus cultures of, 83–4, 101, 205
Atta laevigata, 87
Atta rollenweideri, 87
Atta sexdens, 12, 97, 101, 103, 296, 307
Atta spp., 37, 103, 297; food and foraging of, 86, 87, 88, 90, 97
Atta texana, 89–90, 243
Attinae, 101, 106, 249–50, 327; digestive enzymes of, 237, 238, 241, 242; food of, 86, 87–8, 92, 307; food consumption of, 95, 96; foraging behaviour of, 89–90, 96–8, 103
attractants, insect, 175–6
Azteca spp., 84, 87, 327

bacteria, 183–4, 203, 302, 303; digestive enzymes of, 194–5, 202–3; and symbiosis with termites, 186, 192–3, 197, 204, 283
baiting techniques, 59–61, 91
Basidentitermes potens, 21, 33, 37, 38, 269
Basidiomycetes, 179, 180, 183
Bellicositermes bellicosus rex see *Macrotermes subhyalinus*
Bellicositermes natalensis see *Macrotermes bellicosus*
Bellicositermes spp., 50

Index

biomass, 52, 53, 277–8, 306, 330–1; determinations of, 31–7, 38, 39; and production, 45–6, 48; and respiratory rates, 130–4, 148–50, 157–9, 160–1, 162; of termites in different ecosystems, 267, 268, 269, 273, 280
biomes, of ants, 336–9; *see also* ecosystems
body weight *see* biomass
Bothroponera silvestrii, 334
Bray's association index, frequency of ant species' encounters, 311, 312
Brian's coefficient, for ant population estimations, 321

calcium levels: in ant nests and surrounding soil, 201, 300, 306; in soil, 252, 261, 300; in termite nests, 259, 260, 261
calorific values, determinations of, 37–40
calorimetry, 115–17, 130
Calotermes flavicollis, 43, 75, 200, 204, 206, 220; cannibalism of, 228, 229, 231; digestion of, 79, 185, 187, 189, 213, 225; respiration of, 125, 127, 157, 270, 271; trophallaxis in, 214, 215–16, 217, 218, 219–20
Calotermitidae, 47, 220, 221, 222, 224, 230; food and foraging behaviour, 69, 75, 165, 178
Calotermes minor (*Incisitermes minor*), 178–9
Calotermes spp., 64
Camponotini, 148
Camponotus acvapimensis, 294, 296, 318, 335
Camponotus aethiops, 28, 29
Camponotus compressiscapus, 335
Camponotus congolensis, 335
Camponotus herculeanus, 27, 121, 133, 148; digestive enzymes of, 238, 241, 242
Camponotus herculeanus japonicus, 311, 312
Camponotus inaequalis, 335
Camponotus ligniperda, 241
Camponotus maculatus, 335
Camponotus modoc, 144, 145, 148
Camponotus noveboracensis, 121, 335
Camponotus nylanderi, 27
Camponotus pennsylvanicus, 121, 124, 238, 241, 242, 335
Camponotus santosi, 335
Camponotus spp., 285, 328
cannibalism, 52; of ants, 86, 92, 318, 323–4; of termites, 56, 57, 228–32; *see also* predation
Carebara, spp., 292
carbohydrates, 166–8, 198, 225, 234, 282
carbon levels (organic), 166, 250–1, 262, 300, 301; carbon/nitrogen ratio, 259–60, 282–3
carbon dioxide levels, 113–14, 120, 125–9; *see also* respiration
carnivorous spp., of ants *see* predation
carton, 224–6, 249–50, 258, 259; *see also* faeces
cellulase activity, 194–5, 196
cellulose, 166–8, 169, 173, 182–3, 228; assimilation and digestion by termites, 191–7, 210
α-cellulose, 166, 167
Centromyrmex spp., 289, 292
Cephalotermes rectangularis, 123
Cerapachyini, 86
chromatography, 93, 154
circadian rhythms, of termites, 71–3, 163
Coarctotermes spp., 50
colonies, of ants: density in different ecosystems, 336–9
consumption of food, by ants: ecological efficiencies, 310, 331; and environmental factors, 104–6, 310; field-based estimations of, 95–6; laboratory measurements of, 94–5; primary, 82–5; qualitative aspects of food selection, 82–93; quantitative aspects of food selection, 85–6; and symbiotic fungi, 82–4, 101–2; *see also* feeding habits, of ants
consumption of food, by termites: and energy flow in ecosystems, 273, 276, 277, 278, 279; in the field, 73, 74, 76; under laboratory conditions, 73, 74, 75, 281; in production formulae, 45, 81; *see also* feeding habits, of termites
coprophagy, of termites, 277
Coptotermes acinaciformis, 12, 26, 43, 75, 225; foraging galleries of, 69–70; mound structure and soil composition of, 250, 252, 256, 259, 283
Coptotermes amanii, 75
Coptotermes formosansus, 66, 199, 221; gut symbiotes of, 192, 193, 203; lipid content of, 206–7, 209–10
Coptotermes frenchi, 12, 26
Coptotermes lacteus, 43, 70, 75, 206–7, 225; cannibalism of, 229, 231; mound structure and effect on ecosystem, 250, 252, 259, 283
Coptotermes niger, 68, 180, 181
Coptotermes spp., 57, 66, 69, 225, 264, 269
Cornitermes spp., 63, 249
Crematogaster (*Acrocoelia*) *mimosae*, 23
Crematogaster (*Acrocoelia*) *nigriceps*, 23
Crematogaster dohrni, 334
Crematogaster spp., 318, 328
crop: digestive enzymes in ant, 238, 239, 240–1, 242
Cryptotermes brevis, 66, 177, 200; trophallaxis in, 212, 214, 216, 217, 218
Cryptotermes cavifrons, 75
Cryptotermes dudleyi, 75
Cryptotermes havilandi, 47, 157, 216, 228
Cryptotermes spp., 66
Cubitermes exiguus, 41, 74, 76, 274; biomass of, 31, 32, 33; carbon dioxide evolution of, 120–1; and effect on ecosystems, 162–3; energy flow in, 160–1; oxygen consumption of, 126, 128, 133–4, 137, 279; population density of, 8, 12, 267; respiration of, 112, 121, 128, 129, 156, 270; respiratory data of, 158, 159n, 159–60
Cubitermes fungifaber, 31, 32, 155, 277; trophallaxis in, 216, 218; mounds of, 40, 251, 252, 282
Cubitermes intercalatus, 126
Cubitermes sankurensis, 72, 251, 252; biomass of, 31, 32, 33; population density estimates of, 7–8, 9, 11, 12; respiration of, 133–4, 156, 158, 159n, 160, 270
Cubitermes severus, 9, 40, 41–2, 47
Cubitermes spp., 6, 40, 50, 129, 268, 290; biomass of, 31, 32, 33, 269; and effect on ecosystems, 264; energy flow in, 160; feeding habits of, 57–8, 62–3; mound structure of, 251, 252; respiration of, 154, 158, 159, 269
Cubitermes testaceus, 291

Cubitermes ugandensis, 43, 216–17, 228, 229, 291
Cyphomyrmex costatus, 83, 86
Cyphomyrmex rimosus trinitatis, 83, 86
Cyphomyrmex spp., 82

decomposition, of termites, 283–5
demographic parameters, determinations of, 5–44; biomass and energy content, 30–40; mark-recapture methods of sampling, 26–30; nest sampling methods, 6–14, 25–6; population dynamics, 40–4; soil sampling methods, 15–25
dentition, of termites, 64–5
Dicuspiditermes nemorosus, 270
diet *see* food
digestion and digestive system, of termites, 78–80, 188–209, 63–5; cellulose digestion, 192–5; end-products and metabolic pathways, 195–7; enzyme action, 189–91, 194–5, 213; hemicellulose digestion, 199; lignin degradation, 199–201; lipid digestion, 204–8; morphology of digestive system, 64–5, 185–8; nitrogen metabolism, 201–4; non-symbiotic digestion, 188–9; pH, 187; *see also* nutrient dynamics
digestive enzymes, of ants, 237–43; in different glands, 239–40
diurnal periodicity, of foraging termites, 71–2
diurnal variations, of nest populations, 11
diversity, of termite spp., 247–8
Dolichoderinae, 236, 334
Dolichoderus spp., 328
Dolichoderus quadripunctatus, 324
Dorylinae, 85, 86, 89, 234, 298, 327; adult populations of, 334
Dorylus dentifrons, 291
Dorylus nigricans, 47
Dorylus spp., 290–1, 327
Dorymyrmex emmaricaellus, 335
Drepanotermes rubriceps, 251, 252, 254, 257, 261
Drepanotermes spp., 58, 63, 67, 70, 71

Echinopla spp., 327
Eciton burchelli, 47, 118, 334
Eciton hamatum, 46, 123, 334
Eciton spp., 14, 81, 214, 236, 311
Ecitoninae, 327, 334
ecological parameters, for assessing production, 45–6
economic density, of ants, 91, 99
ecosystems, role of ants in, 293–325, 326–31; as consumers, 306–10; environmental pollution adaptability, 324–5; feeding habits and energy flow, 99–103; as food for other animals, 317–18; importance in different ecosystems, 328–31; and invertebrate fauna, 324–5, 340; nest construction, 293–7; and population of other consumers, 311–17; population regulation, 318–24; soil modification, 297–306; *see also* population energetics
ecosystems, role of termites in, 245–92; diagrammatic representation of, 246, 247; diversity of spp., 247–8; and effect on vegetation, 263–4; effects on fauna, 265; nest systems, 249–53; and population density, 266–70; population energetics, 265–92; return of energy to the ecosystem, 281–92; soil modifications, 253–62
endopolypeptidase *see* protease
energy flow, in ants, 53; energy content, 39–40, 129–30, 140; and feeding habits, 99–103, 152–3; modes of expression of, 129–40; through populations, 309–10, 317–18; respiration, 140–53; and social organization, 117–20; *see also* population energetics, respiration
energy flow, in termites: and effects on ecosystems, 245, 246, 267, 269, 272–3; energy consumption, 277, 278, 279; energy content, 37–8; of individual spp., 265; measurements of, 153–63; modes of expression of, 133, 136–8; and social organization, 117–20; *see also* population energetics, respiration
energy flow, laboratory determinations of: direct, 115–17; indirect, 113–14
environmental pollution, adaptability of ants to, 324–5
Epimyrma stumperi, 335
Eutermellus spp., 50, 269
Eutermes exitiosus, 120
evolution, of ants and termites, 2–3
excreta *see* faeces

faeces, of termites, 64, 200; effect on ecosystem, 257–8, 264, 282–4; production and use of, 80, 204, 224–6, 249, 276, 277
feeding habits, of ants, 81, 92–3; effect on ecosystems, 99–103, 306–10; food storage of, 89–92, 106, 233; and Homoptera mutualism, 85, 100, 313; primary and secondary consumption, 82–6; and symbiotic fungi, 82–4; *see also* foraging, nutrient dynamics
feeding habits, of termites, 55–80; and effect on ecosystems, 248, 263–4; food consumption, 73–7; food stores, 63, 77–8; observations of, (in the field), 57–61, (in the laboratory), 61–3; quantitative aspects of, 73–80; *see also* foraging, nutrient dynamics
food and food selection, by ants, 92, 93, 95; composition of, 82–8, 109–10, 308; effects on ecosystems, 306–10; and Homoptera mutualism, 85, 100, 313; of predatory spp., 85–6, 88; storage, 91–2, 233, 307; and symbiotic fungi, 82–4, 101–2; *see also* foraging, nutrient dynamics
food and food selection, by termites, 65–9; composition of, 55–7, 166–84; consumption rate (in the field), 59–61, 73, 76, 273, 276, 279 (under laboratory conditions), 73, 74, 75, 276, 281; and role in ecosystems, 246, 257–8; storage, 63, 77–8; *see also* foraging, nutrient dynamics
foraging, of ants, 90–1, 92, 95, 108, 121–2; distribution pattern of, 89–90, 96, 311–13; liquid and solid food intake, 96–8; predation, 86, 313–17; by seed-harvesting spp., 107–10
foraging, of termites, 69–73, 163, 246; diurnal periodicity of, 71–2; extent of, 69–71; quantitative assessment of, (in the field), 57–61, (under laboratory conditions), 61–2; seasonal periodicity of, 72–3
Formica aquilonia, 314
Formica cinerea, 294

Index

Formica cunicularia glauca, 294, 295, 298
Formica exsecta, 133, 148, 294, 308
Formica exsectoides, 27, 335
Formica fusca japonica, 311
Formica fusca, 89, 90, 294; digestive enzymes of, 238, 239, 240, 241, 242, 243; population density of, 27, 319, 320, 321; respiratory data, 133, 137, 148, 239
Formica gagatoides, 328
Formica incerta, 335
Formica integra, 238, 240, 241, 242
Formica lemani, 298
Formica lugubris, 298, 314
Formica nigricans, 39
Formica opaciventris, 319, 339
Formica pallidefulva, 335
Formica perpilosa, 133, 136, 147
Formica polyctena, 39, 47, 122–3; digestive enzymes of, 238, 239, 240, 242, 244; energy flow in, 119, 140, 152; food consumption by, 88, 94, 309, 313, 314–15; foraging behaviour of, 91, 96, 104, 307, 308; respiration of, 130–1, 132, 135, 140–2, 145–6, 148; trophallaxis by, 234
Formica pratensis, 133, 137, 148, 311–12, 315
Formica rufa, 39, 40, 47, 52, 93, 104; food consumption of, 88, 90–1, 96, 98, 313; foraging behaviour of, 95, 100, 105–6, 308; nests of, 122–3, 124, 294, 296, 297; population density of, 335, 337; predation by, 86, 100, 314–15
Formica rufa polyctena, 294, 295
Formica rufa rufopratensis, 315
Formica rufibarbis, 41, 298
Formica rufopratensis minor, 234
Formica sanguinea, 41, 237
Formica spp., 39, 40, 142, 237, 296, 318; food and foraging behaviour of, 89, 96, 234; importance in different ecosystems, 328, 329; predation by, 100
Formica subnitens, 308, 335
Formica subpilosa, 311
Formica truncorum yessensis, 294, 311, 312
Formica ulkei, 143, 144, 145, 148, 296–7
Formica yessensis, 119, 122–3
Formicidae, 240
Formicinae, 87, 151, 233, 236, 334–5; digestive enzymes of, 238–42; respiration of, 132–4, 148
fresh weight *see* biomass
Fulleritermes spp., 58
fungi, 200–1, 204–5; and ant symbiosis, 82–4, 95, 101–2, 302, 303, 307; brown and white rot caused by, 179, 180, 181, 182–3, 200, 219; digestive enzymes of, 194–5; wood-decaying, 169, 178–83; *see also* fungus–termite association
fungus combs, of termites, 259, 277–8, 283–4; nutritional role of, 226–8; studies of, 60–1, 63, 64, 68, 80, 205; *see also* fungus–termite association
fungus–termite association, 56, 102, 282, 283; colony respiration of, 111–12; digestive enzymes of fungi, 194–5; and the effects on ecosystems, 163; lipid content of fungi, 204–5; wood-decaying fungi, 178–83, 219–20, 227

galleries, of termites, 69–70, 253, 262
geophagus termites *see* humivorous termites
Gnathamitermes perplexus, 73, 154, 178, 184, 207
Gnathamitermes spp., 110
Gnathamitermes tubiformans, 267, 268
Gonioma spp., 328
granivorous ants, 84, 91, 327, 330
grass-feeding termites, 70, 80, 163; food consumption and digestion of, 65, 78–9, 80, (in the field), 74, 76, 280, (under laboratory conditions), 74, 75; mound structure and soil composition, 251, 252, 258, 259, 262–1
graveyard tests, 59
gut contents, examination of: of ants, 93; of termites, 63–4

habitat, modifications of *see* ecosystems
Harpagoxenus americanus, 335
Harpagoxenus sublaevis, 335
hemicellulose, 166, 167, 199
heteroauxin (3-indolyl acetic acid), 101
Heteroponera monticola, 334
Heterotermes aureus, 61, 73, 74, 75, 76, 267
Heterotermes ferox, 69
Heterotermes indicola, 75, 79, 158, 190, 200, 225
Hodotermes mossambicus, 12—13, 176–7, 208, 224, 225, 232; digestion and assimilation of, 78, 79, 190–1, 199, 200, 231; food consumption of, (in the field), 74, 76, 77, (under laboratory conditions), 74, 75; foraging behaviour of, 58, 62, 66–7, 70, 71–3; as prey, 287, 288; trophallaxis by, 213, 216, 219, 220, 222
Hodotermes spp., 58, 65, 67, 74, 75, 287
Hodotermitidae, 75, 76, 178, 220, 224, 251
Hodotermitinae, 57, 58, 65, 178, 222
Hodotermopsis spp., 65
Homallotermes foraminifer, 270
Homoptera, mutualism with ants, 85, 91, 92, 100, 105–6, 313
Hospitalitermes spp., 57, 58, 71, 288
humidity, in nest systems, 125, 128–9
humivorous termites, 65, 68, 69, 130, 158–9
D-β-hydroxydecanoic acid (myrmicacin), seed germination inhibitor, 91, 101
Hymenoptera, 24
Hypoclinea plagiata, 335, 339
Hypoclinea pustulata, 335, 339
Hypoponera spp., 334
Hypotermes obscuripes, 183–4

imagoes, 48, 207, 222
Incistitermes minor see *Calotermes minor*
Incolitermes spp., 57, 69, 225
3-indolyl acetic acid (heteroauxin), 101
invertase, presence in different ant spp., 238
Iridomyrmex humilis, 235, 244, 287
Iridomyrmex purpureus, 296

labial glands, of ants: enzymes of, 238, 240, 241, 242
Lacessititermes spp., 58
larvae, of ants, 48, 93, 128, 236–7; nutrition of, 81, 92, 234–6, 242–3
larvae, of termites, 33, 51, 158, 207, 213–20
Lasius alienus, 24, 27, 36, 44, 318, 335; colony

Lasius alienus (cont.)
density of, 336, 337; food consumption of, 84, 309; production by, 49, 53, 310, 317; respiration of, 133, 135, 139, 140, 146, 148
Lasius flavus, 92, 133, 148, 297; food of, 85, 309; longevity of, 40, 41, 120, 297; next structure and surrounding soil composition, 129, 298, 300, 301, 302, 303, 336
Lasius fuliginosus, 98, 293
Lasius latipes, 124
Lasius niger, 28, 30, 41, 294, 335, 338; adaptability to environmental pollution, 324, 325; diet and feeding habits of, 85, 89, 92, 100, 309; in ecosystems, 329, 331; nest structure and soil composition, 297–8, 300, 301, 302, 303; production by, 53, 317; respiration of, 133, 137, 146, 148
Lasius spp., 86, 298, 328
Lasius umbratus, 298
Leptogenys latteli, 89
Leptogenys spp., 86
Leptomyxotermes spp., 50
Leptomyrmex spp., 307
Leptothorax acervorum, 328, 335
Leptothorax ambiguus, 335
Leptothorax curvispinosus, 335
Leptothorax duloticus, 335
Leptothorax kutteri, 335
Leptothorax longispinosus, 335
Leptothorax muscorum 335
Leptothorax nylanderi, 335
Leptothorax spp., 85
lestobiosis, 292
lignin, 166, 168–9, 170, 173, 178, 282; degradation by termites, 193, 199–201; fungal effects, on wood content of, 192–3, 228
Liometopum microcephalum, 40
lipase, presence in different ant spp., 242
lipids, 175, 209–10, 213, 234–5, 243–4; digestion by termites of, 204–8
litter bag experiments, for assessing feeding habits of termites, 59, 60
litter-feeding termites, 70, 77–8
longevity, of ant and termite societies, 40–1, 43–4, 45, 120
Lumbricidae, 304–6

Macrotermes bellicosus (*Bellicositermes natalensis*), 259, 284, 289, 291; carbon dioxide and oxygen concentrations of, 120, 125, 128; feeding habits of, 73, 76, 77–8, 289, 291; mounds of, 9, 125–7, 129, 250, 252, 261; respiration of, 111, 154, 158, 269, 270
Macrotermes carbonarius, 270
Macrotermes falciger, 213, 274, 286; nutrient dynamics of, 205, 206, 207, 208, 210
Macrotermes natalensis, 123, 206, 212, 223, 227
Macrotermes spp., 40, 63, 205, 281, 282, 284–5; food consumption by, 76, 77; mound density of, 8, 9, 70; mound structure and soil composition of, 254, 255, 258, 259, 265; as prey, 286, 289, 290, 291; respiration of, 125, 128
Macrotermes subhyalinus (*Bellicositermes bellicositus rex*), 40, 126, 288; feeding habits of, 57, 59, 67, 70, 71–3, 76, 77; mound structure and soil composition of, 250, 252, 254, 255, 261; population density of, 6, 8, 269, 272; production by, 47, 274, 275
Macrotermes ukuzii, 270, 278
Macrotermitinae, 18–19, 213, 273, 274, 292; and effects on ecosystems, 263–4; energy consumption by, 277; feeding habits of, 56, 57, 59, 60–1, 62–3, 194; food consumption rates of, 76, 77, 273, 276; fungus combs of, 63, 64, 80, 226–8, 280, 283; mound structure and soil composition, 249, 250–2, 257, 259, 260, 261–2; population density, biomass and respiration data of, 32, 269, 272, 273; as prey, 287, 289, 290
magnesium levels, in ant nests and surrounding soil, 300, 301, 306
maintenance, 111–17, 145; *see also* energy flow, metabolism, respiration
maltase, presence in different ant spp., 239
mandibular glands, of ants: enzymes of, 238, 240, 241, 242
Marginitermes hubbardi, 154, 201, 202
mark–recapture techniques, 26–30, 49, 53
Mastotermes darwiniensis, 56, 75, 194, 246
Mastotermes spp., 64, 74, 251
Mastotermitidae, 75, 222
mathematical formulae: for energy flow and respiration, 100, 265, 271; for oxygen consumption rates, 128, 130–1, 133, 136, 146, 147; for population estimations, 321; for rate of production, 45–6, 50, 81
maxillary glands, of ant spp.: enzymes of, 238, 240, 241, 242
Megaponera ambigua, 318
Megaponera foetens, 124, 289, 290, 334
Megaponera spp., 285, 290
Melophorus spp., 307, 327
Meranophis spp., 327
Mesoponera ambigua, 334
Mesoponera caffraria, 334
Messor barbarus, 91
Messor capitatus, 40, 243; digestive enzymes of, 237, 238, 239, 241, 242
Messor spp., 17, 84, 239, 240; ecosystems of, 327, 328, 331
metabolism and metabolic rate, 112, 118–20, 122–4, 189; laboratory estimations of, 113–17, 140–53; modes of expression of, 129–40; *see also* energy flow, respiration
metamorphosis, respiration rates of, 130
metathoracic glands, of ants: secretions of, 91, 101
Microcerotermes arboreus, 47
Microcerotermes edentatus, 180, 187–8, 194, 195, 196, 200
Microcerotermes fuscotibialis, 225
Microcerotermes parvulus, 21, 32, 33
Microcerotermes spp., 50, 68–9, 76, 187, 214, 269
Microhodotermes spp., 58, 287
micro-organisms, 169, 192–3, 281, 302, 303, 304–6; symbiotic, 78, 84, 165, 283; *see also* bacteria, fungi, protozoans
Microtermes albopartitus, 19
Microtermes redemanius, 19
Microtermes spp., 50, 214, 269, 270; feeding habits of, 73, 76, 77; fungus combs of, 64, 80, 278;

Index

Microtermes spp (*cont.*)
mound structure and soil composition, 254, 255, 256
Microtermes toumodiensis, 21, 33, 68
midgut, of ant spp: enzymes of, 238, 241, 242
Mollerius spp., 87
Monomorium pharaonis, 47, 89, 237, 334
Monomorium spp., 331
mounds, of ants: moisture content of, 296
mounds, of termites, 9–11, 12–13, 40–1, 70, 122, 294; analysis of, 257, 259–62, 282; and their effect on ecosystems, 246, 263–4; populations of, 266, 267, 268, 269; predation upon, 290–2; and soil profile, 253–5, 261; structure of, 249–53, 282
mouthparts, of termites, 64–5
mucilages, as a component of sound wood, 166, 167
mutilation, by termites, 231–2 *see also* cannibalism
Mycocepurus spp., 82
Myopias spp., 86
Myrmecia gulosa, 335
Myrmecia pilosula, 335
Myrmecia spp., 327
Myrmecia vindex, 335
Myrmeciinae, 335
Myrmecocystus hortideorum, 129
Myrmecocystus mexicanus, 233
Myrmecocystus spp., 91, 233, 307, 327
Myrmica brevinodis, 334
Myrmica emeryana, 334
Myrmica fracticornis, 334
Myrmica laevinodis, 36, 48, 312, 329; food consumption by, 98–9, 309; population dynamics of, 49, 319, 321–3, 334; predation by, 314, 314, 316; production by, 105, 317; *see also Myrmica rubra*
Myrmica punctiventris, 334
Myrmica rubra, 43, 44, 119, 120, 294, 334; digestive enzymes of, 238, 239, 240, 241, 242–3, 244; food consumption of, 88, 92, 94–5, 104–5, 124, 235; production by, 47, 49, 105; respiration of, 133, 137, 141–2, 146, 148
Myrmica ruginodis, 46, 237, 309, 321, 324; population density of, 319, 320, 334
Myrmica rugulosa, 309
Myrmica sabuleti, 49, 53, 117, 317, 334; longevity of, 41, 43–4
Myrmica scabrinodis, 294, 319
Myrmica spp., 36, 41, 43, 51, 297, 318; adaptability to environmental pollution by, 324; in different ecosystems, 328, 329; food consumption of, 94, 96, 100, 329; foraging behaviour of, 89, 297, 308; nest structure and soil composition of, 298, 300, 301, 302, 303, 304; production by, 47, 104, 105, 309, 310
Myrmica sulcinodis, 319
myrmicacin (D-β-hydroxydecanoic acid), 91
Myrmicaria eumenoides, 287
Myrmicinae, 91, 132–4, 149, 233, 236, 334; digestive enzymes of, 238, 240, 241, 242
Myrmicini, 148
Myrmicocrypta buenzlii, 83
Myrmicocrypta ednaella, 86
Myrmicocrypta spp., 82
^{24}Na-labelling, for assessing production rate of worker ants, 49
Nasutitermes costalis, 12, 75, 161, 223, 272; population density of, 267, 268; respiration data of, 127–8, 133, 159, 271, 272, 279
Nasutitermes ephraetae, 79, 159, 200, 225, 228–9
Nasutitermes exitiosus, 43, 207, 231, 282; faecal examination of, 64, 80, 225, 262, 282, 283; food consumption and digestion of, 74, 75, 76, 79, 191, 194; foraging behaviour of, 68–9, 72, 73; mound structure and soil composition of, 250, 252, 259; nest micro-climate of, 123, 126, 129; population density of, 31, 267
Nasutitermes guayanus, 71
Nasutitermes latifrons, 250, 252
Nasutitermes lujae, 213
Nasutitermes magnus, 259
Nasutitermes nigriceps, 203
Nasutitermes ripperti, 75
Nasutitermes spp., 50, 63, 176, 204, 264, 269
Nasutitermes surinamensis, 47
Nasutitermes triodiae, 7, 40, 70, 254, 255, 259
Nasutitermitinae, 57, 58, 65, 194, 217; mound structure of, 249, 251; production and respiration data of, 272, 273
Neivamyrmex carolinensis, 120
Neivamyrmex nigrescens, 334
Neivamyrmex spp., 13
Neotermes castaneus, 75
Neotermes jouleti, 75, 220, 230
nests, 5–6, 12, 40–1, 57–8; micro-climate of, 122–4, 125–9
nest systems, of ants, 293–7, 312–13; chemical properties of, 298, 299, 300, 301; moisture content and temperature of, 294, 296, 328; types of, 293, 295
nest systems, of termites, 246, 290–2; effects on vegetation of, 263–4; structure and composition of, 249–53
nitrogen, 169–75, 227; conservation and recycling in termites, 57, 201–4, 225, 231; levels in termite nest structures, 250–1, 259–60, 282–3; levels in wood, 169–75, 181; soil levels of, 252, 261
Noditermes spp., 50
non-assimilated remains, of ants, 92–3
Novomessor cocherelli, 133, 136, 149, 309; foraging behaviour of, 107–8, 109, 121–2, 308
Novomessor spp., 107, 110
nutrient dynamics, of ants, 233–44; digestive enzymes, 237–43; and fungal contribution to, 83–4; trophallaxis, 234–7; *see also* food
nutrient dynamics, of termites, 165–232; assimilation efficiency, 210–11; cannibalism, 228–32; cellulose digestion, 192–5; digestive enzymes of, 189–91; and effects on ecosystems, 257–8; food source, chemical composition of, 165–77; lignin degradation, 199–201; nitrogen metabolism, 204–8; non-symbiotic digestion, 188–9; premodification by micro-organisms, 177–84; recycling of nutrients, 224–32; trophallaxis, 211–13; utilisation of nutrients, 184–232; *see also* food, fungus–termite association, wood
nymphs, of termites, 52, 53, 157–9, 161, 207; trophallaxis, 214, 215, 216, 218, 220

Odontomachus assiniensis, 333
Odontomachus troglodytes, 333
Odontotermes assmuthi, 207, 285, 286
Odontotermes badius, 205
Odontotermes obesus, 47, 62, 194, 274
Odontotermes latericius, 15
Odontotermes obscuriceps, 250, 252, 259
Odontotermes pauperaris, 68, 289
Odontotermes redemanni, 209, 250, 252
Odontotermes smeathmani, 76, 77, 272, 274, 286; food consumption by, 275, 276
Odontotermes spp., 50, 73, 227, 264, 268, 269; food consumption by, 76, 77; mound structure and soil composition of, 6, 21, 250, 252, 254, 256; as prey, 287, 288, 289; trophallaxis, 213, 214, 281
Odontotermes sudanensis, 9
Oecophylla longinoda, 88
Oecophylla spp., 293, 327
oophagy, 52, 56
Orthotermes spp., 50
oxygen consumption *see* respiration
Oxyopomyrmex spp., 328

^{32}P-labelling: for assessing production rates of worker ants, 49; to assess the rate of honeydew intake by ants, 313; to clarify the function of the ant pharyngeal gland, 244
Paltothyreus spp., 285, 318
Paltothyreus tarsatus, 334
Paraneotermes simplicicornis, 75, 207
Paraneotermes spp., 274
Paratrechina flavipes, 311
pectin substances, as components of sound wood, 166, 167
Pericapritermes spp., 50, 269
pH levels: of carton, 282; in digestive tract of termites, 187; regulation of pH in ant nests, 298; in termite mounds, 259
pharyngeal glands, of ants: digestive enzymes of, 238, 239, 241, 242, 243–4
Pheidole forvida, 311
Pheidole megacephala, 287
Pheidole sculpturata, 314
Pheidole spp., 107, 109, 110, 318
phenyl-acetic acid: in ant metathoracic gland secretions, 101
pheremones, of termites, 66, 176, 187, 204, 218–19, 229–30
phosphorus levels: in the soil of ant nests, 300, 301; in the soil of termite nests, 252, 260, 261
Plagiolepis pygmaea, 47, 235
plant material: changes brought about by, (ants), 304, (termites), 263–4; as a food source for, (ants), 86–8, 308, (termites), 165–7, 177–84; *see also* wood, wood decay
Pogonomyrmex badius, 22, 27, 129, 339; feeding habits of, 99, 100, 234, 307; production by, 48–9, 105, 309, 310, 317; respiration of, 138–9, 149
Pogonomyrmex barbatus, 108, 109, 339
Pogonomyrmex californicus, 28, 107, 108, 122, 334; respiration of, 133, 136, 147, 149
Pogonomyrmex desertorum, 107, 108, 122; respiration of, 133, 136, 147, 149
Pogonomyrmex marcusi, 334
Pogonomyrmex maricopa, 133, 147, 149
Pogonomyrmex occidentalis, 91–2, 297, 298–9, 334, 339; foraging behaviour of, 107, 109, 297, 308; production of, 48, 309, 310, 317
Pogonomyrmex rugosus, 107–8, 122, 133, 149, 308, 309
Pogonomyrmex spp., 84, 107–8, 109, 110, 304, 327
Pogonomyrmex spp., 84, 107, 108, 109, 110, 304
Polyrhachis debilis, 335
Polyrhachis riscosa, 335
Polyrhachis spp., 327
polysaccharides, 166, 167, 213
Ponerinae, 289, 334; food consumption of, 85, 86, 89, 234, 236
population density, determinations of, 6–44; by direct soil sampling, 15–25, 266–8; by mark–recapture, 26–30, 49; nest sampling methods, 6–14, 25–6; population dynamics, 40–4; by quadrat methods, 8–9, 15–24; *see also* population energetics
population energetics, of ants, 40–4, 318–24, 331; adult colony population, 333–5; regulation of colony density, 319–24, 336–9
population energetics, of termites, 40–4, 265–92; population density, 266–70; predation, 285–92; production, 273–6, 276–81; respiration, 271–3; return of energy and nutrients to ecosystem, 281–92; *see also* energy flow
Porotermes adamsoni, 57
Porotermes spp., 64
Porotermitinae, 69
potassium levels, 252, 261, 300, 301, 306; in ant nests, 299; in termite nests, 259, 260, 261
predation, by ants, 85–6, 88, 89, 289, 292; and control of other invertebrates, 313–17; upon termites, 286–90
predation, upon termites, 308; upon alates, 285–6; upon colony brood and nest systems, 290–2; upon foraging sterile castes, 286–90
Prenolepis imparis, 91, 335
Proceratium silaceum, 334
Procubitermes sjöestedti, 21
production, in ant and termite societies, 45–53, 81, 94, 161, 275, 310; of eggs, 46–7; of sexuals, 52–3; of termites, 273–6; of workers, 48–52
Proformica spp., 307, 327, 328
Promirotermes holmgrien, 269
Prorhinotermes simplex, 75
protease (endopolypeptidase): presence of, in different ant spp., 241
protein: in ant food exchange secretions, 234; levels in termite tissue, 201, 210; in sound wood, 166, 173
Protermes spp., 50
protozoans: symbiotic relationship in termite digestion, 190, 192–3, 195, 214, 221, 222; utilization of cellulose, 196–7
Psammotermes hybostoma, 56, 76, 269, 272, 275
Psammotermes spp., 129, 274
Pseudacanthotermes militaris, 21, 33, 69, 76, 77, 269
Pseudacanthotermes spp., 50, 63, 254, 261, 290
Pseudomyrma spp., 84
Pseudomyrmex renefica, 119

Index

Pterotermes occidentis, 184, 201, 202
Pterotermes spp., 273–4
Pyrrhocoris apterus, 206

Q_{10} (respiration rate and temperature), 134–7, 138, 142, 143, 151
quadrat methods, of sampling nest density, 8–9, 12, 15–24
queens, 92, 205, 206, 230; egg production of, 46–7, 221–2, 223; respiration of, (in ants), 130–1, 141–2, 148, (in termites), 158–9; trophallaxis, 213, 214

radioisotope markers and tracers, 26, 49, 244; for assessing the extent of foraging, 28, 91; for assessing honeydew production by Homoptera, 313; for the study of trophallaxis, 95, 211, 218–19, 234
rectal secretions, of ants, 101–2
respiration, 45, 81, 117–20; and abiotic factors, 122–9; of ants, 140–53; carbon dioxide evolution, 113–14, 120, 125–9; of the colony, 111–13, 119; effects of humidity on, 125, 128–9, 147; and intrinsic factors, 120–2; laboratory measurements of, 112, 113–17; oxygen consumption, 113–14, 125–9, 144–5; Q_{10}, respiration rate and temperature correlation, 134–9, 142, 143, 151; rate of, 129–40; of termites, 153–63; *see also* energy flow, respiratory quotient
respiration, of ants, 132–4, 140–53, 310; annual costs of, 138–9; of different castes, 140–7; and effects of temperature, 135–9, 142; laboratory measurements of, 140–53; metabolic costs of, 118–20; respiratory data, 148–50; and seasonal changes of, 143–6
respiration, of termites, 126, 153–63, 275; anaerobic processes of, 153–5; and effects on ecosystems, in terms of energy requirements, 162–3, 246, 269, 270, 271–3
respiratory quotient (RQ), 130, 140–7, 148–50, 151, 152; in termites, 153–5, 157–9, 160, 161, 189, 271
respirometric techniques, 113–14, 143–5
reproductives *see* sexuals
Reticulitermes flavipes, 206–7, 209, 210, 216, 221, 228; food consumption and digestion of, 176, 177, 188, 192, 193, 196; respiratory data of, 157
Reticulitermes hesperus, 127, 221–2, 228, 230
Reticulitermes lucifugus, 43; cannibalism of, 228–9; food consumption and digestion of, 75, 79, 187, 193, 258; lignin degradation by, 200; nutrition of, 220, 222, 225; respiratory data of, 157; trophallactic behaviour of, 216
Reticulitermes spp., 66, 153, 187
Reticulitermes santonensis, 157, 203
Reticulitermes tibialis, 127
Reticulitermes virginicus, 192, 193, 222
Rhinotermitidae, 269; mound structure of, 251; nutrition of, 75, 76, 178, 221, 222; trophallactic behaviour of, 219
Rhinotermitinae, 58
Rhytidoponera convexa, 334
RQ *see* respiratory quotient

salivary secretions: of ants, 293; returned to ecosystem in mound construction, 281; of termites, 52, 213, 275–6
saturation deficit of the atmosphere (humidity) *see* respiration
savannah: biomass, consumption, production and respiration of termites in, 273, 279, 280; population density of termites in, 268, 269
Schedorhinotermes derosus, 58
Schedorhinotermes intermedius actosus, 250, 252
seasonal periodicity of foraging termites, 72–3, 74
seed-harvesting ants, foraging by, 107–10
Sericomyrmex spp., 86, 87
Sericomyrmex urichi, 335
sexuals: of ants, in terms of energy costs, 119; as prey, 318; production of, 42–3, 52–3, 81, 323; of termites, 213, 216, 223, 228, 230
Smithistruma rostrata, 335
societies: of ants and termites compared with vertebrates, 304; variations in composition of, 41–3
sodium levels: in ant nests, 300, 301; in soil, 306
soil-feeding termites: biomass, population density and respiration data, 269, 273; energy consumption and conversion of, 277
soil modifications, by ants, 296–306; area under influence, 297–8; chemical changes, 298, 299, 300, 301, 306; decomposition of organic matter, 299–302; effects on vegetation, 304; and other soil animals, 304–6; physical changes, 298–9
soil modifications, by termites, 253–62; carbon/nitrogen ratio, 258, 259; of chemical properties, 258–62; effects on vegetation, 263–4; of organic matter, 257–8; soil profile changes, 253–5; soil texture, 256–7; subterranean galleries, 262
soil sampling, for nest density estimations, 15–25; quadrat method using pits and trenches, 15–24; soil cores, 24–5
soldiers, 62; production of, 33, 52; respiratory data on termites, 125, 157–8; trophallactic behaviour, of termite, 213–14, 216
Solenopsis geminata, 119
Solenopsis invicta, 16, 233, 235, 334, 339
Solenopsis richteri, 237, 238, 240, 241, 242
Solenopsis saevissima, 120
Sphaerotermes spp., 50
Sphaerotermes sphaerothorax, 194, 197
starch, as a component of sound wood, 166, 167
statistical methods, for density estimations: using line transects, 7–8; using mark-recapture, 26–30; using mound dispersion, 9–11; using quadrat methods, 8–9, 15–24
Stenamma diecki, 334
Stenamma impar, 334
Stenamma meridionale, 334
Stenamma spp., 339
Stenamma westwoodi, 41
Stolotermitinae, 69
stomodeal trophallaxis: of ant larvae, 236–7; of termites, 212, 213–14, 217
swarming, of alates, as an energy loss, 53
subterranean termites: as ant prey, 110; feeding habits of, 59–61; population density of, 266, 267

symbiotes, of termites: bacteria, 183–4, 186, 192–3, 197, 202–3; fungi, wood-decaying, 178–83, 194–5, 200–1, 226–8; nitrogen fixation by, 203; protozoans, 189, 190, 192–3, 195–7, 202–3, 221; vitamin synthesis by, 209
Syntermes spp., 58, 63, 289

TEC *see* total exchangeable cations
Temnothorax recedens, 235
temperature: and effects on the foraging behaviour of ants, 108–9; and effects on respiratory rate, Q_{10}, 134–7
Tenuirostritermes tenuirostris, 71, 223, 228–9
Tenuirostritermes spp., 58
Termes hospes, 185
Termes spp., 50, 225
Termitidae, 50, 65, 223, 224, 272; food consumption by, 75, 76; humus-feeding spp., 57; proctodeal trophallaxis by, 214; symbiotic micro-organisms of, 78, 178; wood-feeding spp., 56, 193–4
Termitinae, 65, 249, 251, 269; foraging behaviour of, 57, 58, 69
Termitomyces spp., 68
termitophily, 217–18, 292
Termitopone commutata, 289
Termopsinae, 65, 69, 181, 222
territory, of foraging ants, 90–1, 91
Tetramorium caespitum, 53; biomass of, 34–5, 36; in different ecosystems, 329, 336, 338; energy content of, 37–9; feeding habits of, 84, 91, 99, 311, 318; metabolic costs of, 118; oral secretions of, 236–7; populations of, 22, 27, 30, 335; respiration of, 133, 137, 149
Tetramorium sericiventre, 314
Tetraponera penzigi, 23
Tetramorium spp., 318
total exchangeable cations (TEC): in ant nests, 299, 300; in termite mounds and surrounding soil, 260–2
Trachymyrmex septentrionalis, 294
Trachymyrmex smithi, 133, 136
Trachymyrmex smithi neomexicanus, 147, 149–50
Trachymyrmex spp., 87
trehalase, presence in different ant spp., 239
Trinervitermes bettonianus, 290
Trinervitermes ebenevianus, 223
Trinervitermes geminatus, 40, 120, 123–4, 163; biomass of, 269; energy flow in, 37, 38, 161–2; feeding habits of, 62, 76, 77–8; foraging behaviour of, 58, 67, 70, 71–3; mound structure and soil composition of, 251, 252; population density of, 32, 33, 41, (estimates of) 8–9, 10, 11, 12, 267, 269; production by, 47, 50–1, 53, 272, 274, 275; respiration of, 112, 121, 128, 269, 272, 275
Trinervitermes occidentalis, 9, 269
Trinervitermes oeconomus, 9, 70, 269
Trinervitermes rhodesiensis, 290
Trinervitermes spp., 50, 264, 274; density of mounds of, 6, 8, 9; feeding habits of, 58, 62, 63, 65, 67, 71; as prey, 287, 290
Trinervitermes togoensis, 9, 10, 269
Trinervitermes trinervius, 9, 10, 275; biomass, population density, and respiratory data, 269; feeding habits of, 67, 73, 187; production of, 272, 274, 275
Trinervitermes trinervoides, 8, 222; digestion of, 191, 194–5, 196, 199; lipid content of, 207, 208
trophallaxis, of termites, 211–23; nutritional functions of, 215–91; of proctodeal food, 214–15, 216, 217; of saliva and stomodeal food, 212, 213–14, 217; social functions of, 219–23
trophic levels, with regard to food consumption: of ants, 81, 306–10; of termites, 55–7
Tumulitermes hastilis, 40, 254, 261, 282
Tumulitermes pastinator, 254, 259, 282
Tumulitermes spp., 63
Tumulitermes tumuli, 63

vegetation *see* plant material
Veromessor pergandei: feeding habits of, 129, 339; food consumption of, 88, 92, 96, 100, 236; foraging behaviour of, 99, 107, 109
Veromessor spp., 84, 107, 327
vitamin requirements, of termites, 209

wood, chemical constituents of, 165–77; carbohydrates, 166–8; lignin, 168–9; lipids, 175; minerals, 175; minor constituents, 175–7; nitrogen, 169–75, 181; and resistance to termite attack, 66–8, 176–7
wood decay, 177–84; abiotic factors of, 177–8; by bacteria, 183–4; by fungi, 178–83; by herbivores and wood-borers, 184
wood-feeding termites: assimilation efficiency, 78–9, 80; feeding habits of, 56, 59, 69, 70, 72, 73; laboratory studies of, 61–3, 64, 65, 74, 75, 80; mound structure and soil composition of, 250, 252, 258, 259
worker populations, of ants, 47, 233; colony density in different ecosystems, 336–9; food consumption and digestion by, 81, 89–90, 92, 93, 242–3; metabolic costs of, 118–19, 120, 139; production and growth of, 48–50, 51–2, 323; respiration of, 125, 130–2, 143–53; trophallaxis by, 234–7
worker populations, of termites: biomass, 32, 33, 133–4; cannibalism of, 229; food consumption and digestion of, 58, 65, 188, 190, 227; lipid content of, 205–7, 210; production and growth of, 48–50, 51–2; respiration data, 133–4, 157–9; trophallaxis of, 213, 214, 216

Zootermopsis angusticollis, 201, 224; cannibalism of, 228, 230–1; digestion and digestive enzymes of, 186, 187, 188–90, 192, 195–6, 197; food consumption of, 67, 75, 201; respiration data of, 136, 157; trophallaxis by, 214, 219
Zootermopsis laticeps, 57, 69
Zootermopsis nevadensis: nitrogen metabolism by, 181, 201, 204; respiration of, 125, 127, 138, 153–4, 157, 271; respiratory rates of, 270; trophallaxis by, 219
Zootermopsis spp.: digestion of, 65, 196, 199, 200, 201; faecal material and carton of, 64, 204, 224; production of, 273–4; respiration of, 153–4